Lina Elmoishedy

WIRELESS COMMUNICATIONS AND NETWORKING

The Morgan Kaufmann Series in Networking
Series Editor, David Clark, M.I.T.

For further information on these books and for a list of forthcoming titles, please visit our Web site at http://www.mkp.com.

WIRELESS COMMUNICATIONS AND NETWORKING

Vijay K. Garg

ELSEVIER

Amsterdam • Boston • Heidelberg
London • New York Oxford
Paris • San Diego • San Francisco
Singapore • Sydney • Tokyo
Morgan Kaufmann is an imprint of Elsevier

MORGAN KAUFMANN PUBLISHERS

Senior Acquisitions Editor	Rick Adams
Publishing Services Manager	George Morrison
Senior Project Manager	Brandy Lilly
Associate Editor	Rachel Roumeliotis
Editorial Assistant	Brian Randall
Cover Design	Alisa Andreola
Composition	diacriTech
Illustration	diacriTech
Copyeditor	Janet Cocker
Proofreader	Janet Cocker
Indexer	Distributech Scientific Indexing
Interior printer	Sheridan Books
Cover printer	Phoenix Color

Morgan Kaufmann Publishers is an imprint of Elsevier.
500 Sansome Street, Suite 400, San Francisco, CA 94111

This book is printed on acid-free paper.

Library of Congress Cataloging-in-Publication Data
Garg, Vijay Kumar, 1938-
   Wireless communications and networking / Vijay K. Garg.–1st ed.
      p. cm.
   Includes bibliographical references and index.
   ISBN-13: 978-0-12-373580-5 (casebound : alk. paper)
   ISBN-10: 0-12-373580-7 (casebound : alk. paper) 1. Wireless communication systems. 2. Wireless LANs.
I. Title.
   TK5103.2.G374 2007
   621.382'1–dc22
                                                                    2006100601

ISBN: 978-0-12-373580-5

For information on all Morgan Kaufmann publications,
visit our Web site at *www.mkp.com* or *www.books.elsevier.com*

Printed in the United States of America
07 08 09 10 11   5 4 3 2 1

Working together to grow
libraries in developing countries

www.elsevier.com | www.bookaid.org | www.sabre.org

ELSEVIER   BOOK AID International   Sabre Foundation

The book is dedicated to my grandchildren—Adam, Devin, Dilan, Nevin, Monica, Renu, and Mollie.

Contents

Contents

The following *Bonus Chapters* can be found on the book's website at
http://books.elsevier.com/9780123735805:

About the Author

Vijay K. Garg has been a professor in the Electrical and Computer Engineering Department at the University of Illinois at Chicago since 1999, where he teaches graduate courses in Wireless Communications and Networking. Dr. Garg was a Distinguished Member of Technical Staff at the Lucent Technologies Bell Labs in Naperville, Illinois from 1985 to 2001. He received his Ph.D. degree from the Illinois Institute of Technologies, Chicago, IL in 1973 and his MS degree from the University of California at Berkeley, CA in 1966. Dr. Garg has co-authored several technical books including five in wireless communications. He is a Fellow of ASCE and ASME, and a Senior Member of IEEE. Dr. Garg is a registered Professional Engineer in the state of Maine and Illinois. He is an Academic Member of the Russian Academy of Transport. Dr. Garg was a Feature Editor of Wireless/PCS Series in IEEE Communication Magazine from 1996–2001.

Preface

During the past three decades, the world has seen significant changes in the telecommunications industry. There has been rapid growth in wireless communications, as seen by large expansion in mobile systems. Wireless communications have moved from first-generation (1G) systems primarily focused on voice communications to third-generation (3G) systems dealing with Internet connectivity and multi-media applications. The fourth-generation (4G) systems will be designed to connect wireless personal area networks (WPANs), wireless local area networks (WLANs) and wireless wide-area networks (WWANs).

With the Internet and corporate intranets becoming essential parts of daily business activities, it has become increasingly advantageous to have wireless offices that can connect mobile users to their enterprises. The potential for technologies that deliver news and other business-related information directly to mobile devices could also develop entirely new revenue streams for service providers.

The 3G mobile systems are expected to provide worldwide access and global roaming for a wide range of services. The 3G WWANs are designed to support data rates up to 144 kbps with comprehensive coverage and up to 2 Mbps for selected local areas. Prior to the emergence of 3G services, mobile data networks such as general packet radio service (GPRS) over time division multiple-access (TDMA) systems and high-speed packet data over IS-95 code-division multiple access (CDMA) systems were already very popular. At the same time, after the introduction of Bluetooth and imode technology in 1998, local broadband and ad hoc wireless networks attracted a great deal of attention. This sector of the wireless networking industry includes the traditional WLANs and the emerging WPANs.

Multi hop wireless ad hoc networks complement the existing WLAN standards like IEEE 802.11a/b/g/n and Bluetooth to allow secure, reliable wireless communications among all possible hand-held devices such as personal digital assistances (PDAs), cell-phones, laptops, or other portable devices that have a wireless communication interface. Ad hoc networks are not dependent on a single point of attachment. The routing protocols for ad hoc networks are designed to self-configure and self-organize the networks to seamlessly create an access point on the fly as a user or device moves.

Provisioning data services over the wireless data networks including ad hoc networks requires smart data management protocols and new transaction models for data delivery and transaction processing, respectively. While personalization of data services is desired, over personalization will have ramifications on scalability of wireless networks? As such, mobile computing not only poses challenges but also opens up an interesting research area. It is redefining existing business models and creating entirely new ones. Envisioning new business processes vis-à-vis the enabling technologies is also quite interesting.

Over the past decade, wireless data networking has developed into its own discipline. There is no doubt that the evolution of wireless networks has had significant impact on our lifestyle. This book is designed to provide a unified foundation of principles for data-oriented wireless networking and mobile communications.

This book is an extensive enhancement to the Wireless & Personal Communications book published by Prentice Hall in 1996, which primarily addressed 2G cellular networks. Since then, wireless technologies have undergone significant changes; new and innovative techniques have been introduced, the focus of wireless communications is increasingly changing from mobile voice applications to mobile data and multimedia applications. Wireless technology and computing have come closer and closer to generating a strong need to address this issue. In addition, wireless networks now include wide area cellular networks, wireless local area networks, wireless metropolitan area networks, and wireless personal area networks. This book addresses these networks in extensive detail. The book primarily discusses wireless technologies up to 3G but also provides some insight into 4G technologies.

It is indeed a challenge to provide an over-arching synopsis for mobile data networking and mobile communications for diverse audiences including managers, practicing engineers, and students who need to understand this industry. My basic motivation in writing this book is to provide the details of mobile data networking and mobile communications under a single cover. In the last two decades, many books have been written on the subject of wireless communications and networking. However, mobile data networking and mobile communications were not fully addressed. This book is written to provide essentials of wireless communications and wireless networking including WPAN, WLAN, WMAN, and WWAN. The book is designed for practicing engineers, as well as senior/first-year graduate students in Electrical and Computer Engineering (ECE), and Computer Science (CS).

The first thirteen chapters of the book focus on the fundamentals that are required to study mobile data networking and mobile communications. Numerous solved examples have been included to show applications of theoretical concepts. In addition, unsolved problems are given at the end of each chapter for practice.

After introducing fundamental concepts, the book focuses on mobile networking aspects with several chapters devoted to the discussion of WPAN, WLAN, WWAN, and other aspects of mobile communications such as mobility management, security, and cellular network planning. Two additional "Bonus" chapters on inter-working between WLAN and WWAN and on 4G systems (along with several helpful appendices) are available free on the book's website at http://books.elsevier.com/9780123735805.

Most of the books in wireless communications and networking appear to ignore the standard activities in the field. I feel students in wireless networking must be exposed to various standard activities. I therefore address important standard activities including 3GPP, 3GPP2, IEEE 802.11, IEEE 802.15 and IEEE 802.16 in the book. This feature of the book is also very beneficial to the professionals who wish to know about a particular standard without going through the voluminous material on that standard.

A unique feature of this book that is missing in most of the available books on wireless communications and networking is to offer a balance between theoretical and practical concepts. This book can be used to teach two semester courses in mobile data networking and mobile communications to ECE and CS students. Chapter 4 may be omitted for ECE students and Chapter 14 for CS students.

The first course—*Introduction to Wireless Communications and Networking* can be offered to senior undergraduate and first year graduate students. This should include first fourteen chapters. Chapters 4 and 14 may be omitted depending on the students' background. The second course—*Wireless Data Networking* should include Chapters 15 through 23. The first course should be a pre-requisite to the second course. The student should be given homework, two examinations, and a project to complete each course. In addition, this book can also be used to teach a comprehensive course in Wireless Data Networking to IT professionals by using Chapters 2, 3, 5, 6, 7, 11, 15, 16, and 18 to 22.

During the preparation of this manuscript my family members were very supportive. I would like to thank my children, Nina, Meena, and Ravi. Also, I appreciate the support given by my wife, Pushpa. In addition, I appreciate the support of the reviewers, Elaine Cheong, Frank Farrante, and Pei Zhang in providing valuable comments on the manuscript. Finally, I am thankful of Rachel Roumeliotis for coordinating the reviews of the manuscript.

Vijay K. Garg
Willowbrook, IL

CHAPTER 1

An Overview of Wireless Systems

1.1 Introduction

The cellular system employs a different design approach than most commercial radio and television systems use [1,2]. Radio and television systems typically operate at maximum power and with the tallest antennas allowed by the regulatory agency of the country. In the cellular system, the service area is divided into cells. A transmitter is designed to serve an individual cell. The system seeks to make efficient use of available channels by using low-power transmitters to allow frequency reuse at much smaller distances. Maximizing the number of times each channel can be reused in a given geographic area is the key to an efficient cellular system design.

During the past three decades, the world has seen significant changes in the telecommunications industry. There have been some remarkable aspects to the rapid growth in wireless communications, as seen by the large expansion in mobile systems. Wireless systems consist of wireless wide-area networks (WWAN) [i.e., cellular systems], wireless local area networks (WLAN) [4], and wireless personal area networks (WPAN) (see Figure 1.1) [17]. The handsets used in all of these systems possess complex functionality, yet they have become small, low-power consuming devices that are mass produced at a low cost, which has in turn accelerated their widespread use. The recent advancements in Internet technology have increased network traffic considerably, resulting in a rapid growth of data rates. This phenomenon has also had an impact on mobile systems, resulting in the extraordinary growth of the mobile Internet.

Wireless data offerings are now evolving to suit consumers due to the simple reason that the Internet has become an everyday tool and users demand data mobility. Currently, wireless data represents about 15 to 20% of all air time. While success has been concentrated in vertical markets such as public safety, health care, and transportation, the horizontal market (i.e., consumers) for wireless data is growing. In 2005, more than 20 million people were using wireless e-mail. The Internet has changed user expectations of what data access means. The ability to retrieve information via the Internet has been "an amplifier of demand" for wireless data applications.

More than three-fourths of Internet users are also wireless users and a mobile subscriber is four times more likely to use the Internet than a nonsubscriber to

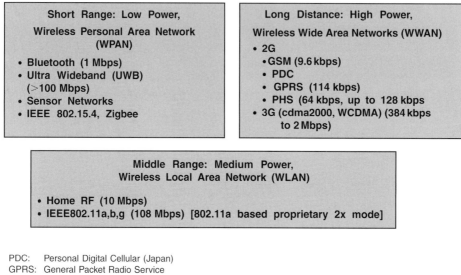

PDC: Personal Digital Cellular (Japan)
GPRS: General Packet Radio Service
PHS: Personal Handy Phone System (Japan)

Figure 1.1   Wireless networks.

mobile services. Such keen interest in both industries is prompting user demand for converged services. With more than a billion Internet users expected by 2008, the potential market for Internet-related wireless data services is quite large.

In this chapter, we discuss briefly 1G, 2G, 2.5G, and 3G cellular systems and outline the ongoing standard activities in Europe, North America, and Japan. We also introduce broadband (4G) systems (see Figure 1.2) aimed on integrating WWAN, WLAN, and WPAN. Details of WWAN, WLAN, and WPAN are given in Chapters 15 to 20.

1.2   First- and Second-Generation Cellular Systems

The first- and second-generation cellular systems are the WWAN. The first public cellular telephone system (first-generation, 1G), called Advanced Mobile Phone System (AMPS) [8,21], was introduced in 1979 in the United States. During the early 1980s, several incompatible cellular systems (TACS, NMT, C450, etc.) were introduced in Western Europe. The deployment of these incompatible systems resulted in mobile phones being designed for one system that could not be used with another system, and roaming between the many countries of Europe was not possible. The first-generation systems were designed for voice applications. Analog frequency modulation (FM) technology was used for radio transmission.

In 1982, the main governing body of the European post telegraph and telephone (PTT), la Conférence européenne des Administrations des postes et des

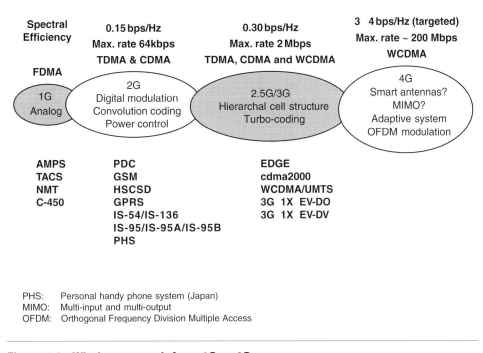

Figure 1.2 Wireless network from 1G to 4G.

télécommunications (CEPT), set up a committee known as Groupe Special Mobile (GSM) [9], under the auspices of its Committee on Harmonization, to define a mobile system that could be introduced across western Europe in the 1990s. The CEPT allocated the necessary duplex radio frequency bands in the 900 MHz region.

The GSM (renamed Global System for Mobile communications) initiative gave the European mobile communications industry a home market of about 300 million subscribers, but at the same time provided it with a significant technical challenge. The early years of the GSM were devoted mainly to the selection of radio technologies for the air interface. In 1986, field trials of different candidate systems proposed for the GSM air interface were conducted in Paris. A set of criteria ranked in the order of importance was established to assess these candidates.

The interfaces, protocols, and protocol stacks in GSM are aligned with the Open System Interconnection (OSI) principles. The GSM architecture is an open architecture which provides maximum independence between network elements (see Chapter 7) such as the Base Station Controller (BSC), the Mobile Switching Center (MSC), the Home Location Register (HLR), etc. This approach simplifies the design, testing, and implementation of the system. It also favors an evolutionary growth path, since network element independence implies that modification to one network element can be made with minimum or no impact on the others. Also, a system operator has the choice of using network elements from different manufacturers.

GSM 900 (i.e., GSM system at 900 MHz) was adopted in many countries, including the major parts of Europe, North Africa, the Middle East, many east Asian countries, and Australia. In most of these cases, roaming agreements exist to make it possible for subscribers to travel within different parts of the world and enjoy continuity of their telecommunications services with a single number and a single bill. The adaptation of GSM at 1800 MHz (GSM 1800) also spreads coverage to some additional east Asian countries and some South American countries. GSM at 1900 MHz (i.e., GSM 1900), a derivative of GSM for North America, covers a substantial area of the United States. All of these systems enjoy a form of roaming, referred to as Subscriber Identity Module (SIM) roaming, between them and with all other GSM-based systems. A subscriber from any of these systems could access telecommunication services by using the personal SIM card in a handset suitable to the network from which coverage is provided. If the subscriber has a multiband phone, then one phone could be used worldwide. This globalization has positioned GSM and its derivatives as one of the leading contenders for offering digital cellular and Personal Communications Services (PCS) worldwide. A PCS system offers multimedia services (i.e., voice, data, video, etc.) at any time and any where. With a three band handset (900, 1800, and 1900 MHz), true worldwide seamless roaming is possible. GSM 900, GSM 1800, and GSM 1900 are second-generation (2G) systems and belong to the GSM family. Cordless Telephony 2 (CT2) is also a 2G system used in Europe for low mobility.

Two digital technologies, Time Division Multiple Access (TDMA) and Code Division Multiple Access (CDMA) (see Chapter 6 for details) [10] emerged as clear choices for the newer PCS systems. TDMA is a narrowband technology in which communication channels on a carrier frequency are apportioned by time slots. For TDMA technology, there are three prevalent 2G systems: North America TIA/EIA/IS-136, Japanese Personal Digital Cellular (PDC), and European Telecommunications Standards Institute (ETSI) Digital Cellular System 1800 (GSM 1800), a derivative of GSM. Another 2G system based on CDMA (TIA/EIA/IS-95) is a direct sequence (DS) spread spectrum (SS) system in which the entire bandwidth of the carrier channel is made available to each user simultaneously (see Chapter 11 for details). The bandwidth is many times larger than the bandwidth required to transmit the basic information. CDMA systems are limited by interference produced by the signals of other users transmitting within the same bandwidth.

The global mobile communications market has grown at a tremendous pace. There are nearly one billion users worldwide with two-thirds being GSM users. CDMA is the fastest growing digital wireless technology, increasing its worldwide subscriber base significantly. Today, there are already more than 200 million CDMA subscribers. The major markets for CDMA technology are North America, Latin America, and Asia, in particular Japan and Korea. In total, CDMA has been adopted by almost 50 countries around the world.

The reasons behind the success of CDMA are obvious. CDMA is an advanced digital cellular technology, which can offer six to eight times the capacity of analog technologies (AMP) and up to four times the capacity of digital technologies such as TDMA. The speech quality provided by CDMA systems is far superior to any other digital cellular system, particularly in difficult RF environments such as dense urban areas and mountainous regions. In both initial deployment and long-term operation, CDMA provides the most cost effective solution for cellular operators. CDMA technology is constantly evolving to offer customers new and advanced services. The mobile data rates offered through CDMA phones have increased and new voice codecs provide speech quality close to the fixed wireline. Internet access is now available through CDMA handsets. Most important, the CDMA network offers operators a smooth migration path to third-generation (3G) mobile systems, [3,5,7,11].

1.3 Cellular Communications from 1G to 3G

Mobile systems have seen a change of generation, from first to second to third, every ten years or so (see Figure 1.3). At the introduction of 1G services, the mobile device was large in size, and would only fit in the trunk of a car. All analog components such as the power amplifier, synthesizer, and shared antenna equipment were bulky. 1G systems were intended to provide voice service and low rate (about 9.6 kbps) circuit-switched data services. Miniaturization of mobile devices progressed before the introduction of 2G services (1990) to the point where the size of mobile phones fell below 200 cubic centimeters (cc). The first-generation handsets provided poor voice quality, low talk-time, and low

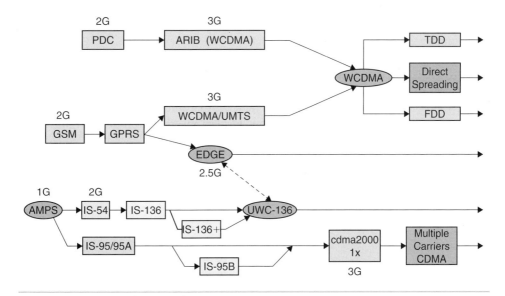

Figure 1.3   Cellular networks (WWAN) evolution from 1G to 3G.

standby time. The 1G systems used Frequency Division Multiple Access (FDMA) technology (see Chapter 6) and analog frequency modulation [8,20].

The 2G systems based on TDMA and CDMA technologies [6] were primarily designed to improve voice quality and provide a set of rich voice features. These systems supported low rate data services (16–32 kbps).

For second-generation systems three major problems impacting system cost and quality of service remained unsolved. These include what method to use for band compression of voice, whether to use a linear or nonlinear modulation scheme, and how to deal with the issue of multipath delay spread caused by multipath propagation of radio waves in which there may not only be phase cancellation but also a significant time difference between the direct and reflected waves.

The swift progress in Digital Signal Processors (DSPs) was probably fueled by the rapid development of voice codecs for mobile environments that dealt with errors. Large increases in the numbers of cellular subscribers and the worries of exhausting spectrum resources led to the choice of linear modulation systems.

To deal with multipath delay spread, Europe, the United States, and Japan took very different approaches. Europe adopted a high transmission rate of 280 kbps per 200 kHz RF channel in GSM [13,14] using a multiplexed TDMA system with 8 to 16 voice channels, and a mandatory equalizer with a high number of taps to overcome inter-symbol interference (ISI) (see Chapter 3). The United States used the carrier transmission rate of 48 kbps in 30 kHz channel, and selected digital advanced mobile phone (DAMP) systems (IS-54/IS-136) to reduce the computational requirements for equalization, and the CDMA system (IS-95) to avoid the need for equalization. In Japan the rate of 42 kbps in 25 kHz channel was used, and equalizers were made optional.

Taking into account the limitations imposed by the finite amount of radio spectrum available, the focus of the third-generation (3G) mobile systems has been on the economy of network and radio transmission design to provide seamless service from the customers' perspective. The third-generation systems provide their users with seamless access to the fixed data network [18,19]. They are perceived as the wireless extension of future fixed networks, as well as an integrated part of the fixed network infrastructure. 3G systems are intended to provide multimedia services including voice, data, and video.

One major distinction of 3G systems relative to 2G systems is the hierarchical cell structure designed to support a wide range of multimedia broadband services within the various cell types by using advanced transmission and protocol technologies. The 2G systems mainly use one-type cell and employ frequency reuse within adjacent cells in such a way that each single cell manages its own radio zone and radio circuit control within the mobile network, including traffic management and handoff procedures. The traffic supported in each cell is fixed because of frequency limitations and little flexibility of radio transmission which is mainly optimized for voice and low data rate transmissions. Increasing traffic leads to costly cellular reconfiguration such as cell splitting and cell sectorization.

The multilayer cell structure in 3G systems aims to overcome these problems by overlaying, discontinuously, pico- and microcells over the macrocell structure with wide area coverage. Global/satellite cells can be used in the same sense by providing area coverage where macrocell constellations are not economical to deploy and/or support long distance traffic.

With low mobility and small delay spread profiles in picocells, high bit rates and high traffic densities can be supported with low complexity as opposed to low bit rates and low traffic load in macrocells that support high mobility. The user expectation will be for service selected in a uniform manner with consistent procedures, irrespective of whether the means of access to these services is fixed or mobile. Freedom of location and means of access will be facilitated by smart cards to allow customers to register on different terminals with varying capabilities (speech, multimedia, data, short messaging).

The choice of a radio interface parameter set corresponding to a multiple access scheme is a critical issue in terms of spectral efficiency, taking into account the ever-increasing market demand for mobile communications and the fact that radio spectrum is a very expensive and scarce resource. A comparative assessment of several different schemes was carried out in the framework of the Research in Advanced Communications Equipment (RACE) program. One possible solution is to use a hybrid CDMA/TDMA/FDMA technique by integrating advantages of each and meeting the varying requirements on channel capacity, traffic load, and transmission quality in different cellular/PCS layouts. Disadvantages of such hybrid access schemes are the high-complexity difficulties in achieving simplified low-power, low-cost transceiver design as well as efficient flexibility management in the several cell layers.

CDMA is the selected approach for 3G systems by the ETSI, ARIB (Association of Radio Industries and Business—Japan) and Telecommunications Industry Association (TIA). In Europe and Japan, Wideband CDMA (WCDMA/UMTS [Universal Mobile Telecommunication Services]) was selected to avoid IS-95 intellectual property rights. In North America, cdma2000 uses a CDMA air-interface based on the existing IS-95 standard to provide wireline quality voice service and high speed data services at 144 kbps for mobile users, 384 kbps for pedestrians, and 2 Mbps for stationary users. The 64 kbps data capability of CDMA IS-95B provides high speed Internet access in a mobile environment, a capability that cannot be matched by other narrowband digital technologies.

Mobile data rates up to 2 Mbps are possible using wide band CDMA technologies. These services are provided without degrading the systems' voice transmission capabilities or requiring additional spectrum. This has tremendous implications for the majority of operators that are spectrum constrained. In the meantime, DSPs have improved in speed by an order of magnitude in each generation, from 4 MIPs (million instructions per second) through 40 MIPs to 400 MIPs.

Since the introduction of 2G systems, the base station has seen the introduction of features such as dynamic channel assignment. In addition, most base stations began making shared use of power amplifiers and linear amplifiers whether or not modulation was linear. As such there has been an increasing demand for high-efficiency, large linear power amplifiers instead of nonlinear amplifiers.

At the beginning of 2G, users were fortunate if they were able to obtain a mobile device below 150 cc. Today, about 10 years later, mobile phone size has reached as low as 70 cc. Furthermore, the enormous increase in very large system integration (VLSI) and improved CPU performance has led to increased functionality in the handset, setting the path toward becoming a small-scale computer.

1.4 Road Map for Higher Data Rate Capability in 3G

The first- and second-generation cellular systems were primarily designed for voice services and their data capabilities were limited. Wireless systems have since been evolving to provide broadband data rate capability as well.

GSM is moving forward to develop cutting-edge, customer-focused solutions to meet the challenges of the 21st century and 3G mobile services. When GSM was first designed, no one could have predicted the dramatic growth of the Internet and the rising demand for multimedia services. These developments have brought about new challenges to the world of GSM. For GSM operators, the emphasis is now rapidly changing from that of instigating and driving the development of technology to fundamentally enable mobile data transmission to that of improving speed, quality, simplicity, coverage, and reliability in terms of tools and services that will boost mass market take-up.

People are increasingly looking to gain access to information and services whenever they want from wherever they are. GSM will provide that connectivity. The combination of Internet access, web browsing, and the whole range of mobile multimedia capability is the major driver for development of higher data speed technologies.

GSM operators have two nonexclusive options for evolving their networks to 3G wide band multimedia operation: (1) they can use General Packet Radio Service (GPRS) and Enhanced Data rates for GSM Evolution (EDGE) [also known as 2.5G] in the existing radio spectrum, and in small amounts of new spectrum; or (2) they can use WCDMA/UMTS in the new 2 GHz bands [12,15,16]. Both approaches offer a high degree of investment flexibility because roll-out can proceed in line with market demand and there is extensive reuse of existing network equipment and radio sites.

The first step to introduce high-speed circuit-switched data service in GSM is by using High Speed Circuit Switched Data (HSCSD). HSCSD is a feature that enables the co-allocation of multiple full rate traffic channels (TCH/F) of GSM into an HSCSD configuration. The aim of HSCSD is to provide a mixture of services with different user data rates using a single physical layer structure. The available capacity of an HSCSD configuration is several times the capacity of a TCH/F, leading to a significant enhancement in data transfer capability.

Ushering faster data rates into the mainstream is the new speed of 14.4 kbps per time slot and HSCSD protocols that approach wire-line access rates of up to 57.6 kbps by using multiple 14.4 kbps time slots. The increase from the current baseline 9.6 kbps to 14.4 kbps is due to a nominal reduction in the error-correction overhead of the GSM radio link protocol, allowing the use of a higher data rate.

The next phase in the high speed road map is the evolution of current short message service (SMS), such as smart messaging and unstructured supplementary service data, toward the new GPRS, a packet data service using TCP/IP and X.25 to offer speeds up to 115.2 kbps. GPRS has been standardized to optimally support a wide range of applications ranging from very frequent transmissions of medium to large data volume. Services of GPRS have been developed to reduce connection set-up time and allow an optimum usage of radio resources. GPRS provides a packet data service for GSM where time slots on the air interface can be assigned to GPRS over which packet data from several mobile stations can be multiplexed.

A similar evolution strategy, also adopting GPRS, has been developed for DAMPS (IS-136). For operators planning to offer wide band multimedia services, the move to GPRS packet-based data bearer service is significant; it is a relatively small step compared to building a totally new 3G network. Use of the GPRS network architecture for IS-136 packet data service enables data subscription roaming with GSM networks around the globe that support GPRS and its evolution. The IS-136 packet data service standard is known as GPRS-136. GPRS-136 provides the same capabilities as GSM GPRS. The user can access either X.25 or IP-based data networks.

GPRS provides a core network platform for current GSM operators not only to expand the wireless data market in preparation for the introduction of 3G services, but also a platform on which to build UMTS frequencies should they acquire them.

GPRS enhances GSM data services significantly by providing end-to-end packet switched data connections. This is particularly efficient in Internet/intranet traffic, where short bursts of intense data communications actively are interspersed with relatively long periods of inactivity. Since there is no real end-to-end connection to be established, setting up a GPRS call is almost instantaneous and users can be continuously on-line. Users have the additional benefits of paying for the actual data transmitted, rather than for connection time.

Because GPRS does not require any dedicated end-to-end connection, it only uses network resources and bandwidth when data is actually being transmitted. This means that a given amount of radio bandwidth can be shared efficiently between many users simultaneously.

The significance of EDGE (also referred to as 2.5G system) for today's GSM operators is that it increases data rates up to 384 kbps and potentially even higher in good quality radio environments that are using current GSM spectrum and carrier structures more efficiently. EDGE will both complement and be an alternative to new WCDMA coverage. EDGE will also have the effect of unifying the GSM, DAMPS, and WCDMA services through the use of dual-mode terminals.

GSM operators who win licenses in new 2 GHz bands will be able to introduce UMTS wideband coverage in areas where early demand is likely to be greatest. Dual-mode EDGE/ UMTS mobile terminals will allow full roaming and handoff from one system to the other, with mapping of services between the two systems. EDGE will contribute to the commercial success of the 3G system in the vital early phases by ensuring that UMTS subscribers will be able to enjoy roaming and interworking globally.

While GPRS and EDGE require new functionality in the GSM network with new types of connections to external packet data networks, they are essentially extensions of GSM. Moving to a GSM/UMTS core network will likewise be a further extension of this network.

EDGE provides GSM operators—whether or not they get a new 3G license—with a commercially attractive solution for developing the market for wide band multimedia services. Using familiar interfaces such as the Internet, volume-based charging and a progressive increase in available user data rates will remove some of the barriers to large-scale take-up of wireless data services. The move to 3G services will be a staged evolution from today's GSM data services using GPRS and EDGE. Table 1.1 provides a comparison of GSM data services.

Table 1.1 Comparison of GSM data services.

Service type	Data unit	Max. sustained user data rate	Technology	Resources used
Short Message Service (SMS)	Single 140 octet packet	9 bps	simplex circuit	SDCCH or SACCH
Circuit-Switched Data	30 octet frames	9.6 kbps	duplex circuits	TCH
HSCSD	192 octet frames	115 kbps	duplex circuits	1-8 TCH
GPRS	1600 octet frames	115 kbps	virtual circuit packet switching	PDCH (1-8 TCH)
EDGE (2.5G)	variable	384 kbps	virtual circuit/ packet switching	1-8 TCH

Note: SDCCH: Stand-alone Dedicated Control Channel; SACCH: Slow Associated Control Channel; TCH: Traffic Channel; PDCH: Packet Data Channel (all refer to GSM logical channels)

The use of CDMA technology began in the United States with the development of the IS-95 standard in 1990. The IS-95 standard has evolved since to provide better voice services and applications to other frequency bands (IS-95A), and to provide higher data rates (up to 115.2 kbps) for data services (IS-95B). To further improve the voice service capability and provide even higher data rates for packet and circuit switched data services, the industry developed the cdma2000 standard in 2000. As the concept of wireless Internet gradually turns into reality, the need for an efficient high-speed data system arises. A CDMA high data rate (HDR) system was developed by Qualcomm. The CDMA-HDR (now called 3G 1X EV-DO, [3G 1X Enhanced Version Data Only]) system design improves the system throughput by using fast channel estimation feedback, dual receiver antenna diversity, and scheduling algorithms that take advantage of multi-user diversity. 3G 1X EV-DO has significant improvements in the downlink structure of cdma2000 including adaptive modulation of up to 8-PSK and 16-quadrature amplitude modulation (QAM), automatic repeat request (ARQ) algorithms and turbo coding. With these enhancements, 3G 1X EV-DO can transmit data in burst rates as high as 2.4 Mbps with 0.5 to 1 Mbps realistic downlink rates for individual users. The uplink design is similar to that in cdma2000. Recently, the 3G 1X EV-Data and Voice (DV) standard was finalized by the TIA and commercial equipment is currently being developed for its deployment. 3G 1X EV-DV can transmit both voice and data traffic on the same carrier with peak data throughput for the downlink being confirmed at 3.09 Mbps.

As an alternative, Time Division-Synchronous CDMA (TD-SCDMA) has been developed by Siemens and the Chinese government. TD-SCDMA uses adaptive modulation of up to quadrature phase shift keying (QPSK) and 8-PSK, as well as turbo coding to obtain downlink data throughput of up to 2 Mbps. TD-SCDMA uses a 1.6 MHz time-division duplex (TDD) carrier whereas cdma2000 uses a 2×1.25 MHz frequency-division duplex (FDD) carrier (2.5 MHz total). TDD allows TD-SCDMA to use the least amount of spectrum of any 3G technologies.

Table 1.2 lists the maximum data rates per user that can be achieved by various systems under ideal conditions. When the number of users increases, and if all the users share the same carrier, the data rate per user will decrease.

One of the objectives of 3G systems is to provide access "anywhere, any time." However, cellular networks can only cover a limited area due to high infrastructure costs. For this reason, *satellite* systems will form an integral part of the 3G networks. Satellite will provide extended wireless coverage to remote areas and to aeronautical and maritime mobiles. The level of integration of the satellite system with the terrestrial cellular networks is under investigation. A fully integrated solution will require mobiles to be dual mode terminals to allow communications with orbiting satellite and terrestrial cellular networks. Low Earth orbit (LEO) satellites are the most likely candidates for providing worldwide coverage. Currently several LEO satellite systems are being deployed to provide global telecommunications.

Table 1.2 Network technology migration paths and their associated data rates.

Technology	Carrier width (MHz)	Duplexing	Multiplexing	Modulation	Max. data rates	End-user data rates
Analog					9.6 kbps	4.8–9.6 kbps
CDPD (1G)					19.2 kbps	about 16 kbps
GSM Circuit Switched Data (2G)	0.20	FDD	TDMA	GMSK	9.6–14.4 kbps	about 12 kbps
GPRS	0.20	FDD	TDMA	GMSK	up to 115.2 kbps (8 channels)	10–56 kbps
EDGE (2.5G)	0.20	FDD	TDMA	GMSK, 8-PSK	384 kbps	about 144 kbps
WCDMA (3G)	5.00	FDD	CDMA	QPSK	2 Mbps (stationary); 384 kbps (mobile)	50 kbps uplink; 150–200 kbps downlink
IS-54/IS-136 TDMA Circuit Switched Data (2G)	0.03	FDD	TDMA	QPSK	14.4 kbps	about 10 kbps
EDGE (2.5G) for North American TDMA system	0.20	FDD	TDMA	GMSK, 8-PSK	64 kbps uplink (initial roll out)	Initial roll out in 2001/2002: 45–50 kbps uplink; 80–90 kbps downlink
					384 kbps	2003: 45–50 kbps uplink; 150–200 kbps downlink

cdma2000 (3G) 1X	1.25	FDD	CDMA	QPSK	153 kbps	90–130 kbps (depending on the number of users and distance from BS)
3G 1X EV-DO (data only)	1.25	FDD	TD-CDMA	QPSK, 8-PSK, 16-QAM	2.4 Mbps	700 kbps
3G 1X EV-DV (data and voice)	1.25	FDD	TD-CDMA	QPSK, 8-PSK, 16-QAM	3–5 Mbps	>1 Mbps
TD-SCDMA	1.60	TDD	TD-CDMA	QPSK, 8-PSK	2 Mbps	1.333 Mbps

Note: FDD = Frequency Division Duplex; TDD = Time Division Duplex; PSK = Phase Shift Keying; QPSK = Quadrature Phase Shift Keying; GMSK = Gaussian Minimum Shift Keying; QAM = Quadrature Amplitude Modulation

1.5 Wireless 4G Systems

4G networks (see Chapter 23) can be defined as wireless ad hoc peer-to-peer networking with high usability and global roaming, distributed computing, personalization, and multimedia support. 4G networks will use distributed architecture and end-to-end Internet Protocol (IP). Every device will be both a transceiver and a router for other devices in the network eliminating the spoke-and-hub architecture weakness of 3G cellular systems. Network coverage/capacity will dynamically change to accommodate changing user patterns. Users will automatically move away from congested routes to allow the network to dynamically and automatically self-balance.

Recently, several wireless broadband technologies [20] have emerged to achieve high data rates and quality of service. Navini Networks developed a wireless broadband system based on TD-SCDMA. The system, named Ripwave, uses beamforming to allow multiple subscribers in different parts of a sector to simultaneously use the majority of the spectrum bandwidth. Beamforming allows the spectrum to be effectively reused in dense environments without having to use excessive sectors. The Ripwave system varies between QPSK, 16 and 64-QAM, which allows the system to burst up to 9.6 Mbps using a single 1.6 MHz TDD carrier. Due to TDD and 64-QAM modulation the Ripwave system is extremely spectrally efficient. Currently Ripwave is being tried by several telecom operators in the United States. The Ripwave Customer Premise Equipment is about the size of a cable modem and has a self-contained antenna. Recently, PC cards for laptops have become available allowing greater portability for the user.

Flarion Technologies is promoting their proprietary Flash-orthogonal frequency-division multiple (OFDM) as a high-speed wireless broadband solution. Flash-OFDM uses frequency hopping spread spectrum (FHSS) to limit interference and allows a reuse pattern of one in an OFDM access environment. Flarion's Flash-OFDM system uses 1.25 MHz FDD carriers with QPSK and 16-QAM modulation. Peak speeds can burst up to 3.2 Mbps with sustained rates leveling off at 1.6 Mbps on the downlink. Flarion has not implemented an antenna enhancement technology that may further improve data rates.

BeamReach is a wireless broadband technology based on OFDM and beamforming. It uses TDD duplexed 1.25 MHz paired carriers. Spread spectrum is used to reduce interference over the 2.5 MHz carriers allowing a frequency reuse of one. Individual users can expect downlink rates of 1.5, 1.2, and 0.8 Mbps using 32-QAM, 16-QAM, and 8-PSK modulation respectively. The aggregate network bandwidth is claimed to be 88 Mbps in 10 MHz of spectrum or 220 Mbps in 24 MHz of spectrum, which equates to a high spectral efficiency of 9 bps/Hz. It should be noted that the system uses either 4 or 6 sectors and these claims are based on those sectoring schemes. For any technology with a reuse number of 1 to achieve 9 bps/Hz per cell with 4 or 6 sectors, the efficiency in each sector would need to be a reasonable 2.3 or 1.5 bps/Hz, respectively.

IPWireless is the broadband technology based upon UMTS. It uses either 5 or 10 MHz TDD carriers and QPSK modulation. The theoretical peak transmission speeds for a 10 MHz deployment is 6 Mbps downlink and 3 Mbps uplink. The IPWireless system only uses QPSK modulation and no advanced antenna technologies. With the inclusion of advanced antenna technologies and the development of High Speed Downlink Packet Access (HSDPA), IPWireless has significant potential.

SOMA networks has also developed a wireless broadband technology based on UMTS. Like UMTS, SOMA's technology uses 5 MHz FHSS carriers. Peak throughput is claimed to be as high as 12 Mbps, making SOMA one of the faster wireless broadband technologies. Table 1.3 compares the wireless broadband technologies and their lowest order modulation data throughput.

1.6   Future Wireless Networks

As mobile networks evolve to offer both circuit and packet-switched services, users will be connected permanently (always on) via their personal terminal of choice to the network. With the development of intelligence in core network (CN), both voice and broadband multimedia traffic will be directed to their intended destination with reduced latency and delays. Transmission speeds will be increased and there will be far more efficient use of network bandwidth and resources. As the number of IP-based mobile applications grows, 3G systems will offer the most flexible access technology because it allows for mobile, office, and residential use in a wide range of public and nonpublic networks. The 3G systems will support both IP and non-IP traffic in a variety of modes, including packet, circuit-switched, and virtual circuit, and will thus benefit directly from the development and extension of IP standards for mobile communications. New developments will allow parameters like quality of service (QoS), data rate, and bit error rate (BER)—vital for mobile operation—to be set by the operator and/or service provider.

Wireless systems beyond 3G (e.g., 4G) will consist of a layered combination of different access technologies:

- Cellular systems (e.g., existing 2G and 3G systems for wide area mobility)
- WLANs for dedicated indoor applications (such as IEEE 802.11a, 802.11b, 802.11g)
- Worldwide interoperability for microwave access (WiMAX) (IEEE 802.16) for metropolitan areas
- WPANs for short-range and low mobility applications around a room in the office or at home (such as Bluetooth)

These access systems will be connected via a common IP-based core network that will also handle interworking between the different systems. The core network will enable inter- and intra-access handoff.

Table 1.3 Non-line of sight (LOS) wireless broadband technologies.

Developer	Technology	Multiplexing	Duplexing	Carrier (MHz)	Modulation	System DL Peak (Mbps)	System DL LOM (Mbps)	Avg. DL efficiency (bps/Hz)
Navini	TD-SCDMA	TD-CDMA	TDD	1.6	QPSK to 64-QAM	8.0	2.0	2.5
IPwireless	TD-WCDMA	TD-CDMA	TDD	5.0	QPSK	3.0	3.0	1.2
Flarion	Flash-OFDM	FHSS OFDM	FDD	1.25	QPSK, 16-QAM	3.2	1.6	1.28
SOMA	UMTS	CDMA	FDD	5.0	QPSK, 16-QAM	12.0	6.0	1.2
Beam Reach	AB-OFDM	DSSS-OFDM	TDD	2.5	8-PSK, 16-, 32-QAM	3.33	2.0	1.6

TDD carriers need one carrier for Tx and Rx, thus efficiency is doubled; BeamReach system throughput includes 6 sectors, thus it was divided by six; LOM—sustained system throughput estimated using lowest order modulation

The peak data rates of 3G systems are around 10 times more than 2G/2.5G systems. The fourth-generation systems may be expected to provide a data rate 10 times higher than 3G systems. User data rates of 2 Mbps for vehicular and 20 Mbps for indoor applications are expected. The fourth-generation systems will also meet the requirements of next generation Internet through compliance with IPv6, Mobile IP, QoS control, and so on.

1.7 Standardization Activities for Cellular Systems

The standardization activities for PCS in North America were carried out by the joint technical committee (JTC) on wireless access, consisting of appropriate groups from within the T1 committee, a unit of the Alliance for Telecommunications Industry Solutions (ATIS), and the engineering committee TR46, a unit of the Telecommunications Industry Association (TIA). The JTC was formed in November of 1992, and its first assignment was to develop a set of criteria for PCS air interfaces. The JTC established seven technical ad hoc groups (TAGs) in March of 1994, one for each of the selected air interface proposals. The TAGs drafted the specifications document for the respective air interface technologies and conducted validation and verification to ensure consistency with the criteria established by the JTC. This was followed by balloting on each of the standards. After the balloting process, four of the proposed standards were adopted as ANSI standards: IS-136-based PCS, IS-95-based PCS, GSM 1900 (a derivative of GSM) and Personal Access Communication System (PACS). Two of the proposed standards—composite CDMA/TDMA and Oki's wide band CDMA—were adopted as trial use standards by ATIS and interim standards by TIA. The Personal Wireless Telecommunications-Enhanced (PWT-E) standard was moved from JTC to TR 46.1 which, after a ballot process, was adopted in March of 1996. Table 1.4 provides a comparison of seven technologies using a set of parameters which include access methods, duplex methods, bandwidth per channel, throughput per channel, maximum power output per subscriber unit, vocoder, and minimum and maximum cell ranges.

The 3G systems were standardized under the umbrella of the International Telecommunications Union (ITU). The main proposals to the ITU International Mobile Telecommunications-2000 (IMT-2000) are: ETSI UMTS Terrestrial Radio Access (UTRA), ARIB WCDMA, TIA cdma2000, and TD-SCDMA. These third-generation systems will provide the necessary quality for multimedia communications. The IMT-2000 requirements are: 384 kbps for full area coverage (144 kbps for fast-moving vehicles between 120 km per hour and 500 km per hour), and 2 Mbps for local coverage. It is, therefore, important to use packet-switched data service to dynamically allocate and release resources based on the current needs of each user.

The ETSI agreed on a WCDMA solution using FDD mode. In Japan, a WCDMA solution was adopted for both TDD and FDD modes. In Korea, two

Table 1.4 Technical characteristics of North American PCS standards.

	TAG-1	TAG-2	TAG-3	TAG-4	TAG-5	TAG-6	TAG-7
Standard	Composite CDMA/TDMA	IS-95 based PCS	PACS	IS-136 based PCS	GSM 1900	PWT-E	Oki's Wide band CDMA
Access	CDMA/TDMA/FDMA	CDMA	TDMA	TDMA	TDMA	TDMA	CDMA
Duplex Method	TDD	FDD	FDD	FDD	FDD	TDD	FDD
Frequency Reuse	3	1	16 × 1	7 × 3	7 × 1, 3 × 3	Portable selected	1
Bandwidth/channel	2.5/5 MHz	1.25 MHz	300 kHz	30 kHz	200 kHz	1 kHz	5, 10, 15 MHz
Throughput/channel (kb/s)	8	8.55/13.3	32	8	13	32	32
Max. Power/subscriber unit	600 mW	200 mW	200 mW	600 mW	0.5–2.0 W	500 mW	500 mW
Vocoder	PHS HCA	CELP	ADPCM	VCELP/ACELP	RPE-LTE ACELP	ADPCM	ADPCM
Max. Cell Range (km)	10.0	50.0	1.6	20.0	35.0	0.15	5.0
Min. Cell Range (km)	0.1	0.05	0–1	0.5	0.5	0.01	0.05

different types of CDMA solutions were proposed—one based on WCDMA similar to that of Europe and Japan and the other similar to the cdma2000 proposed in North America.

A number of groups working on similar technologies pooled their resources. This led to the creation of two standards groups—the third-generation partnership project (3GPP) and 3GPP2. 3GPP works on UMTS, which is based on WCDMA and 3GPP2 works on cdma2000.

The IEEE standard committee 802.11 is responsible for the WLAN standard. There are two IEEE standards committees that are involved in certification of wireless broadband technologies. The 802.16x committee focuses on the wireless metropolitan area network (WMAN) using CDMA and OFDM technologies. 802.16x allows for portability and data rates above 1 Mbps. The newly formed IEEE 802.20 committee, evolved from the 802.16e committee, focuses on mobile wide area network (MWAN). Several key performance requirements include megabit data rates and mobile handoff at speeds of up to 250 km per hour.

The Worldwide Interoperability for Microwave Access (WiMAX) Forum is a nonprofit organization formed by equipment and component suppliers to promote the adoption of IEEE 802.16-compliant equipment by operators of broadband wireless access systems. The organization is working to facilitate the deployment of broadband wireless networks based on IEEE 802.16 standards by helping to ensure the compatibility and interoperability of broadband wireless access equipment.

1.8 Summary

In this chapter, we presented the scope of wireless networks and gave an overview. We briefly discussed 1G, 2G/2.5G, and 3G cellular systems. The advantage of wireless data networking is apparent. Wireless data network users are not confined to the locations of "wired" data jacks, and enjoy connectivity that is less restrictive and therefore well suited to meet the needs of today's mobile users. Wireless network deployment in three service classifications—wireless personal access network (WPAN), wireless local area network (WLAN), and wireless wide area network (WWAN)—was discussed. Today, the core technology behind the wireless service in each of these service classifications is unique and, more important, not an inherently integrated seamless networking strategy. As an example, a user of a personal digital assistant (PDA), such as a PALM (XXX) connecting to the Internet via a WWAN service provider will not be able to directly connect to a WLAN service. Simply stated, they are different services, with different hardware requirements, and have fundamentally different service limitations. In the future, wireless networks have to evolve to provide interoperability of WPAN, WLAN, and WWAN systems.

Problems

1.1   Name the wireless access techniques used in 1G, 2G, and 3G wireless systems.

1.2   What are the three classes of wireless data networking?

1.3   Define the roles of WPAN technology in wireless data networking.

1.4   List the main features of 3G systems.

1.5   What is the role of GPRS in enhancing 2G GSM systems?

1.6   Show how CDMA IS-95 systems are moving to provide 3G services.

1.7   Show how 2G GSM systems are moving to achieve 3G services.

1.8   What are the data rate requirements for 3G systems?

1.9   Define IPWireless technology.

1.10 What are the goals of 4G systems?

References

1. Balston, D. M., and Macario, R. C. V. *Cellular Radio Systems*. Norwood, MA: Artech House, 1993.

2. Balston, D. M. *The Pan-European Cellular Technology*. IEE Conference Publication, 1988.

3. Cai, J., and Goodman, D. J. General packet radio service in GSM. *IEEE Communications Magazine*, vol. 35, no. 10, October 1997, pp. 121–131.

4. Crow, B. P., Widjaja, I., Kim, L. G., and Sakai, P. T. IEEE 802.11 Wireless Local Area Networks. *IEEE Communications Magazine*, September 1998, pp. 116–126.

5. Dasilva, J. S., Ikonomou, D., and Erben, H. European R&D programs on third-generation mobile communications systems. *IEEE Personal Communications*, vol. 4, no. 1, February 1997, pp. 46–52.

6. Dinan, E. H., and Jabbari, B. Spreading codes for direct sequence CDMA and wide band CDMA cellular networks. *IEEE Communications Magazine*, September 1998.

7. The European path towards UMTS. *IEEE Personal Communications*, Special Issue. February 1995.

8. Garg, V. K., and Wilkes, J. E. *Wireless and Personal Communications Systems*. Upper Saddle River, NJ: Prentice Hall, 1996.

9. Garg, V. K., and Wilkes, J. E. *Principles and Applications of GSM*. Upper Saddle River, NJ: Prentice Hall, 1998.

10. Garg, V. K. *CDMA IS-95 and cdma2000*. Upper Saddle River, NJ: Prentice Hall, 2000.

11. 3GPPweb Third generation partnership project website: *http://www.3gpp.org*.

12. 3GPP2web Third generation partnership project-2 website: *http://www.3gpp2.org*.

13. Marley, N. *GSM and PCN Systems and Equipment*. JRC Conference, Harrogate, 1991.

14. Mouly, M., and Pautet, M.-B. *The GSM System for Mobile Communications*. Palaiseau, France, 1992.

15. Nakajma, N. Future mobile communications systems in Japan. *Wireless Personal Communications*, vol. 17, no. 2–3, June 2001, pp. 209–223.

16. Nikula, E., Toshala, A., Dahlman, E., Girard, L., and Klein, A. FRAMES multiple access for UMTS and IMT-2000. *IEEE Personal Communications Magazine*, April 1998, pp. 16–24.

17. Negus, K., Stephens, A., and Lansford, J. Home RF: Wireless networking for the connected home. *IEEE Personal Communications*, February 2000, pp. 20–27.

18. Pirhonen, R., Rautava, T., and Pentinen, J. TDMA convergence for packet data services. *IEEE Personal Communications Magazine*, vol. 6, no. 3, June 1999, pp. 68–73.

19. Rapeli, J. UMTS: Targets, system concepts, and standardization in a global framework. *IEEE Personal Communications*, February 1995.

20. Shafi, M., Ogose, S., and Hattori T. (editors). *Wireless Communication in the 21st Century*. Wiley-Interscience, 2002.

21. Young, W. R. Advanced mobile phone services — Introduction, background and objectives. *Bell Systems Technical Journal*, vol. 58, 1979, pp. 1–14.

CHAPTER 2

Teletraffic Engineering

2.1 Introduction

There are many telephone customers, much larger than the number of available trunks, but not every customer makes or receives a telephone call at the same time. The number of trunks connecting the Mobile Switching Center (MSC) *A* with another MSC *B* are the number of voice pairs used in the connection. One of the most important steps in telecommunication engineering is to determine the number of trunks required on a route or a connection between MSCs. To dimension a route correctly, we must have some idea of its usage, that is, how many subscribers are expected to talk at one time over the route. The usage of a transmission route or an MSC brings us into the realm of *teletraffic engineering*.

The call capacity of an MSC is expressed as the maximum number of originating plus incoming (O + I) calls that can be handled during a high-traffic hour while meeting the dial-tone delay requirements. The call capacity of an MSC depends on the subscriber call mix, feature mix, and equipment configuration. Typically, processor capacity limits the MSC call capacity under specific conditions, but the switching networking, peripheral equipment, trunk terminations, or directory numbers can limit MSC call capacity as well. The call volume offered to an MSC depends on geographical area, class of service mix, and time of day. An MSC is required to process calls while serving a representative supplementary feature (such as call waiting, call forwarding, etc.). The call capacity of an MSC is needed to plan, engineer, and administer its use.

In this chapter we provide definitions of the terms used in teletraffic engineering. We include several numerical examples for calculating traffic. In the latter part of the chapter we focus on the Erlang and Poisson blocking formulas used to calculate the grade of service (GoS) and show their applications in teletraffic engineering. The readers interested in derivation of the blocking formulas should refer to queuing books [2,4,5]. Typical call mixes for systems in different environments are also given. The material in this chapter applies to both wireline and wireless systems.

2.2 Service Level

Service level for telecommunication traffic can be divided into two main areas: the delay in receiving a *dial tone* (for a mobile system, radio signaling delays contribute to dial-tone delay) and the *probability of service denial*. Dial-tone delay is the maximum amount of time a subscriber must wait to hear a dial-tone after removing the handset from the hook. Dial-tone delay has the following characteristics:

- A large number of users compete for a small number of servers (dial-tone connections, dial-tone generators).
- An assumption that the user will wait until a server is available.

Service denial, or the probability that the service trunk will not be available, is similar to dial-tone delay, with several additional characteristics:

- A large number of users compete for a small number of trunks.
- An assumption that no delay will be encountered. The user is either given access to a trunk or is advised by a busy signal or a recording that none are available.
- The user may frequently reinstate the call attempt after receiving a busy signal.

For both cases, the basic measure of performance is either the probability that service delay will exceed some specified value or the probability that the call will be denied or blocked (the *blocking probability*). In a system that drops calls when serving trunks are not available (a *loss system*), the blocking probability represents the performance measure.

2.3 Traffic Usage

A traffic path is a communication channel, time slot, frequency band, line, trunk, switch, or circuit over which individual communications take place in sequence.

Traffic usage is defined by two parameters, *calling rate* and *call holding time* [6].

Calling rate, or the number of times a route or traffic path is used per unit time; more properly defined, the *call intensity* (i.e., calls per hour) per traffic path during busy hour.

Call holding time, or the average duration of occupancy of a traffic path by a call.

The *carried* traffic is the volume of traffic actually carried by a switch, and *offered* traffic is the volume of traffic offered to a switch. The offered load is the sum of the carried load and overflow (traffic that cannot be handled by the switch).

$$\text{Offered load} = \text{carried load} + \text{overflow} \qquad (2.1)$$

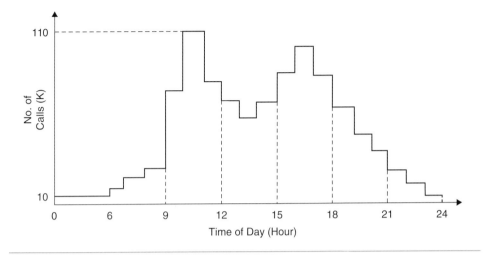

Figure 2.1 Example of voice traffic variation hour by hour.

Figure 2.1 shows a typical hour-by-hour voice traffic variation for an MSC. We notice that the busiest period—the *busy hour* (BH) is between 10 A.M. and 11 A.M. We define the busy hour as the span of time (not necessarily a clock hour) that has the highest average traffic load for the business day throughout the busy season. The peak hour is defined as the clock hour with highest traffic load for a single day.

Since traffic also varies from month to month, we define the average busy season (ABS) as the three months (not necessarily consecutive) with the highest average BH traffic load per access line. Telephone systems are not engineered for maximum peak loads, but for some typical BH load. The blocking probability is defined as the average ratio of blocked calls to total calls and is referred to as the *GoS*.

Example 2.1

If there are 380 seizures (lines connected for service) and 10 blocked calls (lost calls) during the BH, what is the GoS?

Solution

Blocking Probability = (Number of lost calls)/(Total number of offered calls)

$$\text{GoS} = \frac{10}{380 + 10} = \frac{1}{39}$$

2.4 Traffic Measurement Units

Traffic is measured in either Erlangs, percentage of occupancy, centrum (100) call seconds (CCS), or peg count [1].

Traffic intensity is the average number of calls simultaneously in progress during a particular period of time. It is measured either in units of Erlangs or CCS. An average of one call in progress during an hour represents a traffic intensity of 1 Erlang; thus 1 Erlang equals 1×3600 call seconds (36 CCS). The Erlang is a dimensionless number. Traffic intensity can be obtained as:

$$\text{Traffic Intensity} = \frac{\text{(the sum of circuit holding time)}}{\text{(the duration of monitoring period)}} \tag{2.2}$$

$$I = \frac{\sum_{i=1}^{N_c} t_i}{T} = \frac{N_c \bar{t}}{T} = n_c \bar{t} \tag{2.3}$$

where:
I = traffic intensity
T = duration of monitoring period
t_i = the holding time of the ith individual call
N_c = the total number of calls in monitoring period
$\bar{t}$ = average holding time
n_c = number of calls per unit time

Percentage of occupancy. Percentage of time a server is busy.

Peg count. The number of attempts to use a piece of equipment.

The usage (U), peg count (PC) per time period, overflow (O) per period, and average holding time $\bar{t}$ are related as:

$$U = (PC - O) \cdot \bar{t} \tag{2.4}$$

Example 2.2

In a switching office an equipment component with an average holding time of 5 seconds has a peg count of 450 for a one-hour period. Assuming that there was no overflow (i.e., the system handled all calls), how much usage in call-seconds, CCS, and Erlangs has accumulated on the piece of the equipment?

Solution

$$U = (450 - 0) \times \frac{5}{3600} = 0.625 \text{ Erlangs}$$

$$0.625 \text{ Erlangs} \times 36 \text{ CCS/Erlangs} = 22.5 \text{ CCS} = 2250 \text{ call-seconds}$$

Example 2.3

If the carried load for a component is 3000 CCS at 5% blocking, what is the offered load?

Solution

$$\text{Offered load} = \frac{3000}{(1 - 0.05)} \approx 3158 \, \text{CCS}$$

$$\text{Overflow} = (\text{Offered load}) - (\text{Carried load}) = 3158 - 3000 = 158 \, \text{CCS}$$

Example 2.4

In a wireless network each subscriber generates two calls per hour on the average and a typical call holding time is 120 seconds. What is the traffic intensity?

Solution

$$I = \frac{2 \times 120}{3600} = 0.0667 \, \text{Erlangs} = 2.4 \, \text{CCS}$$

Example 2.5

In order to determine voice traffic on a line, we collected the following data during a period of 90 minutes (refer to Table 2.1). Calculate the traffic intensity in Erlangs and CCS.

Table 2.1 Traffic data used to estimate traffic intensity.

Call no.	Duration of call (s)
1	60
2	74
3	80
4	90
5	92
6	70
7	96
8	48
9	64
10	126

Solution

$$\text{Call arrival rate} = \frac{10}{1.5} = 6.667 \text{ calls/hour}$$

Average call holding time:

$$\bar{t} = \frac{(60 + 74 + 80 + 90 + 92 + 70 + 96 + 48 + 64 + 126)}{10} = 80 \text{ seconds per call}$$

$$I = \frac{6.667 \times 80}{3600} = 0.148 \text{ Erlangs} = 5.33 \text{ CCS}$$

Example 2.6

We record data in Table 2.2 by observing the activity of a single customer line during an eight-hour period from 9:00 A.M. to 5:00 P.M. Find the traffic intensity during the eight-hour period, and during busy hour (which occurs between 4:00 P.M. and 5:00 P.M).

Solution

$$\text{Call arrival rate} = \frac{11}{8} = 1.375 \text{ calls/hour}$$

Total call minutes = 3 + 10 + 7 + 10 + 5 + 5 + 1 + 5 + 15 + 34 + 5 = 100 minutes

Table 2.2 Traffic on customer line between 9:00 A.M. and 5:00 P.M.

Call no.	Call started	Call ended	Call duration (min.)
1	9:15	9:18	3.0
2	9:31	9:41	10.0
3	10:17	10:24	7.0
4	10:24	10:34	10.0
5	10:37	10:42	5.0
6	10:55	11:00	5.0
7	12:01	12:02	1.0
8	2:09	2:14	5.0
9	3:15	3:30	15.0
10	4:01	4:35	34.0
11	4:38	4:43	5.0

If statistical stationary condition is assumed, then the calling rate and average holding time could be converted to a one-hour period, with the same result holding for the offered traffic load. That is:

$$\bar{t} = \frac{100}{11} \times \frac{1}{60} = 0.1515 \text{ hours/call}$$

The traffic intensity is:

$$I = 1.375 \times 0.1515 = 0.208 \text{ Erlangs} = 7.5 \text{ CCS}$$

The BH is between 4:00 P.M. and 5:00 P.M. Since there are only two calls between this period, we can write:

Call arrival rate = 2 calls/hour

The average call holding time during BH:

$$\bar{t} = \frac{(34 + 5)}{2} = 19.5 \text{ minutes/call} = 0.325 \text{ hours/call}$$

The traffic load during BH is:

$$I = 2 \times 0.325 = 0.65 \text{ Erlangs} = 23.4 \text{ CCS}$$

Example 2.7

The average mobile user has 500 minutes of use per month; 90% of traffic occurs during work days (i.e., only 10% of traffic occurs on weekends). There are 20 work days per month. Assuming that in a given day, 10% of traffic occurs during the BH, determine the traffic per subscriber per BH in Erlangs.

Solution

Average busy hour usage per subscriber = (minutes of use/month) × (fraction during work day) × (percentage in busy hour)/(work days/month)

Average busy hour usage per subscriber = 500 × 0.9 × 0.1/20 = 2.25 minutes of use per subscriber per busy hour.

$$\text{Traffic per subscriber} = \frac{2.25}{60} = 0.0375 \text{ Erlangs}$$

Example 2.8

If the mean holding time in Example 2.7 is 100 seconds, find the average number of busy hour call attempts (BHCAs).

Solution

BHCAs = (traffic in Erlangs) $\times$ 3600/(mean holding time in seconds)

$$= \frac{0.0375 \times 3600}{100} = 1.35$$

2.5   Call Capacity

Call capacity is defined with respect to a view of the mobile switching center (MSC). In general there are two approaches to view the system—*global view* and *component view* [7].

> **Global view.** The entire MSC is considered to be a single unit. Each request to the MSC for service is counted as an attempt. This approach is applicable to central processors involved in call processing. In the global view, the call volume of interest is expressed as the sum of the originating and incoming (O + I) calls.

> 1. *Originating call (O)* includes the following:
>
>   • Partial dial calls—calls with partial time-outs and abandons
>   • Intra-office calls—all calls that originate from and terminate to the same switch
>   • Outgoing calls—all calls that originate from a line on the switch but terminate to a line on a different switch
>
> 2. *Incoming call (I)* includes the following:
>
>   • Incoming-terminating calls—all calls that terminate to a line on the switch but originate from a different switch
>   • Tandem calls—trunk-to-trunk calls that are routed through the switch
>   • Direct inward dialing (DID)—calls to a PBX system

> **Component view.** The component of interest is considered a subsystem. Each request to the component for service is counted as an attempt. This view applies to principal processors involved in call processing. In the component view, the call volume of interest is expressed as the sum of the originating + terminating (O + T) half-calls

> 1. *Originating half-calls.* One originating half-call is for each originating call, because two peripheral equipment connections are required for a completed call. If the component serves both lines and trunks, incoming and outgoing half-calls are added to the total half-call volume.
>
> 2. *Terminating half-calls.* One terminating half-call is for each incoming-terminating call and each interoffice call.

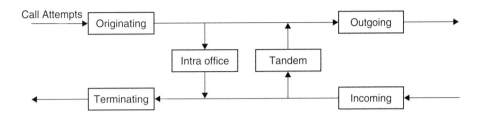

Figure 2.2 Call types in a telephone switch.

It should be noted that false starts and permanent signals (ineffective attempts) are not counted as calls in either case. However, partial dial attempts, attempts receiving busy treatment, and attempts not answered are all considered calls.

The primary determinant of a stored program control system's call handling capacity is the maximum number of calls per hour that the processor can handle while still meeting service criteria. Central processors have high day busy hour (O + I) call capacities, whereas peripheral processors have high day busy hour (O + T) call capacities.

We define the following call types (see Figure 2.2):

- *Originating call (O).* A call placed by a subscriber of the office
- *Terminating call (T).* A call received by a subscriber of the office
- *Outgoing call.* A call going out of the office
- *Incoming call (I).* A call coming into the office from another MSC
- *Intra-office call.* A call that originates from and terminates to the same MSC
- *Tandem calls.* Calls that come in on a trunk from another MSC and go out over a trunk to a different MSC
- *(O + T).* The measure of traffic load on the line side of the MSC both from within the office and from other offices
- *(O + I).* The measure of incoming trunk-circuit traffic load and traffic load on the MSC

Bell Communication Research (Bellcore) [1] suggests the use of standard call mixes given in Tables 2.3 and 2.4 for calculating call capacity of central processors (in terms of (O + I) calls) and peripheral processors (in terms of (O + T) half-calls), respectively. These call mixes do not represent the best and worst case scenarios for call capacity, but they do represent the average or typical case for each environment. These call mixes reflect the differences found in the following traffic environments:

> **Metropolitan (metro).** Class 5 office in a major metropolitan area with a high proportion of business office traffic.
>
> **Single system city.** Class 4 or 5 office serving a medium-sized town with a number of outlying community dial offices homing into it.

Table 2.3 Standard call mixes for central processors (percentage of traffic).

Type	Traffic environments		
	Metro	SSC	Suburban
Ineffective			
False Start	10	12	18
Permanent Signal	1	2	2
Partial Dial			
Abandon	3	4	5
Time-out	1	1	1
Intra office			
Answered	12	24	17
No Answer	2	2	3
Busy	2	4	4
Outgoing			
Answered	34	17	28
No Answer	5	3	4
Busy	5	3	5
Incoming-Terminating			
Answered	27	18	24
No Answer	5	2	5
Busy	4	3	4
Tandem	0	19	0

Suburban. Class 5 office residing in the suburbs of a major city.

Rural. Class 4 office residing in a rural area with a low population density and almost no business traffic.

Table 2.5 shows the typical traffic intensity for metro, suburban, and rural environments in the United States.

Table 2.6 provides an example of a typical traffic distribution for a mobile application in the U.S. metro environment.

2.6 Definitions of Terms

The following terms are used for mobile systems.

Number of calls attempted. The total number of calls attempted (also called number of bids) is the best measure of unconstrained customer demand.

Table 2.4 Standard half-call mixes for peripheral processors controlling subscriber lines (percentage of total traffic).

Type	Traffic environments		
	Metro	SSC	Suburban
Ineffective			
False Start	8	11	14
Permanent Signal	1	2	2
Partial Dial			
Abandon	2	3	4
Time-out	1	1	1
Originating			
Answered	40	37	36
No Answer	6	5	6
Busy	8	6	7
Terminating			
Answered	34	38	33
No Answer	6	4	6
Busy	5	6	7

Table 2.5 Traffic intensity for (O + I).

Environment	(O + I) calls/line
Metro	3.5–4.0
Suburban	2.0–2.5
Rural	1.2–1.5

Table 2.6 Typical traffic distribution in U.S. metro environment for mobile.

Application	Traffic Type	Distribution
Mobile	Mobile to land	60%
	Mobile to mobile	10%
	Land to mobile	30%

Number of calls completed. The number of calls completed in a network sense (i.e., calls reaching ringing tone or being answered), when compared with number of calls attempted, provides a measure of the state of network congestion.

Grade of Service (GoS). (No. of busy hour call attempts) − (No. of busy hour call completed)/(No. of busy hour call attempts).

The number of answered calls is lower than the number of attempted calls, since some calls are bound to encounter either a "subscriber-busy" state or a "ring tone/no reply" condition. The ratio between the number of successful calls (answered) and the total number of call attempts (seizure) is called *answer-seizure ratio* (ASR), whereas *answer-busy ratio* (ABR) is the number of successful calls (answered) and the total number of busy calls. These ratios represent the probability that a user will be able to complete a call on any given attempt.

$$ABR = (\text{No. of calls answered})/(\text{No. of busy calls})$$

$$ASR = (\text{No. of calls answered})/(\text{No. of calls attempted})$$

ABR and ASR are measured over relatively short periods of time (5 to 15 minutes). Both are good indicators of instantaneous network congestion. Lower value of ABR or ASR indicates higher network congestion. However, higher ABR or ASR value does not mean lower network congestion since calls may remain unanswered due to other reasons.

Quality of Service (QoS). QoS has different meanings for different people. For a service provider QoS indicates how satisfied a customer is with the service. As an example, we might find that about half of the time a customer dials a number the call goes awry or the caller does not get a dial tone or is unable to hear what is being said by the calling party. All these have an impact on QoS. QoS is an important factor in many areas of the telecommunications business. Several factors affect QoS. These are:

- Transmission quality (level, crosstalk, echo, etc.)
- Dial-tone delay, and post dial delay
- Grade of service
- Fault incidence and service deficiency
- Adaptation of the system to the subscribers

When QoS requirements are set, one of the two following approaches can be used:

- Design for the maximum permitted impairments in the most unfavorable condition

- Design for a certain range of impairments occurring as a result of a chance combination of elements, such as the majority of a subscriber's opinion being favorable (also known as *statistical design methodology*)

Both approaches have weaknesses. The first may require unnecessary high performance to satisfy those rare unfavorable cases. Some subscribers may be very unhappy with the second approach, which will be considerably more influenced by variability of plant performance. Such variability is observed in areas of signaling and transmission and should be taken into consideration while planning technical areas. The North American Region follows the statistical design approach. In Europe the maximum design impairment method is used, although it is often modified, where, for instance, certain standards will be satisfied for only 95% of the situations.

Example 2.9

We consider a wireless network with following data:

- Total population: 200,000
- Subscriber penetration: 25%
- Average call holding time for mobile-to-land and land-to-mobile subscribers: 100 seconds
- Average calls/hour for mobile-to-land and land-to-mobile subscribers: 3 calls/hour
- Average call holding time for mobile-to-mobile subscribers: 80 seconds
- Average calls/hour for mobile-to-mobile subscribers: 4 calls/hour
- Traffic distribution is:

  Mobile-to-land: 50%

  Land-to-mobile: 40%

  Mobile-to-mobile: 10%

Calculate total traffic in Erlangs. If each MSC can handle 1800 Erlangs traffic, how many MSCs are required to handle the total traffic?

Solution

Traffic generated by a subscriber (mobile-to-land or land-to-mobile):

$$a_{(ML)/(LM)} = \frac{3 \times 100}{3600} = 0.0833 \text{ Erlangs}$$

Traffic generated by a subscriber (mobile-to-mobile):

$$a_{MM} = \frac{4 \times 80}{3600} = 0.0889 \text{ Erlangs}$$

No. of wireless subscribers: $200{,}000 \times 0.25 = 50{,}000$

Total Traffic: $45{,}000 \times 0.0833 + 5{,}000 \times 0.0889 = 4194.5$ Erlangs

No. of MSCs required: $\dfrac{4194.5}{1800} = 2.33 \approx 3$

2.7   Data Collection

Traffic data is collected periodically to access performance of the mobile switching center. The data is used in overall administration, engineering, and maintenance of an office. Traffic measurement reports can also be provided on the customer's local exchange access line or trunk group from the serving central office. Reports are available on a one-week basis, consisting of seven consecutive days or a monthly report that contains a minimum of four consecutive weeks of data. The reports disclose minutes, attempts, overflow, etc. The traffic measurement report is intended to assist customers in designing and administering communications or business activities associated with telephone service. The following data is collected on an MSC [8]:

> **Peg Count:** One peg count for each of the following categories—call attempt, trunk-group seizure attempt, test made for dial tone speed, call queued
>
> **Overflow:** Overflow for each attempt collected for universal tone decoders, trunk groups, etc.
>
> **Traffic Usage:** Measured for trunks, decoders, etc.
>
> **Customer Usage:** Customer usage = Total usage − Maintenance usage

2.8   Office Engineering Considerations

The following steps are often taken in typical office engineering of a wireline or wireless office.

1. MSCs are engineered and administered based on the traffic load during the average busy hour of the busy season.
2. The busy hour is used for the overall administration, engineering, and maintenance of an office.
3. The component busy hour is used to establish trends, make projections, set capacities, and derive future requirements.
4. Dial-tone speed delay is recorded whenever a test call does not receive a dial tone within 3 seconds.
5. Terminating blockage is recorded whenever a terminating call is unable to complete because of a lack of an available path to the called line.
6. Trunk-group busy hour is the time-consistent hour during which maximum trunk-group load occurs. Trunk-group busy hour data is used to provide an adequate trunk base to meet service requirements.

7. Traffic data is collected for one or two weeks by half-hour during all parts of the day that may produce high loads (e.g., 8 A.M. to 11 P.M.).

8. Five days of the week with the heaviest load are determined; this is the *business week* of the office.

9. The hour (on the clock hour or on the half-hour) with the highest total load for the business week is determined; this is the *office busy hour*.

10. Traffic data collected for the busy hour for the months likely to be parts of the year that may produce high loads.

11. The three months, not necessarily consecutive, having the highest busy hour load are determined; this is the busy season.

12. The average load for the busy hour for the busy season's business day is (Average Busy Season per Busy Hour) (ABS/BH)

13. The following approximate relations can be used to estimate the design traffic:

$$(O + T) \text{ call: } (HD)/(ABS) = 1.4 - 1.5$$

$$(O + I) \text{ call: } (HD)/(ABS) = 1.6 - 1.7$$

$$\text{High day (HD) origination attempts per call} = 1.45$$

Example 2.10

Calculate the ABS/BH calling rate and CCS for a switch located in a large metropolitan area. The switch carries 100,000 lines. The distribution of the lines on the switch is as follows:

- Residential lines: 15,000
- Business lines: 80,000
- PBX, WATS, and Foreign Exchange (FX) lines: 5000

The ABS/BH call rates for residential, business, and high-usage customers are 2, 3, and 10 calls per line, respectively. The average call holding times for these customers are 140, 160, and 200 seconds, respectively. Assuming that the HD/ABS for the switch is equal to 1.5, calculate the design call capacity for the switch and design Erlangs.

Solution

Percentage of residential lines: $15,000/100,000 = 0.15 = 15\%$

Percentage of business lines: $80,000/100,000 = 0.80 = 80\%$

Percentage of high usage lines: $5,000/100,000 = 0.05 = 5\%$

CCS per residential line: $2 \times 140/100 = 2.8$

CCS per business line: $3 \times 160/100 = 4.8$

CCS per high-usage line: $10 \times 200/100 = 20$

Calling rate: $2 \times 0.15 + 3 \times 0.8 + 10 \times 0.05 = 3.2$ calls per line

CCS rate $= 2.8 \times 0.15 + 4.8 \times 0.8 + 20 \times 0.05 = 5.26$ CCS per line

Average call holding time per line for the switch: $\frac{5.26}{3.2} \times 100 = 164$ seconds

ABS/BH calls: $3.2 \times 100,000 = 320,000$

ABS/BH usage: $\frac{5.26 \times 100,000}{36} = 14,611$ Erlangs

Design call capacity based on HD: $1.5 \times$ ABS/BH $= 1.5 \times 320,000$
$$= 480,000 \text{ calls}$$

Design Erlangs based on HD: $1.5 \times 14,611 = 21,917$

2.9   Traffic Types

We can classify traffic sources as either infinite or finite. For the infinite traffic sources, the probability of call arrival is constant and does not depend on the occupancy of the system. It also implies an infinite number of call arrivals, each with a small call holding time. When the number of sources offering traffic to a group of trunks or circuits is comparatively small in comparison to the number of circuits, the traffic sources are referred to as finite. We can also conclude that with a finite number of traffic sources the arrival rate is proportional to the number of sources that are not already engaged in sending a call.

The probability distributions of traffic can be classified as smooth, rough, and random. Each traffic distribution can be defined by γ, the variance-to-mean ratio (VMR) given as

$$\gamma = \frac{\sigma^2}{\mu} \tag{2.5}$$

where:

$\sigma^2 =$ variance

$\mu =$ mean

γ is less than 1 for smooth traffic. For rough traffic, γ is greater than 1. When γ is equal to 1, the traffic distribution is random. The Poisson distribution function is an example of random traffic in which VMR $= 1$. Rough traffic tends to be peakier than random or smooth traffic. For a given GoS more circuits are required for rough traffic because of greater spread of the traffic distribution curve.

Smooth traffic behaves like random traffic that has been filtered. The filter is the telephone exchange. The telephone exchange looking out at its subscribers sees call arrivals as random traffic, assuming that the exchange has not been designed for more traffic (over dimensioned). Smooth traffic is the traffic on the telephone exchange outlets. The filtering or limiting of the peakiness is done by call blockage during busy hour. The blocked traffic is actually overflowed to an alternative route. Smooth traffic is characterized by the Bernoulli distribution (refer to Equation 2.6).

If we assume subscribers make calls independent of each other and that each has a probability, p, of being engaged in a conversation, then if n subscribers are examined, the probability that x subscribers will be engaged in a conversation is given as:

$$B(x) = C_x^n p^x (1 - p)^{n - x}; \quad 0 < x < n \tag{2.6}$$

where:

Mean $= np$
Variance $= np(1 - p)$
C_x^n means the number of ways that x entities can be taken n at a time.

The Poisson probability distribution function can be derived from the Bernoulli distribution. If we assume that the number of subscribers, n, is very large and the calling rate per line, h, is low such that mean ($nh = m$) remains constant; allowing n to increase to infinity, then

$$p(x) = \frac{m^x}{x!} e^{-m} \tag{2.7}$$

where:

$x = 0, 1, 2, \ldots, n$
$m = nh$ (mean)
$n =$ number of subscribers
$h =$ calling rate per line

2.10 Blocking Formulas

A call is termed lost or blocked when it cannot be completed because all the connecting equipment is busy, even though the line that the caller wishes to reach may be idle. The probability of blocking is an important parameter in teletraffic engineering. In conventional teletraffic engineering three models are used for handling or dispensing lost calls.

- Blocked Call Held (BCH)
- Blocked Call Cleared (BCC)
- Blocked Call Delayed (BCD)

The BCH concept assumes that the user will immediately reattempt the call on receipt of a congestion signal and will continue to redial. The user hopes to seize connection equipment or a trunk as soon as equipment is available. In the BCH concept, lost calls are held or waiting at the calling user's telephone. The principal traffic formula in North America is based on the BCH concept.

The BCC concept is primarily used in Europe, Asia, and Africa. In this case, the user hangs up and waits for some interval before reattempting the call if the user hears the congestion tone on the first attempt. The *Erlang B* formula is based on this criterion.

The BCD concept assumes that the user is automatically put in a queue and is served when the connection equipment becomes available. The method by which a waiting call is selected from the pool of waiting calls is based on the queue discipline (such as first-come first-served, first-come-last-served etc.). In the queuing system the GoS is defined as the probability of delay. The *Erlang C* formula is based on this concept.

There are several blocking formulas to determine the number of circuits (or trunks) required on a route based on busy hour traffic load. The factors that are used include: call arrivals and holding-time distributions; number of traffic sources; availability; and handling of lost calls. These factors help in determining which formulas to use given a particular set of circumstances.

Erlang B loss formula has been widely used outside the United States. Loss implies the probability of blockage at the switch due to either congestion or all trunks being busy. This is expressed as the GoS (G_B) or probability of finding N channels busy. In the United States the Poisson formula based on the BCH concept is used to determine number of trunks required on a route. This formula is also called the *Molina formula*.

2.10.1   Erlang B Formula

The Erlang B formula [2] is expressed as GoS or probability of finding N channels busy.

The assumptions in the Erlang B formula are:

- Traffic originates from an infinite number of traffic sources independently.
- Lost calls are cleared assuming a zero holding time.
- Number of trunks or service channels is limited.
- Full availability exists.
- Inter-arrival times of call requests are independent of each other.
- The probability of a user occupying a channel (called service time) is based on an exponential distribution.
- Traffic requests are represented by a Poisson distribution implying exponentially distributed call inter-arrival times.

$$G_B = B(N, A) = \frac{A^N/A!}{\sum_{i=0}^{N} (A^i/i!)} \qquad (2.8)$$

where:
> N = number of serving channels
> A = offered load, and
> $B(N, A)$ = blocking probability

2.10.2 Poisson's Formula

The Poisson formula is used for designing trunks on a route for a given GoS [3,9]. It is used in the United States.

The assumptions in Poisson's formula are:

- Traffic originates from an infinite number of independent sources
- Traffic density per traffic source is equal
- Lost calls are held
- A limited number of trunks or service channels exist

$$p_b = e^{-A} \sum_{i=N}^{\infty} \frac{A^i}{i!} \qquad (2.9)$$

where:
> p_b = probability of blocking
> A = offered load, and
> N = number of trunks or service channels

2.10.3 Erlang C Formula

The Erlang C [2] formula assumes that a queue is formed to hold all requested calls that cannot be served immediately. Customers who find all N servers busy join a queue and wait as long as necessary to receive service. This means that the blocked customers are delayed. No server remains idle if a customer is waiting.

The assumptions in the Erlang C formula are:

- Traffic originates from an infinite number of traffic sources independently.
- Lost calls are delayed.
- Number of trunks or service channels is limited.
- The probability of a user occupying a channel (called *service time*) is based on an exponential distribution.
- Calls are served in the order of arrival.

$$C(N, A) = \frac{A^N / [N!(1 - A/N)]}{\sum_{i=0}^{N-1} \frac{A^i}{i!} + \frac{A^N}{N!(1 - A/N)}} \qquad (2.10)$$

where:

    N = number of service channels

    A = offered load, and

    $C(N, A)$ = blocking probability

The Erlang B formula holds even when the load is greater than number of servers ($A > N$) because, unlike the BCD model in which all calls are eventually served, the BCC model allows calls to be lost when all servers are busy. Therefore, the BCC system never becomes unstable.

2.10.4 Comparison of Erlang B and Poisson's Formulas

A comparison between the Erlang B and Poisson's blocking formulas shows that Poisson's formula results in higher blocking than that obtained by the Erlang B formula for a given traffic load.

    For the Erlang B system, the offered load can be divided into the load lost and the carried traffic load A^* (i.e., amount of load serviced by system).

$$A^* = A[1 - B(N, A)] \qquad (2.11)$$

where:

    A = offered load

    N = number of servers

The carried traffic equals that portion of offered traffic load A that is not lost, and $AB[N, A]$ is the lost traffic. In a BCD system, no calls are lost. Thus, the carried load is equal to the offered load. The efficiency, ρ, of a BCC system is defined as the load carried per server (i.e., A^*/N).

2.10.5 Binomial Formula

Since the binomial formula is also used in teletraffic engineering occasionally, we include the information here. More details can be found in reference [3]. The assumptions used in the binomial formula are:

- Traffic originates from a finite number of traffic sources independently.
- Traffic density per traffic source is equal.
- Lost calls are held in the system in a queue.

$$p_b = \left[\frac{s - D}{s} \right]^{s-1} \sum_{i=N}^{s-1} \binom{s-1}{N} \left(\frac{D}{s-D} \right)^i \qquad (2.12)$$

where:

    D = expected traffic density

    p_b = blocking probability

N = number of channels in the group of channels, and
s = number of sources in group sources

Example 2.11

The maximum calls per hour in a mobile cell equals 4000 and the average call holding time is 160 seconds. If the GoS is 2%, find the offered load A. How many service channels are required to handle the load?

Solution

$$A = \frac{4000 \times 160}{3600} = 166.67 \text{ Erlangs (offered load)}$$

Using the Erlang B table in Appendix A; $N = 182$ channels giving 168.3 Erlangs at 2% blocking.

Example 2.12

How many mobile subscribers can be supported with 50 service channels at 2% GoS? Assume the average call holding time equals 120 seconds and the average busy hour call per subscriber is 1.2 calls per hour.

Solution

From the Erlang B table in Appendix A, for 50 channels at 2% blocking, the offered load = 40.26 Erlangs. The carried load will be: $40.26 \times (1 - 0.02)$ = 39.45 Erlangs

$$\text{Average traffic per user} = \frac{1.2 \times 120}{3600} = 0.04 \text{ Erlangs}$$

$$\text{No. of users} = \frac{39.45}{0.04} = 986$$

2.11 Summary

In this chapter we discussed the basic principles of teletraffic engineering that are used to engineer and administer a wireline or wireless switch. We also presented examples to determine the design call capacity of a switching system. We discussed Erlang B and Poisson's blocking formulas for calculating blocking probabilities or the GoS of the system. We concluded the chapter by presenting the Erlang C and binomial formulas that apply to a queued system. Several numerical examples were given to illustrate the applications of various blocking formulas.

Problems

2.1 Define pegcount, busy hour (BH), busy hour call attempts (BHCAs), and grade of service (GoS).

2.2 State the assumptions of the Erlang B formula.

2.3 Define "blocked call clear" and "blocked call held" systems.

2.4 What are the traffic measurements for a switching system?

2.5 What is the average call holding time in seconds of a group of circuits that has accumulated 0.4445 Erlang of usage in an hour, based on 330 call attempts with 10 calls overflowing (i.e., calls not served)?

2.6 A trunk accumulated 0.75 Erlang of usage while 9 calls were carried in an hour with no overflow. What is the average holding time per call in seconds?

2.7 A switching system is designed to support 80,000 subscribers during the BH. Each subscriber generates an average of 1.8 calls/hour during the BH with an average call holding time of 100 seconds. Calculate the offered traffic load to the switching system.

2.8 If there are 60 radio channels in a cell to handle all the calls and the average call holding time is 120 seconds, how many calls are handled in this cell with a GoS of 2%?

2.9 Consider an urban area in which average mobile subscriber has 600 minutes of use (MoU) per month. Eighty percent of traffic occurs during workdays (i.e., only 20% of traffic occurs on weekends). There are 20 workdays per month. Assuming that in a given day, 10% traffic occurs during busy hour, what is the traffic per subscriber in Erlangs?

2.10 Determine the number of subscribers that can be supported by a cell with 63 radio channels. Assume each subscriber generates an average of 2.9 calls per hour with an average call holding time of 110 seconds. Also determine the traffic generated by each subscriber in CCS. Assume GoS = 2%.

2.11 Estimate the number of subscribers that can be supported by a cell with 400 radio channels. Assume each subscriber generates an average of 2.5 calls per hour with a call holding time of 120 seconds. Assume 2% GoS.

2.12 Consider a wireless network with the following data:
- Total population: 300,000
- Subscriber penetration: 40%
- Average call holding time for mobile-to-land and land-to-mobile calls: 100 seconds

- Average call holding time for mobile-to-mobile calls: 80 seconds
- Average calls per hour for mobile-to-land and land-to-mobile: 3
- Average calls per hour for mobile-to-mobile: 4
- Traffic distribution: Mobile-to-land = 50%; Land-to-mobile = 40%; and Mobile-to-mobile = 10%
  a. Calculate the total traffic in Erlangs, and
  b. If each switch can support 3000 Erlangs, how many switches are required in the network?

References

1. Bellcore. "LATA Switching Systems Generic Requirements—Teletraffic Capacity and Environment," Technical Reference, TR-TSY-000517, Issue 3, March 1989.

2. Copper, R. B. *Introduction to Queuing Theory*. New York: North-Holland, 1981.

3. Freeman, R. L. *Telecommunication System Engineering*. New York: John Wiley & Sons, 1989.

4. Kleinrock, L. *Queuing System, Vol. 1*. New York: John Wiley & Sons, 1975.

5. Kleinrock, L. *Queuing System, Vol. 2*. New York: John Wiley & Sons, 1976.

6. Rappaport, T. S. *Wireless Communications*. Upper Saddle River, NJ: Prentice Hall, 1996.

7. Rapp, Y. "Planning of Junction Network in Multiexchange Areas." *Ericsson Technics* 20(1), 1964, pp. 77–130.

8. Rey, R. F. Technical Editor, *Engineering and Operations in Bell System*. Murray Hill, NJ: AT&T Bell Labs, 1984.

9. Sharma, R. L., et al. *Network Systems*. New York: Van Nostrand Reinhold Co., 1982.

CHAPTER 3

Radio Propagation and Propagation Path-Loss Models

3.1 Introduction

Exponential growth of mobile communications has increased interest in many topics in radio propagation. Much effort is now devoted to refine radio propagation path-loss models for urban, suburban, and other environments together with substantiation by field data. Radio propagation in urban areas is quite complex because it often consists of reflected and diffracted waves produced by multipath propagation. Radio propagation in open areas free from obstacles is the simplest to treat, but, in general, propagation over the earth and the water invokes at least one reflected wave.

For closed areas such as indoors, tunnels, and underground passages, no established models have been developed as yet, since the environment has a complicated structure. However, when the environmental structure is random, the Rayleigh model used for urban area propagation may be applied. When the propagation path is on line of sight, as in tunnel and underground passages, the environment may be treated either by the Rician model or waveguide theory. Direct wave models may be used for propagation in a corridor.

In general, radio wave propagation consists of three main attributes: reflection, diffraction and scattering (see Figure 3.1) [2]. *Reflection* occurs when radio wave propagating in one medium impinges upon another medium with different electromagnetic properties. The amplitude and phase of the reflected wave are strongly related to the medium's instrinsic impedance, incident angle, and electric field polarization. Part of the radio wave energy may be absorbed or propagated through the reflecting medium, resulting in a reflected wave that is attenuated.

Diffraction is a phenomenon by which propagating radio waves bend or deviate in the neighborhood of obstacles. Diffraction results from the propagation of wavelets into a shadowy region caused by obstructions such as walls, buildings, mountains, and so on.

Scattering occurs when a radio signal hits a rough surface or an object having a size much smaller than or on the order of the signal wavelength. This causes the

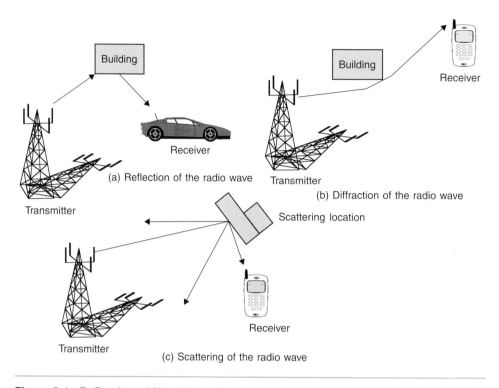

Figure 3.1   Reflection, diffraction and scattering of radio wave.

signal energy to spread out in all directions. Scattering can be viewed at the receiver as another radio wave source. Typical scattering objects are furniture, lamp posts, street signs, and foliage.

In this chapter, our focus is to characterize the radio channel and identify those parameters which distort the information-carrying signal (i.e., base band signal) as it penetrates the propagation medium. The several empirical models used for calculating path-loss are also discussed.

3.2   Free-Space Attenuation

The most simple wave propagation case is that of a direct wave propagation in free space [5,7]. In this special case of line-of-sight (LOS) propagation there are no obstructions due to the earth's surface or other obstacles (see Figure 3.2). We consider radiation from an isotropic antenna. This type of antenna is completely omni-directional, radiating uniformly in all directions. While there is no such thing as a purely isotopic antenna in practice, it is a useful theoretical concept.

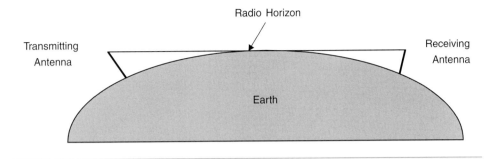

Figure 3.2 Line-of-sight propagation.

The received power, P_r, at the receiving antenna (mobile station), located at a distance, d, from the transmitter (base station) is given for free space propagation [10] as:

$$P_r = P_t \left(\frac{\lambda}{4\pi d} \right)^2 G_b G_m \qquad (3.1)$$

If other losses (not related to propagation) are also present, we can rewrite Equation 3.1 as:

$$\frac{P_r}{P_t} = \left(\frac{\lambda}{4\pi d} \right)^2 \cdot \frac{G_b G_m}{L_0} = \frac{G_b G_m}{L_p L_0} \qquad (3.2)$$

where:

P_r = received power
P_t = transmitted power
λ = wave length $= \dfrac{c}{f} = \dfrac{c \cdot 2\pi}{\omega_c}$
ω_c = carrier frequency in rad./sec
c = velocity of electromagnetic waves in the free space (3×10^8 m/s)
G_b = gain of the transmitting (base station) antenna
G_m = gain of the receiving (mobile) antenna
d = antenna separation distance between transmitter and receiver (i.e., base station and mobile station)
L_0 = other losses expressed as a relative attenuation factor
L_p = $\left[\dfrac{4\pi d}{\lambda} \right]^2$ = free space path loss, often expressed as an attenuation in decibels (dB)

$$= 20 \log \left(\frac{4\pi d}{\lambda} \right) (\text{dB})$$

We can express L_p (dB) in free space as:

$$(L_p)_{\text{free}} = 32.44 + 20\log f + 20\log d \, \text{dB} \tag{3.3}$$

where:

   f = carrier frequency in MHz
   d = separation distance in km ($> 1\,\text{km}$)

It should be noted that the free-space attenuation increases by 6 dB whenever the length of the path is doubled. Similarly, as frequency is doubled, free-space attenuation also increases by 6 dB.

3.3   Attenuation over Reflecting Surface

Free-space propagation is encountered only in rare cases such as satellite-to-satellite paths. In typical terrestrial paths, the signal is partially blocked and attenuated due to urban clutter, trees, and other obstacles. Multipath propagation also occurs due to reflection from the ground. Signals reflected from the ground are fundamentally no different than any other reflected signal. However, they are considered separately because they contribute to the received signal power at virtually all terrestrial receiving locations.

The energy radiated from a transmitting antenna may reach the receiving antenna over several possible propagation paths. For many wireless applications in the 50 to 2000 MHz range, two components of the space wave are of primary concern: energy received by means of the direct wave, which travels a direct path from the transmitter to the receiver, and a ground-reflected wave, which arrives at the receiver after being reflected from the surface of the earth. This is commonly referred to as the two-ray model. Other propagation paths such as sky and surface waves are often neglected.

We assume that the base station and mobile station antenna heights, h_b and h_m, are much smaller compared to their separation, d, and the reflecting earth surface is flat (see Figure 3.3 for direct and reflected waves). The received power at the antenna located at a distance, d, from the transmitter, including other losses, L_0, is given as [7,10] (see Appendix B for derivation):

$$P_r = \left[\frac{h_b h_m}{d^2}\right]^2 \frac{(G_b G_m)}{L_0} P_t \tag{3.4}$$

where:

$$d \gg \overline{d} \text{ and } \overline{d} = \frac{4h_b h_m}{\lambda}$$

Note that under the assumed conditions the received signal level depends only on the transmitted power, antenna heights, and separation distance; there is no frequency dependence. Furthermore, the total attenuation increases by 12 dB when the separation distance is doubled.

The expression for the effects of ground reflections from a flat (or plane) earth provides results that are approximately correct for $\Delta\alpha = \dfrac{2\pi}{\lambda(d_d - d_r)} \leq \dfrac{\pi}{8}$

(see Figure 3.3). The results are not valid for $\Delta\alpha > \pi/8$. When $\Delta\alpha > \pi/8$, the attenuation factor will be:

$$A_{gr} = \sqrt{1 + \rho^2 - 2\rho\cos\Delta\alpha} \qquad (3.5)$$

where:

ρ = reflection coefficient of ground (assumed to be -1)

Example 3.1

With $h_b = 100$ ft, and $h_m = 5$ ft, and a frequency of 881.52 MHz ($\lambda = 1.116$ ft), calculate signal attenuation at a distance equal to 5000 ft. Assume antenna gains are 8 dB and 0 dB for the base station and mobile station, respectively. What are the free-space and reflected surface attenuations? Assume the earth surface to be flat.

Solution

$$\overline{d} = \frac{4h_b h_m}{\lambda} = \frac{4 \times 100 \times 5}{1.116} = 1792 \text{ ft}, \ d > \overline{d}$$

$$G_b = 8\,\text{dB} = 10^{0.8} = 6.3; \quad G_m = 0\,\text{dB} = 1.0$$

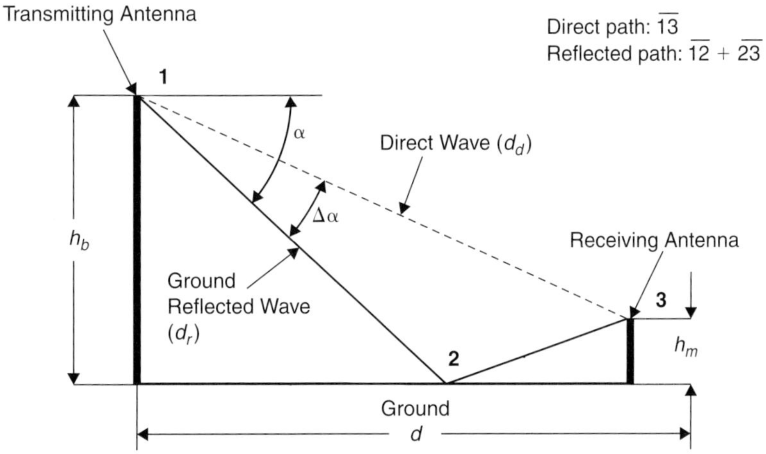

Figure 3.3 Geometry for direct and ground reflected waves.

Free-space attenuation:

$$\frac{P_t}{P_r} = \left(\frac{4\pi d}{\lambda}\right)^2 \cdot \frac{1}{G_b G_m} = \left(\frac{4\pi \times 5000}{1.116}\right)^2 \cdot \frac{1}{6.3 \times 1} = 87\,\text{dB}$$

Attenuation on reflecting surface:

$$\frac{P_t}{P_r} = \left[\frac{d^2}{h_b h_m}\right]^2 \cdot \frac{1}{G_b G_m} = \left[\frac{5000^2}{100 \times 5}\right]^2 \cdot \frac{1}{6.3 \times 1} = 86\,\text{dB}$$

Example 3.2

Consider a base station transmitting to a mobile station in free space. The following parameters relate to this communication system:

- Distance between base station and mobile station: 8000 m
- Transmitter frequency: 1.5 GHz ($\lambda = 0.2$ m)
- Base station transmitting power, $P_t = 10\,\text{W}(10\,\text{dBW})$
- Total system losses: 8 dB
- Mobile receiver noise figure $N_f = 5$ dB
- Mobile receiver antenna temperature = 290 K
- Mobile receiver bandwidth $B_w = 1.25$ MHz
- Antenna gains are 8 dB and 0 dB for the base station and mobile station, respectively.
- Antenna height at the base station and mobile station are 30 m and 3 m, respectively.

Calculate the received signal power at the mobile receiver antenna and signal-to-noise ratio (SNR) of the received signal.

Solution

$$\text{Free space path loss} = -20\log\left[\frac{30 \times 3}{8000^2}\right] = -117\,\text{dB}$$

$$P_r = -117 + 8 + 10 + 0 - 8 = -107\,\text{dBW} = -77\,\text{dBm}$$

$$T_e = 290(3.162 - 1) = 627\,\text{K}$$

$$P_n = 1.38 \times 10^{-23}(627 + 290)(1.25 \times 10^6) = 1.58 \times 10^{-14}\,\text{W} = -138\,\text{dBW}$$

$$\therefore SNR = \frac{P_r}{P_n} = -107 - (-138) = 31\,\text{dB}$$

3.4 Effect of Earth's Curvature

We assumed a flat earth in Section 3.3. In reality, the earth is curved, preventing LOS propagation to great distances. Thus, Equations 3.1 and 3.2 are only approximately correct for distances less than the distance d to the radio horizon (see Figure 3.4). The distance d to the radio horizon can be determined for a terrestrial transmitter as follows.

We consider a circle representing the earth [12]. From Figure 3.4 we write

$$d^2 + R_e^2 = (R_e + h_b)^2$$

$$d^2 = 2R_e h_b + h_b^2$$

where:

R_e = radius of earth
h_b = antenna height

Since $2R_e h_b \gg h_b^2$

$$d \approx \sqrt{2R_e h_b} \tag{3.6}$$

As an example with an antenna height of 600 m and earth's radius of about 6400 km, the distance to the horizon is about 88 km. On average the refractive index of earth's atmosphere is such that the earth's radius appears to be about 33% more than the actual radius. Thus, the effective radius of the earth is often assumed to be 8500 km. The curvature of the earth further affects the propagation of the space

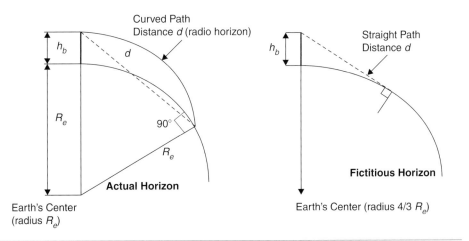

Figure 3.4 Geometry for a spherical earth.

wave since the ground-reflected wave is reflected from a curved surface. Therefore, the energy on a curved surface diverges more than it does from a flat surface and the ground-reflected wave reaching the receiver is weaker than for a flat earth. The divergence factor D that describes this effect is less than unity and is given as:

$$D = \frac{1}{\sqrt{1 + \frac{2}{\left(\frac{R_e(h_b + h_m)^3}{d^2 h_b hm}\right) - 0.5\left(\frac{h_b}{h_m} + \frac{h_m}{h_b}\right)}}}$$ (3.7)

where:
  h_m = mobile antenna height

  D in Equation 3.7 ranges from unity for a small value of d and approaches zero as d approaches the distance to the radio horizon. It can be combined with the ground reflection coefficient so that the attenuation due to ground reflections becomes

$$A_{gr} = \sqrt{1 + (\rho D)^2 - 2\rho D \cos(\Delta\alpha)}$$ (3.8)

where:
  A_{gr} = attenuation factor due to ground reflections
  ρ = reflection coefficient of the earth surface
  D = divergence factor

  In Equation 3.8 the reflection coefficient, ρ, has been modified to account for the divergence factor, D. The effect of the divergence factor is to reduce the effective reflection coefficient of the earth. The equation for calculating the received power in dBm is given as:

$$P_r = P_t + G_b + G_m + 20\log(A_{fs}) + 20\log(A_{gr}) \text{ dBm}$$ (3.9)

where:
  $A_{fs} = [\lambda/(4\pi d)]^2$, free space attenuation

3.5 Radio Wave Propagation

Radio waves propagate [7] through space as travelling electromagnetic waves. The energy of signals exists in the form of electrical and magnetic fields. Both electrical and magnetic fields vary sinusoidally with time. The two fields always exist together because a change in electrical field generates a magnetic field and a change in magnetic field develops an electrical field. Thus there is continuous flow of energy from one field to the other.

Radio waves arrive at a mobile receiver from different directions with different time delays. They combine via vector addition at the receiver antenna to give a resultant signal with a large or small amplitude depending upon whether the incoming waves combine to strengthen each other or cancel each other. As a result, a receiver at one location may experience a signal strength several tens of decibels (dB) different from a similar receiver located only a short distance away. As a mobile receiver moves from one location to another, the phase relationship between the various incoming waves also changes. Thus, there are substantial amplitude and phase fluctuations, and the signal is subjected to *fading*. Figure 3.5 illustrates the fading characteristics of a mobile radio signal.

A steady decrease in the received signal power at a separation distance, *d*, of several kilometers (or miles) occurs. This is the signal attenuation. Attenuation is proportional to the second power of distance in the free space, but can vary to the fourth or fifth power in built-up areas because of reflections and obstacles.

When we focus on a distance of a couple of kilometers, we observe that signal power fluctuates around a mean value and the fluctuations have a somewhat longer period. This is referred to as *long-term* or *slow fading*.

When we concentrate and examine the signal power over a few hundred meters, we find that signal power fluctuates more rapidly. These rapid fluctuations are caused by a local multipath. The phenomenon giving rise to these rapid fluctuations is referred to as *short-term* or *fast fading*.

The slow fading is caused by movement over distances large enough to produce gross variations in the overall path between the base station and the mobile station. Fast fading occurs usually over distances of about half a wavelength. For VHF and

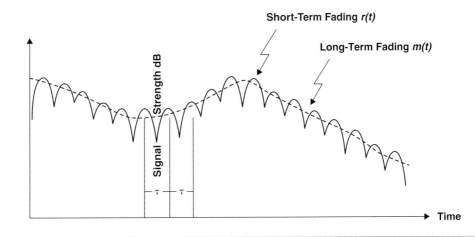

Figure 3.5   Mobile radio signal fading representation.

UHF, a vehicle travelling at 30 miles per hour (mph) can encounter several fast fades in a second. Therefore, the mobile radio signal (see Figure 3.5) contains a short-term fast fading signal superimposed on a long-term slow fading signal (which remains constant over a small area but varies slowly as the mobile receiver moves).

To separate out fast fading from slow fading, the magnitude of the received signal is averaged over a distance on the order of 10 m, and the result is referred to as the small-area average or sector average [1]. The rapid fluctuation in fast fading is a result of small movements of the transmitter, receiver, and surrounding objects. Because fast fading is random, its statistical properties are used to determine system performance. For locations that are heavily shadowed by surrounding buildings, it is typically found that a *Rayleigh distribution* approximates the probability density function (PDF) [8,9,14,18]. For locations where there is one path making a dominant contribution to the received signal, such as when the base station is visible to the mobile station (typically in the indoor environment), the distribution function is typically found to be that of a *Rician distribution* [13,14].

Because of shadowing by buildings and other objects, the average within individual small areas also varies from one small area to the next in an apparently random manner, referred to as the *shadow effect*. Shadow effect is often called *lognormal* fading because its distribution is represented by lognormal distribution. In a moving vehicle, slow fading is observed over a longer time scale than fast fading. Slow fading is the average of received signal power over large transmitter and receiver separation distances. A local mean is computed by averaging signal power over 5 to 40 wavelengths (λ), or separation distance between 40 to 80 fades (where a signal crosses a certain level).

Using a reference distance, d_0, which is the signal attenuation at a standard distance from the antenna (usually $d_0 = 1$ m is used for indoor environment and $d_0 = 1$ km for outdoor environment) the received power, P_r, at distance d is given as:

$$P_r = P_0(d_0/d)^\gamma \tag{3.10}$$

where:

γ = path loss exponent, varying from 2 (free-space) to 5 (urban environment)
P_0 = power at reference distance d_0
P_r = received power that is proportional to $d^{-\gamma}$

The received power under non-line-of-sight propagation conditions can be written as:

$$P_r(d) = 10\log[P_0(d_0)] + 10\gamma\log(d_0/d) \text{ (dBm)} \tag{3.11}$$

The accuracy of P_r can be improved by accounting for a random shadow effect caused by obstructions such as buildings or mountains. Shadow effect is described by a zero-mean Gaussian random variable, X_σ, with standard deviation, σ (dB). Under ideal conditions, it is possible to estimate path loss $L_p(d)$ from the transmitter (P_t) and receiver power (P_r) as $L_p(d) = P_t - P_r$, but this approach ignores the fact that the signal undergoes lognormal fading, which could reduce the received power at any location. Since the fading is long term, no improvements can be expected if the mobile station moves through short distances.

$$\text{Allowable path loss } L_p(d) = \text{Path loss} + \text{Shadow effect } X_\sigma \qquad (3.12)$$

$$L_p(d) = \overline{L_p}(d_0) + 10\gamma \log(d/d_0) + X_\sigma \text{ (dB)} \qquad (3.13)$$

where:
$\overline{L_p}(d_0)$ is known as the 1 m or 1 km loss, or insertion loss that arises due to free-space path loss and antenna inefficiencies (i.e., reference value of path loss).

X_σ is often based on measurement made over a wide range of locations and transmitter-receiver separation. An average value of 8 dB for σ is often used giving X_σ as 10.5 dB.

It should also be noted that whenever relative motion exists between transmitter and receiver, there is a *Doppler shift* in the received signal. The maximum Doppler shift f_m is given as:

$$f_m = \frac{v}{c/f} \qquad (3.14)$$

where:
$\quad c$ = velocity of electromagnetic waves in free space
$\quad v$ = velocity of the moving vehicle
$\quad f$ = frequency of the carrier

In the mobile radio case, the fading and Doppler shift occur as a result of the motion of the receiver through a spatially varying field. Doppler shift also results from the motion of the scattering of the radio waves (e.g., cars, trucks, vegetation). The effect of multipath propagation is to produce a received signal with an amplitude that varies quite substantially with location. At UHF and higher frequencies, the motion of the scattering also causes fading to occur even if the mobile station or handset is not in motion.

The received signal $s(t)$ can be expressed as the product of two components: the signal subjected to long-term fading, $m(t)$, and the signal subjected to short-term fading, $r(t)$, as:

$$s(t) = m(t) \cdot r(t) \qquad (3.15)$$

Example 3.3

Calculate the received power at a distance of 3 km from the transmitter if the path-loss exponent γ is 4. Assume the transmitting power of 4 W at 1800 MHz, a shadow effect of 10.5 dB, and the power at reference distance ($d_0 = 100$ m) of -32 dBm. What is the allowable path loss?

Solution

Using Equation 3.11 and including shadow effect we get

$$P_r(d) = 10\log[P_0(d_0)] + 10\gamma\log(d_0/d) + X_\sigma$$

$$\therefore P_r = -32 + 10 \times 4 \times \log\left(\frac{100}{3000}\right) + 10.5 = -80.5\,\text{dBm}$$

$$\text{Allowable path loss} = P_t - P_r = 36 - (-80.5) = 116.5\,\text{dB}$$

Example 3.4

What is the separation distance between the transmitter and the receiver with an allowable path loss of 150 dB and shadow effect of 10 dB? The path loss in dB is given as:

$$L_p = 133.2 + 43\log d$$

where:
d = separation distance in km

Solution

Using Equation 3.13, we have

$$150 = 133.2 + 40\log d + 10$$

$$\log d = \frac{6.8}{40} = 0.17$$

$$d = 10^{0.17} = 1.48\,\text{km}$$

3.6 Characteristics of a Wireless Channel

The wireless channel is different and much more unpredictable than the wireline channel because of factors such as multipath and shadow fading, Doppler shift, and time dispersion or delay spread. These factors are all related to variability introduced by mobility of the user and the wide range of environmental conditions that are encountered as a result. Multipath delays occur as a transmitted signal is reflected by objects in the environment between a transmitter and a receiver. These objects can be buildings, trees, hills, or even trucks and cars (see Figure 3.6). The reflected signals arrive at the receiver with a random phase offset, since each reflected signal generally follows a different path to reach the user's receiver, resulting in a random signal that fades as the reflections destructively or constructively superimpose on one another. This effectively cancels or adds part of signal energy for brief periods of time. The degree of fading will depend on the delay spread of the reflected signals as embodied by their relative phases, and their relative power.

A mobile radio channel exhibits both time dispersion and frequency dispersion. *Time dispersion* is the distortion to the signal and is manifested by the spreading in time of the modulation symbols. This is caused by *frequency-selective* fading. A channel, which is said to be frequency selective, has many frequency components that take different times to arrive at the receiver and undergo different attenuation levels. The frequency band over which the attenuation remains constant provides a frequency

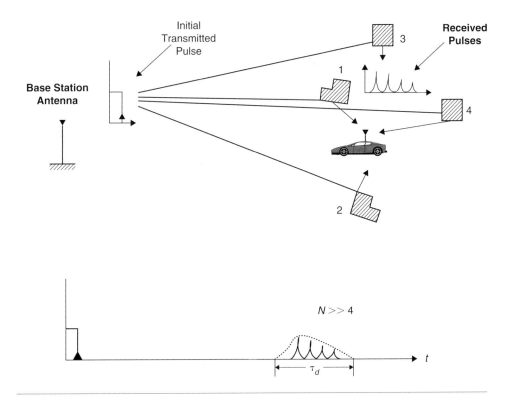

Figure 3.6 Multipath delay.

region where all frequency components behave identically. We identify this frequency band as the *coherence bandwidth* of the channel.

Time dispersion occurs when the channel is band-limited or when the coherence bandwidth of the channel is smaller than the modulation bandwidth. The time dispersion leads to *inter-symbol-interference* (ISI), where the energy from one symbol spills over into another symbol, thereby increasing the bit-error-rate (BER).

In many instances, the fading due to multipath delay will be frequency selective, randomly affecting only a portion of the overall channel bandwidth at a given time. In the case of frequency selective fading, the delay spread exceeds the symbol duration. On the other hand, when there is no dispersion and delay spread is less than the symbol duration, the fading will be *flat*, thereby affecting all frequencies in the signal equally. Flat fading can lead to deep fades of more than 30 to 40 dB.

Doppler shift is the random changes in a channel introduced as a result of a mobile user's mobility. Doppler spread has the effect of shifting or spreading the frequency components of a signal. This is described in terms of frequency dispersion. Like the coherence bandwidth, *coherence time* is defined as the time over which the channel can be assumed to be constant. The coherence time of the channel is the inverse of the Doppler spread. It is the measure of the speed at which channel characteristics change. This determines the rate at which fading occurs. When the channel changes are higher than the modulated symbol rate, fast fading occurs. Slow fading occurs when the channel changes are slower than the symbol rate.

3.6.1 Multipath Delay Spread, Coherence Bandwidth, and Coherence Time

As discussed earlier, the multipath delay spread is the time dispersion characteristic of the channel. Each multipath component is typically associated with different time delays and attenuation, the shortest of which is the LOS path.

We denote the rms delay spread in multipath delay by τ_d and the maximum spread in frequency due to Doppler shift, f_m. We use the coherence bandwidth, which is a range of frequencies over which two frequency components have a strong potential for amplitude correlation, to define whether the channel fading is flat or frequency selective.

The coherence bandwidth (B_c) between two frequency envelopes is given as [16]:

$$B_c \approx \frac{1}{2\pi\tau_d} \tag{3.16}$$

Frequency components of a signal separated by more than B_c will fade independently. A channel is a frequency-selective channel if $B_c < B_w$, where B_w is the signal bandwidth. Frequency selective distortion occurs whenever a signal's spectral components are not all affected equally by the channel. In order to avoid channel-induced ISI distortion, the channel is required to be flat fading by ensuring that $B_c > B_w$. Thus, the channel coherence bandwidth sets an upper limit on the transmission rate that can be used without incorporating an equalizer in the receiver.

The coherence time, T_c, describes the expected time duration over which the impulse response of the channel stays relatively invariant or correlated. The coherence time is approximately inversely proportional to Doppler spread [16,17]

$$T_c \approx \frac{1}{2\pi f_m} \tag{3.17}$$

where:

$f_m = \frac{v}{\lambda}$ = maximum Doppler spread
v = velocity of moving vehicle
λ = wavelength = c/f
f = frequency of carrier
c = speed of electromagnetic wave in free space (3×10^8 m/s)

A rule of thumb for the coherence time value is $T_c = 0.423/f_m$.

If the transmitted symbol interval, T_s, exceeds T_c, then the channel will change during the symbol interval and symbol distortion will occur. In such cases, a matched filter is impossible without equalization and correlator losses occur. A Rayleigh fading signal may change amplitude significantly in the interval T_c.

If the signal symbol interval $T_s \gg T_c$, the channel changes or fades rapidly compared to the symbol rate. This case is called *fast fading* and frequency dispersion occurs, causing distortion. If $T_s \ll T_c$ the channel does not change during the symbol interval. This case is called *slow fading*. Thus, to avoid signal distortion caused by fast fading, the channel must be made to exhibit slow fading by ensuring that the signaling rate exceeds the channel fading rate $T_s < T_c$.

Example 3.5

Assuming the speed of a vehicle is equal to 60 mph (88 ft/sec), carrier frequency, $f_c = 860$ MHz, and rms delay spread $\tau_d = 2\mu$ sec, calculate coherence time and coherence bandwidth. At a coded symbol rate of 19.2 kbps (IS-95) what kind of symbol distortion will be experienced? What type of fading will be experienced by the IS-95 channel?

Solution

$v = 60$ mph ($=88$ ft/sec)

$\lambda = \frac{c}{f} = \frac{9.84 \times 10^8}{860 \times 10^6} = 1.1442$ ft

Maximum Doppler shift $= f_m = \frac{v}{\lambda} = \frac{88}{1.1442} = 77$ Hz

$T_c = \frac{1}{2\pi f_m} = \frac{1}{2\pi \times 77} = 0.0021$ seconds

$$T_s = \frac{10^6}{19,200} = 52\,\mu s$$

The symbol interval is much smaller compared to the channel coherence time. Symbol distortion is, therefore, minimal. In this case fading is slow.

$$\text{Coherence bandwidth} = B_c \approx \frac{1}{2\pi\tau_d} = \frac{1}{2\pi \times 2 \times 10^{-6}} = 79.56\,\text{kHz}$$

This shows that IS-95 is a wide band system in this multipath situation and experiences selective fading only over 6.5% (79.56/1228.8 = 0.0648 ~ 6.5%) of its bandwidth ($B_w = 1228.8\,\text{kHz}$).

3.7   Signal Fading Statistics

As discussed earlier, the rapid variations (fast fading) in signal power caused by local multipaths are represented by *Rayleigh* distribution. The long-term variations in the mean level are denoted by *lognormal* distribution. With a LOS propagation path, the *Rician* distribution is often used for fast fading. Thus, the fading characteristics of a mobile radio signal are described by the following statistical distributions (see Figures 3.7 and 3.8).

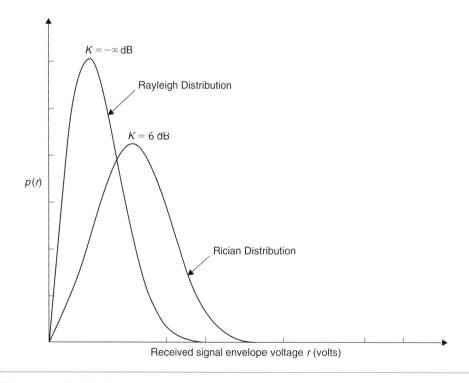

Figure 3.7   Rayleigh and Rician distribution.

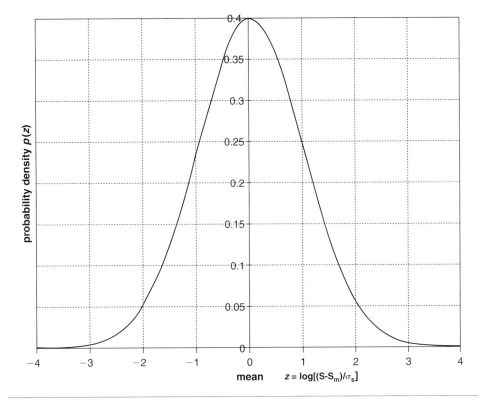

Figure 3.8 Lognormal distribution.

- Rician Distribution
- Rayleigh Distribution
- Lognormal Distribution

3.7.1   Rician Distribution

When there is a dominant stationary (nonfading) signal component present, such as a LOS propagation path, the small-scale fading envelope distribution is Rician. The Rician distribution has a probability density function (PDF) given by:

$$p(r) = \frac{r}{\sigma^2} e^{-\left(\frac{r^2 + A^2}{2\sigma^2}\right)} I_0\left(\frac{Ar}{\sigma^2}\right) \qquad \text{for } A \geq 0, \quad r \geq 0 \qquad (3.18)$$

where:

A = peak amplitude of the dominant signal and I_0 = (…) modified Bessel Function of the first kind and zero order

$r^2/2$ = instantaneous power

σ = standard deviation of the local power

The Rician distribution is often described in terms of a parameter K, known as the Rician factor and is expressed as:

$$K = 10\log\frac{A^2}{2\sigma^2} \text{ dB} \tag{3.19}$$

As $A \to 0$, $K \to \infty$ dB and as the dominant path decreases in amplitude, the Rician distribution degenerates to a Rayleigh distribution (see Figure 3.7).

3.7.2 Rayleigh Distribution

The Rayleigh distribution is used to describe the statistical time-varying nature of the received envelope of a flat fading signal, or the envelope of an individual multipath component. The Rayleigh distribution is given as:

$$p(r) = \frac{r}{\sigma^2}\, e^{-\left(\frac{r^2}{2\sigma^2}\right)} \qquad 0 \le r \le \infty \tag{3.20}$$

where:

σ = rms value of the received signal
$r^2/2$ = instantaneous power
σ^2 = local average power of the received signal before envelope detection

Instead of the distribution of the received envelope, we can also describe a Rayleigh fading signal in terms of the distribution function of its received normalized power.

Let $\Phi = (r^2/2)/\sigma^2$, the instantaneous received power divided by the mean received power. Then $d\Phi = r/\sigma^2 \cdot dr$ and since $p(r)dr$ must be equal to $p(\Phi)d\Phi$, we get

$$p(\Phi) = [p(r)dr]/[(r/\sigma^2)dr] = \frac{(r/\sigma^2)e^{-(r^2/2\sigma^2)}}{(r/\sigma^2)} = e^{-\Phi}, \quad 0 \le \Phi \le \infty \tag{3.21}$$

Equation 3.21 represents a simple exponential density function. One can rightfully say that a flat fading signal is exponentially fading in power.

3.7.3 Lognormal Distribution

Lognormal distribution describes the random shadowing effects which occur over a large number of measurement locations which have the same transmitter and receiver separation, but have different levels of clutter on the propagation path. The signal, $s(t)$, typically follows the Rayleigh distribution but its mean square value or its local mean power is lognormal in dBm with variance equal to σ_s^2. Typically the standard deviation, σ_s equals 8 to 10 dB.

The lognormal distribution is given by (see Figure 3.8):

$$p(S) = \frac{1}{\sqrt{2\pi}\sigma_s} e^{-\left[\frac{(S-S_m)^2}{2\sigma_s^2}\right]}$$

(3.22)

where:
- S_m = mean value of S in dBm
- σ_s = standard deviation of S in dB
- S = 10 log s in dBm
- s = signal power in mW

3.8   Level Crossing Rate and Average Fade Duration

Doppler spread can be used to calculate the *level crossing rate* (LCR), or the average number of times a signal crosses a certain level. The combination of LCR and *average fade duration*, which is the average time over which the signal below a certain level can be used to design the signaling format and a diversity scheme for cellular systems. Let $\rho = R/R_{rms}$ be the value of the specified amplitude level, R, normalized to local rms amplitude of the fading envelope R_{rms}, based on the Rayleigh distribution of the received signal envelope. LCR and average fade duration are given by [1]:

$$LCR = N_R = \sqrt{2\pi} \cdot f_m \cdot \rho e^{-\rho^2} \text{ (crossings per sec)}$$

(3.23)

$$\text{Average fade duration} = \frac{e^{-\rho^2}}{\rho \cdot f_m} \cdot \frac{1}{\sqrt{2\pi}} \text{ (ms)}$$

(3.24)

where:
- f_m = maximum Doppler shift

Example 3.6
Consider a flat Rayleigh fading channel to determine the number of fades per second for $\rho = 1$ and average fade duration, when the maximum Doppler frequency is 20 Hz. What is the maximum velocity of the mobile if the carrier frequency is 900 MHz?

Solution

$$N_R = \sqrt{2\pi} \cdot f_m \cdot \rho e^{-\rho^2} = \sqrt{2\pi} \times 20 \times 1 \times e^{-1} = 18.44 \text{ fades per second}$$

$$\text{Average fade duration} = \frac{e^{-1}}{1 \times 20} \cdot \frac{1}{\sqrt{2\pi}} = 0.0073 \text{ s}$$

$$v = f_m \cdot \frac{c}{f} = 20 \times \frac{3 \times 10^8}{900 \times 10^6} = 6.66 \text{ m/s} = 24 \text{ km/hour}$$

3.9 Propagation Path-Loss Models

Propagation path-loss models [20] play an important role in the design of cellular systems to specify key system parameters such as transmission power, frequency, antenna heights, and so on. Several models have been proposed for cellular systems operating in different environments (indoor, outdoor, urban, suburban, rural). Some of these models were derived in a statistical manner based on field measurements and others were developed analytically based on diffraction effects. Each model uses specific parameters to achieve reasonable prediction accuracy. The long distance prediction models intended for macrocell systems use base station and mobile station antenna heights and frequency. On the other hand, the prediction models for short distance path-loss estimation use building heights, street width, street orientation, and so on. These models are used for microcell systems. When the cell size is quite small (in the range of 10 to 100 m), deterministic models based on ray tracing methods are used. Thus, it is essential to select a proper path-loss model for design of the mobile system in the given environment.

Propagation models are used to determine the number of cell sites required to provide coverage for the network. Initial network design typically is based on coverage. Later growth is engineered for capacity. Some systems may need to start with wide area coverage and high capacity and therefore may start at a later stage of growth.

The coverage requirement along with the traffic requirement relies on the propagation model to determine the traffic distribution, and will offload from an existing cell site to new cell sites as part of a capacity relief program. The propagation model helps to determine where the cell sites should be placed to achieve an optimal location in the network. If the propagation model used is not effective in placing cell sites correctly, the probability of incorrectly deploying a cell site in the network is high.

The performance of the network is affected by the propagation model chosen because it is used for interference predictions. As an example, if the propagation model is inaccurate by 6 dB (provided $S/I = 17$ dB is the design requirement), then the signal-to-interference ratio, S/I, could be 23 dB or 11 dB. Based on traffic conditions, designing for a high S/I could negatively affect financial feasibility. On the other hand, designing for a low S/I would degrade the quality of service.

The propagation model is also used in other system performance aspects including handoff optimization, power level adjustments, and antenna placements. Although no propagation model can account for all variations experienced in real life, it is essential that one should use several models for determining the path losses in the network. Each of the propagation models being used in the industry has pros and cons. It is through a better understanding of the limitations of each of the models that a good RF engineering design can be achieved in a network.

We discuss two widely used empirical models: Okumura/Hata and COST 231 models. The Okumura/Hata model has been used extensively both in Europe and

North America for cellular systems. The COST 231 model has been recommended by the European Telecommunication Standard Institute (ETSI) for use in Personal Communication Network/Personal Communication System (PCN/PCS). In addition, we also present the empirical models proposed by International Mobile Telecommunication-2000 (IMT-2000) for the indoor office environment, outdoor to indoor pedestrian environment, and vehicular environment.

3.9.1 Okumura/Hata Model

Okumura analyzed path-loss characteristics based on a large amount of experimental data collected around Tokyo, Japan [6,11]. He selected propagation path conditions and obtained the average path-loss curves under flat urban areas. Then he applied several correction factors for other propagation conditions, such as:

- Antenna height and carrier frequency
- Suburban, quasi-open space, open space, or hilly terrain areas
- Diffraction loss due to mountains
- Sea or lake areas
- Road slope

Hata derived empirical formulas for the median path loss (L_{50}) to fit Okumura curves. Hata's equations are classified into three models:

1. *Typical Urban*

$$L_{50}(\text{urban}) = 69.55 + 26.16 \log f_c + (44.9 - 6.55 \log h_b) \log d$$

$$- 13.82 \log h_b - a(h_m)(\text{dB}) \tag{3.25}$$

where:

$a(h_m)$ = correction factor (dB) for mobile antenna height as given by:

- For large cities

$$a(h_m) = 8.29[\log(1.54 h_m)]^2 - 11 \qquad f_c \leq 200\,\text{MHz} \tag{3.26}$$

$$a(h_m) = 3.2[\log(11.75 h_m)]^2 - 4.97 \qquad f_c \geq 400\,\text{MHz} \tag{3.27}$$

- For small and medium-sized cities

$$a(h_m) = [1.1 \log(f_c) - 0.7]h_m - [1.56 \log(f_c) - 0.8] \tag{3.28}$$

2. *Typical Suburban*

$$L_{50} = L_{50}(\text{urban}) - 2\left[\left(\log\left(\frac{f_c}{28}\right)^2\right) - 5.4\right]\text{dB} \qquad (3.29)$$

3. *Rural*

$$L_{50} = L_{50}(\text{urban}) - 4.78(\log f_c)^2 + 18.33\log f_c - 40.94\,\text{dB} \qquad (3.30)$$

where:

f_c = carrier frequency (MHz)
d = distance between base station and mobile (km)
h_b = base station antenna height (m)
h_m = mobile antenna height (m)

The range of parameters for which the Hata model is valid is:

$150 \leq f_c \leq 2200\,\text{MHz}$
$30 \leq h_b \leq 200\,\text{m}$
$1 \leq h_m \leq 10\,\text{m}$
$1 \leq d \leq 20\,\text{km}$

3.9.2 Cost 231 Model

This model [19] is a combination of empirical and deterministic models for estimating the path loss in an urban area over the frequency range of 800 MHz to 2000 MHz. The model is used primarily in Europe for the GSM 1800 system.

$$L_{50} = L_f + L_{\text{rts}} + L_{\text{ms}}\,\text{dB} \qquad (3.31)$$

or

$$L_{50} = L_f \text{ when } L_{\text{rts}} + L_{\text{ms}} \leq 0 \qquad (3.32)$$

where:

L_f = free space loss (dB)
L_{rts} = roof top to street diffraction and scatter loss (dB)
L_{ms} = multiscreen loss (dB)

Free space loss is given as:

$$L_f = 32.4 + 20\log d + 20\log f_c\,\text{dB} \qquad (3.33)$$

The roof top to street diffraction and scatter loss is given as:

$$L_{\text{rts}} = -16.9 - 10\log W + 10\log f_c + 20\log \Delta h_m + L_0\,\text{dB} \qquad (3.34)$$

where:

W = street width (m)

$\Delta h_m = h_r - h_m$ m

$L_0 = -10 + 0.354\phi$ $\qquad 0 \le \phi \le 35°$

$L_0 = 2.5 + 0.075(\phi - 35)$ dB $\qquad 35° \le \phi \le 55°$

$L_0 = 4 - 0.114(\phi - 55)$ dB $\qquad 55° \le \phi \le 90°$

where:

ϕ = incident angle relative to the street

The multiscreen (multiscatter) loss is given as:

$$L_{ms} = L_{bsh} + k_a + k_d \log d + k_f \log f_c - 9 \log b \qquad (3.35)$$

where:

b = distance between building along radio path (m)

d = separation between transmitter and receiver (km)

$L_{bsh} = -18 \log(11 + \Delta h_b) \qquad h_b \ge h_r$

$L_{bsh} = 0 \qquad\qquad\qquad\qquad h_b < h_r$

where: $\Delta h_b = h_b - h_r$, h_r = average building height (m)

$k_a = 54 \qquad\qquad\qquad\qquad h_b > h_r$

$k_a = 54 - 0.8 h_b \qquad\qquad d \ge 500\text{m}; \quad h_b \le h_r$

$k_a = 54 - 0.8 \Delta h_b(d/500) \qquad d < 500\text{m}; \quad h_b \le h_r$

Note: Both L_{bsh} and k_a increase path loss with lower base station antenna heights.

$k_d = 18 \qquad\qquad\qquad\qquad h_b < h_r$

$k_d = 18 - \dfrac{15 \Delta h_b}{\Delta h_m} \qquad\qquad h_b \ge h_r$

$k_f = 4 + 0.7(f_c/925 - 1)$ for mid-size city and suburban area with moderate tree density

$k_f = 4 + 1.5\left(\dfrac{f_c}{925} - 1\right)$ for metropolitan area

The range of parameters for which the COST 231 model is valid is:

$800 \le f_c \le 2000\,\text{MHz}$

$4 \le h_b \le 50\,\text{m}$

$1 \le h_m \le 3\,\mathrm{m}$

$0.02 \le d \le 5\,\mathrm{km}$

The following default values may be used in the model:

$b = 20\text{--}50\,\mathrm{m}$

$W = b/2$

$\phi = 90°$

Roof $= 3\,\mathrm{m}$ for pitched roof and $0\,\mathrm{m}$ for flat roof, and

$h_r = 3$ (number of floors) + roof

Example 3.7

Using the Okumura and COST 231 models, calculate the L_{50} path loss for a PCS system in an urban area at 1, 2, 3, 4 and 5 km distance (see Table 3.1). Assume $h_b = 30\,\mathrm{m}$, $h_m = 2\,\mathrm{m}$, and carrier frequency $f_c = 900\,\mathrm{MHz}$. Use the following data for the COST 231 model:

$$W = 15\,\mathrm{m}, \quad b = 30\,\mathrm{m}, \quad \phi = 90°, \quad h_r = 30\,\mathrm{m}$$

COST 231 Model

$$L_{50} = L_f + L_{rts} + L_{ms}$$

$$L_f = 32.4 + 20\log d + 20\log f_c = 32.4 + 20\log d + 20\log 900\,\mathrm{dB}$$

$$L_f = 91.48 + 20\log d\,\mathrm{dB}$$

$$L_{rts} = -16.9 - 10\log W + 10\log f_c + 20\log \Delta h_m + L_0$$

$$\Delta h_m = h_r - h_m = 30 - 2 = 28\,\mathrm{m}$$

Table 3.1 Summary of path losses from COST 231 model.

d (km)	L_f (dB)	L_{rts} (dB)	L_{ms} (dB)	L_{50} (dB)
1	91.49	29.82	9.72	131.03
2	97.50	29.82	15.14	142.46
3	101.03	29.82	18.31	149.16
4	103.55	29.82	20.56	153.91
5	105.47	29.82	22.30	157.59

Note: The table applies to this example only.

$$L_0 = 4 - 0.114(\phi - 55) = 4 - 0.114(90 - 55) = 0$$

$$L_{rts} = -16.9 - 10\log 15 + 10\log 900 + 20\log 28 + 0 = 29.82\,\text{dB}$$

$$L_{ms} = L_{bsh} + k_a + k_d\log d + k_f\log f_c - 9\log b$$

$$k_a = 54 - 0.8h_b = 54 - 0.8 \times 30 = 30$$

$$\Delta h_b = h_b - h_r = 30 - 30 = 0\,\text{m}$$

$$L_{bsh} = -18\log 11 + 0 = -18.75\,\text{dB}$$

$$k_d = 18 - \frac{15\Delta h_b}{\Delta h_m} = 18 - \frac{15 \times 0}{28} = 18$$

$$k_f = 4 + 0.7\left(\frac{f_c}{925} - 1\right) = 4 + 0.7\left(\frac{900}{925} - 1\right) = 3.98 \text{ (for mid-sized city)}$$

$$L_{ms} = -18.75 + 30 + 18\log d + 3.98\log 900 - 9\log 30 = 9.72 + 18\log d\,\text{dB}$$

Okumura/Hata Model

$$L_{50} = 69.55 + 26.16\log f_c + (44.9 - 6.55h_b)\log d - 13.82\log h_b - a(h_m)\,\text{dB}$$

$$a(h_m) = (1.1\log f_c - 0.7)h_m - (1.56\log f_c - 0.8)$$

$$= (1.1\log 1800 - 0.7)(2) - (1.56\log 900 - 0.8) = 1.29\,\text{dB}$$

$$L_{50} = 69.55 + 26.16\log 900 + (44.9 - 6.55\log 30)\log d - 13.82\log 30 - 1.29\,\text{dB}$$

$$= 125.13 + 35.23\log d\,\text{dB (refer to Table 3.2)}$$

The results from the two models are given in Figure 3.9. Note that the calculated path loss with the COST 231 model is higher than the value obtained by the Okumura/Hata model.

Table 3.2 Summary of path losses from Okumura model.

d (km)	L_{50} (dB)
1	125.13
2	135.74
3	141.94
4	142.34
5	145.76

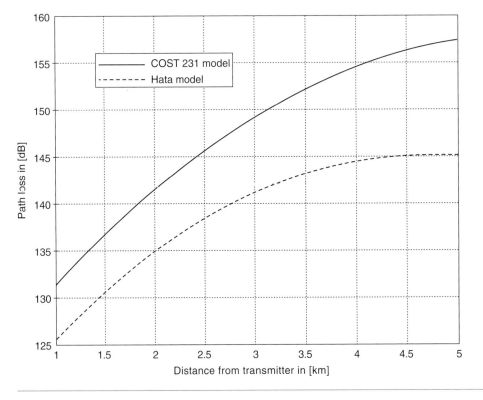

Figure 3.9   Comparison of COST 231 and Hata-Okumura models.

3.9.3   IMT-2000 Models

The operating environments are identified by appropriate subsets consisting of indoor office environments, outdoor to indoor and pedestrian environments, and vehicular (moving vehicle) environments. For narrowband technologies (such as FDMA and TDMA), delay spread is characterized by its rms value alone. However, for wide band technologies (such as CDMA), the strength and relative time delay of the many signal components become important. In addition, for some technologies (e.g., those using power control) the path-loss models must include the coupling between all co-channel propagation links to provide accurate predictions. Also, in some cases, the shadow effect temporal variations of the environment must be modeled. The key parameters of the IMT-2000 propagation models are:

- Delay spread, its structure, and its statistical variation
- Geometrical path loss rule (e.g., $d^{-\gamma}$, $2 \leq \gamma \leq 5$)
- Shadow fading margin
- Multipath fading characteristics (e.g., Doppler spectrum, Rician vs. Rayleigh for envelope of channels)
- Operating radio frequency

Indoor Office Environment

This environment is characterized by small cells and low transmit powers. Both base stations and pedestrian users are located indoors. RMS delay spread ranges from around 35 nsec to 460 nsec. The path loss rule varies due to scatter and attenuation by walls, floors, and metallic structures such as partition and filing cabinets. These objects also produce shadowing effects. A lognormal shadowing with a standard deviation of 12 dB can be expected. Fading characteristic ranges from Rician to Rayleigh with Doppler frequency offsets are determined by walking speeds. Path-loss model for this environment is:

$$L_{50} = 37 + 30 \log d + 18.3 \cdot n^{[(n+2)/(n+1)-0.46]} \, \text{dB} \qquad (3.36)$$

where:
d = separation between transmitter and receiver (m)
n = number of floors in the path

Outdoor to Indoor and Pedestrian Environment

This environment is characterized by small cells and low transmit power. Base stations with low antenna heights are located outdoors. Pedestrian users are located on streets and inside buildings. Coverage into buildings in high power systems is included in the vehicular environment. RMS delay spread varies from 100 to 1800 nsec. A geometric path-loss rule of d^{-4} is applicable. If the path is a line-of-sight on a canyon-like street, the path loss follows a rule of d^{-2}, where there is Fresnel zone clearance. For the region with longer Fresnel zone clearance, a path loss rule of d^{-4} is appropriate, but a range of up to d^{-6} may be encountered due to trees and other obstructions along the path. Lognormal shadow fading with a standard deviation of 10 dB is reasonable for outdoors and 12 dB for indoors. Average building penetration loss of 18 dB with a standard deviation of 10 dB is appropriate. Rayleigh and/or Rician fading rates are generally set by walking speeds, but faster fading due to reflections from moving vehicles may occur sometimes. The following path-loss model has been suggested for this environment:

$$L_{50} = 40 \log d + 30 \log f_c + 49 \, \text{dB} \qquad (3.37)$$

This model is valid for non-line-of-sight (NLOS) cases only and describes the worst-case propagation. Lognormal shadow fading with a standard deviation equal to 10 dB is assumed. The average building penetration loss is 18 dB with a standard deviation of 10 dB.

Vehicular Environment

This environment consists of larger cells and higher transmit power. RMS delay spread from 4 microseconds to about 12 microseconds on elevated roads in hilly or mountainous terrain may occur. A geometric path-loss rule of d^{-4} and lognormal shadow fading with a standard deviation of 10 dB are used in the urban and suburban areas. Building penetration loss averages 18 dB with a 10 dB standard deviation.

In rural areas with flat terrain the path loss is lower than that of urban and suburban areas. In mountainous terrain, if path blockages are avoided by selecting base station locations, the path-loss rule is closer to d^{-2}. Rayleigh fading rates are determined by vehicle speeds. Lower fading rates are appropriate for applications using stationary terminals. The following path-loss model is used in this environment:

$$L_{50} = 40(1 - 4 \times 10^{-2}\Delta h_b)\log d - 18\log(\Delta h_b) + 21\log f_c + 80\,\text{dB} \quad (3.38)$$

where:

Δh_b = base station antenna height measured from average roof top level (m)

Delay Spread

A majority of the time rms delay spreads are relatively small, but occasionally, there are "worst case" multipath characteristics that lead to much larger rms delay spreads. Measurements in outdoor environments show that rms delay spread can vary over an order of magnitude within the same environment. Delay spreads can have a major impact on the system performance. To accurately evaluate the relative performance of radio transmission technologies, it is important to model the variability of delay spread as well as the "worst case" locations where delay spread is relatively large. For each environment IMT-2000 defines three multipath channels: low delay spread, median delay spread, and high delay spread. Channel "A" represents the low delay spread case that occurs frequently; channel "B" corresponds to the median delay spread case that also occurs frequently; and channel "C" is the high delay spread case that occurs only rarely. Table 3.3 provides the rms values of delay spread for each channel and for each environment.

Table 3.3 Rms delay spread (IMT-2000).

	Channel "A"		Channel "B"		Channel "C"	
Environment	τ_{rms} (ns)	% Occurrence	τ_{rms} (ns)	% Occurrence	τ_{rms} (ns)	% Occurrence
Indoor office	35	50	100	45	460	5
Outdoor to indoor and pedestrian	100	40	750	55	1800	5
Vehicular (high antenna)	400	40	4000	55	12,000	5

3.10 Indoor Path-Loss Models

Picocells cover part of a building and span from 30 to 100 meters [13,15]. They are used for WLANs and PCSs operating in the indoor environment. The path-loss model for a picocell is given as:

$$L_p = \overline{L}_p(d_0) + 10\gamma \log d + L_f(n) + X_\sigma \, \text{dB} \tag{3.39}$$

where:

$\overline{L}_p(d_0)$ = reference path loss at the first meter (dB)
γ = path-loss exponent
d = distance between transmitter and receiver (m)
X_σ = shadowing effect (dB)
$L_f(n)$ = signal attenuation through n floors

Indoor-radio measurements at 900 MHz and 1.7 GHz values of L_f per floor are 10 dB and 16 dB, respectively. Table 3.4 lists the values of $\overline{L}_p(d_0)$, $L_f(n)$, γ, and X_σ. Partition dependent losses for signal attenuation at 2.4 GHz are given in Table 3.5.

Table 3.4 Values of $\overline{L}_p(d_0)$, γ, $L_f(n)$ and X_σ in Equation 3.39.

Environment	Residential	Office	Commercial
$\overline{L}_p(d_0)$ (dB)	38	38	38
γ	2.8	3.0	2.2
$L_f(n)$ (dB)	4n	15 + 4(n − 1)	6 + 3(n − 1)
X_σ (dB)	8	10	10

Table 3.5 Partition dependent losses at 2.4 GHz.

Signal attenuation through	Loss (dB)
Window in brick wall	2
Metal frame, glass wall in building	6
Office wall	6
Metal door in office wall	6
Cinder wall	4
Metal door in brick wall	12.4
Brick wall next to metal door	3

Femtocellular systems span from a few meters to a few tens of meters. They exist in individual residences and use low-power devices using Bluetooth chips or Home RF equipment. The data rate is around 1 Mbps. Femtocellular systems use carrier frequencies in the unlicensed bands at 2.4 and 5 GHz. Table 3.6 lists the values of $L_p(d_0)$ and γ for LOS and NLOS conditions.

Example 3.8

In a WLAN the minimum SNR required is 12 dB for an office environment. The background noise at the operational frequency is -115 dBm. If the mobile terminal transmit power is 100 mW, what is the coverage radius of an access point if there are three floors between the mobile transmitter and the access point?

Solution

- Transmit power of mobile terminal $= 10 \log 100 = 20$ dBm
- Receiver sensitivity $=$ background noise $+$ minimum SNR $= -115 + 12$
  $$= -103 \text{ dB}$$
- Maximum allowable path loss $=$ transmit power $-$ receiver sensitivity
  $$= 20 - (-103) = 123 \text{ dB}$$
- $\overline{L}_p(d_0) = 38$ dB, $L_f(n) = 15 + 4(3 - 1) = 23$ dB, $\gamma = 3$, and $X_\sigma = 10$ dB (from Table 3.4)

$$\text{Maximum allowable path loss} = 123 = 38 + 23 + 10 + 30 \log d$$

$$d = 54 \text{ m}$$

3.11 Fade Margin

As we discussed earlier, the local mean signal strength in a given area at a fixed radius, R, from a particular base station antenna is lognormally distributed [7]. The local mean (i.e., the average signal strength) in dB is expressed by a normal random variable with a mean S_m (measured in dBm) and standard deviation σ_s (dB). If S_{min} is the receiver sensitivity, we determine the fraction of the locations (at $d = R$) wherein a mobile would experience a received signal above the receiver sensitivity. The receiver sensitivity is the value that provides an acceptable signal under Rayleigh fading conditions. The probability distribution function for a log-normally distributed random variable is:

$$p(S) = \frac{1}{\sigma_s \sqrt{2\pi}} e^{-[(S - S_m)^2/(2\sigma_s^2)]} \qquad (3.40)$$

Table 3.6 Values of A and γ for femtocellular systems.

Environment	Center frequency (GHz)	Scenario	$\bar{L}_p(d_0)$(dB)	γ
Indoor office	2.4	LOS	41.5	2.0
		NLOS	37.7	3.3
Meeting room	5.1	LOS	46.6	2.22
		NLOS	61.6	2.22
Suburban residence	5.2	LOS (same floor)	47	2 to 3
		NLOS (same floor)		4 to 5
		NLOS & room in higher floor directly above Tx		4 to 6
		NLOS & room in higher floor not directly above Tx		6 to 7

The probability for signal strength exceeding receiver sensitivity $P_{S_{min}}(R)$ is given as

$$P_{S_{min}}(R) = P[S \geq S_{min}] = \int_{S_{min}}^{\infty} p(S)dS = \frac{1}{2} - \frac{1}{2} erf\left(\frac{S_{min} - S_m}{\sigma_s\sqrt{2}}\right) \qquad (3.41)$$

Note: See Appendix D for erf, erfc and Q functions.

Example 3.9

If the mean signal strength and receiver sensitivity are $-100\,$dBm and $-110\,$dBm, respectively and the standard deviation is $10\,$dB, calculate the probability for exceeding signal strength beyond the receiver sensitivity.

Solution

$$P_{S_{min}}(R) = \frac{1}{2} - \frac{1}{2} erf\left(\frac{-110 + 100}{10\sqrt{2}}\right) = 0.5 + 0.5\ erf(0.707) = 0.84$$

Next, we determine the fraction of the coverage within an area in which the received signal strength from a radiating base station antenna exceeds S_{min}. We define the fraction of the useful service area F_u as that area, within an area for which the signal strength received by a mobile antenna exceeds S_{min}. If $P_{S_{min}}$ is the probability that the received signal exceeds S_{min} in an incremental area dA, then

$$F_u = \frac{1}{\pi R^2} \int P_{S_{min}} dA \qquad (3.42)$$

Using the power law we express mean signal strength S_m as

$$S_m = \alpha - 10\gamma \log \frac{d}{R} \qquad (3.43)$$

where α accounts for the transmitter effective radiated power (ERP), receiver antenna gain, feed line losses, etc.

Substituting Equation 3.43 into 3.41 we get:

$$P_{S_{min}} = \frac{1}{2} - \frac{1}{2} \text{erf}\left[\frac{S_{min} - \alpha + 10\gamma\log(d/R)}{\sigma_s\sqrt{2}}\right] \qquad (3.44)$$

Let $a = (S_{min} - \alpha)/(\sigma_s\sqrt{2})$ and $b = (10\gamma\log(d/R))/(\sigma_s\sqrt{2})$

Substituting Equation 3.44 into 3.42, we get

$$\therefore F_u = \frac{1}{2} - \frac{1}{R^2}\int_0^R x\{\text{erf}[a + b\log(x/R)]\}dx \qquad (3.45)$$

Let $t = a + b\log(x/R)$, then

$$\therefore F_u = \frac{1}{2} - \frac{1}{b}e^{(-2a)/b}\int_{-\infty}^a e^{(2t)/b}\text{erf}(t)dt \qquad (3.46)$$

or

$$F_u = \frac{1}{2}\left[1 - \text{erf}(a) + e^{(1-2ab)/b^2}\left(1 - \text{erf}\left[\frac{1-ab}{b}\right]\right)\right] \qquad (3.47)$$

If we choose α such that $S_m = S_{min}$ at $d = R$, then $a = 0$ and

$$F_u = \frac{1}{2}\{1 + e^{1/b^2}[1 - \text{erf}(1/b)]\} \qquad (3.48)$$

Figure 3.10 shows the relation in terms of the parameter σ_s/γ.

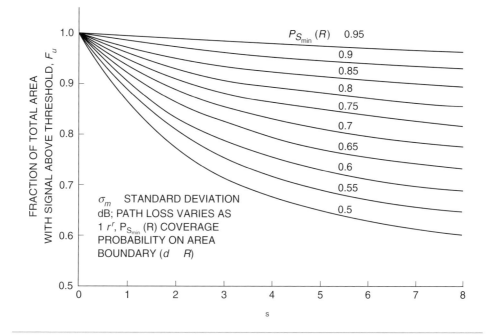

Figure 3.10   Fraction of total area with average power above threshold.

(After: W. C. Jakes, Jr., (editor), *Microwave Mobile Communications.* New York: John Wiley & Sons, 1974, p. 127)

3.12 Link Margin

To consider the losses incurred in transmitting a signal from point *A* to point *B*, we start by adding all the gains and losses in the link to estimate the total overall link performance margin [4].

The receiver power P_r is given as:

$$P_r = \frac{P_t G_t G_r}{Lp} \tag{3.49}$$

where:

P_t is the transmitter power

G_t and G_r are the gains of transmitter and receiver

L_p is the path loss between the transmitter and receiver.

In addition, there are also the effects due to receiver thermal noise, which is generated due to random noise inherent within a receiver's electronics. This increases with temperature. We account for this thermal noise effect with the following:

$$N = kTB_w \tag{3.50}$$

where:

k = Boltzmann's constant $(1.38 \times 10^{-23}\,\text{W/Kelvin-Hz})$
T = temperature in Kelvin
B_w = receiver bandwidth(Hz).

Spectral noise density, N_0, is the ratio of thermal noise to receiver bandwidth

$$N_0 = N/B_w = kT \tag{3.51}$$

Finally, there is an effect on signal-to-noise (SNR) ratio due to the quality of the components used in the receiver's amplifiers, local oscillators (LOs), mixers, etc. The most basic description of a component's quality is its noise figure, N_f, which is the ratio of the SNR at the input of the device versus the SNR at its output. The overall composite effect of several amplifiers' noise figures is cumulative, and can be obtained as:

$$N_{f,\text{total}} = N_{f1} + (N_{f2} - 1)/G_1 + (N_{f3} - 1)/(G_1G_2) + \cdots \tag{3.52}$$

where:

N_{fk} is the noise figure in stage k
G_k = gain of the kth stage.

By combining all the factors, we can develop a relation that allows us to calculate the overall link margin

$$M = \frac{P_t G_t G_r A_g}{N_{f,\text{total}} T_k L_p L_f L_0 F_{\text{margin}}\, R\left(\dfrac{Eb}{N_0}\right)_{\text{reqd}}} \tag{3.53}$$

where:

A_g = gain of receiver amplifier in dB
R = data rate in dB
F_{margin} = fade margin in dB
T_k = noise temperature in Kelvin
$(E_b/N_0)_{\text{reqd}}$ = required value in dB
L_p = path losses in dB
L_f = antenna feed line loss in dB
L_0 = other losses in dB

Expressing Equation 3.53 in dB, we obtain

$$M = P_t + G_t + G_r + A_g - N_{f,\,\text{total}} - T_k - L_p - L_f - L_0 - F_{\text{margin}}$$

$$- R - (E_b/N_0)_{\text{reqd}} \, \text{dB} \qquad\qquad (3.54)$$

Example 3.10

Given a flat rural environment with a path loss of 140 dB, a frequency of 900 MHz, 8 dB transmit antenna gain and 0 dB receive antenna gain, data rate of 9.6 kbps, 12 dB in antenna feed line loss, 20 dB in other losses, a fade margin of 8 dB, a required E_b/N_0 of 10 dB, receiver amplifier gain of 24 dB, noise figure total of 6 dB, and a noise temperature of 290 K, find the total transmit power required of the transmitter in watts for a link margin of 8 dB.

$$k = 10 \log(1.38 \times 10^{-23}) = -228.6 \, \text{dBW}$$

$L_p = 140 \, \text{dB}; \; A_g = 24 \, \text{dB}; \; N_f = 6 \, \text{dB}; \; F_{\text{margin}} = 8 \, \text{dB}; \; G_t = 8 \, \text{dB}; \; G_r = 0 \, \text{dB};$
$L_0 = 20 \, \text{dB}; \; L_{\text{feed}} = 12 \, \text{dB}; \; T = 24.6 \, \text{dB}; \; R = 39.8 \, \text{dB}; \; (E_b/N_0)_{\text{reqd}} = 10 \, \text{dB};$
and M = 8 dB

From Equation (3.54)

$$P_t = M - G_t - G_r - A_g + N_{f,\,\text{total}} + T + k + L_p + L_f + L_0 + F_{\text{margin}} + R$$

$$+ (E_b/N_0)_{\text{reqd}}$$

$$P_t = 8 - 8 - 0 - 24 + 6 + (24.6 - 228.6) + 140 + 12 + 20 + 8 + 39.8 + 10$$

$$= 7.8 \, \text{dBW}$$

$$\therefore P_t = 10^{0.78} \approx 6 \, \text{W}$$

3.13 Summary

In this chapter we discussed propagation and multipath characteristics of a radio channel. The concepts of delay spread that causes channel dispersion and inter-symbol interference were also presented. Since the mathematical modeling of the propagation of radio waves in a real world environment is complicated, empirical models were developed by several authors. We presented these empirical and semi-empirical models used for calculating the path losses in urban, suburban, and rural environments and compared the results obtained with each model. Doppler spread, coherence bandwidth, and time dispersion were also discussed. The forward error correcting algorithms [3] for improving radio channel performances are given in Chapter 8.

Problems

3.1   Define slow and fast fading.

3.2   What is a frequency selective channel?

3.3   Define receiver sensitivity.

3.4   A vehicle travels at a speed of 30 m/s and uses a carrier frequency of 1 GHz. What is the maximum Doppler shift? What is the approximate fade duration?

3.5   A mobile station traveling at 30 km per hour receives a flat Rayleigh fading signal at 800 MHz. Determine the number of fades per second above the rms level. What is the average duration of fade below the rms level? What is the average duration of fade at a level 20 dB below the rms level?

3.6   Find the received power for the link from a synchronous satellite to a terrestrial antenna. Use the following data: height = 60,000 km; satellite transmit power = 4 W; transmit antenna gain = 18 dBi; receive antenna gain = 50 dBi; and transmit frequency = 12 GHz.

3.7   Determine the SNR for the spacecraft that uses a transmitter power of 16 W at a frequency of 2.4 GHz. The transmitter and receiver antenna gain are 28 dBi and 60 dBi, respectively. The distance from the spacecraft to ground is 2×10^{10} m, the effective noise temperature of antenna plus receiver is 14 degrees Kelvin, and a bit rate of 120 kbps. Assume the bandwidth of the system to be half of the bit rate, 60 kHz.

3.8   A base station transmits a power of 10 W into a feeder cable with a loss of cable 10 dB. The transmit antenna has a gain of 12 dBi in the direction of the mobile receiver with a gain of 0 dBi and feeder loss of 2 dB. The mobile receiver has a sensitivity of −104 dBm. (a) Determine the effective isotropic radiated power, and (b) maximum acceptable path loss.

3.9   A receiver in a digital mobile communication system has a noise bandwidth of 200 kHz and requires that its input signal-to-noise ratio should be at least 10 dB when the input signal is −104 dBm. (a) What is the maximum permitted value of the receiver noise figure, and (b) What is the equivalent input noise temperature?

3.10 Calculate the maximum range of the communication system in Problem 8, assuming a mobile antenna height (h_m) of 1.5 m, a base station antenna height (h_b) of 30 m, a frequency equal to 900 MHz and propagation that takes place over a plane earth. Assume base station and mobile station antenna gains to be 12 dBi and 0 dBi, respectively. How will this range change if the base station antenna height is doubled?

3.11 A mobile station traveling at a speed of 60 km/h transmits at 900 MHz. If it receives or transmits data at a rate of 64 kbps, is the channel fading slow or fast?

3.12 The power received at a mobile station is lognormal with a standard deviation of 8 dB. Calculate the outage probability assuming the average received power is −96 dBm and the threshold power is −100 dBm.

3.13 Determine the minimum signal power for an acceptable voice quality at the base station receiver of a GSM system (bandwidth 200 kHz, data rate 271 kbps). Assume the following data: Receiver noise figure = 5 dB, Boltzmann's constant = 1.38×10^{-23} Joules/K, mobile radiated power = 30 dBm, transmitter cable losses = 3 dB, base station antenna gain = 16 dBi, mobile antenna gain = 0 dBi, fade margin = 10.5 dB, and required E_b/N_0 = 13.5 dB. What is the maximum allowable path loss? What is the maximum cell radius in an urban area where a 1 km intercept is 108 dB and the path-loss exponent is 4.2?

3.14 Develop a MATLAB program and obtain a curve for maximum path loss versus cell radius. Test your program using the following data: base station transmit power = 10 W, base station cable loss = 10 dB, base station antenna gain = 8 dBi, base station antenna height = 15 m, mobile station antenna gain = 0 dBi, mobile station antenna height = 1 m, body and matching loss = 6 dB, receiver noise bandwidth = 200 kHz, receiver noise figure = 7 dB, noise density = −174 dBm/Hz, required SNR = 9 dB, building penetration loss = 12 dB, and fade margin = 10 dB.

3.15 In the Bluetooth device with NLOS, S/N required is 10 dB in an indoor office environment. The background noise at the operating frequency is −80 dBm. If the transmit power of the device is 20 dBm, what is its coverage?

References

1. Bertoni, H. L. *Radio Propagation for Modern Wireless Systems*. Upper Saddle River, NJ: Prentice Hall, 2000.

2. Clarke, R. H. "A Statistical Theory of Mobile Radio Reception." *Bell System Technical Journal* 47 (July–August 1968): 957–1000.

3. Forney, G. D. "The Viterbi Algorithm." *Proceedings of IEEE*, vol. 61, no. 3, March 1978, pp. 268–278.

4. Garg, V. K., and Wilkes, J. E. *Wireless and Personal Communications Systems*. Upper Saddle River, NJ: Prentice Hall, 1996.

5. Hanzo, L., and Stefanov, J. "The Pan-European Digital Cellular Mobile Radio System—Known as GSM." Mobile *Radio Communications*. R. Steele, Ed., Chapter 8, London: Prentech Press, 1992.

6. Hata, M. "Empirical Formula for Propagation Loss in Land Mobile Radio Services." *IEEE Transactions Vehicular Technology*, vol. 29, no. 3, 1980.

7. Jakes, W. C., Ed., *Microwave Mobile Communications*. New York: John Wiley, 1974.

8. Lecours, M., Chouinard, I. Y., Delisle, G. Y., and Roy, J. "Statistical Modeling of the Received Signal Envelope in a Mobile Radio Channel." *IEEE Transactions Vehicular Technology*, vol. VT-37, 1988, pp. 204–212.

9. Lee, William, C. Y. *Mobile Communications Engineering*. New York: John Wiley, 1989.

10. Mark, J. W., and Zhuang, W. *Wireless Communications and Networking*. Upper Saddle River, NJ: Prentice Hall, 2003.

11. Okumura, Y., et al. "Field Strength and Its Variability in VHF and UHF Land Mobile Radio Service." Review Electronic Communication Lab 16, no. 9–10, 1968.

12. Parsons, D. *The Mobile Radio Propagation Channel*. Chichester, West Sussex, England: John Wiley, 1996.

13. Rappaport, T. S. *Wireless Communications*. Upper Saddle River, NJ: Prentice Hall, 1996.

14. Sampei, Seiichi. *Applications of Digital Wireless Technologies to Global Wireless Communications*. Upper Saddle River, NJ: Prentice Hall, 1997.

15. Seidel, S. Y., and Rappaport, T. S. "914 MHz Path Loss Prediction Models for Indoor Wireless Communications in Multifloor Buildings." *IEEE Transactions, Antenna & Propagation*, 40(2), February 1992.

16. Sklar, B. "Rayleigh Fading Channels in Mobile Digital Communication Systems Part I: Characterization." *IEEE Communications Magazine*, vol. 35, no. 9, September 1997, pp. 136–146.

17. Sklar, B. "Rayleigh Fading Channels in Mobile Digital Communication Systems Part II: Mitigation." *IEEE Communications Magazine*, vol. 35, no. 9, September 1997, pp. 148–155.

18. Turin, G. L., et al. "A Statistical Model of Urban Multipath Propagation." *IEEE Transactions Vehicular Technology*, vol. VT-21, 1972, pp. 1–9.

19. Walfisch, J., and Bertoni, H. L. "A Theoretical Model of UHF Propagation in Urban Environment." *IEEE Transactions, Antennas & Propagation*, AP-36:1788–1796, October 1988.

20. Weissberger, M. A. "An Initial Critical Summary of Models for Predicting the Attenuations of Radio Waves." ESD-TR-81–101, Electromagnetic Compatibility Analysis Center, Annapolis, MD, July 1982.

CHAPTER 4

An Overview of Digital Communication and Transmission

4.1 Introduction

The basic part of any digital communication system is the communication channel. This is the physical medium that carries information bearing signals from the *source* of the information to the *sink*. In a radio system the communication channel is the propagation of radio waves in free space (see Figure 4.1). As discussed in Chapter 3, radio waves in free space are subjected to fading. In nearly all communication systems some equipment is required to convert the information-bearing signal into a suitable form for transmission over the communication channel and then back into a form that is comprehensible to the end-user. This equipment is the *transmitter* and *receiver*. The receiver does not only perform the inverse translation to the transmitter, but it also has to overcome the distortions and disturbances (see Chapter 3) that occur over the communication channel. Thus, it is often more difficult to design the receiver than the transmitter. Speech coding, forward-error-correcting (FEC) coding, bit-interleaving, diversity, equalization, and modulation play important roles in a communication system, particularly in a radio system (see Chapters 7, 8, and 9).

The transmitter for a radio system consists of antenna, RF section, encoder, and modulator. An *antenna* converts the electrical signal into a radio wave propagating in free space. The *RF section* of the transmitter generates a signal of sufficient power at the required frequency. It typically consists of a power amplifier, a local oscillator, and an up-converter. However, generally the RF section only amplifies and frequency-converts a signal (see Figure 4.2).

At the input of the transmitter the user interface interacts and converts the information into a suitable digital data stream. The information source can be analog (such as speech) or discrete (such as data). Analog information is converted into digital information through the use of sampling and quantization. Sampling, quantization, and encoding techniques are called formatting and source coding.

The *source encoder* and *modulator* bridge the gap between the digital data and electrical signal required at the input to the RF section. The encoder converts the data stream into a form that is more resistant to the degradations introduced

Figure 4.1 Overview of a communication system.

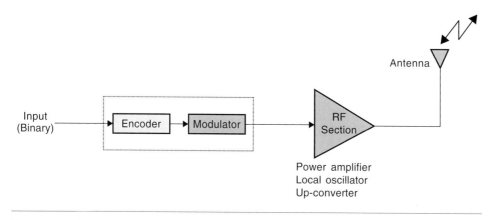

Figure 4.2   Structure of the transmitter for a radio system.

by the communication channel (see Chapter 3). The modulator takes a sine wave at the required carrier frequency and modifies the signal characteristics. We may regard the *encoder/modulator* as a single subsystem that maps data presented to it by the user interface onto a modulated RF carrier for subsequent processing, amplification, and transmission by the RF section. The *demodulator/decoder* takes the received RF signal and performs the inverse mapping back to the data stream for onward transmission.

The encoder/modulator determines the bandwidth occupied by the transmitted signal. The demodulator/decoder determines the quality of the resulting service, in terms of bit-error-rate (BER), availability and delay. These subsystems also determine the robustness of the communication system to channel impairments due to the RF subsystem (such as phase noise and nonlinearity) and RF channel (such as multipath dispersion and fading, discussed in Chapter 3). Therefore, the correct choice of the coding/modulation scheme is vital for efficient operation of the whole system. In addition to coding/modulation, other means for improving signal quality in a wireless system include the speech coding, bit-interleaving, equalization, and diversity [4,6,9,10,14,16].

The basic objective of this chapter is to familiarize the readers who are not exposed to digital communications. The chapter may be omitted by the students who have been exposed to digital communications course(s). Details of techniques

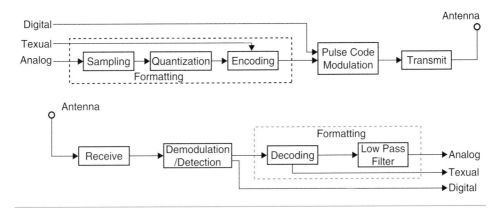

Figure 4.3 Formatting, transmission, and reception of baseband signals.

such as speech/channel coding, bit-interleaving, modulation, diversity, and the antenna used for improving signal quality are discussed in Chapters 7, 8, and 9.

This chapter focuses only on the performance parameters of coding and the modulation scheme. It outlines the OSI layers, types of data services, and discusses briefly transmission media to familiarize the readers.

4.2 Baseband Systems

Source information may contain either analog, textual, or digital data. *Formatting* involves sampling, quantization, and encoding. It is used to make the message compatible with digital processing. *Transmit formatting* transforms source information into digital symbols. When data *compression* is used in addition to formatting, the process is referred to as *source coding*. Figure 4.3 shows a functional diagram that primarily focuses on the formatting and transmission of baseband (information bearing) signals. The receiver with a detector followed by a signal decoder performs two main functions: (1) does reverse operations performed in the transmitter, and (2) minimizes the effect of channel noise for the transmitted symbol.

4.3 Messages, Characters, and Symbols

During digital transmission the characters are first encoded into a sequence of bits, called a *bit stream* or *baseband signal*. Groups of b bits form a finite symbol set or word $M\ (=2^b)$ of such symbols [14,17]. A system using a symbol set size of M is referred to as an *M-ary system*. The value of b or M is an important initial choice in the design of any digital communication system. For $b = 1$, the system is called a *binary system*, the size of symbol set M is 2, and the modulator uses two different waveforms to represent the binary "1" and the binary "0" (see Figure 4.4). In this case, the symbol rate and the bit rate are the same. For $b = 2$, the system is called

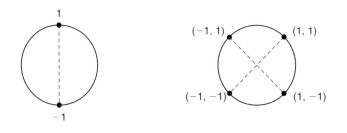

Figure 4.4   Binary and quatenary systems.

quatenary or *4-ary* ($M = 4$) system. At each symbol time, the modulator uses one of the four different waveforms that represents the symbol (see Figure 4.4).

4.4   Sampling Process

The analog information is transformed into a digital format. The process starts with *sampling* the waveform to produce a *discrete pulse-amplitude-modulated waveform* (see Figure 4.5) [6,16,17]. The *sampling process* is usually described in time domain. This operation is basic to digital signal processing and digital communication. Using the sampling process, we convert the analog signal into corresponding sequences of samples that are usually spaced uniformly in time. The sampling process can be implemented in several ways, the most popular being the *sample-and-hold operation*. In this operation, a switch and storage mechanism (such as a transistor and a capacitor, or shutter and a filmstrip) form a sequence of samples of the continuous input waveform. The output of the sampling process is called *pulse amplitude modulation* (PAM) (see Section 4.6) because the successive output intervals can be described as a sequence of pulses with amplitudes derived from the input waveform samples. The analog waveform can be retrieved from a PAM waveform by simple low-pass filtering provided we choose the sampling rate properly. The ideal form of sampling is called *instantaneous sampling*.

We sample the signal $s(t)$ instantaneously at a uniform rate, f_s, once every T_s ($1/f_s$) seconds. Thus, we obtain:

$$s_\delta(t) = \sum_{n=-\infty}^{\infty} s(nT_s)\delta(t - nT_s) \tag{4.1}$$

where:

   $s_\delta(t)$ is the ideal sampled signal
   $\delta(t - nT_s)$ is the delta function positioned at time $t = nT_s$

A delta function is closely approximated by a rectangular pulse of duration Δt and amplitude $s(nT_s)/\Delta t$; the smaller Δt the better the approximation.

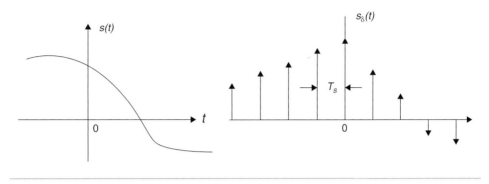

Figure 4.5 Sampling process.

We determine the Fourier transform of the ideal sampled signal $s_\delta(t)$ by using the duality property of the Fourier transform and the fact that a delta function is an even function [6] as:

$$s_\delta(t) \Longleftrightarrow f_s \sum_{m=-\infty}^{\infty} S(f - mf_s) \tag{4.2}$$

where:

$S(f)$ is the Fourier transform of the original signal $s(t)$ and f_s is the sampling rate.

Equation 4.2 states that the process of uniformly sampling a continuous-time signal of finite energy results in a periodic spectrum with a period equal to the sampling rate.

Taking the Fourier transform of both sides of Equation 4.1 and noting that the Fourier transform of the delta function $\delta(t - nT_s)$ is equal to $e^{-j2\pi nfT_s}$, we get

$$S_\delta(f) = \sum_{n=-\infty}^{\infty} s(nT_s)e^{-j2\pi nfT_s} \tag{4.3}$$

Equation 4.3 is called the *discrete-time Fourier transform*. It is the complex Fourier series representation of the periodic frequency function $S_\delta(f)$, with the sequence of samples $\{s(nT_s)\}$ defining the coefficients of the expansion.

From Equation 4.2, we see that the Fourier transform of $s_\delta(t)$ may also be expressed as:

$$S_\delta(f) = f_s S(f) + f_s \sum_{m=-\infty, m \neq 0}^{\infty} S(f - mf_s) \tag{4.4}$$

Next, we consider a continuous-time signal $s(t)$ of finite energy and infinite duration. The signal is strictly band-limited with no frequency component higher than B_w Hz. This implies that the Fourier transform $S(f)$ of the signal $s(t)$ has the property that $S(f)$ is zero for $|f| \geq B_w$. If we choose the sampling period $T_s = 1/(2B_w)$, then the corresponding spectrum is given as:

$$S_\delta(f) = \sum_{n=-\infty}^{\infty} s\left(\frac{n}{2B_w}\right) e^{-\left(\frac{j\pi nf}{B_w}\right)} \tag{4.5}$$

Using the following two conditions:

1. $S(f) = 0$ for $|f| \geq B_w$ and
2. $f_s = 2B_w$

We find from Equation 4.4

$$S(f) = \frac{1}{2B_w} S_\delta(f), \quad -B_w < f < B_w \tag{4.6}$$

Substituting Equation 4.5 into Equation 4.6, we can write

$$\therefore S(f) = \frac{1}{2B_w} \sum_{n=-\infty}^{\infty} s\left(\frac{n}{2B_w}\right) e^{-\left(\frac{j\pi nf}{B_w}\right)}, \quad -B_w < f < B_w \tag{4.7}$$

Thus, if the sample value $s(n/2B_w)$ of a signal $s(t)$ is specified for all n, then the Fourier transform $S(f)$ of the signal is uniquely determined by using the discrete-time Fourier transform of Equation 4.7. Because $s(t)$ is related to $S(f)$ by inverse Fourier transform, it follows that the signal $s(t)$ is itself uniquely determined by the sample values $s(n/2B_w)$ for $-\infty < n < \infty$. In other words, the sequence $\{s(n/2B_w)\}$ has all the information contained in $s(t)$.

We state the sampling theorem for *band-limited* signals of finite energy in two parts, which apply to the transmitter and receiver of a pulse modulation system, respectively.

1. A band-limited signal of finite energy with no frequency components higher than B_w Hz is completely described by specifying the values of signals at instants of time separated by $1/(2B_w)$ seconds.

2. A band-limited signal of finite energy with no frequency components higher than B_w Hz, may be completely recovered from a knowledge of its samples taken at the rate of $2B_w$ samples per second.

This is known as the *uniform sampling theorem*. The sampling rate of $2B_w$ samples per second for a signal bandwidth B_w Hz, is called the *Nyquist rate* and $1/(2B_w)$ second is called the *Nyquist interval* [8].

We discussed the sampling theorem by assuming that signal $s(t)$ is strictly band limited. In practice, however, an information-bearing signal is not strictly band limited, with the result that some degree of undersampling is encountered. Consequently, some aliasing is produced by the sampling process. *Aliasing* refers to the phenomenon of a high-frequency component in the spectrum of the signal seemingly taking on the identity of a lower frequency in the spectrum of its sampled version.

4.4.1 Aliasing

Figure 4.6 shows the part of the spectrum that is aliased due to *undersampling* [13]. The aliased spectral components represent ambiguous data that can be retrieved only under special conditions. In general, the ambiguity is not resolved and ambiguous data appears in the frequency band between $(f_s - f_m)$ and f_m, where f_m is the maximum frequency and f_s is the sampling rate.

In Figure 4.7 we use the higher sampling rate f_s' to eliminate the aliasing by separating the spectral replicas. Figures 4.8 and 4.9 show two ways to eliminate

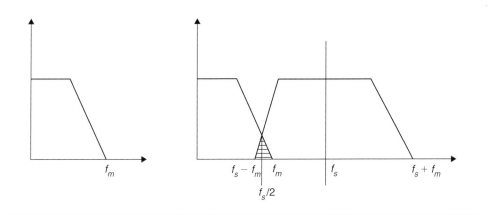

Figure 4.6   Sampled signal spectrum.

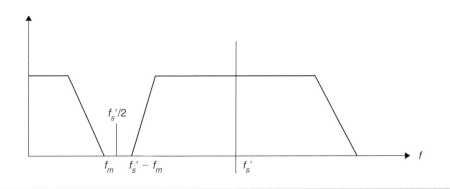

Figure 4.7   Higher sampling rate to eliminate aliasing.

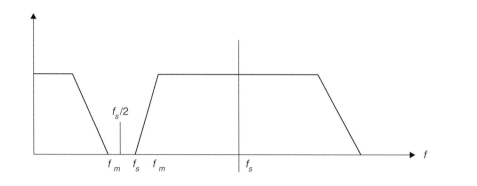

Figure 4.8   Pre-filtering to eliminate aliasing.

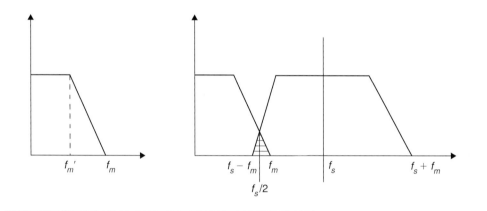

Figure 4.9   Post-filter to eliminate aliasing portion of the spectrum.

aliasing using anti-aliasing filters. The analog signal is *prefiltered* so that the new maximum frequency f_m is less than or equal to $f_s/2$. Thus there are no aliasing components seen in Figure 4.8, since $f_s > 2f_m'$. Eliminating aliasing terms prior to sampling is good engineering practice. When the signal structure is well known, the aliased terms can be eliminated after sampling with a linear pass filter (LPF) operating on the sampled data. In this case the aliased components are removed by *post-filtering* after sampling. The filter cutoff frequency f_m' removes the aliased components; f_m' needs to be less than $(f_s - f_m)$. It should be noted that filtering techniques for eliminating the aliased portion of the spectrum will result in a loss of some of the signal information. For this reason, the sample rate, cutoff bandwidth, and filter type selected for a particular signal bandwidth are all interrelated.

Realizable filters require a non-zero bandwidth for the transition between the passband and the required out-of-band attenuation. This is called the *transition bandwidth*. To minimize the system sample rate, we want the anti-aliasing filter to have a small transition bandwidth. Filter complexity and cost increase sharply with a narrower transition bandwidth, so a trade-off is required between the cost of a small transition bandwidth and the costs of the higher sampling rate, which are those of more storage and higher transition rates.

In many systems the answer is to make the transition bandwidth 10 to 20% of the signal bandwidth. If we account for a 20% transition bandwidth of the anti-aliasing filter, we have an engineering version of the Nyquist sampling rate: $f_s \geq 2.2f_m$.

Example 4.1

What should be the sampling rates to produce a high-quality digitalization of a 20-kHz bandwidth music signal?

Solution

A sampling rate, of ≥ 44 ksps should be used.

The sampling rate for a compact disc digital audio player = 44.1 ksps and for a studio-quality audio player = 48 ksps are used.

4.4.2 Quantization

In Figure 4.10, each pulse is expressed as a level from a finite number of predetermined levels; each level can be represented by a symbol from a finite alphabet. The pulses in Figure 4.10 are called *quantized samples*. When the sample values are quantized to a finite set, this format can interface with a digital system. After quantization, the analog waveform can still be recovered, but not precisely; improved reconstruction fidelity of the analog waveform can be achieved by increasing the number of quantization levels (requiring increased system bandwidth).

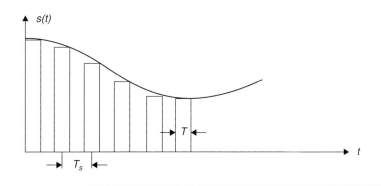

Figure 4.10 Flat-top quantization.

4.4.3 Sources of Error

The quantization process introduces an error defined as the difference between the input and output signal. The error is referred to as *quantization noise*. Other sources of error are due to sampling and channel condition [2,13].

Quantization Noise

The distortion inherent in quantization is a round-off or truncation error. The process of encoding the pulse amplitude modulation (PAM) waveform into a quantized waveform involves discarding some of the original analog information. This distortion is called *quantization noise*; the amount of such noise is inversely proportional to the number of quantization levels used in the process.

Quantizer Saturation

The quantizer allocates L levels to the task of approximating the continuous range of inputs with a finite set of outputs. The range of inputs for which the difference between the input and output is small is called the *operating range* of the converter. If the input exceeds this range, the difference between the input and output becomes large and we say that the converter is operating in *saturation mode*. Saturation errors are more objectionable than quantizing noise. Generally, saturation is avoided by use of automatic gain control, which effectively extends the operating range of the converter.

Timing Jitter

If there is a slight jitter in the position of the sample, the sampling is no longer uniform. The effect of the jitter is equivalent to frequency modulation of the baseband signal. If the jitter is random, a low-level wide band spectral contribution is induced whose properties are very close to those of the quantizing noise. *Timing jitter can be controlled with good power supply isolation and stable clock reference.*

Channel Noise

Thermal noise, interference from other users, and interference from circuit switching transients can cause errors in detecting the pulses carrying the digitized samples. Channel-induced errors can degrade the reconstructed signal quality quite significantly. The rapid degradation of the output signal quality with channel-induced errors is called a *threshold* effect.

Inter-Symbol Interference

A bandwidth-limited channel spreads a pulse waveform passing through it. When the channel bandwidth is much greater than the pulse bandwidth, the spreading of the pulse will be slight. When the channel bandwidth is close to the signal bandwidth, the spreading will exceed a symbol duration and cause signal pulses to overlap. This overlapping is called *inter-symbol interference* (ISI) (see Chapter 3).

ISI causes system degradation (higher bit-error-rate); it is a particularly insidious form of interference because raising the signal power to overcome interference will not improve the error performance.

4.4.4 Uniform Quantization

We consider a uniform quantization process as shown in Figure 4.11. With the quantizer input having zero mean, and the quantizer being symmetric, the quantizer output and quantization error will also have zero mean.

Since the mean of quantization error e is zero, its variance will be:

$$\sigma^2 = \int_{-q/2}^{q/2} e^2 p(e)\,de = \int_{-q/2}^{q/2} e^2 \left(\frac{1}{q}\right) de = \frac{q^2}{12} = \text{Average quantization noise power}$$

where:

$p(e)$ = probability of error = $1/q$ (for uniform distribution of quantization error)

σ^2 = variance of the average quantization noise power

L = number of quantization levels.

$$\text{Signal power} = (V_p)^2 = \left(\frac{V_{pp}}{2}\right)^2 = \left(\frac{Lq}{2}\right)^2 = \frac{L^2 q^2}{4} \tag{4.8}$$

where:

V_{pp} = peak-to-peak voltage range

$$\text{SNR} = \frac{\text{Signal power}}{\text{Average quantization noise power}} = \frac{(V_p)^2}{\sigma^2}$$

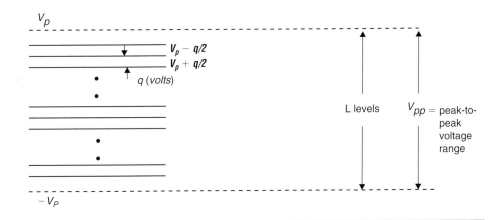

Figure 4.11   Uniform quantization.

$$\therefore \left(\frac{S}{N}\right)_q = \frac{(L^2 q^2)/4}{q^2/12} = 3L^2 \tag{4.9}$$

With $L = 2^R$, R = bits per sample, we obtain

$$\sigma^2 = \frac{q^2}{12} = \left(\frac{2V_p}{2^R}\right)^2 /12 = \frac{1}{3}(V_p)^2 (2^{-2R})$$

Let P denote the average power of the message signal $s(t)$, then

$$\left(\frac{S}{N}\right)_q = \frac{P}{\sigma^2} = \left[\frac{3P}{(V_p)^2}\right](2^{2R}) \tag{4.10}$$

The output signal-to-noise ratio (SNR) of the quantizer increases exponentially with increasing number of bits per sample, R. An increase in R requires a proportionate increase in the channel bandwidth.

In the limit as $L \to \infty$, the signal approaches the PAM format (with no quantization error) and signal-to-quantization noise ratio is infinite. In other words, with an infinite number of quantization levels, there is no quantization error.

Example 4.2

We consider a full-load sinusodial modulating signal of amplitude A, which uses all representation levels provided. If the average signal power corresponds to a load of 1 Ohm, determine the SNR for $L = 32$, 64, 128, and 256.

Solution

$$P = \frac{A^2}{2}$$

$$\sigma^2 = \frac{1}{3}A^2 2^{-2R}$$

$$(\text{SNR})_q = \frac{\frac{A^2}{2}}{\left(\frac{1}{3}A^2 2^{-2R}\right)} = \frac{3}{2}(2^{2R}) = 1.8 + 6R \,\text{dB}$$

$L = 2^R$	R (bits)	$(\text{SNR})_q$(dB)
32	5	31.8
64	6	37.8
128	7	43.8
256	8	49.8

4.5 Voice Communication

For most voice communication, very low speech volumes predominate: 50% of the time, the voltage characterizing detected speech energy is less than 1/4 of the rms value of the voltage. Large amplitude values are relatively rare: only 15% of the time does the voltage exceed the rms value. Uniform quantization would be wasteful for speech signals; many of the quantizing steps would rarely be used. In a system that uses equally spaced quantization levels, the quantization noise is the same for all signal magnitudes because noise depends on the step size of quantization. The uniform quantization results in poor signal to quantization noise ratios for low-amplitude signals. Nonuniform quantization can provide better quantization of the weak signals and coarse quantization of the strong signals. Thus, in the case of nonuniform quantization, quantization noise can be made proportional to signal magnitude. The effect is to improve overall SNR by reducing the noise for predominant weak signals, at the expense of an increase in noise for rarely occurring signals. The nonuniform form of quantization is used to make the SNR a constant for all signals within the input range.

Nonuniform quantization is achieved by first distorting the original signal with a logarithmic *compression* characteristic, and then using a uniform quantizer. For small magnitude signals the compression characteristic has a much steeper slope than large magnitude signals. Thus, a given signal change at small magnitudes will carry the uniform quantizer through more steps than the same change at large magnitudes. The compression characteristic effectively changes the distribution of the input signal magnitude. By compression, the low amplitudes are scaled up while the high amplitudes are scaled down. After compression, the distorted signal is used as input to the uniform quantizer. Thus, we achieve nonuniform quantization. There are two compression algorithms commonly used, the μ-law and the *A*-law [6,10].

We use a device called *expander* at the receiver with characteristics complementary to the compression. It is used so that the overall transmission is not distorted. The whole process (compression and expansion) is called *companding* [7]. The μ - and *A*-law compression characteristics as given below are used.

• **μ-Law** compression characteristic used in North America is given as:

$$|v_{\text{out}}| = \frac{\log[1 + \mu|v_{\text{in}}|]}{\log\{1 + \mu\}} \tag{4.11}$$

where:
   v_{in} and v_{out} are the normalized input and output voltages, respectively
   μ = a positive constant
   $\mu = 0$ represents uniform quantization, $\mu = 255$ is used in North America

- *A*-**Law** compression characteristic used in Europe and most of the rest of the world is given as:

$$|v_{out}| = \frac{A|v_{in}|}{1 + \log A} \quad 0 \le |v_{in}| \le \frac{1}{A} \tag{4.12a}$$

$$|v_{out}| = \frac{1 + \log A|v_{in}|}{1 + \log A} \quad \frac{1}{A} \le |v_{in}| \le 1 \tag{4.12b}$$

where *A* is the positive constant. $A = 87.6$ is the standard value used in Europe. The case of uniform quantization corresponds to $A = 1$.

4.6 Pulse Amplitude Modulation (PAM)

Pulse amplitude modulation [6] is a process that represents a continuous analog signal with a series of discrete analog pulses in which the amplitude of the information signal at a given time is coded as a binary number. PAM is now rarely used, having been largely superseded by pulse code modulation (PCM).

Two operations involved in the generation of the PAM signal are:

1. Instantaneous sampling of the message signal $s(t)$ every T_s seconds, where $f_s = 1/T_s$ is selected according to the sampling theorem.

2. Lengthening the duration of each sample obtained to some constant value T.

These operations are jointly referred to as *sample and hold*. One important reason for intentionally lengthening the duration of each sample is to avoid the use of an excessive channel bandwidth, since bandwidth is inversely proportional to pulse duration.

The Fourier transform of the rectangular pulse $h(t)$ is given as (see Figure 4.12):

$$S(f) = T \text{sinc}\,(fT)e^{-j2\pi fT} \tag{4.13}$$

Using flat-top sampling of an analog signal with a sample-and-hold circuit such that the sample has the same amplitude for its whole duration introduces *amplitude distortion* as well as a *delay*. This effect is similar to the variation in transmission frequency that is caused by the finite size of the scanning aperture in television. The distortion caused by the use of PAM to transmit an analog signal is called the *aperture effect*. The distortion may be corrected by use of an *equalizer* (see Figure 4.13). The equalizer decreases the in-band loss of the reconstruction filter as the frequency increases in such a manner to compensate for the aperture effect. The amount of equalization required in practice is usually small. For $T/T_s \le 0.1$, the amplitude distortion is less than 0.5%, in which case the need for equalization may be omitted altogether.

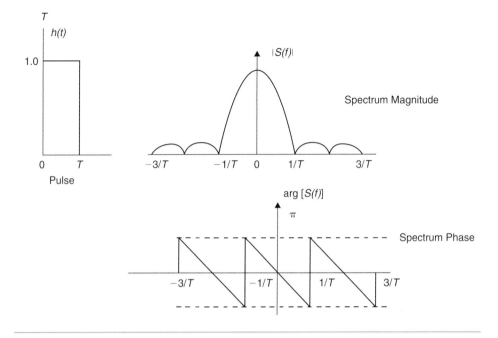

Figure 4.12   Rectangular pulse and its spectrum.

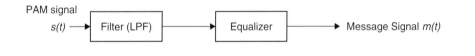

Figure 4.13   An equalizer application.

Example 4.3

Twenty-four voice signals are sampled uniformly and then time-division multi-plexed. The sampling operation is flat-top samples with $1\,\mu s$ duration. The multi-plexing operation includes provision for synchronization by adding an extra pulse of sufficient amplitude and also $1\,\mu s$ duration. The highest frequency components of each voice signal is $3.4\,kHz$.

   a. Assuming a sampling rate of $8\,kHz$, calculate the spacing between succes-sive pulses of the multiplexed signal.

   b. Repeat your calculations using the engineering version of Nyquist rate sampling.

Solution

a. $T_s = \dfrac{1}{8000} = 125\,\mu s$

25 channels (24 voice channels + 1 synchronization channel), time allocated for each channel = 125/25 = 5 μs. Since the pulse duration is 1 μs, the time between pulses is (5 − 1) = 4 μs

b. Nyquist rate = 7.48 kHz (2.2 × 3.4)

$$T_s = \frac{1}{7480} = 134\,\mu s$$

$$T_c = \frac{134}{25} = 5.36\,\mu s$$

The time between pulses is (5.36 − 1) = 4.36 μs.

4.7   Pulse Code Modulation

Pulse code modulation (PCM) [13] is a digital scheme for transmitting analog data. It converts an analog signal into digital form. Using PCM, it is possible to digitize all forms of analog data, including full-motion video, voice, music, telemetry, etc.

To obtain a PCM signal from an analog signal at the source (transmitter) of a communications circuit, the analog signal is sampled at regular time intervals. The sampling rate is several times the maximum frequency of the analog signal. The instantaneous amplitude of the analog signal at each sample is rounded off to the nearest of several specific, predetermined levels (quantization). The number of levels is always a power of 2. The output of a pulse code modulator is a series of binary numbers, each represented by some power of 2 bits. At the destination of the communications circuit, the pulse code modulator converts the binary numbers back into pulses having the same quantum levels as those in the modulator. These pulses are further processed to restore the original analog waveform.

When pulse modulation is applied to a binary symbol, the resulting binary wave form is called a *pulse code modulation* waveform. When pulse modulation is applied to a nonbinary symbol, the resulting waveform is called *M-ary pulse modulation* waveform. Each analog sample is transmitted into a PCM word consisting of groups of *b* bits. The PCM word size can be described by the number of quantization levels that are used for each sample. The choice of the number of quantization levels, or bits per sample, depends on the magnitude of quantization distortion that one is willing to tolerate with the PCM format.

In North America and Japan, PCM samples the analog waveform 8000 times per second and converts each sample into an 8-bit number, resulting in a 64 kbps data stream. The sample rate is twice the 4 kHz bandwidth required for a toll-quality voice conversion.

Differential (or Delta) PCM (DPCM) encodes the PCM values as differences between the current and the previous value. For audio, this type of encoding reduces the number of bits required per sample by about 25% compared to PCM. Adaptive Differential PCM (ADPCM) is a variant of DPCM that varies the size of the quantization step to allow further reduction of the required bandwidth for a given signal-to-noise ratio.

Example 4.4

The information in an analog waveform with a maximum frequency of $f_m = 3\,\text{kHz}$ is transmitted over an M-ary PCM system, where the number of pulse levels is $M = 2^5 = 32$. The quantization distortion is specified not to exceed $\pm 1\%$ of the peak-to-peak analog signal.

    a. What is the minimum number of bits/sample or bits/PCM word that should be used?

    b. What is the minimum sampling rate, and what is the resulting transmission rate?

    c. What is the PCM pulse or symbol transmission rate?

Solution

$$|e| \le pV_{pp}$$

where:

    p = the fraction of the peak-to-peak analog voltage, V_{pp}

    $|e|$ = the magnitude of quantization distortion error specified as a fraction p of the peak-to-peak analog voltage V_{pp}

$$|e_{max}| = \frac{V_{pp}}{2L}$$

$$\therefore \left(\frac{V_{pp}}{2L} \le pV_{pp} \right)$$

$$2^R = L \ge \frac{1}{2p}$$

where R is the number of bits required to represent quantization levels L

$$2^R \ge \frac{1}{2 \times 0.01} = 50$$

$$\therefore (R \ge 5.64) \quad \text{use } R = 6$$

$$f_s = 2f_m = 6000 \text{ samples per second}$$

$$f_s = 6 \times 6000 = 36\,\text{kbps}$$

$$M = 2^b = 32$$

$$b = 5 \text{ bits per symbol}$$

$$R_s = \frac{36{,}000}{5} = 7200 \text{ symbols/sec.}$$

4.8   Shannon Limit

For a Gaussian channel (with additive white Gaussian noise [AWGN]) a theoretical limit to the maximum data rate (R_b) that can be transmitted in a channel with a given bandwidth (B_w) is given by the Shannon theorem as [11,12]:

$$C \leq B_w \log_2\left(1 + \frac{S}{N}\right) \tag{4.14}$$

$$\frac{C}{B_w} \leq \log_2\left(1 + \frac{E_b}{N_0} \cdot \frac{R_b}{B_w}\right) \tag{4.15}$$

where:

E_b/N_0 = signal-to-noise density ratio

At channel capacity $R_b = C$

Let $C/B_w \equiv \eta$ (spectral efficiency of the channel), then Equation 4.15 can be written as:

$$\therefore \frac{E_b}{N_0} \geq \frac{2^\eta - 1}{\eta} \tag{4.16}$$

where:

C = channel capacity (bits per second)

S = signal strength

N = noise power

N_0 = noise density

In the limit, as $\eta \to 0$, we get

$$\frac{E_b}{N_0} \geq \lim_{\eta \to 0}\left(\frac{2^\eta - 1}{\eta}\right) \geq \lim_{\eta \to 0}(2^\eta \cdot \log_e 2) \geq \log_e 2 \geq 0.6931 \geq -1.592\,\text{dB} \tag{4.17}$$

Figure 4.14 shows η versus E_b/N_0 for the Gaussian channel.

The Shannon capacity bound states that there exists a coding/modulation scheme that achieves, over the AWGN channel, an arbitrarily low probability of bit-error, P_b provided that

$$R_b < C(\text{SNR}) \tag{4.18}$$

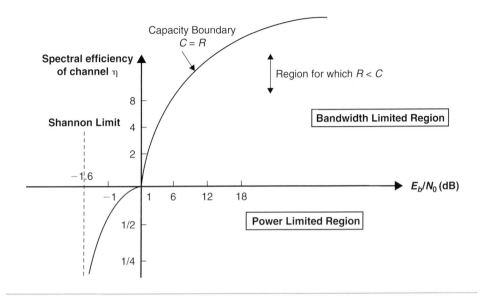

Figure 4.14 η versus E_b/N_0.

4.9 Modulation

Baseband signals are generated at low rates, therefore these signals are modulated onto a radio frequency carrier for transmission. Baseband signal $s(t)$ is complex, and can be represented mathematically as

$$s(t) = a(t)e^{j\phi(t)} \tag{4.19}$$

where:

$a(t)$ is the amplitude and $\phi(t)$ is the phase

Assuming a sampling rate the same as the Nyquist rate, the low-pass reconstruction filter extends from $-f_m$ to f_m (see Figure 4.15). The maximum frequency f_m of $s(t)$ is an approximate measure of its bandwidth. The Fourier transform of $s(t)$ is given by

$$S(f) = \int_{-\infty}^{\infty} s(t)e^{-j2\pi ft}\,dt \tag{4.20}$$

A functional block diagram of a generic modulation procedure for signal $s(t)$ is given in Figure 4.16.

$$x(t) = \text{Real}\,[s(t)A_c e^{j2\pi f_c t}] = A_c a(t)\cos[2\pi f_c t + \phi(t)] \tag{4.21}$$

$$x(t) = A_c a(t)\cos\phi(t)\cos(2\pi f_c t) - A_c a(t)\sin\phi(t)\sin(2\pi f_c t) \tag{4.22}$$

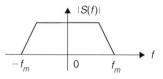

Figure 4.15   Power spectral density (PSD).

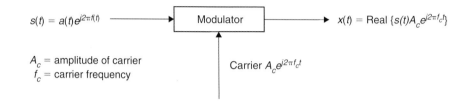

Figure 4.16   Functional block diagram of a generic modulator.

The modulation can be classified as *linear* modulation or nonlinear modulation. A modulation process is *linear* when $a(t)\cos\phi(t)$ and $a(t)\sin\phi(t)$ are linearly related to the message information signal. Examples of linear modulation are *amplitude modulation*, where the modulating signal affects only the amplitude of the modulated signal (i.e., $\phi(t)$ is constant for any t), and *phase modulation*, where the modulating signal affects only the phase of the modulated signal (i.e., when $\phi(t)$ is a constant over each signaling (symbol) interval and $a(t)$ is constant for any t).

The modulation process is *nonlinear* when the modulating signal $s(t)$ affects the frequency of the modulated signal. The definition of a nonlinear system is that superposition does not apply. The modulation process is nonlinear whether or not the amplitude of the modulating signal is a function of time. We consider a frequency modulation process and let $a(t) = a$ for any t. Then, the nonlinear modulated signal is $x(t) = aA_c\cos[2\pi f_c t + \phi(t)]$, where $\phi(t)$ is the integral of a frequency function.

Selection of modulation and demodulation schemes is based on spectral efficiency, power efficiency, and fading immunity. During the late 1970s and early 1980s, constant envelope modulation schemes were used for cellular systems to achieve a high-power efficient terminal with a C class amplifier. As a result, Gaussian minimum shift keying (GMSK) (see Chapter 9) is the widely used modulation scheme in the GSM and DECT systems.

In the mid-1980s, when cellular systems' capacity became a serious problem, developments of linear modulations with two bits per second per Hz (bps/Hz) transmission capability were initiated. To apply a linear modulation in a wireless communication system, we need high spectral efficiency as well as high-power efficiency at the same time. π/4-quadrature phase shift keying (QPSK) (see Chapter 9) was used in the Japanese and North American digital cellular and personal systems. More details of modulation schemes used in various wireless systems are given in Chapter 9.

4.10   Performance Parameters of Coding and Modulation Scheme

The most important parameter of a coding and modulation scheme is the bandwidth requirement, which is determined by the spectrum of the modulated signal usually presented as a plot of power spectral density (PSD) against frequency (see Figure 4.17) [1]. Ideally, the PSD should be zero outside the band occupied. However, in practice this can never be achieved, and the spectrum extends to infinity beyond the band. This is either because of the inherent characteristics of the modulation scheme, or because of the practical implementation of filters. Hence, we must define the bandwidth, B_w, such that the signal power falling outside the band is below a specified threshold. In practice, this threshold is determined by the tolerance of the system to adjacent channel interference. The coding and modulation selection should be based on the following factors:

- bit error rate probability, P_b
- bandwidth efficiency, η

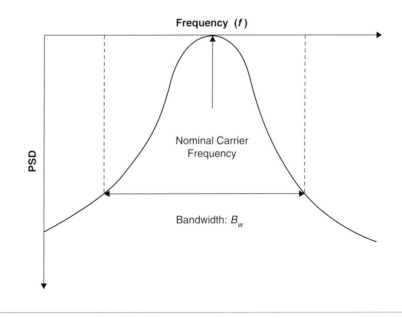

Figure 4.17   PSD versus frequency.

- signal-to-noise density ratio, E_b/N_0 (E_b is the energy per bit and N_0 is the noise density)
- complexity of transmitter/receiver

The bandwidth efficiency (or spectrum efficiency), η, of a coding and modulation scheme determines the bandwidth requirement. This is defined as the information bit rate, R_b, per unit bandwidth occupied, and is measured in bits/sec/Hz (bps/Hz).

$$\eta = \frac{R_b}{B_w} \tag{4.23}$$

An ideal coding and modulation system should provide a small P_b with a high bandwidth efficiency (η) and a low signal-to-noise density ratio (E_b/N_0). The information rate, R_b, is related to the number of waveforms, M, used by the modulator and the duration of these waveforms, T_s

$$R_b = \frac{\log_2 M}{T_s} \tag{4.24}$$

The average power used by the modulator is $P = E_s/T_s$, where E_s is the average energy of the modulator signals. Each signal carries a $\log_2 M$ information bit.

$$\therefore P = \frac{E_b \log_2 M}{T_s} = E_b R_b \tag{4.25}$$

The signal-to-noise ratio (SNR) is the ratio between the average signal power and the average noise power over the signal bandwidth

$$SNR = \frac{P}{N_0 B_w} = \left(\frac{E_b}{N_0}\right) \cdot \left(\frac{R_b}{B_w}\right) \tag{4.26}$$

Equation 4.26 shows that SNR is the product of (E_b/N_0) and (R_b/B_w), the bandwidth (or spectral) efficiency of a modulation scheme. Table 4.1 lists the bandwidth efficiencies of several 2G wireless systems.

Table 4.1 Bandwidth efficiencies of wireless systems.

Wireless System	R_b (kb/s)	B_w (kHz)	R_b/B_w (bps/Hz)
CT2	72	100	0.72
GSM	270.8	200	1.354
IS-54	48.6	30	1.62
PDC	42.0	25	1.68
IS-95-CDMA	1228	1230	0.998

For high SNR, the error probability can be closely approximated by a complementary error function $\text{erfc}[d_{\min}/(2\sqrt{N_0})]$, where $d_{\min}$ is the minimum Euclidean distance between any two elements of the modulator signal set [1]. It is related to the power efficiency (γ) of a modulation scheme. The parameter γ expresses how efficiently a modulation scheme uses the available signal energy to generate minimum Euclidean distance. γ is defined as:

$$\gamma = \frac{d^2{}_{\min}}{4E_b} \tag{4.27}$$

Equation 4.27 provides an approximation to error probability that is asymptotically tight for large SNR.

$$P_b \approx \frac{M-1}{2}\text{erfc}\left(\sqrt{\gamma \frac{E_b}{N_0}}\right) \tag{4.28}$$

Thus, for high signal-to-noise ratios, a modulation scheme is better if its power efficiency is greater. Table 4.2 lists bandwidth and power efficiency of several *M-ary* modulation schemes.

The BER performance of a coding and modulation scheme is an important parameter. The shape of E_b/N_0 versus BER (P_b) curves depends on the channel and the modulation scheme. Ideally, the BER of the service offered to the user should be zero, but it is not possible in practice. Hence, we must specify a BER to obtain the required E_b/N_0. There is an inherent trade-off between γ and the E_b/N_0 requirement of a coding and modulation scheme. The greater the bandwidth efficiency, the greater the required E_b/N_0 is likely to be. Hence, it is usually possible to increase the capacity of a communication system for a given bandwidth allocation by increasing the signal power.

Table 4.2 Bandwidth and power efficiency of *M-ary* modulation scheme [1].

Modulation	R_b/B_w	γ
Pulse Amplitude Modulation (PAM)	$2 \log_2 M$	$3 \log_2 M/(M^2-1)$
Phase Shift Keying (PSK)	$\log_2 M$	$\sin^2(\pi/M) \log_2 M$
Quadrature Amplitude Modulation (QAM)	$\log_2 M$	$3 \log_2 M/(2M-1)$
Frequency Shift Keying (FSK)	$2 \log_2 M/M$	$1/2 \log_2 M$

4.11 Power Limited and Bandwidth-Limited Channel

The objective of a general communication system design is to use transmitted power and channel bandwidth as efficiently as possible. In most communication channels, one resource may be more important than another. We may therefore classify communication channels as *power-limited* or *bandwidth-limited*.

For a power-limited channel, the desired system performance is obtained with the smallest possible power. One choice is the use of standard error control codes [5] which increase the power efficiency by adding extra bits to the transmitted symbol sequence. This procedure requires the modulator to operate at a higher data rate and, hence, requires a larger bandwidth.

In a bandwidth-limited channel, increased efficiency in both power and frequency utilization is obtained by selecting an integrated coding and modulation solution, where higher-order modulation schemes (e.g., 8-PSK instead of 4-PSK; see Chapter 9) are combined with low-complexity coding schemes. Trellis codes [9] for bandwidth-limited channels result from the treatment of modulation and coding as a combined entity rather than as two separate operations. The trellis-coded-modulation (TCM) solution combines the choice of a modulation scheme with that of a convolutional code, while the receiver, instead of performing demodulation and decoding in two separate steps, combines the two operations into one.

Table 4.3 summarizes some of the energy savings (coding gains) in dB that can be obtained by doubling the constellation size and using TCM. These are considered for coded 8-PSK (relative to uncoded 4-PSK) and for coded 16-QAM (relative to uncoded 8-PSK). These gains can be achieved only for high signal-to-noise ratios, and decrease as the signal-to-noise ratios decrease. The complexity of decoder/demodulator is proportional to the number of states.

Table 4.3 Coding gain of TCM [1].

No. of states	Coding gain (dB) (8-PSK)	Coding gain (dB) (16-QAM)
4	3.0	4.4
8	3.6	5.3
16	4.1	6.1
32	4.6	6.1
64	4.8	6.8
128	5.0	7.4
256	5.4	7.4

4.12 Nyquist Bandwidth

As discussed earlier, Nyquist showed that the theoretical minimum bandwidth (Nyquist bandwidth) required for the baseband transmission of R_s symbols per second without inter-symbol interference (ISI) is $R_s/2$ Hz. In practice, the Nyquist minimum bandwidth is increased by about 10 to 20% because of the constraints of real filters. Thus, typical baseband digital communication throughput is reduced from the ideal two symbols per Hz to a range of about 1.8 to 1.6 symbols per Hz. Assuming Nyquist filtering at baseband, the required bandwidth is related to the symbol rate by

$$B_w = \frac{(1 + \beta)}{T_s} = R_s(1 + \beta) \tag{4.29}$$

where:

β = roll-off factor (may vary between 0 and 1). This allows a trade-off between bandwidth and desirable time-domain properties of the signal. Very low values are difficult to realize in practice because they require very rapid roll-off in the filter response and because of the ringing in time-domain response.

From the set of M symbols, the coding system assigns to each symbol b bits, where $M = 2^b$. The number of bits per symbol will be:

$$b = \log_2 M$$

$$R_b = bR_s$$

$$\therefore R_s = \frac{R_b}{b} = \frac{R_b}{\log_2 M} \tag{4.30}$$

For signaling at a fixed symbol rate as b is increased, the data rate R_b is increased. In the case of M-PSK, increasing b results in an increased bandwidth efficiency η. In other words, with the same system bandwidth, we can transmit M-PSK signals at an increased data rate and hence at an increased η.

The effective duration T_b of each bit in terms of the symbol duration, T_s, or the symbol rate, R_s, is

$$T_b = \frac{1}{R_b} = \frac{T_s}{b} = \frac{T_s}{\log_2 M}$$

$$\therefore \frac{R_b}{B_w} = \frac{\log_2 M}{B_w T_s} = \frac{1}{B_w T_b} \text{ bits/s/Hz} \tag{4.31}$$

From Equation 4.31, we can observe that a digital communication system becomes more bandwidth efficient as its $B_w T_b$ product is decreased. Thus, signals with small $B_w T_b$ product are often used with a *bandwidth-limited* system. For example the Global System of Mobile (GSM) communication uses GMSK modulation with a $B_w T_b$ product equal to 0.3, where B_w is the 3 dB bandwidth of a Gaussian filter (see Chapter 9). For uncoded bandwidth-limited systems, the objective is to maximize the transmitted information rate within the allowable bandwidth, at the expense of a signal-to-noise density ratio (E_b/N_0) (while maintaining a specified value of BER).

For the case of *power-limited* systems, power is scarce but system bandwidth is available. The following trade-offs are possible: (1) improved BER at the expense of bandwidth for a fixed signal-to-noise density ratio (E_b/N_0); or (2) reduction in signal-to-noise density ratio (E_b/N_0) at the expense of bandwidth for a fixed BER. A natural modulation choice for a power-limited system is *M*-FSK (see Chapter 9).

Using Equations 4.29 and 4.31, we get

$$\frac{R_b}{B_w} = \frac{\log_2 M}{(1 + \beta)} \text{ bit/s/Hz} \tag{4.32}$$

The *M*-PSK modulation is a bandwidth-efficient scheme. As *M* increases in value, R_b/B_w also increases. Thus *M*-PSK modulation can achieve improved bandwidth efficiency at the cost of increased signal to noise density ratio (E_b/N_0).

For *M*-FSK, the Nyquist minimum bandwidth is given as

$$B_w = \frac{M(1 + \beta)}{T_S} = MR_s(1 + \beta) \tag{4.33}$$

From Equations 4.32 and 4.30, we get

$$\frac{R_b}{B_w} = \frac{\log_2 M}{M(1 + \beta)} \tag{4.34}$$

For simplicity, the modulation choice is limited to constant-envelope types — either *M*-PSK or noncoherent orthogonal *M*-FSK. Thus, for an uncoded system, if the channel is bandwidth limited, *M*-PSK is selected, and if the channel is power limited, *M*-FSK is selected. Note that, when we consider error-correction coding, the selection of a modulation scheme is not so simple because there may exist coding schemes which can provide power-bandwidth trade-off more effectively than would be possible using any *M-ary* modulation scheme.

Example 4.5

In a digital communication system, the received power to noise-power spectral density (S/N_0) is 53 dB-Hz, the required data rate is 9.6 kbps and the available bandwidth is 4.8 kHz. The required BER performance P_b is $\leq 10^{-5}$. What is the design choice of the modulation scheme if no error-correction coding is used?

Solution

Since the required data rate of 9.6 kbps is more than the available bandwidth of 4.8 kHz, the channel is bandwidth-limited.

$$\frac{E_b}{N_0}(\text{dB}) = \frac{S}{N_0}(\text{dB} \cdot \text{Hz}) - R(\text{dB})$$

$$\therefore \frac{E_b}{N_0} = 53 - 10\log 9600 = 20.78 \, \text{dB}$$

Try the 8-PSK modulation scheme

$$P_s \le (\log_2 M) \times P_b \le (\log_2 8) \times 10^{-5} \le 3 \times 10^{-5}$$

$$\frac{E_s}{N_0} = (\log_2 8) \times \frac{E_b}{N_0} = 3 \times 20.78 = 62.34$$

$$P_s(8) = 2Q\left[\sqrt{2\left(\frac{E_s}{N_0}\right)} \cdot \sin\left(\frac{\pi}{8}\right)\right] = 2Q(4.28) = 2.2 \times 10^{-5} < 3 \times 10^{-5} \, (\text{see Chapter 9})$$

Therefore, use 8-PSK modulation.

Example 4.6

In a digital communication system, the signal-to-noise spectral density ratio is 48 dB-Hz, the available bandwidth is equal to 45 kHz, and the data rate is 9.6 kbps. The required BER performance, P_b, is 10^{-5}. What is the design choice of the modulation scheme without an error-correction coding?

Solution

In this case, the channel is not bandwidth limited since the available bandwidth of 45 kHz is more than adequate to support the required data rate of 9.6 kbps.

$$\frac{E_b}{N_0}(\text{dB}) = \frac{S}{N_0}(\text{dB} \cdot \text{Hz}) - R(\text{dB})$$

$$\therefore \frac{E_b}{N_0} = 48 - 10\log 9600 = 6.61 \, \text{dB}$$

We try the 16-FSK modulation scheme

$$\therefore \frac{E_s}{N_0} = (\log_2 16) \times \frac{E_b}{N_0} = 4 \times 6.61 = 26.44$$

$$P_s(M) = \frac{M-1}{2}e^{\frac{-E_s}{2N_0}} = \frac{15}{2}e^{-13.22} = 1.36 \times 10^{-5} \, (\text{see Chapter 9})$$

For orthogonal signaling

$$P_s \leq \frac{2^M - 1}{2^{M-1}} \cdot P_b \leq \frac{65{,}535}{32{,}768} \times 10^{-5} \leq 2 \times 10^{-5} \text{ (see Chapter 9)}$$

We can meet the given specifications for this power-limited channel with a 16-FSK modulation scheme without any error-correction coding.

4.13 OSI Model

Coding and modulation are the primary tasks of the OSI physical layer, which adapts the information transmitted to it from the higher layers into a form that can be transmitted over the physical medium. Some functions of coding and modulation, such as error control [15] and multiple access, appear higher up the OSI stack. Table 4.4 summarizes network-related functions performed by the first three OSI layers.

4.13.1 OSI Upper Layers

The role of the OSI upper layers (Transport, Session, Presentation, and Application) are summarized as follows [3,14]:

The *transport layer* segments the messages into packets of acceptable sizes and performs the reassembly at the destination. It may multiplex many low-rate transmissions onto one virtual circuit or divide a high-rate transmission into parallel virtual circuits. The transport layer controls transmission errors and requests retransmissions of packets corrupted by transmission errors. In addition, the flow may be controlled by some mechanism to prevent one host from sending data faster than the destination host can handle.

The *session layer* sets up the call and takes care of the authentication of the user and of billing. The session layer supervises the synchronization (packet

Table 4.4 Functions of first four OSI layers.

OSI layers	Function
Network Layer	Routing
Data Link Layer	
• *Logical Link Control* (*LLC*)	Error Control
• *Media Access Control* (*MAC*)	Multiple Access protocols
Physical Layer	Modulation, Forward Error Correction (FEC), Encryption, Equalization, Synchronization
Physical Medium	Radio Propagating Mechanism

numbering) and the recovery in case of failures. The session layer closes the session at the end of transmission.

The *presentation layer* asks the session layer to set up a call. It specifies the destination's name and type of transmission (e.g., datagram, high priority). The presentation layer translates between the local syntax used by the application process and transfer syntax. The presentation layer also performs the required encryption and data compression.

The *application layer* provides information transfer services for user application programs. The user interacts with the application layer through a user interface. The application layer is composed of Specific Application Service Elements (SASEs) that use the services of Common Application Service Elements (CASEs). A CASE establishes are association between SASEs and may include an Association Control Service Element (ACSE), a Remote Operation Service Element (ROSE), and the Commitment Concurrency and Recovery (CCR) element.

4.14 Data Communication Services

End-to-end communication services are classified as either *synchronous (sync)* or *asynchronous (async)* [2]. A sync communication service delivers a bit stream with a fixed delay and a given bit error rate. The voice communication is an example of the sync communication service. The sync delivery of a 64 kbps voice bit stream can be implemented by dividing the bit stream into packets that are received with random delays and are stored in a buffer to hold the bits until they are delivered. This implementation of a sync transmission service is called *packetized-voice*. In packetized-voice, a buffer is used to absorb the random fluctuations in the packet transmission delays. Another implementation of the sync transmission of the bit stream is to use a dedicated coaxial cable that propagates the bits one after the other, all with the same delay.

In an async communication service, the bit stream to be transferred is divided into packets. The packets are received by the destination with varying delays, and a fraction of them may not be received correctly at the destination. An async communication service is evaluated by its QoS. The QoS deals with parameters, such as the packet error rate, delay, throughput, reliability, and security of the communication. There are two classes of async communication services: *connection-oriented* and *connectionless*. A connection-oriented communication service delivers the packet in sequence, i.e., in the correct order and confirms the delivery. Depending on the QoS requirements, the delivery may be guaranteed to be error-free. Thus, connection-oriented service looks from end-to-end like a dedicated link, which may be noiseless or noisy. A connectionless communication service delivers the packets individually. The packets can be delivered out of order, and some may contain errors and others may be lost. Some connectionless services provide an acknowledgment (ACK) of the correctly delivered packets. The connectionless services are similar to the mail service provided by the post office: letters may be delivered out of order; normal mail delivery does not guarantee or acknowledge the delivery. Another class of connection-oriented communication service is also used in some applications. It is called *expedited data*, and corresponds to a potentially faster delivery of certain packets, usually by making them jump to the head of the queues of packets that are waiting to be transmitted.

Communication services are implemented by transporting bits over the network. One essential objective of the bit transport is *connectivity*, in which one network user should be able to exchange information with many other users. It should be possible to route the bits of one user to any one of a large number of other users. The property to vary the path followed by the bits is called *switching*. Three basic methods used for switching bits in communication networks are:

- circuit-switching
- virtual-circuit packet switching
- datagram packet switching

In circuit-switching, the switch connects transmission paths to establish a circuit between transmitter and receiver. Circuit-switching is quite suitable for continuous data transmission services. It is a sync service.

A packet-switched network uses another scheme. The nodes of the network, *packet-switching nodes*, play a role similar to that of switches in a circuit-switched network. Packet-switched networks can use two different methods for selecting the path followed by the packets: *virtual-circuit* (VC) and *datagram*. In the VC transport, the different packets that are part of the same information transfer are sent along the same path. The packets follow one another as if they were using a dedicated circuit even though they may be interleaved with other packet streams. Some implementations of VC perform an error control on each link between successive nodes. Thus, not only are the packets delivered in sequence by each node to the next node along the path, but they are also transmitted without errors. This is implemented by each node checking the correctness of the packets it receives and asking the previous node along the path to retransmit incorrect packets. VC packet-switching does not need a buffer at the destination. VC packet-switching is a sync service.

Since multiple virtual circuits may exist between the source-destination pair, routing cannot be done on the basis of the source-destination address only. Data packets must carry an indication of VC identification as well. Routing is done on the basis of explicit route number and destination address. An explicit routing table at each node associates an appropriate outgoing transmission group with the destination address and explicit route number. By changing the explicit route number for a given destination, a new path will be followed. This introduces alternative route capability. If a link or node along the path becomes inoperative, any session using that path can be reestablished on an explicit route by bypassing the failed element. Explicit routes can also be assigned on the basis of type of traffic, type of physical media along the path (satellite or terrestrial, for example), or other criteria. Routes could also be listed on the basis of cost, the smallest-cost route being assigned first, then the next-smallest-cost route, and so forth.

In datagram packet-switching, the bits are grouped as packets. Each packet is labeled with the address of its destination. The packets are routed independently of one another and arrive at destination out-of-sequence. The datagram packet switching requires buffers at the source and destination. In datagram packet switching networks, each network node keeps a complete (global) topological database that is updated regularly as topological changes occur. Generally, the routing philosophy of datagram networks is to route packets (datagram) along paths of minimum time delay. The datagram is an async service.

4.15 Multiplexing

Multiplexing [2] refers to a variety of techniques that are used to make an efficient use of transmission facility. In many cases, the capacity of transmission facility exceeds the requirements for the transfer of data between two devices. That capacity can be shared among multiple transmitters by multiplexing a number of signals onto the same medium. In this case, the actual transmission path is called a *circuit* or *link*, and the portion of capacity devoted to each pair of transmitter/receivers is called a *channel*.

There are three types of multiplexing techniques: frequency-division multiplexing (FDM), synchronous time-division multiplexing (TDM), and improved synchronous TDM. FDM is the most widespread. TDM is commonly used for multiplexing digitized voice streams. The third type improves the efficiency of synchronous TDM. It is known by various names:

- Statistical TDM
- Asynchronous TDM
- Intelligent TDM

Figure 4.18 shows the concept of multiplexing. We have *m* inputs to a multiplexer. The multiplexer is connected by a single data link to a demultiplexer.

Figure 4.18 Multiplexing.

The link carries m separate channels of data. The multiplexer combines data from m input lines and transmits over a higher-capacity data link. The demultiplexer accepts the combined data stream, separates the data according to channel, and delivers them to the appropriate output lines.

4.16   Transmission Media

The transmission medium [2] is the physical path between the transmitter and receiver in a communication system. The characteristics and quality of transmission depend on the nature of the signal and the nature of the medium.

Two general ranges of frequencies are of interest. Microwave frequencies cover a range of about 2 to 40 GHz. At these frequencies, highly directional beams are possible. Microwave is quite suitable for point-to-point transmission. Signals in the frequency range 30 MHz to 1 GHz are referred to as radio waves. Omnidirectional transmission is used and signals at these frequencies are often employed for broadcast applications.

The following are the commonly used transmission mediums used in wired and wireless communications:

- Twisted pair
- Coaxial cable
- Fiber optics
- Terrestrial microwave
- Radio
- Satellite

A *twisted pair* contains two insulated copper wires arranged in a regular spiral pattern. A wire pair acts as a single communication link. Typically, a number of these pairs are bundled together into a cable by wrapping them in a tough protective sheath. Over longer distances, cables may contain hundreds of pairs. The twisting of the individual pairs minimizes electromagnetic interference between the pairs. The wires in a pair have a thickness of 0.016 to 0.036 in. The twisted pair is by far the most common medium for both analog and digital data transmission. It is the backbone of the telephone system as well as the workhorse for intra-building communications.

A *coaxial cable* is simply a transmission line consisting of a pair made up of an inner conductor surrounded by a grounded outer conductor, which is held in a concentric configuration by a dielectric. Systems have been designed to use coaxial cable as a transmission medium with a capability of transmitting an FDM configuration from 120 to 10,800 voice channels. Community antenna television (CATV) systems use single cable for transmitting a bandwidth of the order of 300 MHz.

One of the advantages of coaxial cable systems is reduced noise accumulation when compared to radio links. A coaxial cable system is attractive for television transmission or other video applications.

Fiber optics as a transmission medium has a comparatively unlimited bandwidth. It has excellent attenuation properties—as low as 0.25 dB/km. A major advantage of fiber optics compared to coaxial cable is that no equalization is needed. Also repeaters separation of the order of 10 to 1000 times that of coaxial cable for equal transmission bandwidths can be used. Other advantages of fiber optics are:

- Electromagnetic immunity
- Ground loop elimination
- Security
- Small size and light weight
- Expansion capabilities require change out of electronics only, in most cases.

Fiber optics uses three wavelength bands: around 800, 1300, and 1600 nm or near-visible infrared. Fiber optics has analog transmission applications, particularly for video/TV.

The primary use of *terrestrial microwaves* is in long-haul telecommunications service as an alternative to coaxial cable for transmitting television and voice. Microwaves can support high data rates over long distances. The microwave requires far fewer amplifiers or repeaters than coaxial cable for the same distance, but requires line-of-sight transmission. One of the potential uses for terrestrial microwaves is to provide digital data transmission in small regions (radius < 10 km). Another common use of microwaves is for short point-to-point links between buildings. It can also be used as a data link between local area networks.

The principal difference between *radio* and microwaves is that radio is omnidirectional and microwaves are focused. Radio does not require dish-shaped antennas, and the antennas need not be rigidly mounted to precise alignment. Radio covers VHF and some UHF bands: 30 MHz to 1 GHz. A primary source of impairment for radio waves is multipath interference. Reflection from land, water, and natural or human-made objects can create multiple paths between antennas. This effect is frequently evident when TV reception displays multiple images as a plane passes by.

Satellite communication is nothing more than a radio link communication using one or two RF repeaters located at great distances from terminal earth stations. If the range from an earth antenna to a satellite is the same as the satellite altitude, the round-trip delay is around 500 ms, which is more than that encountered in conventional terrestrial systems. Thus, one major problem is propagation time and the resulting echo on telephone circuits. To reply for packet transmission systems affects the delay and requires careful selection of telephone signaling systems, otherwise the call setup time may become excessive.

The equatorial orbit is filling with geostationary satellites and RF interference from one satellite system to another is increasing. The most desirable frequency bands for commercial satellite communication are in the spectrum 1000–10,000 MHz.

4.17   Transmission Impairments

The nominal voice channel occupies the band 0–4 kHz. The CCITT (ITU) voice channel occupies the band 300–3400 Hz. There are four other parameters that are used to characterize the voice channel:

1. Attenuation distortion
2. Phase distortion
3. Level (signal power level)
4. Noise and SNR

4.17.1   Attenuation Distortion

A communication channel suffers various types of distortion, i.e., the output signal from the channel is distorted in some manner such that it is not an exact replica of the input. Attenuation distortion can be avoided if all frequencies within the passband are subjected to exactly the same loss (or gain). One type of distortion is referred to as *attenuation distortion* and is the result of imperfect amplitude-frequency response.

4.17.2   Phase Distortion

A voice channel may be regarded as a band-pass filter. A signal takes a finite time to pass through a filter. This time is a function of the velocity of propagation, which varies with the medium involved. The velocity also tends to vary with frequency because of the electrical characteristics associated with it.

The finite time a signal takes to pass through the total extension of a voice channel or any network is called *delay*. Absolute delay is the delay a signal experiences while passing through the channel at a reference frequency. The propagation time is different for different frequencies with the wave front of one frequency arriving prior to the wave front of another in the passband. Thus, the phase is shifted or distorted. If the phase shift is uniform with respect to frequency, a modulated signal will not be distorted, but if the phase shift is nonlinear with respect to frequency, the output signal is distorted compared to the input.

4.17.3   Level

System levels are used to engineer a communication system. A zero test-level point is established. A zero test-level is the point at which the test-tone level should be 0 dBm. From zero test-level other points may be shown using dBr (decibel reference). The dBm and dBr are related as:

$$dBm = dBm0 + dBr \tag{4.35}$$

For example, a value of -30 dBm at a -20 dBr point corresponds to a reference level of -10 dBm0. A -10 dBm0 signal introduced at 0 dBr point (zero test-level) has an absolute signal level of -10 dBm.

4.17.4 Noise and SNR

Noise

Noise can be put into the following categories:

- Thermal noise
- Inter modulation (IM) noise
- Crosstalk
- Impulse noise

Thermal noise occurs in all transmission media and all communication equipment. It occurs due to random electron motion and is characterized by a uniform distribution of energy over the frequency spectrum with a Gaussian distribution of levels. The white noise refers to the average uniform spectral distribution of energy with respect to frequency. Thermal noise is directly proportional to bandwidth (B_w) and temperature (T). The amount of thermal noise to be found in 1 Hz of bandwidth in an actual device is given as:

$$N_0 = kT \text{ (W/Hz)} \tag{4.36}$$

where:

  k = Boltzmann's constant (1.3803×10^{-23} J/K)
  T = absolute temperature (K)
  At room temperature $T = 290$ K

$$N_0 = 1.3803 \times 290 \times 10^{-23} = -204 \text{ dBW/Hz} = -174 \text{ dBm/Hz}$$

The noise power at temperature T with bandwidth B_w is given as:

$$N = kTB_w$$

$$N = -198.6 \text{ dBW} + 10 \log T + 10 \log B_w \tag{4.37}$$

Inter modulation (IM) noise is the result of the presence of IM products. If two signals with frequencies f_1 and f_2 are passed through a nonlinear device or medium, the result will be IM products that are spurious frequency components. These components may be present either inside or outside the band of interest for the device. IM noise may result from a number of causes:

- Improper level setting. If the level of input to a device is too high, the device is driven into its nonlinear operating region
- Improper alignment causing a device to function nonlinearly

- Nonlinear envelope
- Device malfunction

Crosstalk refers to unwanted coupling between signal paths. Crosstalk is caused by (1) electrical coupling between transmission media, (2) poor control of frequency response, and (3) the nonlinear performance in an analog multiplex system. There are two types of crosstalk:

- *Intelligible*—where at least four words are intelligible to the listener from extraneous conversations in a seven-second period (for voice applications)
- *Unintelligible*—crosstalk resulting from any other form of disturbing effects of one channel on another.

Impulse noise is noncontinuous and consists of irregular pulses or noise spikes of short duration and of relatively high amplitude. These spikes are often called "hits." Impulse-noise degrades voice telephony only marginally, if at all; however, it may seriously degrade error rate on data or other digital circuits.

SNR

SNR is expressed in decibels (dB) — the amount by which a signal level exceeds the noise within a specified bandwidth.

$$\frac{S}{N}(\text{dB}) = (\text{Signal level})(\text{dBm}) - (\text{Noise level})(\text{dBm}) \tag{4.38}$$

The following are the suggested SNR values for various services:

- Voice: 40 dB*
- Video: 45 dB*
- Data ~15 dB, based on specified BER and modulation type.

Note: The values marked by * are based on customer input.

4.18 Summary

In this chapter we presented the essentials of digital communications. This chapter may be skipped by those readers who have been exposed to digital communications. The chapter outlines the process of converting an analog signal to a digital signal. The chapter presented the modulation and error correcting schemes that are used in wireless communications systems. It also discussed data communication services and defined circuit-switched and packet-switched services. Finally, it looks at the most common types of transmission media, and the behavior of signals on those media.

The readers who wish to learn more about digital communication theory should refer to several excellent books listed in the reference section (for example [6] and [7]). More details about modulation schemes used in wireless communications can be found in Chapter 9.

Problems

4.1 Define linear and nonlinear modulation schemes.

4.2 What are the power limited and bandwidth limited channels?

4.3 Define the Nyquist bandwidth.

4.4 A digital signaling system is required to operate at 9.6 kbps. If a signal element encodes a 4-bit word, what is the minimum bandwidth of the channel?

4.5 Given a channel with intended capacity of 20 Mbps, the bandwidth of the channel is 3 MHz. What signal-to-noise ratio is required to achieve this capacity?

4.6 The receiver in a communication system has a received signal power of -134 dBm, a received noise power spectral density of -174 dBm/Hz, and a bandwidth of 2000 Hz. What is the maximum rate of error-free information for the system?

4.7 Find the minimum required bandwidth for the baseband transmission of an 8-level PAM pulse sequence having a data rate, R_b, of 9600 bps. The system characteristics consist of a raised-cosine spectrum with a 50% excess bandwidth.

4.8 Find the BER probability, P_e, of an M-PSK system for 14.4 kbps signal, amplitude is 15 mv, and noise density, N_0 is 10^{-9} W/Hz.

4.9 A noncoherent orthogonal M-FSK system carries 3 bits per symbol. The system is designed for $E_b/N_0 = 6$ dB. What is the maximum bit error rate probability? Find the bandwidth efficiency of the system.

4.10 A noncoherent orthogonal M-FSK system carries 4 bits per symbol. The system is designed to have a maximum probability of symbol-error of 10^{-6}. What is the required E_b/N_0? What is the bandwidth efficiency?

References

1. Burr, Alister. *Modulation and Coding for Wireless Communications*. Upper Saddle River, NJ: Prentice Hall, 2001.

2. Freeman, R. L. *Telecommunication System Engineering*. New York: John Wiley and Sons, 1989.

3. Garg, V. K., and Wilkes, J. E. *Principles & Applications of GSM*. Upper Saddle River, NJ: Prentice Hall, 1999.

4. Glover, I. A., and Grant, P. M. *Digital Communications*. Upper Saddle River, NJ: Prentice Hall, 1998.

5. Hamming, R.W. Error detecting and correcting codes. *Bell System Technical Journal*, 29:147–160, 1950.

6. Haykin, S. *Communication Systems*. New York: John Wiley and Sons, 2001.

7. Lathi, B. P. *Modern Digital and Analog Communications Systems*. 2nd Edition. Holt, Reinhart and Winston, 1989.

8. Nyquist, H. Certain topics in telegraph transmission theory. *AIEE Transactions*, 47: 617–644, 1928.

9. Proakis, J. G. *Digital Communications*. 3rd Edition. McGraw-Hill, 1995.

10. Rappaport, T. S. *Wireless Communications: Principles and Practice*, 2nd Edition. Upper Saddle River, NJ: Prentice Hall, 2000.

11. Shannon, C. E. A mathematical theory of communication, Part 1. *Bell System Technical Journal*, 27:379, 1948.

12. Shannon, C. E. A mathematical theory of communication, Part 2. *Bell System Technical Journal*, 27:623, 1948.

13. Skalar, B. *Digital Communication—Fundamental and Applications*. Englewood Cliffs, NJ: Prentice Hall, 1988.

14. Stallings, W. *Data and Computer Communications*. New York: Macmillan Publishing Company, 1988.

15. Sweeney, P. *Error Control Coding: An Introduction*. Prentice Hall, 1991.

16. Taub, H., and Schilling, D. L. *Principles of Communications Systems*. New York: McGraw Hill, 1999.

17. Ziemer, R. E., and Peterson, R. L. *Introduction to Digital Communication*. New York: Macmillan Publishing Company, 1992.

CHAPTER 5

Fundamentals of Cellular Communications

5.1 Introduction

In this chapter, we present the concept of a cellular system and discuss the fundamentals of cellular communications. We develop a relationship between reuse ratio (q) and cluster size or reuse factor (N) for hexagonal cell geometry, as well as study cochannel interference for omnidirectional and sectorized cells. We also discuss cell splitting and segmentation procedures used in cellular systems.

5.2 Cellular Systems

Most commercial radio and television systems are designed to cover as much area as possible. These systems typically operate at maximum power and with the tallest antennas allowed by the Federal Communications Commission (FCC). The frequency used by the transmitter cannot be reused again until there is enough geographical separation so that one station does not interfere significantly with another station assigned to that frequency. There may even be a large region between two transmitters using the same frequency where neither signal is received.

The cellular system takes the opposite approach [1,3,4,5,9,11–14]. It seeks to make an efficient use of available channels by employing low-power transmitters to allow frequency reuse at much smaller distances (see Figure 5.1). Maximizing the number of times each channel may be reused in a given geographic area is the key to an efficient cellular system design.

Cellular systems are designed to operate with groups of low-power radios spread out over the geographical service area. Each group of radios serve mobile stations located near them. The area served by each group of radios is called a *cell*. Each cell has an appropriate number of low-power radios to communicate within the cell itself. The power transmitted by the cell is chosen to be large enough to communicate with mobile stations located near the edge of the cell. The radius of each cell may be chosen to be perhaps 28 km (about 16 miles) in a start-up system with relatively few subscribers, down to less than 2 km (about 1 mile) for a mature system requiring considerable frequency reuse.

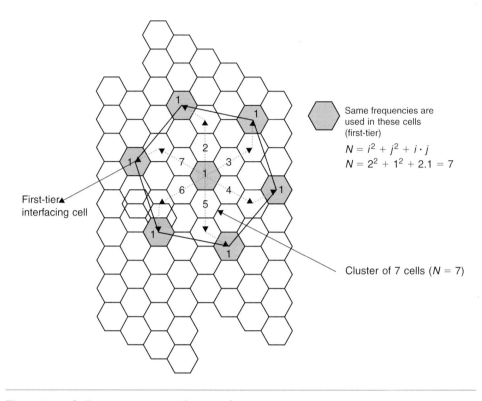

Figure 5.1   Cell arrangement with reuse factor.

As the traffic grows, new cells and channels are added to the system. If an irregular cell pattern is selected, it would lead to an inefficient use of the spectrum due to its inability to reuse frequencies because of cochannel interference. In addition, it would also result in an uneconomical deployment of equipment, requiring relocation from one cell site to another. Therefore, a great deal of engineering effort would be required to readjust the transmission, switching, and control resources every time the system goes through its development phase. The use of a regular cell pattern in a cellular system design eliminates all these difficulties.

In reality, cell coverage is an irregularly shaped circle. The exact coverage of the cell depends on the terrain and many other factors. For design purposes and as a first-order approximation, we assume that the coverage areas are regular polygons. For example, for omnidirectional antennas with constant signal power, each cell site coverage area would be circular. To achieve full coverage without dead spots, a series of regular polygons are required for cell sites. Any regular polygon such as an equilateral triangle, a square, or a hexagon can be used for cell design. The hexagon is used for two reasons: a hexagonal layout requires fewer cells and, therefore, fewer transmitter sites, and a hexagonal cell layout is less expensive compared to square and triangular cells. In practice, after the polygons are drawn on a map of the coverage area, radial lines are drawn and the signal-to-noise ratio (SNR) calculated for various directions using the propagation models (discussed in Chapter 3), or using appropriate computer programs [2,6–8]. For the remainder of this chapter, we assume regular polygons for coverage areas even though in practice that is only an approximation.

5.3 Hexagonal Cell Geometry

We use the u-v axes to calculate the distance D between points C_1 and C_2 (see Figure 5.2). The u-v axes are chosen so that u-axis passes through the centers of the hexagons. C_1 and C_2 are the centers of the hexagonal cells with coordinates (u_1, v_1) and (u_2, v_2) [11,12].

$$D = \left\{ (u_2 - u_1)^2 (\cos 30°)^2 + [(v_2 - v_1) + (u_2 - u_1)(\sin 30°)]^2 \right\}^{1/2} \qquad (5.1)$$

$$= \left\{ (u_2 - u_1)^2 + (v_2 - v_1)^2 + (v_2 - v_1)(u_2 - u_1) \right\}^{1/2} \qquad (5.2)$$

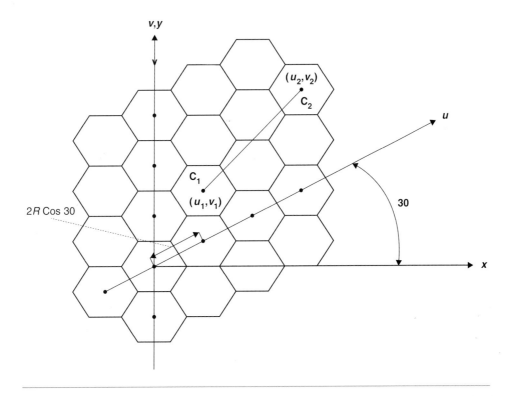

Figure 5.2 Coordinate system.

If we assume $(u_1, v_1) = (0,0)$, or the origin of the coordinate system is the center of a hexagonal cell, and restrict (u_2, v_2) to be a positive integer valued (i, j) then normalized distance D_{norm} from Equation 5.2 can be written as:

$$D_{norm} = [i^2 + j^2 + ij]^{1/2} \tag{5.3}$$

The normalized distance between two adjacent cells is unity for $(i = 1, j = 0)$ or $(i = 0, j = 1)$. The actual center-to-center distance (D) between two adjacent hexagonal cells is $2R\cos 30°$ or $\sqrt{3}R$ where R is the center-to-vertex distance.

We assume the size of all the cells is roughly the same. As long as the cell size is fixed, and each cell transmits the same power, cochannel interference will be independent of the transmitted power of each cell. The cochannel interference is a function of q where $q = D/R$. But D (distance between any two cells) is a function of N_I and S/I. N_I is the number of cochannel-interfering cells in the first tier (see Figure 5.3, *Note*: N_I is 6; the number of cochannel-interfering cells in the second tier will be 12) and S/I is the received signal-to-interference ratio at the desired mobile receiver. We neglect the effects of cochannel-interfering cells in the second, third, and higher tiers because their contributions are much smaller compared to the first tier ($<1\%$ of the total interference is caused by cells in the second and higher tiers).

Assuming that the first cell is centered at the origin ($u = 0$, $v = 0$), the distance between any two cells will be:

$$D = (D_{norm})(\sqrt{3}R) \tag{5.4a}$$

Using Equation 5.3 cochannel separation (D) (Figure 5.4) will be:

$$D^2 = 3R^2(i^2 + j^2 + ij) \tag{5.4b}$$

Since the area of a hexagon is proportional to the square of the distance between the center and vertex, the area enclosed by the large hexagon is:

$$A_{large} = k[3R^2(i^2 + j^2 + ij)] \tag{5.5}$$

where k is a constant

Similarly the area enclosed by the small hexagon is given as:

$$A_{small} = k(R^2) \tag{5.6}$$

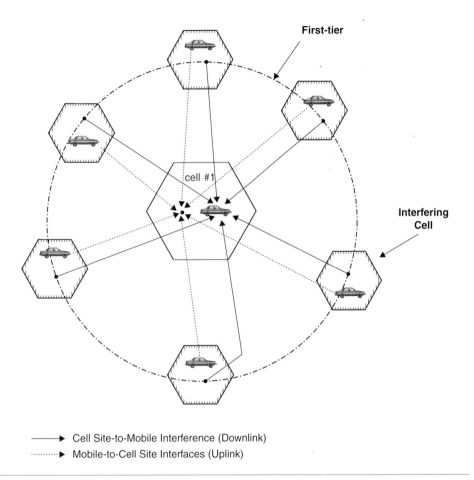

First-tier

cell #1

Interfering Cell

→ Cell Site-to-Mobile Interference (Downlink)
⋯⋯▶ Mobile-to-Cell Site Interfaces (Uplink)

Figure 5.3 Cochannel interference with omnidirectional cell site.

Comparing Equation 5.5 and Equation 5.6, we can write

$$\frac{A_{\text{large}}}{A_{\text{small}}} = 3(i^2 + j^2 + ij) = \frac{D^2}{R^2} \tag{5.7}$$

From symmetry, we can see that the large hexagonal encloses the center cluster of N cells plus one-third of the number of cells associated with six other peripheral hexagons. Thus, the total number of cells enclosed in the large hexagon is equal to $N + 6(N/3) = 3N$. Since the area is proportional to the number of cells, $A_{\text{large}} = 3N$ and $A_{\text{small}} = 1$.

$$\frac{A_{\text{large}}}{A_{\text{small}}} = 3N \tag{5.8}$$

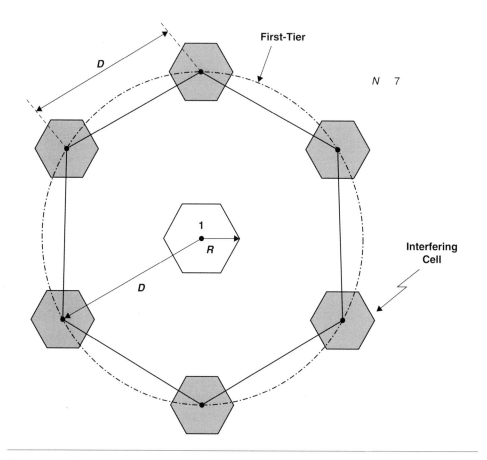

Figure 5.4 Six effective interfering cells in tier 1 of cell 1.

Substituting Equation 5.8 into Equation 5.7 we get:

$$N = i^2 + j^2 + ij \tag{5.9}$$

$$\therefore \frac{D^2}{R^2} = 3N \tag{5.10}$$

$$\frac{D}{R} = q = \sqrt{3N} \tag{5.11}$$

where:

    q = reuse ratio (refer to Figure 5.5)
    N = cluster size (see Figure 5.1) or reuse factor

Table 5.1 lists the values of q for different values of N.

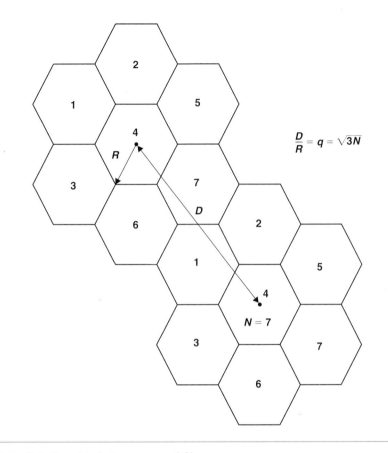

$$\frac{D}{R} = q = \sqrt{3N}$$

Figure 5.5 Relationship between q and N.

Equation 5.11 is important because it affects the traffic-carrying capacity of a cellular system and the cochannel interference. By reducing q, the number of cells per cluster is reduced. If total RF channels are constant, then the number of channels per cell is increased, thereby increasing the system capacity. On the other hand, cochannel interference is increased with small q. The reverse is true when q is increased—an increase in q reduces cochannel interference and also the traffic capacity of the cellular system.

Table 5.1 shows that a 2-cell, 5-cell, etc., reuse pattern does not exist. However, the basic assumption in Equation 5.9 is that all six first-tier interferers are located at the same distance from the desired cell. Asymmetrical reuse arrangements, where interferers are located at various distances, do allow for 2-cell, 5-cell, etc., reuse.

Table 5.1 Cochannel reuse ratio (*q*) vs. frequency reuse pattern (*N*).

i	*j*	*N*	*q = D/R*
1	0	1	1.73
1	1	3	3.00
2	0	4	3.46
2	1	7	4.58
3	0	9	5.20
2	2	12	6.00
3	1	13	6.24
4	0	16	6.93
3	2	19	7.55
4	1	21	7.94
4	2	28	9.17

Example 5.1

We consider a cellular system in which total available voice channels to handle the traffic are 960. The area of each cell is 6 km^2 and the total coverage area of the system is 2000 km^2. Calculate (a) the system capacity if the cluster size, N (reuse factor), is 4 and (b) the system capacity if the cluster size is 7. How many times would a cluster of size 4 have to be replicated to cover the entire cellular area? Does decreasing the reuse factor N increase the system capacity? Explain.

Solution

- Total available channels = 960
- Cell area = 6 km^2
- Total coverage area = 2000 km^2

  $N = 4$
- Area of a cluster with reuse $N = 4$: $4 \times 6 = 24$ km^2
- Number of clusters for covering total area with N equals 4 = 2000/24 = 83.33 ~ 83
- Number of channels per cell = 960/4 = 240
- System capacity = $83 \times 960 = 79,680$ channels

  $N = 7$

- Area of cluster with reuse $N = 7$: $7 \times 6 = 42\,\text{km}^2$
- Number of clusters for covering total area with N equals $7 = 2000/42 = 47.62 \sim 48$
- Number of channels per cell $= 960/7 = 137.15 \sim 137$
- System capacity $= 48 \times 960 = 46{,}080$ channels

It is evident when we decrease the value of N from 7 to 4, we increase the system capacity from 46,080 to 79,680 channels. Thus, decreasing the reuse factor (N) increases the system capacity.

5.4 Cochannel Interference Ratio

The S/I ratio at the desired mobile receiver is given as [5,10]:

$$\frac{S}{I} = \frac{S}{\sum\limits_{k=1}^{N_I} I_k} \tag{5.12}$$

where:

I_k = the interference due to the kth interferer

N_I = the number of interfering cells in the first tier.

In a fully equipped hexagonal-shaped cellular system, there are always six cochannel-interfering cells in the first tier (i.e., $N_I = 6$, see Figure 5.1). Most of the cochannel interference results from the first tier. Contribution from second and higher tiers amounts to less than 1% of the total interference and, therefore, it is ignored. Cochannel interference can be experienced both at the cell site and the mobile stations in the center cell. In a small cell system, interference will be the dominating factor and thermal noise can be neglected. Thus the S/I ratio can be given as:

$$\frac{S}{I} = \frac{1}{\sum\limits_{k=1}^{6} \left(\dfrac{D_k}{R}\right)^{-\gamma}} \tag{5.13}$$

where:

$2 \leq \gamma \leq 5$ = the propagation path loss, and γ depends upon the terrain environment (see Chapter 3)

D_k = the distance between mobile and kth interfering cell

R = the cell radius

If we assume D_k is the same for the six interfering cells for simplification, or $D = D_k$, then Equation 5.13 becomes:

$$\frac{S}{I} = \frac{1}{6(q)^{-\gamma}} = \frac{q^\gamma}{6} \tag{5.14}$$

where:

$q = D/R$, reuse ratio

$$\therefore q = \left[6\left(\frac{S}{I}\right) \right]^{1/\gamma} \tag{5.15}$$

Substituting for q in Equation 5.15 from Equation 5.11 we get:

$$N = \frac{1}{3}\left[6\left(\frac{S}{I}\right) \right]^{2/\gamma} \tag{5.16}$$

Example 5.2

Consider the advanced mobile phone system in which an S/I ratio of 18 dB is required for the accepted voice quality. What should be the reuse factor for the system? Assume $\gamma = 4$. What will be the reuse factor of the Global System of Mobile (GSM) system in which an S/I of 12 dB is required?

Solution

Using Equation 5.16, we get

$$N_{AMP} = \frac{1}{3}[6(10^{1.8})]^{2/4} = 6.486 \approx 7$$

and

$$N_{GSM} = \frac{1}{3}\left[6(10^{1.2})\right]^{2/4} = 3.251 \approx 4$$

Example 5.3

Consider a cellular system with 395 total allocated voice channel frequencies. If the traffic is uniform with an average call holding time of 120 seconds and the call blocking during the system busy hour is 2%, calculate:

1. The number of calls per cell site per hour (i.e., call capacity of cell)
2. Mean S/I ratio for cell reuse factor equal to 4, 7, and 12.

Assume omnidirectional antennas with six interferers in the first tier and a slope for path loss of 40 dB/decade ($\gamma = 4$).

Solution

For a reuse factor $N = 4$, the number of voice channels per cell site $= 395/4 = 99$. Using the Erlang-B traffic table (see Appendix A) for 99 channels with 2% blocking, we find a traffic load of 87 Erlangs. The carried load will be $(1 - 0.02) \times 87 = 85.26$ Erlangs.

$$\therefore \frac{N_{call} \times 120}{3600} = 85.26$$

$$N_{call} = 2558 \text{ calls/hour/cell}$$

Using Equation 5.16, we get

$$4 = \frac{1}{3}\left[6\left(\frac{S}{I}\right)\right]^{2/4}$$

$$\therefore \frac{S}{I} = 24 \ (13.8 \text{ dB})$$

The results for $N = 7$ and $N = 12$ are given in Table 5.2.

It is evident from the results that, by increasing the reuse factor from $N = 4$ to $N = 12$, the mean S/I ratio is improved from 13.8 to 23.3 dB. However, the call capacity of cell (i.e., calls per hour per cell) is reduced from 2558 to 724 calls per hour.

Table 5.2 Cell reuse factor vs. mean S/I ratio and call capacity of cell.

N	Voice channels per cell	Calls per hour per cell (N_{call})	Mean S/I (dB)
4	99	2558	13.8
7	56	1349	18.7
12	33	724	23.3

Example 5.4

Consider a GSM system with a one-way spectrum of 12.5 MHz and channel spacing of 200 kHz. There are three control channels per cell, and the reuse factor is 4. Assuming an omnidirectional antenna with six interferers in the first tier and a slope of path loss of 40 dB/decade, calculate the number of calls per hour per cell site with 2% blocking during the system busy hour and an average call holding time of 120 seconds. The GSM uses eight voice channels per RF channel.

Solution

$$\therefore \text{No. of voice channels per cell site} = \frac{12.5 \times 10^6 \times 8}{200 \times 10^3 \times 4} - 3 = 122$$

Using the Erlang-B traffic table for 122 channels with 2% blocking, we find a traffic load of 110 Erlangs. The carried traffic load will be $(1 - 0.02) \times 110 = 107.8$

$$\therefore \frac{N_{call} \times 120}{3600} = 107.8$$

$$N_{call} = 3234 \text{ calls/hour/cell}$$

Using Equation 5.16 we calculate the mean S/I ratio as 13.8 dB.

5.5 Cellular System Design in Worst-Case Scenario with an Omnidirectional Antenna

In the previous section we showed that with a seven-reuse pattern ($N = 7$), the value of $q = 4.6$ (see Table 5.1) is adequate for a normal interference condition [5]. Let us reexamine the seven-cell reuse pattern and consider the worst case in which the mobile unit is located at the cell boundary as shown in Figure 5.6. In this situation, the mobile unit receives the weakest signal from its own cell and is subjected to strong interference from all the interfering cells in the first tier. The distances from the six interfering cells are given in Figure 5.6.

The S/I ratio can be given as:

$$\frac{S}{I} = \frac{R^{-\gamma}}{2(D-R)^{-\gamma} + 2D^{-\gamma} + 2(D + R)^{-\gamma}} \qquad (5.17)$$

Using the path-loss exponent $\gamma = 4$ and $D/R = q$, we rewrite Equation 5.17 as:

$$\frac{S}{I} = \frac{1}{2(q-1)^{-4} + 2q^{-4} + 2(q+1)^{-4}} \qquad (5.18)$$

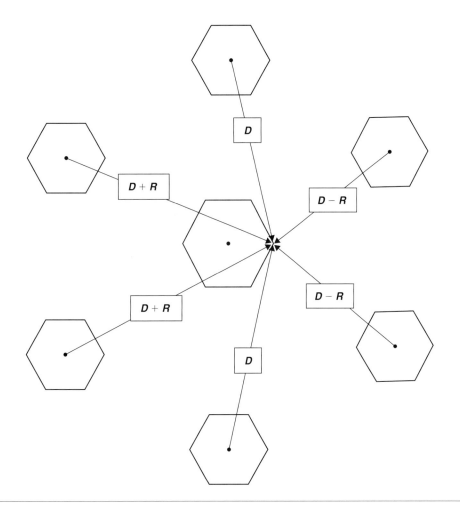

Figure 5.6 Worst-case scenario for cochannel interference.

where:

    $q = 4.6$ for a normal seven-cell reuse pattern ($N = 7$).

Substituting $q = 4.6$ in Equation 5.18, we get $S/I = 54.3$ or $17.3\,\text{dB}$. For a conservative estimate, if we use the shortest distance $(D-R)$ then

$$\frac{S}{I} = \frac{1}{6(q - 1)^{-4}} = \frac{1}{6(3.6)^{-4}} = 28 \text{ or } 14.47\,\text{dB} \tag{5.19}$$

In a real situation, because of imperfect cell-site locations and the rolling nature of the terrain configuration, the S/I ratio is often less than $17.3\,\text{dB}$. It could be $14\,\text{dB}$

or lower. Such conditions may occur in heavy traffic. Therefore, the cellular system should be designed around the *S/I* ratio of worst case. If we consider the worst case for a seven-cell reuse ($N = 7$) pattern, we conclude that $q = 4.6$ is not enough in an omnidirectional cell system. In an omnidirectional cell system, $N = 9$ ($q = 5.2$) or $N = 12$ ($q = 6.0$) cell reuse pattern would be a better choice. These cell reuse patterns would provide the *S/I* ratio of 19.78 and 22.54 dB, respectively.

5.6 Cochannel Interference Reduction

In the case of increased call traffic, the frequency spectrum should be used efficiently. We should avoid increasing the number of cells N in a frequency reuse pattern. As N increases, the number of frequency channels assigned to a cell is reduced, thereby decreasing the call capacity of the cell [7].

Instead of increasing N, we use either a directional antenna arrangement (sectorization) to reduce cochannel interference or perform cell splitting to subdivide a congested cell into smaller cells. In case of a sectorized cell, each cell is divided into three or six sectors and uses three or six directional antennas at the base station to reduce the number of cochannel interferers (refer to Figures 5.7 and 5.8). Each

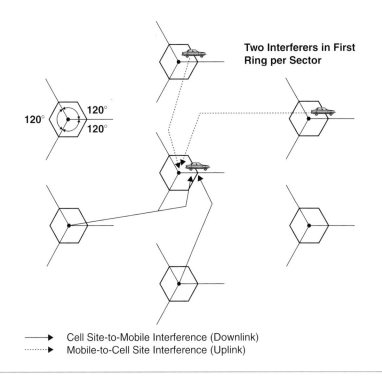

Figure 5.7   Cochannel interference with 120° sectorized cells.

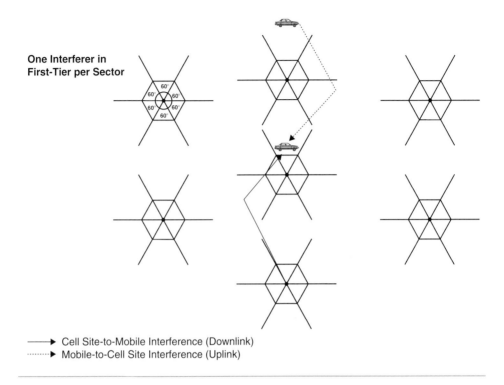

**One Interferer in
First-Tier per Sector**

→ Cell Site-to-Mobile Interference (Downlink)
┈▸ Mobile-to-Cell Site Interference (Uplink)

Figure 5.8   Cochannel interference with 60° sectorized cells.

sector is assigned a set of channels (frequencies) (either 1/3 or 1/6 of the frequencies of the omnidirectional cell).

5.7 Directional Antennas in Seven-Cell Reuse Pattern

5.7.1 Three-Sector Case

We consider the worst case in which the mobile unit is at position M (see Figure 5.9). In this situation, the mobile receives the weakest signal from its own cell and fairly strong interference from two interfering cells 1 and 2. Because of the use of directional antennas, the number of interfering cells is reduced from six to two. At point M, the distances between the mobile unit and the two interfering cells are D and $(D + 0.7R)$, respectively. The S/I ratio in the worst case with $\gamma = 4$ will be:

$$\frac{S}{I} = \frac{R^{-4}}{D^{-4} + (D + 0.7R)^{-4}} \tag{5.20}$$

$$\frac{S}{I} = \frac{1}{q^{-4} + (q + 0.7)^{-4}} \tag{5.21}$$

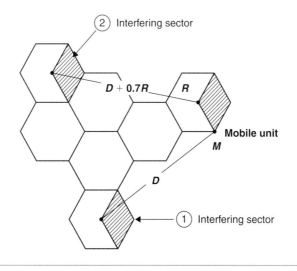

Figure 5.9   Worst-case interference with 120° sectorized cells.

Using $q = 4.6$ in Equation 5.21 for $N = 7$, we get $S/I = 285$ or 24.5 dB. The S/I for a mobile unit served by a cell site with 120° directional antenna exceeds 18 dB in the worst case. It is evident from Equation 5.21 that the use of a directional antenna is helpful in reducing cochannel interference. In a real situation, under heavy traffic, the S/I could be up to 6 dB weaker than in Equation 5.21 due to irregular terrain configurations and imperfect site locations. The resulting 18.5 dB S/I is still adequate.

5.7.2   Six-Sector Case

In this case the cell is divided into six sectors by using a 60° beam width directional antenna. In this case, only one interference can occur. The worst case S/I ratio with $\gamma = 4$ will be (see Figure 5.10):

$$\frac{S}{I} = \frac{R^{-4}}{(D + 0.7R)^{-4}} = (q + 0.7)^4 \tag{5.22}$$

For $q = 4.6$ ($N = 7$), Equation 5.22 gives $S/I = 789$ or 29 dB. This indicates a further reduction in cochannel interference. Using the argument that was used for the three-sector case and subtracting 6 dB from 29 dB, the remaining 23 dB is still more than adequate. Under heavy traffic, the 60° sector configuration can be used to reduce cochannel interference. However, with the six-sector configuration, the call capacity of the cell is decreased.

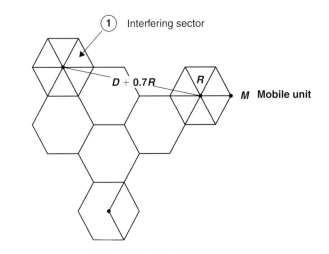

Figure 5.10 Worst-case interference with 60° sectorized cells.

Example 5.5

We consider a cellular system with 395 total allocated voice channels of 30 kHz each. The total available bandwidth in each direction is 12.5 MHz. The traffic is uniform with the average call holding time of 120 seconds, and call blocking during the system busy hour is 2%.

Calculate:

    a. The calls per hour per cell site

    b. The mean S/I ratio

    c. The spectral efficiency in Erlang/km²/MHz

For a cell reuse factor $N = 4, 7$, and 12, respectively, and for omnidirectional (see Example 5.3) 120° and 60° antenna systems, calculate the call capacity. Assume that there are 10 mobiles/km² with each mobile generating traffic of 0.02 Erlangs. The slope of path loss is $\gamma = 40$ dB per decade. The mean S/I ratio is given as:

$$\text{Mean } \frac{S}{I} = \gamma \log \sqrt{3N} - 10 \log m$$

where:

    γ = slope of path loss (dB/decade)

    m = number of interferers in the first tier

    $m = 6$ for omnidirectional system

    $m = 2$ for 120° sectorized system

    $m = 1$ for 60° sectorized system

Solution

We consider only the first tier interferers and neglect the effects of cochannel interference from the second and other higher tiers.

The traffic carried per cell site $= V \times t \times A_c = V \times t \times 2.6R^2$ Erlangs

where:

V = number of mobiles per km^2

t = traffic in Erlangs per mobile

A_c = area of hexagonal cell (i.e., $A_c = 2.6R^2$)

The traffic carried per cell site $= 10 \times 0.02 \times 2.6R^2 = 0.52R^2$ Erlangs

The spectral efficiency $= \dfrac{\text{Traffic per cell} \times N_c}{B_w \times A}$ Erlangs/km^2/MHz

where:

N_c = number of cells in the system (i.e., $A/2.6R^2$) and

A = area of the system

The spectral efficiency $= \dfrac{\text{Traffic per cell}}{2.6R^2 \times B_w}$ Erlangs/km^2/MHz

We will demonstrate the procedure for calculating results in one row in Table 5.3; the remaining calculations can be performed without any difficulty.

120° Sectorized Cell

a. $N = 7$

b. Number of voice channels per sector $= 395/(7 \times 3) \approx 19$

c. Offered traffic load per sector from Erlang-B tables $= 12.3$ Erlangs

d. Offered traffic load per cell site $= 3 \times 12.3 = 36.9$ Erlangs

e. Carried traffic load per cell site $= (1 - 0.02) \times 36.9 = 36.2$ Erlangs

$$\text{No. of calls per hour per cell site} = \frac{36.2 \times 3600}{120} = 1086$$

$$\therefore R = \sqrt{\frac{36.2}{0.52}} = 8.3\,\text{km}$$

$$\text{Spectral efficiency} = \frac{36.2}{2.6 \times 12.5 \times 8.3^2} = 0.016\,\text{Erlang/km}^2/\text{MHz}$$

$$\text{Mean } S/I = 40\log\sqrt{21} - 10\log2 = 26.44 - 3.01 = 23.43\,\text{dB}$$

Table 5.3 Omni vs. sectorized cellular system performance.

System	N	Channels per sector	Offered load/ cell (E)	Carried load/ cell (E)	Calls/ hour/ cell (cell capacity)	Required cell radius R (km)	Spectral efficiency E/km^2/ MHz	Mean S/I (dB)
Omni	4	99	87.0	85.3	2558	12.8	0.016	13.8
	7	56	45.9	45.0	1349	9.3	0.016	18.7
	12	33	24.6	24.1	724	6.8	0.016	23.3
120° Sector	4	33	73.8	72.3	2169	11.8	0.016	18.6
	7	19	36.9	36.2	1086	8.3	0.016	23.4
	12	11	17.5	17.2	516	5.8	0.016	28.1
60° Sector	3	22	84.2	82.6	2477	12.6	0.016	19.1
	4	17	64.2	62.9	1887	11.0	0.016	21.6
	7	9	26.0	25.5	765	7.0	0.016	26.4
	12	6	13.7	13.4	402	5.1	0.016	31.1

From the results in Table 5.3, we can draw the following conclusions:

1. Sectorization reduces cochannel interference and improves the mean S/I ratio for a given cell reuse factor. However, it reduces cell capacity since the channel resource is distributed more thinly among various sectors.

2. Since a sectorized cellular system has fewer cochannel interferers, it is possible to reduce the cluster size.

3. An omnidirectional cellular system requires a cluster size of 7, while a 120° sectorized system requires a cluster size of 4 and a 60° sectorized system requires a cluster size of 3 for a desired mean S/I ratio of approximately 18 dB.

5.8 Cell Splitting

After a cellular system has been in operation, traffic will grow in the system and will require that additional channels be made available. The cell with heavy traffic is split into smaller cells (see Figure 5.11) [10]. This is done in such a way that cell areas, or the individual component coverage areas of the cellular system, are

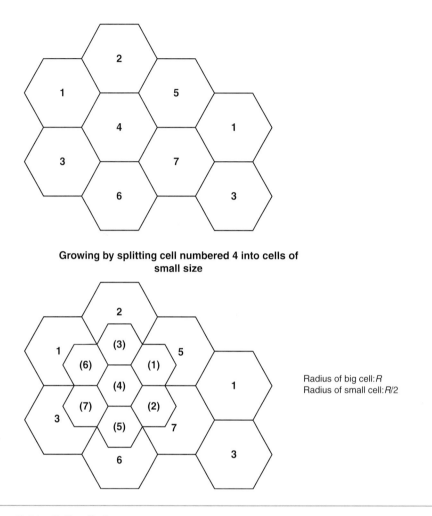

Growing by splitting cell numbered 4 into cells of small size

Radius of big cell: *R*
Radius of small cell: *R*/2

Figure 5.11   Cell splitting.

further divided to yield yet more cell areas. The splitting of cell areas by adding new cells provides for an increasing amount of channel reuse and, hence, increasing system capacity.

When cell splitting occurs, the designer should minimize changes in the system. The value of the reuse factor, N, is important with respect to the type of split. The following splitting patterns can be used for various values of N:

- For $N = 3$, use 4:1 cell splitting.
- For $N = 4$, use 3:1 cell splitting.

- For $N = 7$, use 3:1 or 4:1 cell splitting.
- For $N = 9$, use 4:1 cell splitting.

The 4:1 cell splitting works as follows. When the new cell is located on the border between two existing cell areas, the new cell will cover an area with a radius of one-half that of the larger cell areas from which it was split. Thus, the area covered by the new cell is one-fourth of the area of the old cell; hence, this is called 4:1 splitting. For example, if the old cell had a radius of 8 km, the new cell would have a radius of 4 km.

The 3:1 cell splitting works as follows. When the new cell is located on the corner between the cell areas covered by three existing cells, the radius of the cell area covered by the new cell would be $1/\sqrt{3}$ of the radius of the cell areas covered by the existing cell sites. Thus, the new coverage area is only one-third of the larger cell's coverage area; hence, this is called 3:1 splitting.

Decreasing cell radii imply that cell boundaries will be crossed more often. This will result in more handoffs per call and a higher processing load per subscriber. Simple calculations show that a reduction in a cell radius by a factor of four will produce about a tenfold increase in the handoff rate per subscriber. Since the call processing load tends to increase geometrically with the increase in the number of subscribers, with cell splitting the handoff rate will increase exponentially. Therefore, it is desirable to perform a cost-benefit analysis to compare the overall cost of cell splitting versus other available alternatives to handle increased traffic load.

The large cell with radius R is split into cells with radius $R/2$. Let d be the separation between the transmitter and receiver, and d_0 be the distance from the transmitter to a close-in reference point. P_0 is the power received at the close-in reference point. The average received power, P_r, is given by:

$$P_r = P_0\left(\frac{d}{d_0}\right)^{-\gamma} \tag{5.23}$$

Expressing Equation 5.23 in decibels, we have:

$$P_r \text{ (dBm)} = P_0(\text{dBm}) - 10\gamma\log\left(\frac{d}{d_0}\right) \tag{5.24}$$

P_{t1} and P_{t2} are the transmit power of the large cell (radius R) base station and small cell (radius $R/2$) base station, respectively. The received power, P_r at the large cell boundary is proportional to $P_{t1}R^{-\gamma}$, and P_r at the small cell boundary is proportional to $P_{t2}(R/2)^{-\gamma}$. On the basis of equal power, we get

$$P_{t1}R^{-\gamma} = P_{t2}\left(\frac{R}{2}\right)^{-\gamma} \tag{5.25}$$

or

$$\frac{P_{t1}}{P_{t2}} = 4\gamma \qquad\qquad (5.26)$$

For $\gamma = 4$, $P_{t1}/P_{t2} = 16$ (12 dB). Thus, with cell splitting, where the radius of the new cell is one-half of that of the old cell, we achieve a 12 dB reduction in the transmit power.

5.9 Adjacent Channel Interference (ACI)

Signals which are adjacent in frequency to the desired signal cause adjacent channel interference [7,8]. ACI is brought about primarily because of imperfect receiver filters which allow nearby frequencies to move into the pass band, and nonlinearity of the amplifiers. The ACI can be reduced by: (1) using modulation schemes which have low out-of-band radiation; (2) carefully designing the bandpass filter at the receiver front end; and (c) assigning adjacent channels to different cells in order to keep the frequency separation between each channel in a given cell as large as possible. The effects of ACI can also be reduced using advanced signal processing techniques that employ equalizers.

5.10 Segmentation

Sometimes engineers have to add an additional cell at less than the reuse distance without using a complete cell splitting process. This method might be used to fill in a coverage gap in the system. This can result in cochannel interference. The most straightforward method to avoid an increase in cochannel interference is simply not to reuse them. Segmentation divides a channel group into segments of mutually exclusive channel frequencies. Then, by assigning different segments to particular cell sites, cochannel interference between these cell sites can be avoided. The disadvantage of segmentation is that the capacity of the segmented cells is lower than the unsegmented cell.

When a cellular system is growing, there may be cells of different radii in the same region of the coverage area. This also can result in cochannel interference. By dividing the radii at the cell site into two separate serving groups, one for the larger (overlaid) cell and other for the smaller (underlaid) cell, the interference can be minimized (see Figure 5.12). The radii for the primary serving group serve the underlaid cell, and the radii of the secondary serving group are used to serve mobiles in the overlaid cell areas. As traffic in the smaller cells grows, more and more channels are removed from the secondary serving group and assigned to the primary group until the secondary group (and its larger cell) disappears.

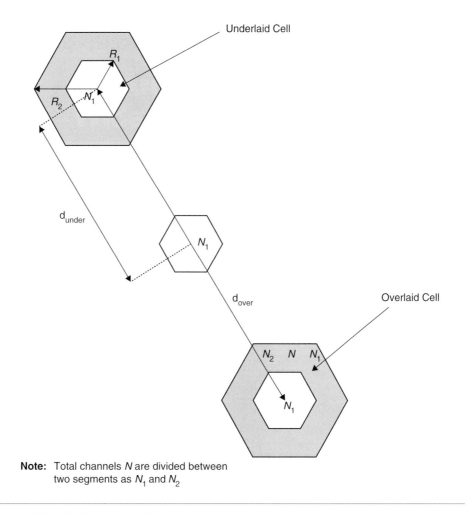

Note: Total channels *N* are divided between
two segments as N_1 and N_2

Figure 5.12 Cell segmentation.

5.11 Summary

In this chapter, we developed a relationship between the reuse ratio (q) and cell cluster size (N) for the hexagonal geometry. Cochannel interference ratios for the omnidirectional and sectorized cell were derived. A numerical example was given to demonstrate that, for a given cluster size, sectorization yields a higher *S/I* ratio, but reduces spectral efficiency. However, it is possible to achieve a higher spectral efficiency by reducing the cluster size in a sectorized system without lowering the *S/I* ratio below the minimum requirement. We concluded the chapter by discussing cell splitting, adjacent channel interference, and segmentation.

Problems

5.1 Why is cell splitting needed? Define 4:1 and 3:1 cell splitting.

5.2 Explain segmentation in a cellular system. Why is it required?

5.3 What is adjacent channel interference? How can it be minimized?

5.4 The *S/I* ratio was calculated by neglecting the interference from cells other than in the first tier. Calculate the amount of interference from the second tier of cells. Is it reasonable to neglect this interference?

5.5 Consider a GSM system with a one-way spectrum of 12.5 MHz and channel spacing of 200 kHz. There are three control channels per cell and the reuse factor is 4. Assuming an omnidirectional antenna with six interferers in the first tier and slope for path loss of 45 dB/decade, calculate the number of calls per cell site per hour with 2% blocking during system busy hour and an average call holding time of 120 seconds. What is the *S/I* ratio?

5.6 Repeat Problem 5.5 with 3-sector and 6-sector antenna systems. Discuss your results with an omnidirectional antenna and provide comments.

5.7 Compare the spectral efficiency of the GSM system with the total access communication system (TACS) using the following data:

- GSM channel spacing: 200 kHz
- TACS channel spacing: 25 kHz
- The required *S/I* for the GSM: 12 dB
- The required *S/I* for the TACS: 16 dB
- The total available one-way spectrum: 25 MHz
- The number of control channels for GSM per cell: 3
- The number of control channels for TACS per cell: 6
- The reuse factor for GSM: 4
- The reuse factor for TACS: 7
- Total coverage area: 10,000 km^2

5.8 A signal-to-interference ratio of 16 dB is required for a satisfactory forward link performance of a cellular system. What is the frequency reuse factor and cluster size that should be used for maximum capacity if the path loss is 40 dB/decade? Assume that there are six cochannel interferers in the first tier and all of them are the same distance from the mobile. Neglect interference from higher tiers.

5.9 A suburb has an area of 1500 square miles and is covered by a cellular system that uses a seven reuse pattern. Each cell has a radius of four miles and the city is allocated 50 MHz of spectrum of a full duplex channel bandwidth of 60 kHz. Assuming a GoS of 2% for an Erlang-B system and a traffic load per user of 0.03 Erlangs, calculate:

- The number of cells in the service area
- The number of channels per cell
- The maximum carried traffic
- Traffic intensity of each cell
- Total number of users that can be served for 2% GoS
- Number of mobiles per channel

References

1. AT&T Technical Center, "Cellular System Design and Performance Engineering I," CC 1400, version 1.12, 1993.

2. Chen, G. K. Effects of Sectorization on Spectrum Efficiency of Cellular Radio Systems. *IEEE Transactions on Vehicular Technology,* 41, no. 3 (August 1992): 217–225.

3. Dersh, U., and Braun, W. A Physical Mobile Radio Channel Model. Proceedings of IEEE Vehicular Technology Conference. May 1991, 289–294.

4. French, R. C. The effects of Fading and Shadowing on Channel Reuse in Mobile Radio. *IEEE Transactions on Vehicular Technology,* 28 (August 1979).

5. Lee, W. C. Y. Elements of Cellular Radio System. *IEEE Transactions on Vehicular Technology,* 35 (May 1986): 48–56.

6. Lee, W. C. Y. Spectrum Efficiency and Digital Cellular. Presented at 38th IEEE Vehicular Technology Conference, Philadelphia, June 1988.

7. Lee, W. C. Y., *Mobile Cellular Telecommunications System.* New York: McGraw-Hill, 1989.

8. Lee, W. C. Y. Spectrum Efficiency in Cellular. *IEEE Transactions on Vehicular Technology,* 38 (May 1989): 69–75.

9. MacDonald, V. H. The Cellular Concept. *Bell System Technical Journal,* 58, no. 1 (January 1979): 15–41.

10. Mark, J. W., and Weihua, Zhuang. *Wireless Communications and Networking.* Upper Saddle River, NJ: Prentice Hall, 2003.

11. Mehrotra, A. *Cellular Radio Analog and Digital System.* Boston: Artech House, 1994.

12. Rappaport, T. S. *Wireless Communication.* Upper Saddle River, NJ: Prentice Hall, 2002.

13. Whitehead, J. F. Cellular System Design: An Emerging Engineering Discipline. *IEEE Communications Society Magazine,* 24, no. 2 (February 1986): 8–15.

14. Young, W. R. Advanced Mobile Phone Service: Introduction, Background, and Objectives. *Bell System Technical Journal,* 58, no. 1 (January 1979): 1–14.

CHAPTER 6

Multiple Access Techniques

6.1 Introduction

In Chapter 5 we discussed that cellular systems divide a geographic region into cells where mobile units in each cell communicate with the cell's base station. The goal in the design of a cellular system is to be able to handle as many calls as possible in a given bandwidth with the specified blocking probability (reliability).

Multiplexing deals with the division of the resources to create multiple channels. Multiplexing can create channels in frequency, time, etc., and the corresponding terms are then frequency division multiplexing (FDM), time division multiplexing (TDM), etc. [1,3]. Since the amount of spectrum available is limited, we need to find ways to allow multiple users to share the available spectrum simultaneously. Shared access is used to implement a multiple access scheme when access by many users to a channel is required [13,14,15]. For example, one can create multiple channels using TDM, but each of these channels can be accessed by a group of users using the ALOHA multiple access scheme [8,9]. The multiple access schemes can be either reservation-based or random.

Multiple access schemes allow many users to share the radio spectrum. Sharing the bandwidth efficiently among users is one of the main objectives of multiple access schemes [16,17]. The variability of wireless channels presents both challenges and opportunities in designing multiple access communications systems. Multiple access strategy has an impact on robustness and interference levels generated in other cells. Therefore, multiple access schemes are designed to maintain orthogonality and reduce interference effects [10].

Multiple access schemes can be classified as *reservation-based* multiple access (e.g., FDMA, TDMA, CDMA) [4,5] and *random* multiple access (e.g., ALOHA, CSMA) (see Figure 6.1) [9,23]. If data traffic is continuous and a small transmission delay is required (for example in voice communication) reservation-based multiple access is used. The family of reservation-based multiple access includes frequency division multiple access (FDMA), time division multiple access (TDMA), and code division multiple access (CDMA) [6,7,12,21,22]. In many wireless systems for voice communication, the control channel is based on random multiple access and the communication channel is based on FDMA, TDMA, or CDMA. The reservation-based multiple access technique has a disadvantage in that once the channel is assigned, it remains idle if the user has nothing to transmit,

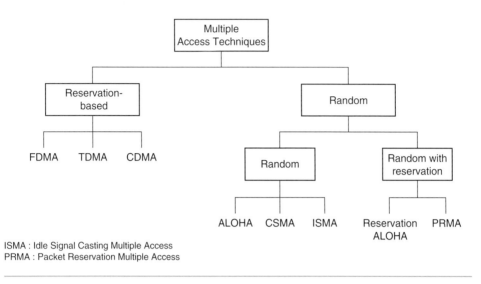

Figure 6.1 Multiple access schemes.

while other users may have data waiting to be transmitted. This problem is critical when data generation is random and has a high peak-rate to average-rate ratio. In this situation, random multiple access is more efficient, because a communication channel is shared by many users and users transmit their data in a random or partially coordinated fashion. ALOHA and carrier sense multiple access (CSMA) are examples of random multiple access [8]. If the data arrives in a random manner, and the data length is large, then random multiple access combined with a reservation protocol will perform better than both random- and reservation-based schemes.

We first focus on the reservation-based multiple access schemes including narrowband channelized and wideband nonchannelized systems for wireless communications. We discuss access technologies—FDMA, TDMA, and CDMA. We examine FDMA and TDMA from a capacity, performance, and spectral efficiency viewpoint. As networks have evolved, the demand for higher capacities has encouraged researchers and system designers to examine access schemes that are even more spectrally efficient than TDMA. Therefore, we also examine the CDMA system. Work in standards bodies around the world indicates that the 3G/4G wireless systems are evolving to wideband CDMA (WCDMA) to achieve high efficiencies and high access data rates. The later part of the chapter is devoted to the discussion of random multiple access schemes.

6.2 Narrowband Channelized Systems

Traditional architectures for analog and digital wireless systems are channelized [6,11]. In a channelized system, the total spectrum is divided into a large number of relatively narrow radio channels that are defined by carrier frequency. Each radio channel consists of a pair of frequencies. The frequency used for transmission from the base station to the mobile station is called the *forward channel* (down-link channel) and the frequency used for transmission from the mobile station to the base station is called the *reverse channel* (uplink channel). A user is assigned both frequencies for the duration of the call. The forward and reverse channels are assigned widely separated frequencies to keep the interference between transmission and reception to a minimum.

A narrowband channelized system demands precise control of output frequencies for an individual transmitter. In this case, the transmission by a given mobile station occurs within the specified narrow bandwidth to avoid interference with adjacent channels. The tightness of bandwidth limitations plays a dominant role in the evaluation and selection of modulation technique. It also influences the design of transmitter and receiver elements, particularly the filters which can greatly affect the cost of a mobile station.

A critical issue with regulators and operators around the world is how efficiently the radio spectrum is being used. Regulatory bodies want to encourage competition for cellular services. Thus for a given availability of bandwidth, more operators can be licensed. For a particular operator, a more efficient technology can support more users within the assigned spectrum and thus increase profits.

When we examine efficiencies of various technologies, we find that each system has made different trade-offs in determining the optimum method for access. Some of the parameters that are used in the trade-off are bandwidth per user, guard bands between channels, frequency reuse among different cells in the system, the signal-to-noise and signal-to-interference ratio, the methods of channel and speech coding, and the complexity of the system.

The first-generation analog cellular systems showed signs of capacity saturation in major urban areas, even with a modest total user population. A major capacity increase was needed to meet future demand. Several digital techniques were deployed to solve the capacity problem of analog cellular systems. There are two basic types of systems whereby a fixed spectrum resource is partitioned and shared among different users [13,16]. The channels are created by dividing the total system bandwidth into frequency channels through the use of FDM and then further dividing each frequency channel into time channels through the use of TDM. Most systems use a combination of FDMA and TDMA.

6.2.1 Frequency Division Duplex (FDD) and Time Division Duplex (TDD) System

Many cellular systems (such as AMP, GSM, DAMP, etc.) use *frequency division duplex* (*FDD*) in which the transmitter and receiver operate simultaneously on different frequencies. Separation is provided between the downlink and uplink channels to avoid the transmitter causing self interference to its receiver. Other precautions are also needed to prevent self interference, such as the use of two antennas, or alternatively one antenna with a duplexer (a special design of RF filters protecting the receiver from the transmit frequency). A duplexer adds weight, size, and cost to a radio transceiver and can limit the minimum size of a subscriber unit.

A cellular system can be designed to use one frequency band by using *time division duplex* (*TDD*). In TDD a bidirectional flow of information is achieved using the simplex-type scheme by automatically alternating in time the direction of transmission on a single frequency. At best TDD can only provide a quasi-simultaneous bidirectional flow, since one direction must be off while the other is using the frequency. However, with a high enough transmission rate on the channel, the off time is not noticeable during conversations, and with a digital speech system, the only effect is a very short delay.

The amount of spectrum required for both FDD and TDD is the same. The difference lies in the use of two bands of spectrum separated by the required bandwidth for FDD, whereas TDD requires only one band of frequencies but twice the bandwidth. It may be easier to find a single band of unassigned frequencies than finding two bands separated by the required bandwidth.

With TDD systems, the transmit time slot and the receiver time slot of the subscriber unit occur at different times. With the use of a simple RF switch in the subscriber unit, the antenna can be connected to the transmitter when a transmit burst is required (thus disconnecting the receiver from the antenna) and to the receiver for the incoming signal. The RF switch thus performs the function of the duplexer, but is less complex, smaller in size, and less costly. TDD uses a burst mode scheme like TDMA and therefore also does not require a duplexer. Since the bandwidth of a TDD channel is twice that of a transmitter and receiver in an FDD system, RF filters in all the transmitters and receivers for TDD systems must be designed to cover twice the bandwidth of FDD system filters.

6.2.2   Frequency Division Multiple Access

The FDMA is the simplest scheme used to provide multiple access. It separates different users by assigning a different carrier frequency (see Figure 6.2). Multiple users are isolated using bandpass filters. In FDMA, signals from various users are assigned different frequencies, just as in an analog system. Frequency guard bands are provided between adjacent signal spectra to minimize crosstalk between adjacent channels.

The advantages and disadvantages of FDMA with respect to TDMA or CDMA are:

Advantages
1. Capacity can be increased by reducing the information bit rate and using an efficient digital speech coding scheme (See Chapter 8) [20].

2. Technological advances required for implementation are simple. A system can be configured so that improvements in terms of a lower bit rate speech coding could be easily incorporated.

3. Hardware simplicity, because multiple users are isolated by employing simple bandpass filters.

Disadvantages

1. The system architecture based on FDMA was implemented in first-generation analog systems such as advanced mobile phone system (AMPS) or total access communication system (TACS). The improvement in capacity depends on operation at a reduced signal-to-interference (S/I) ratio. But the narrowband digital approach gives only limited advantages in this regard so that modest capacity improvements could be expected from the allocated spectrum.

2. The maximum bit-rate per channel is fixed and small, inhibiting the flexibility in bit-rate capability that may be a requirement for computer file transfer in some applications in the future.

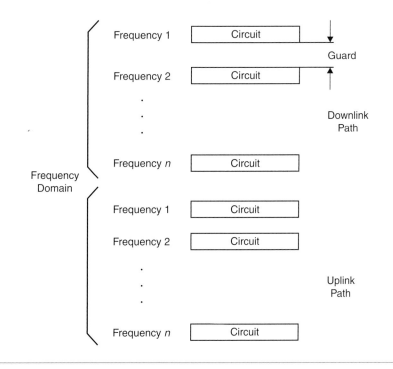

Figure 6.2 FDMA/FDD channel architecture.

3. Inefficient use of spectrum, in FDMA if a channel is not in use, it remains idle and cannot be used to enhance the system capacity.

4. Crosstalk arising from adjacent channel interference is produced by non-linear effects.

6.2.3 Time Division Multiple Access

In a TDMA system, each user uses the whole channel bandwidth for a fraction of time (see Figure 6.3) compared to an FDMA system where a single user occupies the channel bandwidth for the entire duration (see Figure 6.2) [2]. In a TDMA system, time is divided into equal time intervals, called *slots*. User data is transmitted in the slots. Several slots make up a frame. Guard times are used between each user's transmission to minimize crosstalk between channels (see Figure 6.4). Each user is assigned a frequency and a time slot to transmit data. The data is transmitted via a radio-carrier from a base station to several active mobiles in the downlink. In the reverse direction (uplink), transmission from mobiles to base stations is time-sequenced and synchronized on a common frequency for TDMA. The preamble carries the address and synchronization information that both the base station and mobile stations use for identification.

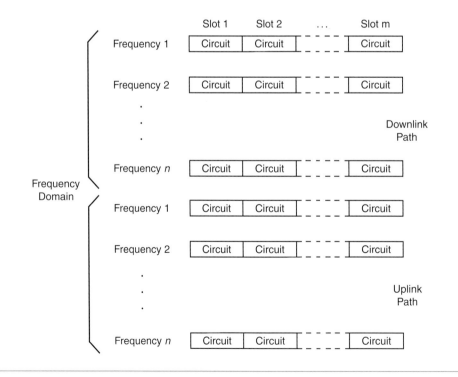

Figure 6.3 TDMA/FDD channel architecture.

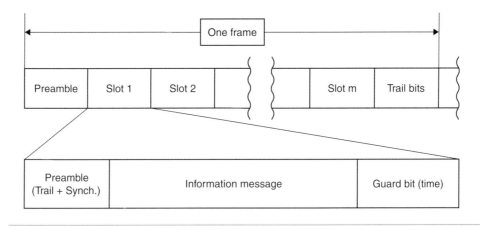

Figure 6.4 TDMA frame.

In a TDMA system, the user can use multiple slots to support a wide range of bit rates by selecting the lowest multiplexing rate or multiple of it. This enables supporting a variety of voice coding techniques at different bit rates with different voice qualities. Similarly, data communications customers could use the same kinds of schemes, choosing and paying for the digital data rate as required. This would allow customers to request and pay for a bandwidth on demand.

Depending on the data rate used and the number of slots per frame, a TDMA system can use the entire bandwidth of the system or can employ an FDD scheme. The resultant multiplexing is a mixture of frequency division and time division. The entire frequency band is divided into a number of duplex channels (about 350 to 400 kHz). These channels are deployed in a frequency-reuse pattern, in which radio-port frequencies are assigned using an autonomous adaptive frequency assignment algorithm. Each channel is configured in a TDM mode for the downlink direction and a TDMA mode for the uplink direction.

The advantages and disadvantages of TDMA are:

Advantages

1. TDMA permits a flexible bit rate, not only for multiples of the basic single channel rate but also submultiples for low bit rate broadcast-type traffic.

2. TDMA offers the opportunity for frame-by-frame monitoring of signal strength/bit error rates to enable either mobiles or base stations to initiate and execute handoffs.

3. TDMA, when used exclusively and not with FDMA, utilizes bandwidth more efficiently because no frequency guard band is required between channels.

4. TDMA transmits each signal with sufficient guard time between time slots to accommodate time inaccuracies because of clock instability, delay spread, transmission delay because of propagation distance, and the tails of signal pulse because of transient responses.

Disadvantages

1. For mobiles and particularly for hand-sets, TDMA on the uplink demands high peak power in transmit mode, that shortens battery life.

2. TDMA requires a substantial amount of signal processing for matched filtering and correlation detection for synchronizing with a time slot.

3. TDMA requires synchronization. If the time slot synchronization is lost, the channels may collide with each other.

4. One complicating feature in a TDMA system is that the propagation time for a signal from a mobile station to a base station varies with its distance to the base station.

6.3 Spectral Efficiency

An efficient use of the spectrum is the most desirable feature of a mobile communications system. To realize this, a number of techniques have been proposed or already implemented. Some of these techniques used to improve spectral efficiency are reducing the channel bandwidth, information compression (low-rate speech coding), variable bit rate codec (see Chapter 8), improved channel assignment algorithms (dynamic channel assignment), and so on [11,17,19]. Spectral efficiency of a mobile communications system shows how efficiently the spectrum is used by the system. Spectral efficiency of a mobile communications system depends on the choice of a multiple access scheme. The measure of spectral efficiency enables one to estimate the capacity of a mobile communications system.

The overall spectral efficiency of a mobile communications system can be estimated by knowing the *modulation* and the *multiple access* spectral efficiencies separately [16].

6.3.1 Spectral Efficiency of Modulation

The spectral efficiency with respect to modulation is defined as [16]:

$$\eta_m = \frac{\text{(Total Number of Channels Available in the System)}}{\text{(Bandwidth)(Total Coverage Area)}} \quad (6.1\text{a})$$

$$\eta_m = \frac{\dfrac{B_w}{B_c} \times \dfrac{N_c}{N}}{B_w \times N_c \times A_c} \quad (6.1\text{b})$$

$$\eta_m = \frac{1}{B_c \times N \times A_c} \text{ Channels/MHz/km}^2 \qquad (6.1c)$$

where:

η_m = modulation efficiency (channels/MHz/km^2)
B_w = bandwidth of the system (MHz)
B_c = channel spacing (MHz)
N_c = total number of cells in the covered area
N = frequency reuse factor of the system (or cluster size)
A_c = area covered by a cell (km^2)

Equation 6.1c indicates that the spectral efficiency of modulation does not depend on the bandwidth of the system. It only depends on the channel spacing, the cell area, and the frequency reuse factor, N. By reducing the channel spacing, the spectral efficiency of modulation for the system is increased, provided the cell area (A_c) and reuse factor (N) remain unchanged. If a modulation scheme can be designed to reduce N then more channels are available in a cell and efficiency is improved.

Another definition of spectral efficiency of modulation is *Erlangs/MHz/km^2*

$$\eta_m = \frac{\text{Maximum Total Traffic Carried by System}}{(\text{System Bandwidth})(\text{Total Coverage Area})} \qquad (6.2a)$$

$$\eta_m = \frac{\text{Total Traffic Carried by } \left(\dfrac{B_w/B_c}{N}\right) \text{ Channels}}{B_w A_c} \qquad (6.2b)$$

By introducing the trunking efficiency factor, η_t in Equation 6.2b (<1, it is a function of the blocking probability and number of available channels per cell), the total traffic carried by the system is given as:

$$\eta_m = \frac{\eta_t \left(\dfrac{B_w/B_c}{N}\right)}{B_w A_c} \qquad (6.2c)$$

$$\eta_m = \frac{\eta_t}{B_c N A_c} \qquad (6.2d)$$

where:

η_t is a function of the blocking probability and the total number of available channels per cell $\left[\dfrac{B_w/B_c}{N}\right]$

Based on Equation 6.2d we can conclude:

1. The voice quality will depend on the frequency reuse factor, N, which is a function of the signal-to-interference (S/I) ratio of the modulation scheme used in the mobile communications system (see Chapter 5).

2. The relationship between system bandwidth, B_w, and the amount of traffic carried by the system is nonlinear, i.e., for a given percentage increase in B_w, the increase in the traffic carried by the system is more than the increase in B_w.

3. From the average traffic per user (*Erlang/user*) during the busy hour and *Erlang/MHz/km²*, the capacity of the system in terms of *users/km²/MHz* can be obtained.

4. The spectral efficiency of modulation depends on the blocking probability.

Example 6.1

In the GSM800 digital channelized cellular system, the one-way bandwidth of the system is 12.5 MHz. The RF channel spacing is 200 kHz. Eight users share each RF channel and three channels per cell are used for control channels. Calculate the spectral efficiency of modulation (for a dense metropolitan area with small cells) using the following parameters:

- Area of a cell = 8 km^2
- Total coverage area = 4000 km^2
- Average number of calls per user during the busy hour = 1.2
- Average holding time of a call = 100 seconds
- Call blocking probability = 2%
- Frequency reuse factor = 4

Solution

Number of 200 kHz RF channels $= \dfrac{12.5 \times 1000}{200} = 62$

Number of traffic channels $= 62 \times 8 = 496$

Number of signaling channels per cell $= 3$

Number of traffic channels per cell $= \dfrac{496}{4} - 3 = 121$

Number of cells $= \dfrac{4000}{8} = 500$

With 2% blocking for an omnidirectional case, the total traffic carried by 121 channels (using Erlang-B tables) = 108.4 (1.0 − 0.02) = 106.2 Erlangs/cell or 13.28 Erlangs/km²

$$\text{Number of calls per hour per cell} = \frac{106.2 \times 3600}{100} = 3823, \text{calls/hour/km}^2 =$$

$$\frac{3823}{8} = 477.9\,\text{calls/hour/km}^2$$

$$\text{Max. number of users/cell/hour} = \frac{3823}{1.2} = 3186, \text{users/hour/channel} = \frac{3186}{121} =$$

26.33

$$\eta_m = \frac{(\text{Erlangs per cell}) \times \text{no. of cells}}{B_w \times \text{Coverage Area}} = \frac{106.2 \times 500}{12.5 \times 4000} = 1.06\,\text{Erlangs/MHz/km}^2$$

6.3.2   Multiple Access Spectral Efficiency

Multiple access spectral efficiency is defined as the ratio of the total time or frequency dedicated for traffic transmission to the total time or frequency available to the system. Thus, the multiple access spectral efficiency is a dimensionless number with an upper limit of unity.

In FDMA, users share the radio spectrum in the frequency domain. In FDMA, the multiple access efficiency is reduced because of guard bands between channels and also because of signaling channels. In TDMA, the efficiency is reduced because of guard time and synchronization sequence.

FDMA Spectral Efficiency

For FDMA, multiple access spectral efficiency is given as:

$$\eta_a = \frac{B_c N_T}{B_w} \le 1 \tag{6.3}$$

where:

η_a = multiple access spectral efficiency
N_T = total number of traffic channels in the covered area
B_c = channel spacing
B_w = system bandwidth

Example 6.2

In a first-generation AMP system where there are 395 channels of 30 kHz each in a bandwidth of 12.5 MHz, what is the multiple access spectral efficiency for FDMA?

Solution

$$\eta_a = \frac{30 \times 395}{12.5 \times 1000} = 0.948$$

TDMA Spectral Efficiency

TDMA can operate as wideband or narrowband. In the wideband TDMA, the entire spectrum is used by each individual user. For the wideband TDMA, multiple access spectral efficiency is given as:

$$\eta_a = \frac{\tau M_t}{T_f} \qquad (6.4)$$

where:

τ = duration of a time slot that carries data
T_f = frame duration
M_t = number of time slots per frame

In Equation 6.4 it is assumed that the total available bandwidth is shared by all users. For the narrowband TDMA schemes, the total band is divided into a number of sub-bands, each using the TDMA technique. For the narrowband TDMA system, frequency domain efficiency is not unity as the individual user channel does not use the whole frequency band available to the system. The multiple access spectral efficiency of the narrowband TDMA system is given as:

$$\eta_a = \left(\frac{(\tau M_t)}{T_f} \right) \left(\frac{(B_u N_u)}{B_w} \right) \qquad (6.5)$$

where:

B_u = bandwidth of an individual user during his or her time slot
N_u = number of users sharing the same time slot in the system, but having access to different frequency sub-bands

6.3.3   Overall Spectral Efficiency of FDMA and TDMA Systems

The overall spectral efficiency, η, of a mobile communications system is obtained by considering both the modulation and multiple access spectral efficiencies

$$\eta = \eta_m \eta_a \qquad (6.6)$$

Example 6.3

In the North American Narrowband TDMA cellular system, the one-way bandwidth of the system is 12.5 MHz. The channel spacing is 30 kHz and the total number of voice channels in the system is 395. The frame duration is 40 ms, with six time slots per frame. The system has an individual user data rate of 16.2 kbps in which the speech with error protection has a rate of 13 kbps. Calculate the multiple access spectral efficiency of the TDMA system.

Solution

The time slot duration that carries data: $\tau = \left(\dfrac{13}{16.2}\right)\left(\dfrac{40}{6}\right) = 5.35\,\text{ms}$

$T_f = 40\,\text{ms}$, $M_t = 6$, $N_u = 395$, $B_u = 30\,\text{kHz}$, and $B_w = 12.5\,\text{MHz}$

$$\eta_a = \frac{5.35 \times 6}{40} \times \frac{30 \times 395}{12500} = 0.76$$

The overhead portion of the frame $= 1.0 - 0.76 = 24\%$

Capacity and Frame Efficiency of a TDMA System

Cell Capacity

The cell capacity is defined as the maximum number of users that can be supported simultaneously in each cell.

The capacity of a TDMA system is given by [16]:

$$N_u = \frac{\eta_b \mu}{v_f} \times \frac{B_w}{RN} \tag{6.7}$$

where:

$N_u =$ number of channels (mobile users) per cell
$\eta_b \;=$ bandwidth efficiency factor (<1.0)
$\mu \;\;=$ bit efficiency ($= 2$ bit/symbol for QPSK, $= 1$ bit/symbol for GMSK as used in GSM)
$v_f \;=$ voice activity factor (equal to one for TDMA)
$B_w =$ one-way bandwidth of the system
$R \;\;=$ information (bit rate plus overhead) per user
$N \;\;=$ frequency reuse factor

$$\text{Spectral efficiency } \eta = \frac{N_u \times R}{B_w} \text{ bit/sec/Hz} \tag{6.8}$$

Example 6.4

Calculate the capacity and spectral efficiency of a TDMA system using the following parameters: bandwidth efficiency factor $\eta_b = 0.9$, bit efficiency (with QPSK) $\mu = 2$, voice activity factor $v_f = 1.0$, one-way system bandwidth $B_w = 12.5\,\text{MHz}$, information bit rate $R = 16.2\,\text{kbps}$, and frequency reuse factor $N = 19$.

Solution

$$N_u = \frac{0.9 \times 2}{1.0} \times \frac{12.5 \times 10^6}{16.2 \times 10^3 \times 19}$$

$N \;\; = 73.1$ (say 73 mobile users per cell)

$$\text{Spectral efficiency } \eta = \frac{73 \times 16.2}{12.5 \times 1000} = 0.094 \text{ bit/sec/Hz}$$

Efficiency of a TDMA Frame

We refer to Figure 6.4 that shows a TDMA frame. The number of overhead bits per frame is:

$$b_0 = N_r b_r + N_t b_p + (N_t + N_r)b_g \tag{6.9}$$

where:

N_r = number of reference bursts per frame
N_t = number of traffic bursts (slots) per frame
b_r = number of overhead bits per reference burst
b_p = number of overhead bits per preamble per slot
b_g = number of equivalent bits in each guard time interval

The total number of bits per frame is:

$$b_T = T_f \times R_{rf} \tag{6.10a}$$

where:

T_f = frame duration
R_{rf} = bit rate of the RF channel

$$\text{Frame efficiency } \eta = (1 - b_0/b_T) \times 100\% \tag{6.10b}$$

It is desirable to maintain the efficiency of the frame as high as possible.

The number of bits per data channel (user) per frame is $b_c = RT_f$, where R = bit rate of each channel (user).

$$\text{No. of channels/frame} \quad N_{CF} = \frac{\text{(Total Data Bits)/(frame)}}{\text{(Bits per Channel)/(frame)}}$$

$$N_{CF} = \frac{\eta R_{rf} T_f}{R T_f} \tag{6.11a}$$

$$N_{CF} = \frac{\eta R_{rf}}{R} \tag{6.11b}$$

Equation 6.11b indicates the number of time slots per frame.

Example 6.5
Consider the GSM TDMA system with the following parameters:

$N_r = 2$
$N_t = 24$ frames of 120 ms each with eight time slots per frame
$b_r = 148$ bits in each of 8 time slots
$b_p = 34$ bits in each of 8 time slots
$b_g = 8.25$ bits in each of 8 time slots
$T_f = 120$ ms
$R_{rf} = 270.8333333$ kbps
$R = 22.8$ kbps

Calculate the frame efficiency and the number of channels per frame.

Solution

$$b_0 = 2 \times (8 \times 148) + 24 \times (8 \times 34) + 8 \times 8.25 = 10,612 \text{ bits per frame}$$

$$b_T = 120 \times 10^{-3} \times 270.8333333 \times 10^3 = 32,500 \text{ bits per frame}$$

$$\eta = \left(1 - \frac{10612}{32500}\right) \times 100 = 67.35\%$$

$$\text{Number of channels/frame} = \frac{0.6735 \times 270.8333333}{22.8} = 8$$

The last calculation, with an answer of 8 channels, confirms that our calculation of efficiency is correct.

6.4 Wideband Systems

In wideband systems, the entire system bandwidth is made available to each user, and is many times larger than the bandwidth required to transmit information. Such systems are known as *spread spectrum* (SS) systems. There are two fundamental types of spread spectrum systems: (1) direct sequence spread spectrum (DSSS) and (2) frequency hopping spread spectrum (FHSS) [3,26].

In a DSSS system, the bandwidth of the baseband information carrying signals from a different user is spread by different codes with a bandwidth much larger than that of the baseband signals (see Chapter 11 for details). The spreading codes used for different users are orthogonal or nearly orthogonal to each other. In the DSSS, the spectrum of the transmitted signal is much wider than the spectrum associated with the information rate. At the receiver, the same code is used for despreading to recover the baseband signal from the target user while suppressing the transmissions from all other users (see Figure 6.5).

One of the advantages of the DSSS system is that the transmission bandwidth exceeds the coherence bandwidth (see Chapter 3). The received signal, after despreading (see Chapter 11 for details), resolves into multiple signals with different time delays. A Rake receiver (see Chapter 11) can be used to recover the multiple time

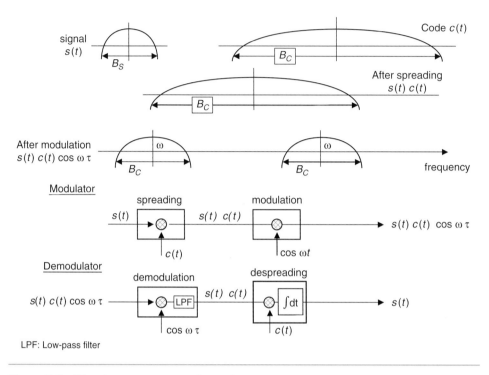

Figure 6.5   Direct sequence spread spectrum.

delayed signals and combine them into one signal to provide a time diversity with a lower frequency of deep fades. Thus, the DSSS system provides an inherent robustness against mobile-channel degradations. Another potential benefit of a DSSS system is the greater resistance to interference effects in a frequency reuse situation. Also, there may be no hard limit on the number of mobile users who can simultaneously gain access. The capacity of a DSSS system depends upon the desired value of E_b/I_0 instead of resources (frequencies or time slots) as in FDMA or TDMA systems.

Frequency hopping (FH) is the periodic changing of the frequency or the frequency set associated with transmission (see Figure 6.6). If the modulation is M-*ary* frequency-shift-keying (MFSK) (see Chapter 9 for details), two or more frequencies are in the set that change at each hop. For other modulations, a single center or carrier frequency is changed at each hop.

An FH signal may be considered a sequence of modulated pulses with pseudo-random carrier frequencies. The set of possible carrier frequencies is called the hop set. Hopping occurs over a frequency band that includes a number of frequency channels. The bandwidth of a frequency channel is called the *instantaneous bandwidth* (B_I). The bandwidth of the frequency band over which the hopping occurs is called the total *hopping bandwidth* (B_H). The time duration between hops is called the *hop duration* or hopping period (T_H).

Frequency

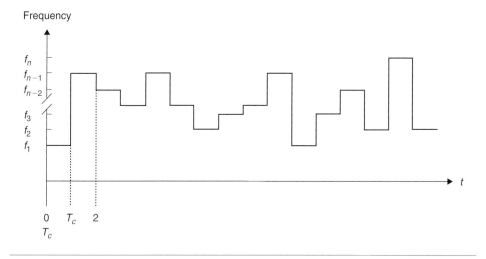

Figure 6.6   Frequency hopping spread spectrum system.

Frequency hopping can be classified as fast or slow. *Fast frequency hopping* occurs if there is frequency hop for each transmitted symbol. Thus, fast frequency hopping implies that the hopping rate equals or exceeds the information symbol rate. *Slow frequency hopping* occurs if two or more symbols are transmitted in the time interval between frequency hops.

Frequency hopping allows communicators to hop out of frequency channels with interference or to hop out of fades. To exploit this capability, error-correcting codes, appropriate interleaving, and disjoint frequency channels are nearly always used. A frequency synthesizer is required for frequency hopping systems to convert a stable reference frequency into the various frequency of hop set.

Frequency hopping communicators do not often operate in isolation. Instead, they are usually elements of a network of frequency hopping systems that create mutual multiple-access interference. This network is called a frequency-hopping multiple-access (FHMA) network.

If the hoppers of an FHMA network all use the same M frequency channels, but coordinate their frequency transitions and their hopping sequence, then the multiple-access interference for a lightly loaded system can be greatly reduced compared to a non-hopped system. For the number of hopped signals (M_h) less than the number of channels (N_c), a coordinated hopping pattern can eliminate interference. As the number of hopped signals increases beyond N_c, then the interference will increase in proportion to the ratio of the number of signals to the number of channels. In the absence of fading or multipath interference, since there is no interference suppression system in frequency hopping, for a high channel loading the performance of a frequency hopping system is no better than a non-hopped system. Frequency hopping systems are best for light channel loadings in the presence of conventional non-hopped systems.

When fading or multipath interference is present, the frequency hopping system has better error performance than a non-hopped system. If the transmitter hops to a channel in a fade, the errors are limited in duration since the system will shortly hop to a new frequency where the fade may not be as deep.

Network coordination for frequency hopping systems are simpler to implement than that for DSSS systems because the timing alignments must be within a fraction of a hop duration, rather than a fraction of a sequence chip (narrow pulse). In general, frequency hopping systems reject interference by trying to avoid it, whereas DSSS systems reject interference by spreading it. The interleaving and error-correcting codes that are effective with frequency hopping systems are also effective with DSSS systems.

The major problems with frequency hopping systems with increasing hopping rates are the cost of the frequency synthesizer increases and its reliability decreases, and synchronization becomes more difficult.

In theory, a wideband system can be overlaid on existing, fully loaded, narrowband channelized systems (as an example, the IS-95 CDMA system overlays on existing AMPS [FDMA]). Thus, it may be possible to create a wideband network right on top of the narrowband cellular system using the same spectrum.

6.5   Comparisons of FDMA, TDMA, and DS-CDMA

The DSSS approach is the basis to implementation of the direct sequence code division multiple access (DS-CDMA) technique introduced by Qualcom. The DS-CDMA has been used in commercial applications of mobile communications. The primary advantage of DS-CDMA is its ability to tolerate a fair amount of interfering signals compared to FDMA and TDMA that typically cannot tolerate any such interference(Figure 6.7). As a result of the interference tolerance of CDMA, the problems of frequency band assignment and adjacent cell interference are greatly simplified. Also, flexibility in system design and deployment are significantly improved since interference to others is not a problem. On the other hand, FDMA and TDMA radios must be carefully assigned a frequency or time slot to assure that there is no interference with other similar radios. Therefore, sophisticated filtering and guard band protection is needed with FDMA and TDMA technologies. With DS-CDMA, adjacent microcells share the same frequencies whereas with FDMA/TDMA it is not feasible for adjacent microcells to share the same frequencies because of interference. In both FDMA and TDMA systems, a time-consuming frequency planning task is required whenever a network changes, whereas no such frequency planning is needed for a CDMA network since each cell uses the same frequencies.

Capacity improvements with DS-CDMA also result from voice activity patterns during two-way conversations, (i.e., times when a party is not talking) that cannot be cost-effectively exploited in FDMA or TDMA systems. DS-CDMA radios can, therefore, accommodate more mobile users than FDMA/TDMA radios

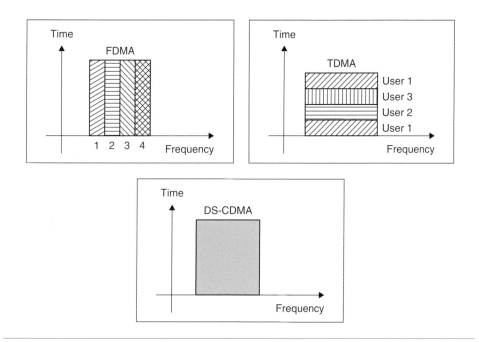

Figure 6.7 Comparison of multiple access methods.

on the same bandwidth. Further capacity gains for FDMA, TDMA, and CDMA can also result from antenna technology advancement by using directional antennas that allow the microcell area to be divided into sectors. Table 6.1 provides a summary of access technologies used for various wireless systems.

Table 6.1 Access technologies for wireless system.

System	Access technology	Mode of operation	Frame rate (kbps)
North American IS-54 (Dual Mode)	TDMA/FDD FDMA/FDD	Digital/ Analog FM	48.6 —
North American IS-95 (Dual Mode)	DS-CDMA/FDD FDMA/FDD	Digital/ Analog FM	1228.8 —
North American IS-136	TDMA/FDD	Digital	48.6
GSM (used all over world)	TDMA/FDD	Digital	270.833
European CT-2 Cordless	FDMA/TDD	Digital	72.0
DECT Cordless	TDMA/TDD	Digital	1152.0

6.6 Capacity of a DS-CDMA System

The capacity of a DS-CDMA system depends on the processing gain, G_p (a ratio of spreading bandwidth, B_w, and information rate, R), the bit energy-to-interference ratio, E_b/I_0, the voice duty cycle, v_f, the DS-CDMA omnidirectional frequency reuse efficiency, η_f, and the number of sectors, G, in the cell-site antenna.

The received signal power at the cell from a mobile is $S = R \times E_b$. The signal-to-interference ratio is

$$\frac{S}{I} = \frac{R}{B_w} \times \frac{Eb}{I_0} \tag{6.12}$$

where:

    E_b = energy per bit
    I_0 = interference density

In a cell with N_u mobile transmitters, the number of effective interferers is $N_u - 1$ because each mobile is an interferer to all other mobiles. This is valid regardless of how the mobiles are distributed within the cell since automatic power control (APC) is used in the mobiles. The APC operates such that the received power at the cell from each mobile is the same as for every other mobile in the cell, regardless of the distance from the center of the cell. APC conserves battery power in the mobiles, minimizes interference to other users, and helps overcome fading.

In a hexagonal cell structure, because of interference from each tier, the S/I ratio is given as (see Chapter 5):

$$\frac{S}{I} = \frac{1}{(N_u - 1) \times [1 + 6 \times k_1 + 12 \times k_2 + 18 \times k_3 + \cdots]} \tag{6.13}$$

where:

    N_u = number of mobile users in the band, B_w
    k_i, $i = 1, 2, 3, \ldots$ = the interference contribution from all terminals in individual cells in tiers 1, 2, 3, etc., relative to the interference from the center cell. This loss contribution is a function of both the path loss to the center cell and the power reduction because of power control to an interfering mobile's own cell center.

If we define a frequency reuse efficiency, η_f, as in Equation 6.14a, then E_b/I_0 is given by Equation 6.15.

$$\eta_f = \frac{1}{[1 + 6 \times k_1 + 12 \times k_2 + 18 \times k_3 + \cdots]} \tag{6.14a}$$

$$\frac{S}{I} = \frac{\eta_f}{(N_u - 1)} \tag{6.14b}$$

$$\frac{E_b}{I_0} = \frac{B_w}{R} \times \frac{\eta_f}{(N_u - 1)} \tag{6.15}$$

This equation does not include the effect of background thermal and spurious noise (i.e., ρ) in the spreading bandwidth B_w. Including this as an additive degradation term in the denominator results in a bit energy-to-interference ratio of:

$$\frac{E_b}{I_0} = \frac{B_w}{R} \times \frac{\eta_f}{(N_u - 1) + \rho/S} \tag{6.16}$$

Note that from Equation 6.16 the capacity of the DS-CDMA system is reduced by ρ/S which is the ratio of background thermal plus spurious noise to power level.

For a fixed $G_p = B_w/R$, one way to increase the capacity of the DS-CDMA system is to reduce the required E_b/I_0, which depends upon the modulation and coding scheme. By using a powerful coding scheme, the E_b/I_0 ratio can be reduced, but this increases system complexity. Also, it is not possible to reduce the E_b/I_0, ratio indefinitely. The only other way to increase the system capacity is to reduce the interference. Two approaches are used: one is based on the natural behavior of human speech and the other is based on the application of the sectorized antennas. From experimental studies it has been found that typically in a full duplex 2-way voice conversation, the duty cycle of each voice is, on the average, less than 40%. Thus, for the remaining period of time the interference induced by the speaker can be eliminated. Since the channel is shared among all the users, noise induced in the desired channel is reduced due to the silent interval of other interfering channels. It is not cost-effective to exploit the voice activity in the FDMA or TDMA system because of the time delay associated with reassigning the channel resource during the speech pauses. If we define v_f as the voice activity factor (<1), then Equation 6.16 can be written as:

$$\frac{E_b}{I_0} = \frac{\eta_f}{v_f} \times \frac{B_w}{R} \times \frac{1}{(N_u - 1) + \rho/S} \tag{6.17a}$$

$$(N_u - 1) + \frac{\rho}{S} = \left[\frac{\eta_f}{v_f}\right] \times \left[\frac{B_w}{R}\right] \times \left[\frac{I_0}{E_b}\right] \tag{6.17b}$$

The equation to determine the capacity of a DS-CDMA system should also include additional parameters to reflect the bandwidth efficiency factor, the capacity degradation factor due to imperfect power control, and the number of sectors in the cell-site antenna. Equation 6.17b is augmented by these additional factors to provide the following equation for DS-CDMA capacity at one cell:

$$N_u = \frac{\eta_f \eta_b c_d \lambda}{v_f} \times \frac{B_w}{R \times (E_b/I_0)} + 1 - \frac{\rho}{S} \qquad (6.18a)$$

Equation 6.18a can be rewritten as Equation 6.18b by neglecting the last two terms.

$$N_u = \frac{\eta_f \eta_b c_d \lambda}{v_f} \times \frac{B_w}{R \times (E_b/I_0)} \qquad (6.18b)$$

where:

η_f = frequency reuse efficiency <1
η_b = bandwidth efficiency factor <1
c_d = capacity degradation factor to account for imperfect APC <1
v_f = voice activity factor <1
B_w = one-way bandwidth of the system
R = information bit rate plus overhead
E_b = energy per bit of the desired signal
E_b/I_0 = desired energy-to-interference ratio (dependent on quality of service)
λ = efficiency of sector-antenna in cell ($< G$, number of sectors in the cell-site antenna)

For digital voice transmission, E_b/I_0 is the required value for a bit error rate (BER) of about 10^{-3} or better, and η_f depends on the quality of the diversity. Under the most optimistic assumption, $\eta_f < 0.5$. The voice activity factor, v_f is usually assumed to be less than or equal to 0.6. E_b/I_0 for a BER of 10^{-3} can be as high as 63 (18 dB) if no coding is used and as low as 5 (7 dB) for a system using a powerful coding scheme. The capacity degradation factor, c_d will depend on the implementation but will always be less than 1.

Example 6.6
Calculate the capacity and spectral efficiency of the DS-CDMA system with an omnidirectional cell using the following data:

- bandwidth efficiency $\eta_b = 0.9$
- frequency reuse efficiency $\eta_f = 0.45$

- capacity degradation factor $c_d = 0.8$
- voice activity factor $v_f = 0.4$
- information bit rate $R = 16.2\,\text{kbps}$
- $E_b/I_0 = 7\,\text{dB}$
- one-way system bandwidth $B_w = 12.5\,\text{MHz}$

Neglect other sources of interference.

Solution

$E_b/I_0 = 5.02\ (7\,\text{dB})$

$$N_u = \frac{0.45 \times 0.9 \times 0.8 \times 1}{0.4} \times \frac{12.5 \times 10^6}{16.2 \times 10^3 \times 5.02}$$

$N_u = 124.5\ (\text{say } 125)$

The spectral efficiency, $\eta = \dfrac{125 \times 16.2}{12.5 \times 10^3} = 0.162\,\text{bits/sec/Hz}$

In these calculations, an omnidirectional antenna is assumed. If a three sector antenna (i.e., $G = 3$) is used at a cell site with $\lambda = 2.6$, the capacity will be increased to 325 mobile users per cell, and spectral efficiency will be $0.421\,\text{bits/sec/Hz}$.

6.7 Comparison of DS-CDMA vs. TDMA System Capacity

Using Equations 6.7 and 6.18b with $v_f = 1$ (no voice activity) for TDMA and $\lambda = 1.0$ (omnidirectional cell) for DS-CDMA the ratio of the cell capacity for the DS-CDMA and TDMA systems is given as:

$$\frac{N_{\text{CDMA}}}{N_{\text{TDMA}}} = \frac{c_d N \eta_f}{E_b/I_0} \times \frac{1}{v_{f_{\text{cdma}}}} \times \frac{1}{\mu} \times \frac{R_{\text{TDMA}}}{R_{\text{CDMA}}} \tag{6.19}$$

Example 6.7

Using the data given in Examples 6.4 and 6.6, compare the capacity of the DS-CDMA and TDMA omnidirectional cell.

Solution

$$\frac{N_{\text{CDMA}}}{N_{\text{TDMA}}} = \frac{0.8 \times 19 \times 0.45}{5.02} \times \frac{1}{0.4} \times \frac{1}{2} \times \frac{16.2}{16.2} = 1.703$$

6.8   Frequency Hopping Spread Spectrum with M-*ary* Frequency Shift Keying

The FHSS system uses M-*ary* frequency shift keying modulation (MFSK) and involves the hopping of the carrier frequency in a random manner. It uses MFSK, in which $b = \log_2 M$ information bits determine which one of M frequencies is to be used [19]. The portion of the M-*ary* signal set is shifted pseudo-randomly by the frequency synthesizer over a hopping bandwidth, B_{ss}. A typical block diagram is shown in Figure 6.8.

In a conventional MFSK system, the data symbol is modulated on a carrier whose frequency is pseudo-randomly determined. The frequency synthesizer produces a transmission tone based on simultaneous dictates of the pseudo-noise (PN) code (see Chapter 11) and the data. At each frequency hop time a PN generator feeds the frequency synthesizer a frequency word (a sequence of L chips), which dictates one of $2L$ symbol-set positions. The FH bandwidth, B_{ss}, and the minimum frequency spacing between consecutive hop positions, Δf, dictate the minimum number of chips required in the frequency word.

Example 6.8

A hopping bandwidth, B_{ss}, of 600 MHz and a frequency step size, Δf, of 400 Hz are used. What is the minimum number of PN chips that are required for each frequency word?

Solution

$$\text{Number of tones contained in } B_{ss} = \frac{B_{ss}}{\Delta f} = \frac{600 \times 10^6}{400} = 1.5 \times 10^6$$

$$\text{Minimum number of chips required} = \left\lceil \log_2(1.5 \times 10^6) \right\rceil = 20 \text{ chips}$$

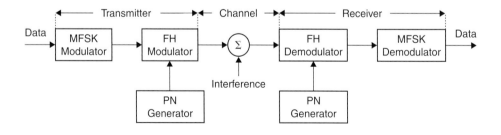

Figure 6.8   Frequency hopping using MFSK.

6.9 Orthogonal Frequency Division Multiplexing (OFDM)

In this section we briefly introduce OFDM. For more details readers should refer to [19]. OFDM uses three transmission principles, multirate, multisymbol, and multicarrier. OFDM is similar to frequency division multiplexing (FDM). OFDM distributes the data over a large number of carriers that are spaced apart at precise frequencies. The spacing provides the orthogonality in this technique, which prevents the demodulator from seeing frequencies other than their own.

Multiple Input, Multiple Output-OFDM (MIMO-OFDM) uses multiple antennas to transmit and receive radio signals. MIMO-OFDM allows service providers to deploy a broadband wireless access system that has non-line-of-sight (NLOS) functionality. MIMO-OFDM takes advantage of the multipath properties of the environment using base station antennas that do not have LOS. The MIMO-OFDM system uses multiple antennas to simultaneously transmit data in small pieces to the receiver, which can process the data flow and put it back together. This process, called *spatial multiplexing*, proportionally boosts the data transmission speed by a factor equal to the number of transmitting antennas. In addition, since all data is transmitted both in the same frequency band and with separate spatial signatures, this technique utilizes spectrum efficiently. VOFDM (vector OFDM) uses the concept of MIMO technology.

We consider a data stream operating at R bps and an available bandwidth of $N\Delta f$ centered at f_c. The entire bandwidth could be used to transmit a data stream, in which case the bit duration would be $1/R$. By splitting the data stream into N substreams using a serial-to-parallel converter, each substream has a data rate of R/N and is transmitted on a separate subcarrier, with spacing between adjacent subcarriers of Δf (see Figure 6.9). The bit duration is N/R. The advantage of OFDM is that on a multiple channel the multipath is reduced relative to the symbol interval by a ratio of $1/N$ and thus imposes less distortion in each modulated symbol. OFDM overcomes inter-symbol interference (ISI) in a multipath environment. ISI has a greater impact at higher data rates because the distance between bits or symbols is smaller. With OFDM, the data rate is reduced by a factor of N, which increases the symbol duration by a factor of N. Thus, if the symbol duration is T_s for the source stream, the duration of OFDM signals is NT_s. This significantly reduces the effect of ISI. As a design criterion, N is selected so that NT_s is significantly greater than τ_{rms} (rms delay spread) of the channel. With the use of OFDM, it may not be necessary to deploy an equalizer. OFDM is an ideal solution for broadband communications, because increasing the data rate is simply a matter of increasing the number of subcarriers. To avoid overlap between consecutive symbols, a time guard is enforced between the transmissions of two OFDM pulses that will reduce the effective data rate. Also, some subcarriers are devoted to synchronization of signal, and some are reserved for redundancy.

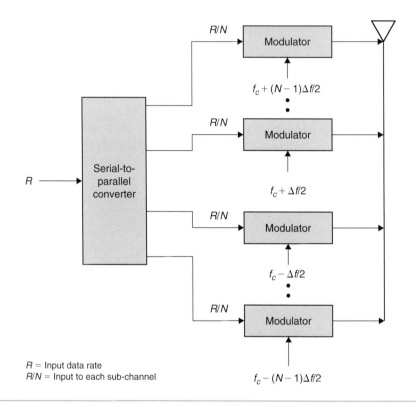

Figure 6.9   Orthogonal frequency division multiplexing (OFDM).

The most important feature of OFDM is the orthogonal relationship between the subcarrier signals. Orthogonality allows the OFDM subcarriers to overlap each other without interference. OFDM uses FH to create a spread spectrum system. FH has several advantages over DSSS, for example, no near-far problem, easier synchronization, less complex receivers, and so on.

In the OFDM the input information sequence is first converted into parallel data sequences and each serial/parallel converter output is multiplied with spreading code. Data from all subcarriers is modulated in baseband by inverse fast Fourier transform (IFFT) and converted back into serial data. The guard interval is inserted between symbols to avoid ISI caused by multipath fading and finally the signal is transmitted after RF up-conversion. At the receiver, after down-conversion, the m-subcarrier component corresponding to the received data is first coherently detected with FFT and then multiplied with gain to combine the energy of the received signal scattered in the frequency domain (see Figure 6.10).

Wireless Local Area Networks (WLAN) development is ongoing for wireless point-to-point and point-to-multipoint configurations using OFDM technology.

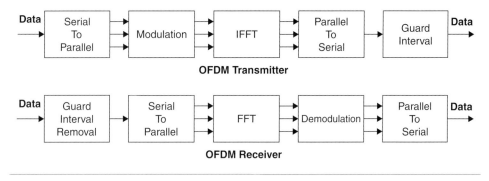

Figure 6.10   IEEE 802.11 a Transmit and Receive OFDM.

In a supplement to the IEEE 802.11 standard, the IEEE 802.11 working group published IEEE 802.11a, which outlines the use of OFDM in the 5.8 GHz band.

The basic principal of operation is to divide a high-speed binary signal to be transmitted into a number of lower data rate subcarriers. There are 48 data subcarriers and 4 pilot subcarriers for a total of 52 subcarriers. Each lower data rate bit stream is used to modulate a separate subcarrier from one of the channels in the 5 GHz band. Prior to transmission the data is encoded using convolutional code (see Chapter 8) of rate, $R = 1/2$ and bit interleaved for the desired data rate. Each bit is then mapped into a complex number according to the modulation type and subdivided in 48 data subcarriers and 4 pilot subcarriers. The subcarriers are combined using an IFFT and transmitted. At the receiver, the carrier is converted back to a multicarrier lower data rate form using FFT. The lower data subcarriers are combined to form a high rate data unit.

6.10 Multicarrier DS-CDMA (MC-DS-CDMA)

Future wireless systems such as a fourth-generation (4G) system will need flexibility to provide subscribers with a variety of services such as voice, data, images, and video. Because these services have widely differing data rates and traffic profiles, future generation systems will have to accommodate a wide variety of data rates. DS-CDMA has proven very successful for large-scale cellular voice systems, but there are concerns whether DS-CDMA will be well-suited to non-voice traffic. The DS-CDMA system suffers inter-symbol interference (ISI) and multi-user interference (MUI) caused by multipath propagation, leading to a high loss of performance.

With OFDM, the time dispersive channel is seen in the frequency domain as a set of parallel independent flat subchannels and can be equalized at a low complexity. There are potential benefits to combining OFDM and DS-CDMA. Basically the frequency-selective channel is first equalized in the frequency domain using the

OFDM modulation technique. DS-CDMA is applied on top of the equalized channel, keeping the orthogonality properties of spreading codes. The combination of OFDM and DS-CDMA is used in MC-DS-CDMA. MC-DS-CDMA [4,5,12,25] marries the best of the OFDM and DS-CDMA world and, consequently, it can ensure good performance in severe multipath conditions. MC-DS-CDMA can achieve very large average throughput. To further enhance the spectral efficiency of the system, some form of adaptive modulation can be used.

Basically, three main designs exist in the literature, namely, MC-CDMA, MC-DS-CDMA, and multitone (MT)-CDMA. In MC-CDMA, the spreading code is applied across a number of orthogonal subcarriers in the frequency domain. In MC-DS-CDMA, the data stream is first divided into a number of substreams. Each substream is spread in time through a spreading code and then transmitted over one of a set of orthogonal subcarriers. In MT-CDMA the system undergoes similar operations as MC-DS-CDMA except that the different subcarriers are not orthogonal after spreading. This allows higher spectral efficiencies and longer spreading codes; however, different substreams interfere with one other. The MC-DS-CDMA transmitter spreads the original data stream over different orthogonal subcarriers using a given spreading code in the frequency domain.

6.11   Random Access Methods

So far we have discussed the reservation-based schemes, now we focus on random-access schemes [8]. When each user has a steady flow of information to transmit (for example, a data file transfer or a facsimile transmission), reservation-based access methods are useful as they make an efficient use of communication resources. However, when the information to be transmitted is bursty in nature, the reservation-based access methods result in wasting communication resources. Furthermore, in a cellular system where subscribers are charged based on a channel connection time, the reservation-based access methods may be too expensive to transmit short messages. Random-access protocols provide flexible and efficient methods for managing a channel access to transmit short messages. The random-access methods give freedom for each user to gain access to the network whenever the user has information to send. Because of this freedom, these schemes result in contention among users accessing the network. Contention may cause collisions and may require retransmission of the information. The commonly used random-access protocols are pure ALOHA, slotted-ALOHA, and CSMA/CD. In the following section we briefly describe details of each of these protocols and provide the necessary throughput expressions.

6.11.1   Pure ALOHA

In the pure ALOHA [18,23] scheme, each user transmits information whenever the user has information to send. A user sends information in packets. After sending a packet, the user waits a length of time equal to the round-trip delay for an acknowledgment (ACK) of the packet from the receiver. If no ACK is received, the packet is assumed to be lost in a collision and it is retransmitted with a randomly selected delay to avoid repeated collisions.* The normalized throughput S (average new packet arrival rate divided by the maximum packet throughput) of the pure ALOHA protocol is given as:

$$S = Ge^{-2G} \tag{6.20}$$

where G = normalized offered traffic load

From Equation 6.20 it should be noted that the maximum throughput occurs at traffic load $G = 50\%$ and is $S = 1/2e$. This is about 0.184. Thus, the best channel utilization with the pure ALOHA protocol is only 18.4%.

6.11.2 Slotted ALOHA

In the slotted-ALOHA [23] system, the transmission time is divided into time slots. Each time slot is made exactly equal to packet transmission time. Users are synchronized to the time slots, so that whenever a user has a packet to send, the packet is held and transmitted in the next time slot. With the synchronized time slots scheme, the interval of a possible collision for any packet is reduced to one packet time from two packet times, as in the pure ALOHA scheme. The normalized throughput S for the slotted-ALOHA protocol is given as:

$$S = Ge^{-G} \tag{6.21}$$

where G = normalized offered traffic load

The maximum throughput for the slotted ALOHA occurs at $G = 1.0$ (Equation 6.21) and it is equal to $1/e$ or about 0.368. This implies that at the maximum throughput, 36.8% of the time slots carry successfully transmitted packets. The best channel utilization with the slotted ALOHA protocol is 36.8%—twice the pure ALOHA protocol.

---

*It should be noted that the protocol on CDMA access channels as implemented in TIA IS-95-A is based upon the pure ALOHA approach. The mobile station randomizes its attempt for sending a message on the access channel and may retry if an acknowledgment is not received from the base station. For further details, one should reference Section 6.6.3.1.1.1 of TIA IS-95-A.

6.11.3   Carrier Sense Multiple Access (CSMA)

The carrier sense multiple access (CSMA) [8,18] protocols have been widely used in both wired and wireless LANs. These protocols provide enhancements over the pure and slotted ALOHA protocols. The enhancements are achieved through the use of the additional capability at each user station to sense the transmissions of other user stations. The carrier sense information is used to minimize the length of collision intervals. For carrier sensing to be effective, propagation delays must be less than packet transmission times. Two general classes of CSMA protocols are nonpersistent and p-persistent.

- **Nonpersistent CSMA:** A user station does not sense the channel continuously while it is busy. Instead, after sensing the busy condition, it waits for a randomly selected interval of time before sensing again. The algorithm works as follows: if the channel is found to be idle, the packet is transmitted; or if the channel is sensed busy, the user station backs off to reschedule the packet to a later time. After backing off, the channel is sensed again, and the algorithm is repeated again.

- **p-persistent CSMA:** The slot length is typically selected to be the maximum propagation delay. When a station has information to transmit, it senses the channel. If the channel is found to be idle, it transmits with probability p. With probability q = 1 − p, the user station postpones its action to the next slot, where it senses the channel again. If that slot is idle, the station transmits with probability p or postpones again with probability q. The procedure is repeated until either the frame has been transmitted or the channel is found to be busy. If the station initially senses the channel to be busy, it simply waits one slot and applies the above procedure.

- **1-persistent CSMA:** 1-persistent CSMA is the simplest form of the p-persistent CSMA. It signifies the transmission strategy, which is to transmit with probability 1 as soon as the channel becomes idle. After sending the packet, the user station waits for an ACK, and if it is not received within a specified amount of time, the user station waits for a random amount of time, and then resumes listening to the channel. When the channel is again found to be idle, the packet is retransmitted immediately.

For more details, the reader should refer to [18].

The throughput expressions for the CSMA protocols are:

- **Unslotted nonpersistent CSMA**

$$S = \frac{Ge^{-aG}}{G(1 + 2a) + e^{-aG}} \tag{6.22}$$

- Slotted nonpersistent CSMA

$$S = \frac{aGe^{-aG}}{1 - e^{-aG} + a} \qquad (6.23)$$

- Unslotted 1-persistent CSMA

$$S = \frac{G[1 + G + aG(1 + G + (aG)/2)]e^{-G(1 + 2a)}}{G(1 + 2a) - (1 - e^{-aG}) + (1 + aG)e^{-G(1 + a)}} \qquad (6.24)$$

- Slotted 1-persistent CSMA

$$S = \frac{Ge^{-G(1 + a)}[1 + a - e^{-aG}]}{(1 + a)(1 - e^{-aG}) + ae^{-G(1 + a)}} \qquad (6.25)$$

where:
S = normalized throughput
G = normalized offered traffic load
a = τ/T_p
τ = maximum propagation delay
T_p = packet transmission time

Example 6.9

We consider a WLAN installation in which the maximum propagation delay is 0.4 sec. The WLAN operates at a data rate of 10 Mbps, and packets have 400 bits. Calculate the normalized throughput with: (1) an unslotted nonpersistent, (2) a slotted persistent, and (3) a slotted 1-persistent CSMA protocol.

Solution

$$T_p = \frac{400}{10} = 40\,\mu s$$

$$a = \frac{\tau}{T_p} = \frac{0.4}{40} = 0.01$$

$$G = \frac{40 \times 10^{-6} \times 10 \times 10^6}{400} = 1$$

- Slotted nonpersistent:

$$S = \frac{0.01 \times 1 \times e^{-0.01}}{1 - e^{-0.01} + 0.01} = 0.495$$

- Unslotted nonpersistent:

$$S = \frac{1 \times e^{-0.01}}{(1 + 0.02) + e^{-0.01}} = 0.493$$

- Slotted 1-persistent:

$$S = \frac{e^{-1.01}(1 + 0.01 - e^{-0.01})}{(1 + 0.01)(1 - e^{-0.01}) + 0.01e^{-1.01}} = 0.531$$

6.11.4   Carrier Sense Multiple Access with Collision Detection

A considerable performance improvement in the basic CSMA protocols can be achieved by means of the *carrier sense multiple access with collision detection* (CSMA/CD) technique. The CSMA/CD protocols are essentially the same as those for CSMA with addition of the collision-detection feature. Similar to CSMA protocols, there are nonpersistent, 1-persistent, and p-persistent CSMA/CD protocols. More details about CSMA/CD protocols can be found in [27].

When a CSMA/CD station senses that a collision has occurred, it immediately stops transmitting its packets and sends a brief jamming signal to notify all stations of this collision. Collisions are detected by monitoring the analog waveform directly from the channel. When signals from two or more stations are present simultaneously, the composite waveform is distorted from that of a single station. This is manifested in the form of larger than normal voltage amplitude on the cable. In the Ethernet the collision is recognized by the transmitting station, which goes into a retransmission phase based on an exponential random backoff algorithm.

The normalized throughput for unslotted nonpersistent and slotted nonpersistent CSMA/CD is given as:

Unslotted nonpersistent CSMA/CD

$$S = \frac{Ge^{-aG}}{Ge^{-aG} + bG(1 - e^{-aG}) + 2aG(1 - e^{-aG}) + (2 - e^{-aG})} \tag{6.26}$$

where b = jamming signal length

Slotted nonpersistent CSMA/CD

$$S = \frac{aGe^{-aG}}{aGe^{-aG} + b(1 - e^{-aG} - aGe^{-aG}) + a(2 - e^{-aG} - aGe^{-aG})} \qquad (6.27)$$

While these collision detection mechanisms are a good idea on a wired local area network (LAN), they cannot be used on a wireless local area network (WLAN) environment for two main reasons:

- Implementing a collision detection mechanism would require the implementation of a full duplex radio capable of transmitting and receiving at the same time—an approach that would increase the cost significantly.
- In a wireless environment we cannot assume that all stations hear each other (which is the basic assumption of the collision detection scheme), and the fact that a station wants to transmit and senses the medium as free does not necessarily mean that the medium is free around the receiver area.

6.11.5 Carrier Sense Multiple Access with Collision Avoidance (CSMA/CA)

IEEE 802.11 uses a protocol known as *carrier sense multiple access with collision avoidance* (CSMA/CA) or distributed coordination function (DCF). CSMA/CA attempts to avoid collisions by using *explicit packet acknowledgment* (ACK), which means an ACK packet is sent by the receiving station to confirm that the data packet arrived intact.

The CSMA/CA protocol works as follows. A station wishing to transmit senses the medium, if the medium is busy (i.e., some other station is transmitting) then the station defers its transmission to a later time. If no activity is detected, the station waits an additional, randomly selected period of time and then transmits if the medium is still free. If the packet is received intact, the receiving station issues an ACK frame that, once successfully received by the sender, completes the process. If the ACK frame is not detected by the sending station, either because the original data packet was not received intact or the ACK was not received intact, a collision is assumed to have occurred and the data packet is transmitted again after waiting another random amount of time. The CSMA/CA provides a way to share access over the medium. This explicit ACK mechanism also handles interference and other radio-related problems very effectively. However, it does add some overhead to 802.11 that 802.3 does not have, so that an 802.11 WLAN will always have slower performance than the equivalent Ethernet LAN (802.3).

The CSMA/CA protocol is very effective when the medium is not heavily loaded since it allows stations to transmit with minimum delay. But there is always a chance of stations simultaneously sensing the medium as being free and transmitting at the same time, causing a collision. These collisions must be identified, so that the media access control (MAC) layer can retransmit the packet by itself and not by the upper layers, which would cause significant delay. In particular, the hidden node and exposed node problems should be addressed by MAC. Both of them give rise to many performance problems including throughput degradtion, unfair throughput distribution, and throughput instability (see Chapter 18 for details).

The IEEE 802.11 uses a collision avoidance (CA) mechanism together with a positive ACK. The MAC layer of a station wishing to transmit senses the medium. If the medium is free for a specified time (called *distributed inter-frame space (DIFS)*), then the station is able to transmit the packet; if the medium is busy (or becomes busy during the DIFS interval) the station defers using the *exponential backoff algorithm*.

This scheme implies that, except in cases of very high network congestion, no packets will be lost, because retransmission occurs each time a packet is not acknowledged. This entails that all packets sent will reach their destination in sequence.

The IEEE 802.11 MAC layer provides cyclic redundancy check (*CRC*) *checksum* and *packet fragmentation*. Each packet has a CRC checksum calculated and attached to ensure that the data was not corrupted in transmit. Packet fragmentation is used to segment large packets into smaller units when sent over the medium. This is useful in very congested environments or when interference is a factor, since large packets have a better chance of being corrupted. This technique reduces the need for retransmission in many cases and improves overall wireless network performance. The MAC layer is responsible for reassembling fragments received, rendering the process transparent to higher-level protocols. The following are some of the reasons it is preferable to use smaller packets in a WLAN environment.

- Due to higher BER of a radio link, the probability of a packet getting corrupted increases with packet size.
- In case of packet corruption (either due to collision or interference), the smaller the packet, the less overhead it needs to retransmit.

A simple *stop-and-wait* algorithm is used at the MAC sublayer. In this mechanism the transmitting station is not allowed to transmit a new fragment until one of the following happens:

- Receives an ACK for the fragment, or
- Decides that the fragment was retransmitted too many times and drops the whole frame.

Exponential backoff scheme is used to resolve contention problems among different stations wishing to transmit data at the same time. When a station goes into the backoff state, it waits an additional, randomly selected number of time slots known as a contention window (in 802.11b a slot has a $20\,\mu s$ duration and the random number must be greater than 0 and smaller than a maximum value referred to as a contention window (CW)). During the wait, the station continues sensing the medium to check whether it remains free or if another transmission begins. At the end of its window, if the medium is still free the station can send its frame. If during the window another station begins transmitting data, the backoff counter is frozen and counting down starts again as the channel returns to the idle state.

There is a problem related to the CW dimension. With a small CW, if many stations attempt to transmit data at the same time it is possible that some of them may have the same backoff interval. This means that there will continuously be collisions, with serious effects on network performance. On the other hand, with a large CW, if a few stations wish to transmit data they will likely have long back-off delays resulting in degradation of network performance. The solution is to use an exponentially growing CW size. It starts from a small value ($CW_{min} = 31$) and doubles after each collision, until it reaches the maximum value CW_{max} ($CW_{max} = 1023$). In 802.11 the backoff algorithm is executed in three cases:

1. When the station senses the medium busy before the first transmission of a packet
2. Before each retransmission
3. After a successful transmission

This is necessary to avoid a single host wanting to transmit a large quantity of data and occupying the channel for too long, denying access to all other stations. The backoff mechanism is not used when the station decides to transmit

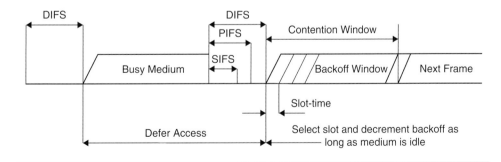

Figure 6.11 CSMA/CA in IEEE 802.11b.

a new packet after an idle period and the medium has been free for more than a distributed inter-frame space (DIFS) (see Figure 6.11).

To support time-bounded services, the IEEE 802.11 standard defines the point coordinate function (PCF) to let stations have priority access to the wireless medium, coordinated by a station called point coordinate (PC). The PCF has higher priority than DIFS, because it may start transmissions after a shorter duration than DIFS; this time space is called PCF inter frame space (PIFS), which is 25 μs for IEEE 802.11 and larger than SIFS.

$$\text{The transmission time for a data frame} = \left(\text{PLCP} + \frac{D}{R}\right) \mu\text{s}$$

where:

   PLCP = the time required to transmit the physical layer convergence protocol
           (PLCP)
   D = the frame size
   R = the channel bit rate

$$\text{CSMA/CA packet transmission time} = \text{BO} + \text{DIFS} + 2\text{PLCP} + \frac{D}{R} + \text{SIFS} + \frac{A}{R} \ \mu\text{s}$$

where:

   A = the ACK frame size
   BO = the backoff time
   DIFS = the distributed inter-frame space
   SIFS = the short inter-frame space
   PIFS = the point coordinate interframe space

6.12   Idle Signal Casting Multiple Access

In the CSMA scheme, each terminal must be able to detect the transmissions of all other terminals. However, not all packets transmitted from different terminals can be sensed, or terminals may be hidden from each other by buildings or some other obstacles. This is known as the *hidden terminal problem*, which severely degrades the throughput of the CSMA. The idle signal casting multiple access (ISMA) system transmits an idle/busy signal from the base station to indicate the presence or absence of another terminal's transmission. The ISMA and CSMA are basically the same. In the CSMA, each terminal must listen to all other terminals, whereas in the ISMA, each terminal is informed from the base station of the other terminals' transmission. Similar to CSMAs, there are nonpersistent ISMAs and 1-persistent ISMAs.

6.13   Packet Reservation Multiple Access

Packet reservation multiple access (PRMA) allows a variety of information sources to share the same communication channel and obtains a statistical multiplexing effect. In PRMA, time is divided into frames, each of which consists of a fixed number of time slots. For voice terminals, voice activity detection is adopted. The voice signal comprises a sequence of talk spurts. At the beginning of a talk spurt, the terminal transmits the first packet based on slotted ALOHA. Once the packet is transmitted successfully, that terminal is allowed to use the same time slots in the succeeding frames (reservation is made). The reservation is kept until the end of the talk spurt. The status "reserved" or "unreserved" of each slot is broadcast from the base station.

6.14 Error Control Schemes for Link Layer

Error control schemes for the link layer are used to improve the performance of mobile communication systems [8]. Several automatic repeat request (ARQ) schemes are used. At the physical layer of wireless mobile communication systems, error detection and correction techniques such as forward error correction (FEC) schemes are used. For some of the data services, higher layer protocols use ARQ schemes to enable retransmission of any data frames in which an error is detected. The ARQ schemes are classified as follows [23]:

- **Stop and Wait:** The sender transmits the first packet numbered 0 after storing a copy of that packet. The sender then waits for an ACK numbered 0, (ACK0) of that packet. If the ACK0 does not arrive before a time-out, the sender transmits another copy of the first packet. If the ACK0 arrives before a time-out, the sender discards the copy of the first packet and is ready to transmit the next packet, which it numbers 1. The sender repeats the previous steps, with numbers 0 and 1 interchanged. The advantages of the Stop and Wait protocol are its simplicity and its small buffer requirements. The sender needs to keep only the copy of the packet that it last transmitted, and the receiver does not need to buffer packets at the data link layer. The main disadvantage of the Stop and Wait protocol is that it does not use the communication link very efficiently.

The total time taken to transmit a packet and to prepare for transmitting the next one is

$$T = T_p + 2T_{\text{prop}} + 2T_{\text{proc}} + T_a \qquad (6.28)$$

where:

T = total time for transmission time
T_p = transmission time for a packet
T_{prop} = propagation time of a packet or an ACK
T_{proc} = processing time for a packet or an ACK
T_a = transmission time for an ACK

The protocol efficiency without any error is:

$$\eta(0) = \frac{T_p}{T} \qquad (6.29)$$

If p is the probability that a packet or its ACK is corrupted by transmission errors, and a successful transmission of a packet and its ACK takes T seconds and occurs with probability $1 - p$, the protocol efficiency for full duplex (FD) is given as:

$$\eta_{FD} = \frac{(1 - p)T_p}{(1 - p)T + pT_p} \qquad (6.30)$$

- **Selective Repeat Protocol (SRP):** In case of the SRP, only the selected packets are retransmitted. The data link layer in the receiver delivers exactly one copy of every packet in the correct order. The data link layer in the receiver may get the packets in the wrong order from the physical layer. This occurs, for example, when transmission errors corrupt the first packet and not the second one. The second packet arrives correctly at the receiver before the first. The data link layer in the receiver uses a buffer to store the packets that arrive out of order. Once the data link layer in the receiver has a consecutive group of packets in its buffer, it can deliver them to the network layer. The sender also uses a buffer to store copies of the unacknowledged packets. The number of the packets which can be held in the sender/receiver buffer is a design parameter.

Let W = the number of packets which the sender and receiver buffers can each hold and SRP = number of packets in modulo 2W. The protocol efficiency without any error and with a packet error probability of p is given as:

$$\eta(0) = \min\left\{\frac{WT_p}{T}, 1\right\} \qquad (6.31)$$

For very large W, the protocol efficiency is

$$\eta(p) = 1 - p \qquad (6.32)$$

where:
  WT_p = time-out

$$\eta(p) = \frac{2 + p(W - 1)}{2 + p(3W - 1)} \qquad (6.33)$$

SRP is very efficient, but it requires buffering packets at both the sender and the receiver.

- **Go-Back-N (GBN):** The Go-Back-N protocol allows the sender to have multiple unacknowledged packets without the receiver having to store packets. This is done by not allowing the receiver to accept packets that are out of order. When a time-out timer expires for a packet, the transmitter resends that packet and all subsequent packets. The Go-Back-N protocol improves on the efficiency of the Stop and Wait protocol, but is less efficient than SRP. The protocol efficiency for full duplex is given as:

$$\eta_{FD} = \frac{1}{1 + \left(\dfrac{p}{1-p}\right)W} \tag{6.34}$$

- **Window-control Operation Based on Reception Memory (WORM) ARQ:** In digital cellular systems, bursty errors occur by multipath fading, shadowing, and handoffs. The typical bit-error rate fluctuates from 10^{-1} to 10^{-6}. Therefore, the conventional ARQ schemes do not operate well in a digital cellular system. WORM ARQ has been suggested for control of dynamic error characteristics. It is a hybrid scheme that combines SRP GBN protocol. GBN protocol is chosen in the severe error condition whereas SRP is selected in the normal error condition.

- **Variable Window and Frame Size GBN and SRP [24]:** Since wireless systems have bursty error characteristics, the error control schemes should have a dynamic adaptation to a bursty channel environment. The SRP and GBN with variable window and frame size have been proposed to improve error control in wireless systems. Table 6.2 provides the window and frame size for different BER. If the error rate increases, the window and frame size are decreased. In the case of the error rate being small, the window and frame size are increased. The optimum threshold values of BER, window and frame size were obtained through computer simulation.

Table 6.2 Bit-error rate versus window and size.

Bit-error rate (BER)	Window size (W)	Frame size (bits)
BER $\leq 10^{-4}$	32	172
$10^{-4} < $ BER $ < 10^{-3}$	8	80
$10^{-3} < $ BER $ < 10^{-2}$	4	40
$10^{-2} < $ BER	2	16

Example 6.10

We consider a WLAN in which the maximum propagation delay is 4 sec. The WLAN operates at a data rate of 10 Mbps. The data and ACK packets are of 400 and 20 bits, respectively. The processing time for a data or ACK packet is 1 sec. If the probability p that a data packet or its ACK can be corrupted during transmission is 0.01, find the data link protocol efficiency with (1) Stop and Wait protocol—full duplex, (2) SRP with window size W = 8, and (3) Go-Back-N protocol with window size W = 8.

Solution

$$T_p = \frac{400}{10} = 40\,\mu s$$

$$T_a = \frac{20}{10} = 2\,\mu s$$

$$T_{\text{prop}} = 4\,\mu s$$

$$T_{\text{proc}} = 1\,\mu s$$

$$T = 40 + 2 \times 4 + 2 \times 1 + 2 = 52\,\mu s$$

Stop and Wait:

$$\eta = \frac{(1 - 0.01) \times 40}{(1 - 0.01) \times 52 + 0.01 \times 40} = 0.763$$

SRP:

$$\eta = \frac{2 + 0.01(8 - 1)}{2 + 0.01(24 - 1)} = 0.954$$

GBN:

$$\eta = \frac{1}{1 + 8\left(\dfrac{0.01}{1 - 0.01}\right)} = 0.925$$

6.15 Summary

The chapter described the access technologies used in wireless communications including reservation-based multiple access and random multiple access. FDMA,

TDMA, and CDMA technologies were discussed and their advantages and disadvantages were listed. Illustrated examples were given to show calculations for determining the capacity of TDMA and CDMA systems. Brief descriptions of the FDD, TDD, TDM/TDMA, and TDM/TDMA/FDD approaches were also given. Since packet networks are an important part of wireless networks, we briefly stated the characteristics of the access methods in common use and defined their throughput equations. The common packet protocols such as ALOHA, slotted ALOHA, and Carrier Sense Multiple Access (CSMA) were discussed. We also presented the methods used to control errors for data link protocols.

Problems

6.1 In a proposed TDMA cellular system, the one-way bandwidth of the system is 40 MHz. The channel spacing is 30 kHz and total voice channels in the system are 1333. The frame duration is 40 ms divided equally between six time slots. The system has an individual user data rate of 16.2 kbps in which the speech with error protection has a rate of 13 kbps. Calculate the efficiency of the TDMA system. What is the efficiency of the system with 20, 60, 80 and 100 MHz?

6.2 Recompute the capacity of the GSM system in Example 5.1 when a sectorized system is used. With sectorization, there are 12 channel sets of 39 channels each with three sets assigned at each cell, one for each sector.

6.3 In the IS-54 (TDMA/FDD), the frame duration is 40 ms. The frame contains six time slots. The transmit bit rate is 48.6 kbps. Each time slot carries 260 bits of user information. The total number of 30 kHz voice channels available is 395 and the total system bandwidth is 12.5 MHz. Calculate the access efficiency of the system.

6.4 Calculate the capacity and spectral efficiency (η) of the IS-54 system using the following parameters: $\eta_b = 0.96$, $\mu = 2$ (i.e., $\pi/4$-DQPSK), voice activity factor $v_f = 1.0$, information bit rate = 19.5 kbps, frequency reuse factor = 7 and system bandwidth = 12.5 MHz.

6.5 Calculate the cell capacity and spectral efficiency of a GSM system using the following data: (1) bandwidth efficiency factor = 1, (2) bit efficiency (with GMSK modulation) = 1, (3) voice activity factor = 1, (4) one-way bandwidth of the system = 10 MHz, (4) information bit rate per frame = 270.83 kbps, (5) number of users per frame = 8, and (6) frequency reuse factor = 4.

6.6 Consider a CDMA system that uses QPSK modulation and convolutional coding. The system has a bandwidth of 1.25 MHz and transmits data at 9.6 kbps. Find the number of users that can be supported by the system and bandwidth efficiency. Assume a three-sector antenna system with an effective gain of 2.6, power control efficiency = 90%, and frequency reuse efficiency of 66.67%. A bit-error rate of 10^{-3} is required.

6.7 A QPSK/DSSS WLAN is designed to transmit in the 902- to 928-MHz ISM band. The symbol transmission rate is 0.25 Megasymbols per second. An orthogonal code with eight symbols is used. A bit-error rate of 10^{-5} is required. How many users can be supported by the WLAN? A three-sector antenna with gain = 2.6 is used. Assume frequency reuse efficiency of 66.67% and power control efficiency of 90%. What is the bandwidth efficiency?

6.8 A WLAN accommodates 50 stations running the same application. The transmission rate per station is 2 Mbps and the stations use slotted ALOHA protocol. The total traffic produced by the stations is assumed to form a Poisson process. What is the maximum throughput in Erlangs? What is the maximum throughput in Mbps? What is the maximum throughput in Mbps for each station?

6.9 Consider a WLAN installation in which maximum propagation delay is $0.5\,\mu s$. The WLAN operates at a data rate of 12 Mbps, and each packet is 600 bits. Calculate the throughput with (1) an unslotted nonpersistent, (2) a slotted persistent, and (3) a slotted 1-persistent CSMA protocol.

6.10 Consider a WLAN in which the maximum propagation delay is $5\,\mu s$. The WLAN operates at a data rate of 12 Mbps. The data and ACK packet are 600 and 24 bits, respectively. The processing time for data or ACK packet is $2\,\mu s$. If the probability p that a data packet or its ACK can be corrupted during transmission is 1%, find the data link protocol efficiency with (1) Stop-and-Wait protocol, full duplex, (2) SRP with window size W = 12, and (3) Go-Back-N protocol with window size W = 12.

References

1. Bates, R. J. *Wireless Network Communication.* New York: McGraw-Hill, Inc., 1994.

2. Bell Communications Research. Generic Framework Criteria for version 1.0 Wireless Access Communication System (WACS), FA-NWT-001318. Piscataway, NJ, June 1992.

3. Calhoun, G. *Digital Cellular Radio.* Boston: Artech House, 1988.

4. Chen, Q., Sousa, E. S., and Pasupathy, S. Performance of a coded multi-carrier DS-CDMA in multipath fading channels. *Kluwer Journal of Wireless Personal Communications*, vol. 2, 1995, pp. 167–183.

5. Fazel, K. Performance of CDMA/OFDM for mobile communication system. *IEEE Proceedings of ICUPC*, vol. 2, October 1993, pp. 975–979.

6. Garg, V. K., and Wilkes, J. E. *Wireless and Personal Communications.* Upper Saddle River, NJ, Prentice Hall, 1996.

7. Gilhousen, S., et al. Increased Capacity Using CDMA for Mobile Satellite Communication. *IEEE Journal on Selected Areas in Communications*, vol. 8, no. 4, May 1990.

8. Hammond, J. L., and O'Reilly, J. P. *Performance Analysis of Local Computer Networks.* Reading, MA: Addison-Wesley Publishing Company, 1986.

9. Habab, I. M., Kavehrad, M., and Sundberg, C.-E. W. ALOHA with Capture Over Slow and Fast Fading Radio Channels with Coding and Diversity. *IEEE Journal of Selected Areas of Communications*, vol. 6, 1988, pp. 79–88.

10. Jacobs, I. M., et al. Comparison of CDMA and FDMA for the MobileStar System. Proceedings of Mobile Satellite Conference, Pasadena, CA, May 3–5, 1988, pp. 283–290.

11. Jellamy, J. C., *Digital Telephony.* Second Edition. New York: John Wiley & Sons, Inc., 1990.

12. Kondo, S., and Milstein, L. Performance of multi-carrier DS-CDMA systems. *IEEE Transactions on Communications*, vol. 44, no. 2, February 1996, pp. 238–246.

13. Lee, W. C. Y. Spectrum efficiency in cellular. *IEEE Transactions Vehicular Technology*, vol. 38, May 1989, pp. 69–75.

14. Lee, W. C. Y. Overview of cellular CDMA. *IEEE Transactions Vehicular Technology*, vol. 40, no. 2, May 1991, pp. 291–302.

15. Lee, W. C. Y. *Mobile Communications Design Fundamentals.* Second Edition. New York: John Wiley & Sons, Inc., 1993.

16. Mehrotra, A. *Cellular Radio—Analog & Digital Systems.* Boston: Artech House, 1994 ISBN 0-89006-731-7.

17. Parsons, D., and Gardiner, J. G. *Mobile Communication Systems.* New York: Halsted Press, 1988.

18. Pahlavan, K., and Levesque, A. H. *Wireless Information Networks.* New York: John Wiley and Sons, Inc., 1995.

19. Skalar, B. *Digital Communications—Fundamentals and Applications.* Englewood Cliffs: Prentice Hall, 1988.

20. Torrieri, J. *Principles of Secure Communication Systems.* Second Edition. Boston: Artech House, 1992.

21. Viterbi, A. J. When not to spread spectrum-A sequel. *IEEE Communications Magazine*, vol. 23, April 1985, pp. 12–17.

22. Viterbi, A. J., and Padovani, Roberto. Implications of mobile cellular CDMA. *IEEE Communication Magazine*, vol. 30, no. 12, 1992, pp. 38–41.

23. Walrand, J. *Communications Networks: A First Course.* Homewood, IL: Irwin, 1991.

24. Woo, Ill and Cho, Dong-Ho. A Study on the Performance Improvements of Error Control Schemes in Digital Cellular DS/CDMA Systems. *IEICE Transactions Communications*, vol. E77-B, no. 7, July 1994.

25. Yee, N., Linnartz, J -P., and Feltweis, G. Multi-carrier CDMA in indoor wireless radio networks. *IEEE Proceedings of PIMRC*, vol. 1, September 1993, pp. 109–113.

26. Ziemer, E., and Peterson, R. L. *Introduction to Digital Communications.* New York: Macmillan Publishing Company, 1992.

27. Keiser, G. E. *Local Area Networks.* New York: McGraw-Hill Publishing Company, 1989.

CHAPTER 7

Architecture of a Wireless Wide-Area Network (WWAN)

7.1 Introduction

A wireless network does not operate in isolation; it uses the services of public switched telephone networks (PSTNs) to make or receive calls from wireline users. A number of functions is required to support the services and facilities in a wireless wide-area network (WWAN). The basic subsystems of the WWAN are: radio station subsystem (RSS), networking and switching subsystem (NSS), and operational and maintenance subsystem (OMSS) (see Figure 7.1). The radio subsystem is responsible for providing and managing transmission paths between the user equipment and the NSS. This includes management of the radio interface between the user equipment and the rest of the WWAN system. The NSS has the responsibility of managing communications and connecting user equipment to the relevant networks or other users. The NSS is not in direct

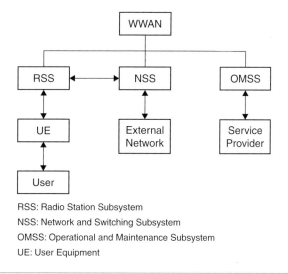

RSS: Radio Station Subsystem
NSS: Network and Switching Subsystem
OMSS: Operational and Maintenance Subsystem
UE: User Equipment

Figure 7.1   Model of a WWAN system.

contact with the user equipment, nor is the radio subsystem in direct contact with external networks. The user equipment, radio subsystem, and NSS form the operational part of the WWAN system. The OMSS provides the means for a service provider to control them. Figure 7.1 shows the model for the WWAN system. In the WWAN, interaction between the subsystems can be grouped into two main parts [3]:

- **Operational part**: External Networks ⇔ NSS ⇔ RSS ⇔ UE ⇔ User
- **Control and maintenace part**: OMSS ⇔ Service Provider

The operational part provides transmission paths and establishes them. The control and maintenance parts interact with the traffic-handling activity of the operational part by monitoring and modifying it to maintain or improve functions.

In this chapter, we present the architecture of a WWAN and discuss subsystem entities and their roles. We also look into the frame and channel structure and point out the role of different logical channels used in the GSM. We then describe how information is processed in the GSM. We conclude the chapter with a brief description of services available in the GSM900. More detailed discussion of WWANs will be given in Chapter 15.

7.2 WWAN Subsystem Entities

Figure 7.2 shows the functional entities of a WWAN and their logical interaction. A brief description of these functional entities is provided below [1,10,11].

7.2.1 User Equipment

The user equipment (UE) consists of the physical equipment used by the subscriber to access a WWAN for offered telecommunication services. Functionally, the UE includes a mobile terminal and, depending on the services it can support, various terminal equipment and combinations of terminal equipment and terminal adaptor (TA) functions (TA acts as a gateway between the terminal equipment and mobile terminal) (see Figure 7.3). Various types of mobile stations, such as a vehicle-mounted station, portable station, or handheld station, are used [7]. Basically, a mobile station can be divided into two parts. The first part contains the hardware and software to support radio and man-machine interface functions and is available at retail stores to buy or rent. The second part contains terminal/user-specific data in the form of a smart card (subscriber identity module (SIM) card), which can effectively be considered a sort of logical terminal. The SIM card plugs into the first part of the mobile station and remains in it for the duration of use. Without the SIM card, the mobile station is not associated with any user and cannot make or receive calls (except possibly an emergency call if the network allows). The SIM card is issued by the mobile service provider after

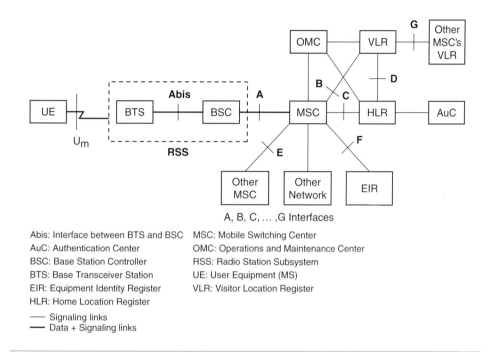

A, B, C, ... ,G Interfaces

Abis: Interface between BTS and BSC   MSC: Mobile Switching Center
AuC: Authentication Center   OMC: Operations and Maintenance Center
BSC: Base Station Controller   RSS: Radio Station Subsystem
BTS: Base Transceiver Station   UE: User Equipment (MS)
EIR: Equipment Identity Register   VLR: Visitor Location Register
HLR: Home Location Register

—— Signaling links
▬▬ Data + Signaling links

Figure 7.2 WWAN reference model.

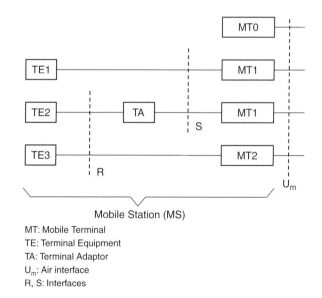

Mobile Station (MS)

MT: Mobile Terminal
TE: Terminal Equipment
TA: Terminal Adaptor
U_m: Air interface
R, S: Interfaces

Figure 7.3 Functional model of a mobile station.

subscription. This type of SIM-card mobility is analogous to terminal mobility, but it also provides a personal-mobility-like service within the WWAN.

A mobile station has a number of identities, including the international mobile subscriber identity (IMSI), the temporary mobile subscriber identity (TMSI), the international mobile equipment identity (IMEI), and integrated services of digital network (ISDN) number. The IMSI is embodied in the SIM. SIM contains all subscriber-related information stored on the user's side of the radio interface.

- *International Mobile Subscriber Identity:* The IMSI is assigned to a mobile station at subscription time. It uniquely identifies a given mobile station. IMSI is transmitted over the radio interface only if necessary. IMSI contains 15 digits and has:

  1. Mobile Country Code (MCC)—3 digits (home country)
  2. Mobile Network Code (MNC)—2 digits (home network)
  3. Mobile Subscriber Identification Number (MSIN)
  4. National Mobile Subscriber Identity Number (NMSI)

- *Temporary Mobile Subscriber Identity:* The TMSI is assigned to a mobile station by the visitor location register (VLR). The TMSI uniquely identifies a mobile station within the area controlled by a given visitor location register. A maximum of 32 bits can be used for TMSI.

- *International Mobile Equipment Identity:* The IMEI uniquely identifies the mobile station equipment. It is assigned by the equipment manufacturer. The IMEI contains 15 digits and carries:

  1. Type of Approval Code—6 digits
  2. Final Assembly Code—2 digits
  3. Serial Number—6 digits
  4. Spare—1 digit

The *Subscriber Identity Module* (SIM) contains: IMSI, authentication key (K_i), subscriber information, access control class, cipher key (K_c), TMSI, information about additional services, location area identity (LAI) and forbidden networks list.

7.2.2   Radio Station Subsystem

The radio station subsystem (RSS) is the physical equipment that provides radio coverage to prescribed geographical areas, known as cells. It contains equipment required to communicate with the user equipment. Functionally, an RSS consists of a control function performed by the base station controller (BSC) and a transmitting/receiving function carried out by the base station transceiver (BTS) system. The BTS is the radio transmission/receiving equipment and covers a cell. An RSS can serve several cells and can have multiple base station transceivers.

The base station transceiver contains the transcoder rate adapter unit (TRAU). In the GSM TRAU, the speech encoding and decoding is carried out, as well as the rate adaptation function for data. In certain situations TRAU is located between the base station controller (BSC) and the mobile switching center (MSC) to gain an advantage of a more-compressed transmission between the BTS and the TRAU. Interface between the BTS and BSC is A_{bis}. The interface between the user equipment and radio station subsystem is air interface (U_m).

7.2.3 Network and Switching Subsystem

The NSS includes the main switching functions of the WWAN, databases required for the subscribers, and mobility management. Its main role is to manage the communications between the WWAN and other network users [8]. Within the NSS, the switching functions are performed by the mobile switching center (MSC). Subscriber information relevant to provisioning of service is kept in the home location register (HLR). The other database in the NSS is the visitor location register (VLR), which maintains data required for mobility management.

The MSC performs the necessary switching functions required for the user equipment located in an associated geographical area, called an MSC area. The MSC monitors the mobility of its subscribers and manages necessary resources required to handle and update the location registration procedures and to carry out the handoff functions. The MSC is involved in the interworking functions to communicate with other networks such as PSTN and ISDN. The interworking functions of the MSC depend upon the type of the network to which it is connected and the type of service to be performed. The call routing, call control, and echo control functions are also performed by the MSC.

The HLR is the functional unit used for management of mobile subscribers. The number of home location registers in a network varies with the characteristics of the network. Two types of information are stored in the HLR: subscriber information and part of the mobile information to allow incoming calls to be routed to the MSC for the particular mobile. Any administrative action (such as changes in service profile, etc.) by a service provider on subscriber data is carried out in the HLR. The HLR stores IMSI, MS ISDN number, VLR address, and subscriber data (e.g., supplementary services, etc.).

The VLR is linked to one or more MSCs. The VLR is a functional unit that stores subscriber information, such as location area, when the subscriber is located in the area covered by the VLR. When a roaming user enters an MSC area, the MSC informs the associated VLR about the UE; the UE goes through a registration procedure that includes the following steps:

- The VLR recognizes that the UE is from another network.
- If roaming is allowed, the VLR finds the UE's HLR in its home network.

- The VLR constructs a global title from the IMSI to allow signaling from the VLR to the UE's HLR via the PSTN/ISDN networks.
- The VLR generates a mobile subscriber roaming number (MSRN) that is used to route incoming calls to the UE.
- The MSRN is sent to the UE's HLR.

The information included in the VLR is:

1. MSRN
2. TMSI
3. LAI where the UE has been registered
4. data related to supplementary services
5. MS ISDN number
6. IMSI
7. HLR address or global title (GT)
8. local UE identity, if used

The NSS contains more than MSCs, HLRs, and VLRs. In order to set up a call, the call is first routed to a gateway switch, referred to as the gateway MSC (GMSC). The GMSC is responsible for collecting the location information and routing the call to the MSC through which the subscriber can obtain service at that instant (i.e., the visited MSC). The GMSC first finds the right HLR from the directory number of the subscriber and interrogates it to obtain user information. The GMSC has an interface with external networks for which it provides a gateway function. It also has an interface with the signaling system 7 (SS7) network for interworking with other NSS entities.

7.2.4   Operation and Maintenance Subsystem (OMSS)

The OMSS is responsible for handling system security based on the validation of identities of various telecommunications entities. These functions are performed in the authentication center (AuC) and equipment identity register (EIR). The AuC is accessed by the HLR to determine whether a UE will be granted service. The EIR provides UE information used by the MSC. The EIR maintains a list of legitimate, fraudulent, or faulty UEs.

The OMSS is also in charge of remote operation and maintenance of the network. Functions are monitored and controlled in the OMSS. The OMSS may have one or more network management centers (NMCs) to centralize network control.

The operations and maintenance center (OMC) is the functional entity through which a service provider monitors and controls the system. The OMC provides a single point for maintenance personnel to maintain the entire system. One OMC can serve multiple MSCs.

7.2.5   Interworking and Interfaces

Necessary interfaces are required to achieve an optimum interworking between different entities of the network. The use of the SS7 between the MSC and VLR and between the MSC and HLR allows transmission of both call control signals and other information. The corresponding signaling capabilities are supported by the mobile application part (MAP) in SS7 defined in the particular network standards. The interface labels on the reference model (Figure 7.2—A, A_{bis}, B, C, D, E, F, G, and U_m) corresponds to interfaces between network nodes. Each interface is specified in network standards along with their corresponding procedures. For example the GSM recommendations in the 09 series cover interworking procedures between a public land mobile network (PLMN) and other networks [9].

7.3   Logical Channels

A WWAN uses a variety of channels in which the information is carried. In GSM, these channels are separated into *physical channels* and *logical channels* [10,11]. The physical channels are determined by the time slots, whereas the logical channels are determined by the information carried within the physical channels. It can be further summarized by saying that several recurring time slots on a carrier constitute a physical channel. These are then used by different logical channels to transfer information. These channels may either be used for user data (payload) or signaling to enable the system to operate.

In GSM logical channels (see Figure 7.4) are used to carry traffic and control information. There are three types of logical channels: *traffic channels* (TCHs), *control channels* (CCHs), and the *cell broadcast channel* (CBCH). The traffic channels are used to transmit user information (speech or data). The control channels are used to transmit control and signaling information. The cell broadcast channel is used to broadcast user information from a service center to the mobile stations listening to a given cell area. It is a unidirectional (downlink-only, base station to mobile station), point-to-multipoint channel used for a short-information message service. Some special constraints are imposed on the design of the CBCH because of the requirement that this channel can be listened in parallel with the broadcast control channel (BCCH) information and the paging messages.

The control channels consist of broadcast channel (BCH), common control channel (CCCH), and dedicated control channel (DCCH). The broadcast channels are used for such functions as correcting mobile frequencies, frame synchronization, control channel structure, and so on. They are point-to-multipoint, downlink-only channels. These channels consist of BCCH, frequency correction channel (FCCH), and synchronization channel (SCH).

The BCCH is used to send cell identities, organization information about common control channels, cell service available, and so on. The frequency correction channel is used to transmit frequency correction data bursts that contain the set of

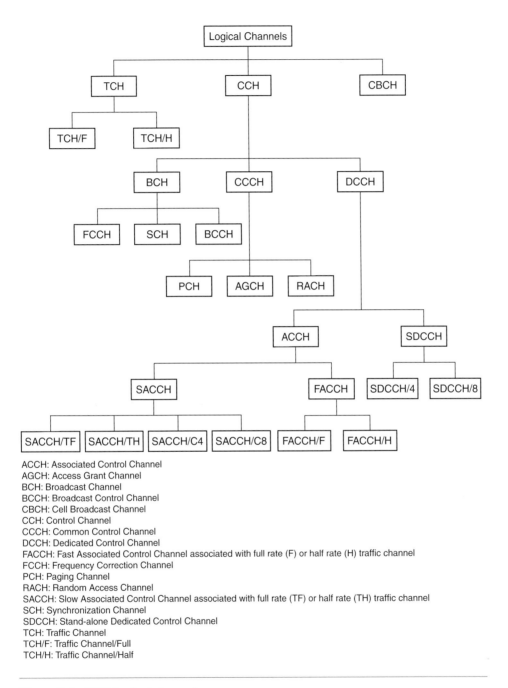

ACCH: Associated Control Channel
AGCH: Access Grant Channel
BCH: Broadcast Channel
BCCH: Broadcast Control Channel
CBCH: Cell Broadcast Channel
CCH: Control Channel
CCCH: Common Control Channel
DCCH: Dedicated Control Channel
FACCH: Fast Associated Control Channel associated with full rate (F) or half rate (H) traffic channel
FCCH: Frequency Correction Channel
PCH: Paging Channel
RACH: Random Access Channel
SACCH: Slow Associated Control Channel associated with full rate (TF) or half rate (TH) traffic channel
SCH: Synchronization Channel
SDCCH: Stand-alone Dedicated Control Channel
TCH: Traffic Channel
TCH/F: Traffic Channel/Full
TCH/H: Traffic Channel/Half

Figure 7.4   GSM logical channel.

all "0." This gives a constant frequency shift of the RF carrier that can be used by the mobile station for frequency correction. The synchronization channel (SCH) is used for time synchronization of the mobile stations. The data on this channel include frame number as well as the base station identity code (BSIC) required by mobile stations when measuring base station signal strength.

The CCCHs include paging channel (PCH), access grant channel (AGCH) and random access channel (RACH). The CCCHs are point-to-multipoint downlink-only channels that are used for paging and access. The PCHs are used to page mobile stations. The mobile stations need to listen for paging during certain times. The AGCHs are downlink-only channels used to assign mobiles to stand-alone dedicated control channels (SDCCHs) for initial assignment. The RACHs are uplink-only channels used by mobile stations for transmitting their requests for dedicated connections to the network.

There are two types of DCCHs: SDCCH and associated control channel (ACCH). The SDCCHs are bidirectional, point-to-point channels that are used for service request, subscriber authentication, ciphering initiation, equipment validation, and assignment to a traffic channel (TCH). The ACCHs are bidirectional, point-to-point channels that are associated with a given TCH and SDCCH. These channels are used to send out-of-band signaling and control data between the mobile station and the base station. Examples of their use are to send signal strength measurements from a mobile station to the base station or to send transmission timing information from the base station to the mobile station. The associated control channels are further divided as slow associated control channels (SACCHs) and fast associated control channels (FACCHs). Figure 7.4 shows all GSM logical channels. Table 7.1 lists briefly the role of each logical channel. For more details, refer to the GSM 04.03, 05.01, and 05.2 series recommendations [4,5].

7.4 Channel and Frame Structure

We discuss channel and frame structure with reference to the GSM900 system that uses FDMA/TDMA (see Chapter 6) and a bandwidth of 25 MHz. The frequency band used for uplink (i.e., from the mobile station to the base station) is 890 to 915 MHz, whereas for the downlink (i.e., from the base station to the mobile station) is 935 to 960 MHz. The GSM900 has 124 RF channels, each with a bandwidth of 200 kHz. For a given RF channel the uplink (UL_f) and downlink (DL_f) frequency can be obtained from Equations 7.1 and 7.2, respectively:

$$UL_f = 890.2 + 0.20(N - 1)\,\text{MHz} \tag{7.1}$$

$$DL_f = 935.2 + 0.20(N - 1)\,\text{MHz} \tag{7.2}$$

Table 7.1 Role of logical channels in GSM.

Logical channel	Role of the logical channel
TCH/F	Full rate traffic channel to carry payload at 22.8 kbps
TCH/H	Half rate traffic channel to carry payload at 11.4 kbps
BCCH	Broadcast network information
SCH	Synchronization of the mobile stations
FCCH	Used for frequency correction
AGCH	Acknowledges channel request from mobile and allocates an SDCCH
FACCH	For time critical signaling over TCH (e.g., for handoff signaling); traffic burst is stolen for a full signaling burst
SACCH	TCH in-band signaling, e.g., for link monitoring
SDCCH	For signaling exchange, e.g., during call setup, registration/location updates
FACCHs	FACCH for SDCCH. The SDCCH burst is stolen for a full signaling burst.
SACCHs	SDCCH in-band signaling, e.g., for link monitoring

where:

$N = 1, 2, \ldots, 124$

When the mobile station is assigned a channel, an RF channel and a time slot are also assigned. RF channels are assigned in frequency pairs—one for the uplink and the other for the downlink. Each pair of RF channels supports up to eight simultaneous voice calls (see Figure 7.5). Thus, the GSM can support up to 992 simultaneous users with a full-rate speech coder. This number is doubled to 1984 users with the half-rate speech coder.

In GSM, the RF carriers are divided in time using a TDMA scheme. The fundamental unit of time is called a burst (time slot) period and it lasts for 0.577 ms. Eight of these burst periods are grouped into a TDMA frame. The frame lasts for 4.615 ms and it forms the basic unit for the definition of logical channels. One physical channel is one burst period allocated in each TDMA frame. There are different types of frames that are transmitted to carry different data. The frames are organized into multiframes, superframes, and hyperframes to provide overall synchronization.

The GSM multiframe is 120 ms. It contains 26 frames of 8 bursts. The structure of a GSM hyperframe, superframe, multiframe, frame, and burst is given in Figure 7.6. A burst carries 156.25 bits. The same format is used for the uplink and downlink transmission with various burst types as shown in Figure 7.7. In a normal burst, two user information groups of 58 bits account for most of the transmission time in a time slot (57 bits carry user data, while the H bit is used to distinguish speech from other transmissions). Twenty-six training (T) bits are

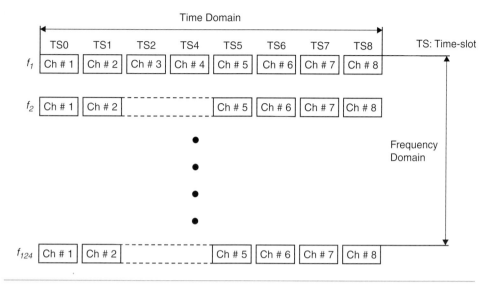

Figure 7.5 GSM FDMA/TDMA structure.

used in the middle of the time slot. The time slot starts and ends with three tail bits. The time slot also contains 8.25 guard (G) bits.

7.5 Basic Signal Characteristics

The GSM system operates using frequency division duplex (FDD) and as a result, paired frequency bands are required for the up- and downlink transmissions. The frequency separation is dependent upon the band in use (for GSM900, it is 45 MHz). The RF carrier is modulated using Gaussian minimum shift keying (GMSK) (see Chapter 9 for details). The GMSK occupies a relatively narrow bandwidth, and it has a constant power level.

The data transported by the RF carrier serves up to eight different users under the basic system. Even though the full data rate on the carrier is about 270.83 kbps, some of this supports the management overhead, and therefore the data rate allocated to each time slot is only 22.8 kbps. In addition to error correction, the problem of interference, fading and the like must be overcome. This means that the available data rate for transporting the digitally encoded speech is 13 kbps for the basic vocoder.

7.6 Speech Processing

Two major steps are taken in transmitting and receiving information over a digital radio link: information processing and modulation processing. Information processing deals with the preparation of the basic information signals so that they are

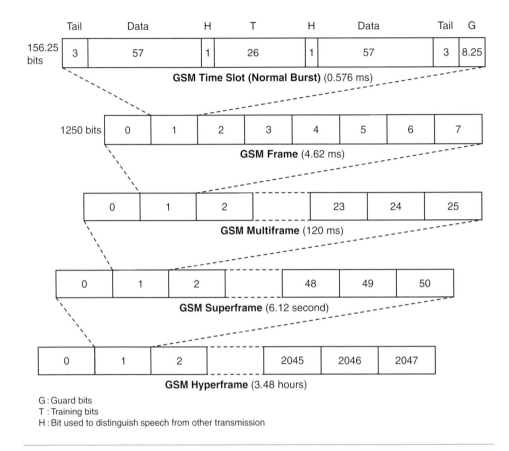

Figure 7.6 Physical structure for GSM hyperframe, superframe, multiframe, frame, and time slot.

protected and converted into a form that the radio link can handle. Information processing involves transcoding, channel coding, bit-interleaving, encrypting, and multiplexing. Modulation processing involves the physical preparation of the signal to carry information on an RF carrier [6].

Each digital radio link process in the transmitting path has its peer in the receiving path (see Figure 7.8). The delay equalization process in the receiving path is required to compensate for the spread in time delays resulting from the multipath propagation (see Chapter 3). It can be part of the demodulation process. However, it should be emphasized that it is required when delay spreads are significant compared to the information bit duration. In the GSM the transmitted bit duration is about 37 µsec, and a delay spread of about 6 to 8 µsec is quite common. The delay spread problem becomes critical and is difficult to solve with the higher transmission bit rate.

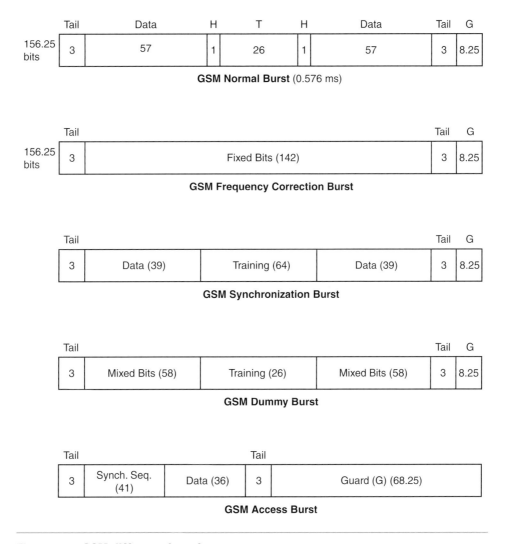

Figure 7.7 GSM different time slot structures.

In the GSM, the analog speech from the mobile station is passed through a low-pass filter to remove the high-frequency contents from the speech. The speech is sampled at the rate of 8000 samples per second, uniformly quantized (see Chapter 4) to 2^{13} (8192) levels and coded using 13 bits per sample. This results in a digital information stream at a rate of 104 kbps. At the base station, the speech signal is digital (64 kbps), which is first transcoded from the A-law (see Chapter 4) 8-bit samples into 13-bit samples corresponding to a linear representation of amplitudes. This results in a digital information stream at a rate of 104 kbps.

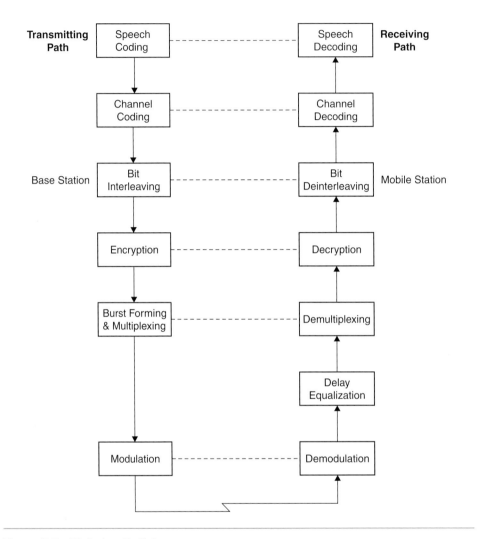

Figure 7.8 Digital radio link process.

The 104-kbps digital signal stream is fed into the speech encoder (see Figure 7.9) which then transcodes the speech into a 13-kbps stream. The full-rate speech encoder takes a 2080 bit block from the 13-bit transcoder every 20 ms (i.e., 160 samples) and produces 36 "filter parameters" bits over the 20 ms period, 9 long-term prediction (LTP) bits every 5 ms, and 47 regular pulse excited-linear prediction coding (RPE-LPC) bits every 5 ms (refer to Table 7.2). Thus, 260 bits are generated every 20 ms. Of these 260 bits, 182 are classified as class I bits related to the excitation signal, whereas the remaining 78 are class II bits related to the parameters of linear prediction coding (LPC) and LTP filters. The class I bits

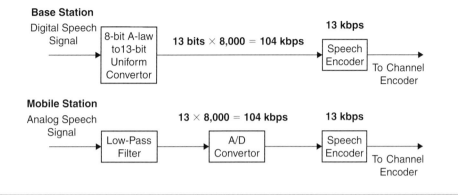

Figure 7.9   Speech processing in GSM.

Table 7.2 Details of class I and class II bits in GSM vocoder.

	Bits per 5 ms	Bits per 20 ms
Linear Prediction Coding (LPC) filter		36
Long Term Prediction (LTP) filter	9	36
Excitation signal	47	188
Total		260
Class I		182 (class Ia = 50, class Ib = 132)
Class II		78

are further divided into class Ia (50 bits) and class Ib (132 bits). The class Ia bits are the most significant bits which are used to generate 3 cyclic redundancy check (CRC) bits. The 3 CRC bits along with 4 tail bits are added to the 182 class I bits before they are passed through the half-rate convolutional codec to produce twice as many bits output as there are bits input (i.e., $2 \times 189 = 378$). The 78 class II bits remain uncoded and are bypassed (see Figure 7.10). The total 456 bits (i.e., $378 + 78$) are fed to the bit interleaver. Since 456 bits are generated during 20 ms, the user data rate is $456/0.02 = 22.8$ kbps. This includes 13 kbps raw data and 9.8 kbps of parity, tail, and channel coding.

Another vocoder (the enhanced full rate (EFR)) (see Chapter 8) has been added in response to the poor voice quality perceived by the users of the RPE-LPC vocoder. This new vocoder gives much better voice quality. It uses the algebraic code excitation linear prediction (ACELP) compression technology (see Chapter 8) to provide a significant improvement in voice quality compared to the original RPE-LPC. There is also a half-rate vocoder (11.4 kbps). Although this vocoder

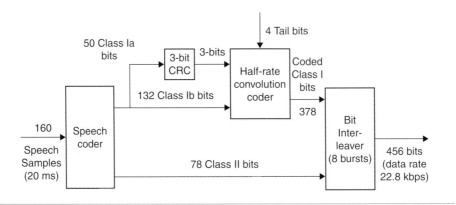

Figure 7.10   Channel coding for full-rate speech in GSM.

gives more inferior voice quality compared to full-rate vocoders, it does allow for an increase in network capacity. It is used in some instances when network loading is very high to handle all the calls.

Of the 456 bits, 57 at a time are interleaved with 57 other bits from an adjacent data block to form a data burst of 114 bits. At this stage, 42.25 overhead bits are added to the data burst to carry it in a time slot (see Figure 7.11). Bit interleaving is used to reduce the adverse effects of fading by preventing entire blocks of bits from being destroyed by a signal fade. Interleaved data is passed through the GMSK modulator where it is filtered by a Gaussian filter before applying it to the modulator. The modulated data passes through a duplexer switch where filtering is provided between the transmitted and the received signal. On the receiving side the signal is demodulated and deinterleaved before the error correction is applied to the recovered bits.

7.7   Power Levels in Mobile Station

A variety of power levels are allocated by the GSM standard, the lowest being only 800 mW (29 dBm). As mobiles may only transmit for one-eighth of the time, i.e., for their allocated slot, which is one of eight, the average power level is an eighth of the maximum power.

To reduce the levels of transmitted power additionally and hence the levels of interference, mobiles are able to step the power down in increments of 2 dB from the maximum to a minimum of 20 mW (13 dBm). The mobile station measures the signal strength or signal quality (based on bit-error-rate), and passes the information to BTS and hence to BSC, which ultimately decides if and when the power level should be changed.

A further power-saving and interference-reducing facility is the discontinuous transmission (DTx) capability (an optional feature in GSM) that is incorporated

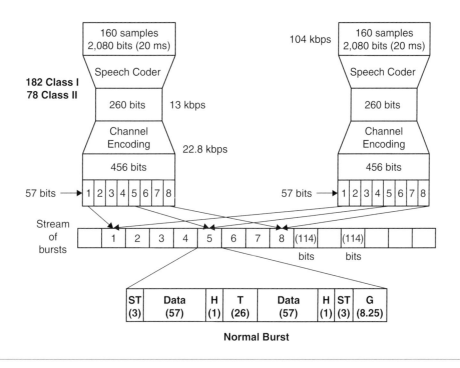

Figure 7.11   Speech coding in GSM (full rate).

within specifications. It is particularly useful because there are long pauses in speech, for example, when the person using the mobile phone is listening, and during these periods there is no need to transmit a signal. The most important element of DTx is the voice activity detector. It must distinguish between voice and noise inputs, a task that is not trivial.

7.8   GSM Public Land Mobile Network Services

GSM offers users good voice quality, call privacy, and network security. The SIM card provides the security mechanism for GSM. SIM cards are like credit cards and identify the user to the GSM network. They can be used with any GSM handset, providing phone access, ensuring delivery of appropriate services to that user, and automatically billing the subscribers's network usage to the home network.

Roaming agreements have been established between most GSM service providers in different countries, allowing subscribers to roam between networks and have access to the same services no matter where they travel.

Of major importance is GSM's potential for delivering enhanced services requiring multimedia communication: voice, image, and data. Several mobile service providers offer free voice mailboxes and phone answering services to subscribers.

The basic telecommunication services provided by the GSM public land mobile network (PLMN) are divided into three main groups: *bearer services*, *teleservices*, and *supplementary services* [2]. Bearer services give the subscriber the capacity required to transmit appropriate signals between certain access points (i.e., user-network interfaces). Table 7.3 provides a summary of these services and compares them with service available with integrated services of digital network (ISDN).

Teleservices provide the subscriber with necessary capabilities, including terminal equipment functions, to communicate with other subscribers. The GSM teleservices are:

- Speech transmission—telephony, emergency call
- Short message services—mobile terminating point-to-point, mobile-originating point-to-point, cell broadcast
- Message handling and storage services
- Videotex access
- Teletext transmission
- Facsimile transmission

A summary of the teleservices is given in Table 7.4. A comparison is made between the GSM and ISDN teleservices.

Supplementary services modify or supplement basic telecommunications services and are offered with or in association with basic telecommunications services. Table 7.5 summarizes the supplementary services and compares them to the supplementary services available with ISDN.

The GSM system was designed as a 2G cellular communications system. One of the basic aims was to provide a system that would enable greater capacity to be achieved than 1G analog systems. GSM achieved this by using a digital TDMA approach. By adopting TDMA more users could be accommodated within the

Table 7.3 A comparison of bearer services supported by GSM and ISDN.

Service	GSM	ISDN
Data Service	yes	yes
Alternate speech/data	yes	yes
Speech followed by data	yes	yes
Clear 3.1-kHz audio	yes	yes
Unrestricted digital information (UDI)	yes	yes
Packet Assembler/Disassembler (PAD)	yes	no
3.1-kHz external to PLMN	yes	no
Others	no	yes

available spectrum. In addition to this, ciphering of digitally encoded speech was used to retain privacy.

Table 7.4 A comparison of teleservices supported by GSM and ISDN.

Service	GSM	ISDN
Circuit switch (telephony)	yes	yes
Emergency call	yes	yes
Short message point-to-point	yes	yes
Short message cell broadcast	yes	yes
Alternate speech/facsimile group 3	yes	yes
Automatic facsimile group 3 service	yes	yes
Voice-band modem (3.1-kHz audio)	yes	yes
Messaging teleservices	yes	no
Paging teleservices	yes	no
Others	no	yes

Table 7.5 A comparison of supplementary services supported by GSM and ISDN.

Service	GSM	ISDN
Call Number ID Presentation	yes	yes
Call Number ID Restriction	yes	yes
Connected Number ID Presentation	yes	yes
Connected Number ID Restriction	yes	yes
Malicious Call Identification	yes	yes
Call Forwarding Unrestricted	yes	yes
Call Forwarding Mobile Busy	yes	yes
Call Forwarding No Reply	yes	yes
Call Forwarding Mobile Not Reachable	yes	yes
Call Transfer	yes	yes
Call Waiting	yes	yes
Call Hold	yes	yes
Completion of Call to Busy Subscriber	yes	yes
3-Party Service	yes	yes
Conference Calling	yes	yes

(continued)

Table 7.5 (*continued*)

Service	GSM	ISDN
Closed User Group	yes	yes
Multiparty	yes	yes
Advice of Charge	yes	yes
Reverse Charging	yes	yes
Flexible Altering	no	yes
Mobile Access Hunting	yes	yes
Freephone	yes	no
Barring All Originating Calls	yes	yes
Barring Outgoing Calls	yes	yes
Barring Outgoing International Calls	yes	yes
Barring All International Calls	yes	yes
Barring Outgoing International Calls—except Home	yes	yes
Barring Incoming Calls when Roaming	yes	yes
Do Not Disturb	Barring	yes
Message Waiting Notification	SMS	yes
Preferred Language Service	no	yes
Remote Feature Control	no	yes
Selective Call Acceptance	Barring	yes
Voice Privacy	Encryption	yes
Priority Access & Channel Assignment	no	yes
Password Call Acceptance	no	yes
Subscriber PIN Intercept	Barring	yes
Subscriber PIN Access	no	yes
Voice Mail Retrieval	no	yes
Others	no	yes

7.9 Summary

The use of GSM has grown steadily since it was first deployed in 1991. It is now the most widely used mobile phone system in the world. GSM reached the 1 billion subscriber point in 2004, and continues to grow in popularity.

In this chapter, we presented an overview of GSM900—a typical WWAN—and discussed network architecture, logical channels, frame structures, speech processing, power control, and services. We will discuss enhancements to the GSM in Chapter 15, including general packet radio services (GPRS) and universal mobile telecommunication services (UMTS), which are derived from the GSM.

Problems

7.1 What are the three subsystems that are used in the WWAN?

7.2 Define the role of a VLR in the WWAN architecture.

7.3 Define the role of an HLR in the WWAN architecture.

7.4 Why are the IMSI, TMSI, and IMEI used in a WWAN?

7.5 Define the three services available in the GSM.

7.6 What is the role of a speech codec in the speech processing of a WWAN?

7.7 What is a bit-interleaver? Why it is used in a WWAN?

7.8 Why do you need an equalizer in the GSM?

7.9 Why are convolution codecs used in wireless networks?

7.10 Why are so many logical channels used in the GSM?

References

1. GSM Specification Series 1.02–1.06, "GSM Overview, Glossary, Abbreviation, Service Phases."

2. GSM Specification Series 2.01–2.88, "GSM Services and Features."

3. GSM Specification Series 3.01–3.88, "GSM PLMN Functions, Architecture, Numbering and Addressing Procedures."

4. GSM Specification Series 4.01–4.88, "MS-BSS Interface."

5. GSM Specification Series 5.01–5.10, "Radio Link."

6. GSM Specification Series 6.01–6.32, "Speech Processing."

7. GSM Specification Series 7.01–7.03, "Terminal Adaptation."

8. GSM Specification Series 8.01–8.60, "BSS-MSC Interface, BSC-BTS Interface."

9. GSM Specification Series 9.01–9.11, "Network Interworking, MAP."

10. Mouly, M., and Pautet, M. *The GSM System for Mobile Communications*. Palaiseau, France, 1992.

11. Garg, V. K., and Wilkes, J. E. *Wireless and Personal Communications System*. Upper Saddle River, NJ: Prentice Hall, 1996.

CHAPTER 8

Speech Coding and Channel Coding

8.1 Introduction

In Chapter 7 we introduced speech and channel coding with reference to the GSM. In this chapter we provide detailed descriptions of speech and channel coding schemes. We first discuss speech coding methods and attributes of a speech codec. (The speech codec is also called a voice codec or vocoder. It is a hardware circuit that converts the spoken word into digital code and vice versa.) These are then followed by a brief discussion of linear-prediction-based analysis-by-synthesis (LPAS) method. We also discuss the QCELP, EVRC, EFR, and AMR codecs that are used in WWAN systems. We then focus on channel coding, concentrating on three channel coding schemes—convolutional code, Reed-Solomon (R-S) block code, and turbo code. The convolutional codes have been used in direct sequence spread spectrum CDMA (IS-95) and R-S and turbo code are the proposed channel coding schemes for cdma2000 and WCDMA.

To achieve reliable communication with minimum possible signal power is the main objective of a communications system engineer. Speech coding is used to save bandwidth and improve bandwidth efficiency whereas channel coding is employed to improve signal quality and reduce bit-error-rate (BER). The idea of using a channel coding scheme is to recover from errors that occur during transmission over the communication channel. The channel coding strategy is aimed at allowing the transmitter to use minimum possible signal power in accomplishing the design objective of providing a specified error rate.

8.2 Speech Coding

The speech quality of codecs operating at a fixed bit rate is largely determined by the worst-case speech segments, i.e., those that are the most difficult to code at the given rate. Variable rate coding can provide a given level of speech quality at an average bit rate that is substantially less than the bit rate that would be required by an equivalent quality fixed rate codec [1–4].

The original CDMA codec known as Qualcomm code-excited linear predictive (QCELP), IS 96A was an 8 kbps code-excited linear predictive (CELP) variable rate codec designed for use in the 900 MHz digital cellular band [11,12]. The desire for improved voice quality spurred the TIA to begin working on a 13 kbps CELP variable rate codec to provide higher quality voice transmissions. The 13 kbps CDMA codec takes advantage of the higher data rate (14.4 kbps as compared to 9.6 kbps for the 8 kbps codec) to improve speech quality. It has produced mean opinion scores (MOS) close to toll-quality voice, the benchmark used for comparison. Unfortunately, system capacity and cell coverage are reduced by the higher data rate codec.

8.2.1   Speech Coding Methods

Speech coding is the process for reducing the bit rate of digital speech representation for transmission or storage, while maintaining a speech quality that is acceptable for the application. Speech coding methods are classified as waveform coding, source coding, and hybrid coding. The following sections will explain these concepts.

In Figure 8.1, the bit rate is plotted on a logarithmic axis versus speech quality classes of "poor to excellent" corresponding to the five-point MOS scale values of 1 to 5, defined by the International Telecommunications Union (ITU). It may be noted that for low complexity and low delay, a bit rate of 32 to 64 kbps is required. This suggests the use of waveform codecs. However, for low bit rate of 4 to 8 kbps, hybrid codecs should be used (see Figure 8.1). These types of codecs tend to be complex with high delay [5–8].

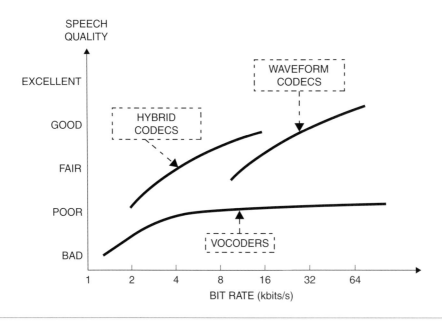

Figure 8.1   Quality of service versus bit rate.

8.2.2 Speech Codec Attributes

Speech quality as produced by a codec is a function of transmission bit rate, complexity, delay, and bandwidth. Therefore, when considering speech codecs it is essential to consider all of these attributes and their interactions. For example, low bit rate codecs tend to have more delay compared to the higher bit rate codecs. They are generally more complex to implement and often have lower speech quality than the higher bit rate codecs. The following are the speech codec attributes.

Transmission Bit Rate

Since the speech codec shares the communications channel with other data, the peak bit rate should be as low as possible so as not to use a disproportionate share of the channel. The codecs below 64 kbps are primarily developed to increase the capacity of circuit multiplication equipment used for narrow bandwidth links. For the most part, they are fixed bit rate codecs, meaning they operate at the same rate regardless of the input. In the variable bit rate codecs, network loading and voice activity determine the instantaneous rate assigned to a particular voice channel. Any of the fixed rate speech codecs can be combined with a voice activity detector (VAD) and made into a simple two-state variable bit rate system. The lower rate could be either zero or some low rate needed to characterize slowly changing background noise characteristics. Either way, the bandwidth of the communications channel is only used for active speech.

Delay

The delay of a speech codec can have a great impact on its suitability for a particular application. For a one-way delay of conversation greater than 300 ms, the conversation becomes more like a half-duplex or push-to-talk experience, rather than an ordinary conversation. The components of total system delay include: (1) frame size, look ahead, multiplexing delay, (2) processing delay for computations, and (3) transmission delay.

Most low bit rate speech codecs process a frame of speech data at a time. The speech parameters are updated and transmitted for every frame. In addition, to analyze the data properly it is sometimes necessary to analyze data beyond the frame boundary; hence, before the speech can be analyzed it is necessary to buffer a frame's worth of data. The resulting delay is referred to as *algorithmic delay*. This delay component cannot be reduced by changing the implementation, but all other delay components can. The second major contribution for delay comes from the time taken by the encoder to analyze the speech and the decoder to reconstruct the speech. This part of the delay is referred to as *processing delay*. It depends on the speed of the hardware used to implement the coder. The sum of the algorithmic and processing delays is called the *one-way codec delay*. The third component of delay is due to transmission. It is the time taken for an entire frame of data to be transmitted from the encoder to the decoder. The total of the three delays is the *one-way system delay*. In addition, frame interleaving delay adds an additional frame delay to the total transmission delay. Frame interleaving is necessary to combat channel fading (discussed in Chapter 3), and is part of the channel coding process.

Complexity

Speech codecs are implemented on special purpose hardware, such as digital signal processing (DSP) chips. DSP attributes are the computing speed in millions of instructions per second (MIPS), random access memory (RAM), and read only memory (ROM). For a speech codec, the system designer makes a choice about how much of these resources are to be allocated to the speech codec. Speech codecs using less than 15 MIPS are considered low-complexity codecs; those requiring 30 MIPS or more are thought of as high-complexity codecs. More complexity results in higher costs and greater power usage; for portable applications, greater power usage means reduced time between battery recharges or use of larger batteries, which means more expense and weight.

Quality

Of all the attributes, quality has the most dimensions. In many applications there are large amounts of background noise (car noise, street noise, office noise, etc.). How well does the codec perform under these adverse conditions? What happens when there are channel errors during transmission? Are the errors detected or undetected? If undetected, the codec must perform even more robustly than when it is informed that entire frames are in error. How good does the codec sound when speech is encoded and decoded twice? All these questions must be carefully evaluated during the testing phase of a speech codec. The speech quality is often based on the five-point MOS scale as defined by the International Telecommunication Union-Technical (ITU-T).

8.2.3   Linear-Prediction-Based Analysis-by-Synthesis (LPAS)

In a linear predictive codec attempt is made to synthesize certain features of the voice message (time waveform). The linear-prediction-based analysis-by-synthesis (LPAS) methods provide efficient speech coding at rates between 4 and 16 kbps. In LPAS speech codecs [10], the speech is divided into frames, each of about 20 ms length, for which the coefficients of the linear predictor (LP) are computed. The resulting LP filter predicts each sample from a set of previous samples.

In LPAS codecs, the residual signal is quantized on a subframe-by-subframe basis (there are commonly 2 to 8 subframes per frame). The resulting quantized signal forms the excitation signal for the LP synthesis filter. For each subframe, a criterion is used to select the best excitation signal from a set of trial excitation signals. The criterion compares the original speech signal with trial reconstructed speech signals. Because of the synthesis implicit in the evaluation criterion, the method is called *analysis-by-synthesis* coding. Various representations of excitation have been used. For lower bit rates, the most efficient representation is achieved by using vector quantization. For each subframe, the excitation signal is selected from a multitude of vectors, which are stored in a codebook. (The codebook consists of a set of stochastic excitation signals, ensembles of zero-mean Gaussian noise.) The index of the best matching vector is transmitted. At the receiver this vector is retrieved from the same codebook. The resulting excitation signal is filtered through the LP synthesis filter to produce the reconstructed speech. Linear prediction analysis-by-synthesis codecs using a codebook approach are commonly known as *code-excited linear predictive* (CELP) codec.

The parametric coding is used for those aspects of the speech signal which are well understood. Parametric codecs are traditionally used at low bit rates. A proper understanding of a speech signal and its perception is essential to obtaining good speech quality with a parametric codec. The waveform matching procedure is employed for those aspects of speech which are not well understood. Waveform matching constraints are relaxed for those aspects which can be replaced by parametric models without degrading the quality of the reconstructed speech.

A parameter which is well understood in parametric coding is the pitch of the speech signal. Satisfactory pitch estimation procedures are available. Piecewise linear interpolation of the pitch does not degrade speech quality. Pitch period is typically determined once every 20 ms and linearly interpolated between the updates. The challenge is to generalize the LPAS method such that its matching accuracy becomes independent of the synthetic pitch-period contour used. This is done by determining a time wrap of speech signal such that its pitch-period contour matches the synthetic pitch-period contour. The time wraps are determined by comparing a multitude of time-wrapped original signals with a synthesized signal. This coding scheme is called the generalized analysis-by-synthesis method and is referred to as *Relaxed CELP* (*RCELP*). The generalization relaxes the waveform-matching constraints without affecting speech quality.

8.2.4 Waveform Coding

In general, waveform codecs are designed to be independent of signal. They map the input waveform of the encoder into a facsimile-like replica of it at the output of the decoder. Coding efficiency is quite modest. The coding efficiency can be improved by exploiting some statistical signal properties, if the codec parameters are optimized for most likely categories of input signals, while still maintaining good quality for other types of signals as well. The waveform codecs are further subdivided into *time domain waveform* codec and *frequency domain waveform* codec.

Time Domain Waveform Coding

The well-known representation of speech signal using time domain waveform coding is the A-law (in Europe) or μ-law (in North America) companded pulse code modulation (PCM) at 64 kbps (see Chapter 4). Both use nonlinear companding characteristics to give a near-constant signal-to-noise ratio (SNR) over the total input dynamic range.

The ITU G.721, 32 kbps adaptive differential PCM (ADPCM) codec is an example of a time domain waveform codec. More flexible counterparts of the G.721 are the G.726 and G.727 codecs. The G.726 codec is a variable-rate arrangement for bit rates between 16 and 40 kbps. This may be advantageous in various networking applications to allow speech quality and bit rate to be adjusted on the basis of the instantaneous requirement. The G.727 codec uses core bits and enhancement bits in its bit stream to allow the network to drop the enhancement bits under restricted channel capacity conditions, while benefiting from them when the network is lightly loaded.

In differential codecs a linear combination of the last few samples is used to generate an estimate of the current one, which occurs in the adaptive predictor. The resultant difference signal (i.e., the prediction residual) is computed and encoded by the adaptive quantizer with a lower number of bits than the original signal, since it has a lower variance than the incoming signal. For a sampling rate of 8000 samples per second, an 8-bit PCM sample is represented by a 4-bit ADPCM sample to give a transmission rate of 32 kbps.

Time domain waveform codecs encode the speech signal as a full-band signal and map it into as close a replica of the input as possible. The difference between various coding schemes is their way of using prediction to reduce the variance of the signal to be encoded in order to reduce the number of bits necessary to represent the encoded waveform.

Frequency Domain Waveform Coding

In the frequency domain waveform codecs the input signal undergoes some short-time spectral analysis. The signal is split into a number of frequency-domain sub-bands. The individual sub-band signals are then encoded by using different numbers of bits to fulfill the quality requirements of that band based on its prominence. The various schemes differ in their accuracies of spectral analysis and in the bit allocation principle (fixed, adaptive, semi-adaptive). Two well-known representatives of this class are sub-band coding (SBC) and adaptive transform coding (ATC).

8.2.5   Vocoders

Vocoders are parametric digitizers which use certain properties of the human speech production mechanism. Human speech is produced by emitting sound pressure waves which are radiated primarily from the lips, although, with some sounds, significant energy emanates also from the nostrils, throat, etc. In human speech, the air compressed by the lungs excites the vocal cord in two typical modes. When generating voice sounds, the vocal cord vibrates and generates quasi-periodic voice sounds. In the case of lower energy unvoiced sounds, the vocal cord does not participate in voice production and the source acts like a noise generator. The excitation signal is then filtered through the vocal apparatus, which behaves like a spectral shaping filter. This can be described adequately by an all-pole transfer function that is constituted by the spectral shaping action of the vocal tract, lip radian characteristics, etc.

In the case of vocoders, instead of producing a close replica of an input signal at the output, an appropriate set of source parameters is generated to characterize the input signal sufficiently close for a given period of time. The following steps are used in this process:

1. Speech signal is partitioned in segments of 5 to 20 ms.

2. Speech segments are subjected to spectral analysis to produce the coefficients of the all-zero analysis filter to minimize the prediction residual energy. This process is based on the computation of the speech autocorrelation coefficients and then using either matrix inversion or iterative scheme.

3. The corresponding source parameters are specified. The excitation parameters as well as filter coefficients are quantized and transmitted to the decoder to synthesize a replica of the original signal by exciting the all-pole synthesis filter.

The quality of this type of scheme is predetermined by the accuracy of the source model rather than the accuracy of the quantization of the parameters. The speech quality is limited by the fidelity of the source model used. The main advantage of vocoders is their low bit rate, with the penalty of relatively low, synthetic speech quality. Vocoders can be classified into the frequency domain and time domain subclasses. However, frequency domain vocoders are generally more effective than time domain vocoders.

8.2.6 Hybrid Coding

Hybrid coding is an attractive trade-off between waveform coding and vocoder, both in terms of speech quality and transmission bit rate, although generally at the price of higher complexity. They are also referred to as analysis-by-synthesis (ABS) codecs.

The most recent international and regional speech coding standards belong to a class of LPAS codecs. This class of codecs includes ITU G723.1, G.728 (low-delay (LD) CELP, 16 kbps), and G.729 and all the current digital cellular standards including:

- European global system for mobile communications (GSM), full-rate, half-rate, enhanced full-rate (EFR), and adaptive multiple rate (AMR) codec
- North American full-rate, half-rate, and enhanced full-rate for time division multiple access (TDMA) IS-136 and code division multiple access (CDMA) IS-95 systems
- Japanese public digital cellular (PDC) full-rate and half-rate

In an LPAS coder, the decoded speech is generated by filtering the signal produced by the excitation generator through both a long-term (LT) predictor synthesis filter and a short-term (ST) predictor synthesis filter. The excitation signal is found by minimizing the mean-squared error over a block of samples. The error signal is the difference between the original and decoded signal. It is weighted by filtering it through a weighting filter. Both ST and LT predictors are adapted over time. Since the analysis procedure (encoder) includes the synthesis procedure (decoder), the description of the encoder defines the decoder. The short-term synthesis filter models the short-term correlations (spectral envelope) in the speech signal. This is an all-pole filter. The predictor coefficients are determined from the speech signal using linear prediction (LP) techniques. The coefficients of the short-term predictor are adapted in time, with rates varying from 30 to as high as 400 times per second.

The long-term predictor filter models the long-term correlations (fine spectral structure) in the speech signal. Its parameters are a delay and gain coefficient. For periodic signals, delay corresponds to the pitch period (or possibly an integral number of pitch periods). The delay is random for non-periodic signals. Typically, the long-term predictor coefficients are adapted at rates varying from 100 to 200 times per second.

An alternative structure for the pitch filter is the *adaptive codebook*. In this case, the long-term synthesis filter is replaced by a codebook that contains the previous excitation at different delays. The resulting vectors are searched, and the one that provides the best result is selected. In addition, an optimal scaling factor is determined for the selected vector. This representation simplifies the determination of the excitation for delays smaller than the length of excitation frames.

Code-excited linear predictive (CELP) codecs use another approach to reduce the number of bits per sample. Both encoder and decoder store the same collection of codes (C) of possible length L in a codebook. The excitation for each frame is described completely by the index to an appropriate vector. This index is found by an exhaustive search over all possible codebook vectors, using the one that gives the smallest error between the original and decoded signals. To simplify the search it is common to use a gain-shape codebook in which the gain is searched and quantized separately. Further simplifications are obtained by populating the codebook vectors with a multipulse structure. By using only a few non-zero unit pulses in each codebook vector, efficient search procedures are derived. This partitioning of excitation space is referred to as an *algebraic codebook*. The excitation method is known as *algebraic codebook excited linear prediction (ACELP)*.

8.3   Speech Codecs in European Systems

8.3.1   GSM Enhanced Full-Rate (EFR)

GSM EFR is the same as a US1 vocoder with 8-PSK modulation and is a 12.2 kbps vocoder. It is used in GSM1900. The distribution of bits in a GSM EFR vocoder is given in Table 8.1.

The IS-136 (TDMA) system uses the vector self-excited linear predictor (VSELP) codec. The VSELP algorithm uses a codebook with a predefined structure to reduce the number of computations. The output of the VSELP codec for IS-136 is 7.95 kbps. It produces a speech frame every 20 ms containing 159 bits. Recently, the algebraic codebook excited prediction (ACELP) codec (IS-641 vocoder) was selected to replace the VSELP codec in IS-136. This codec has an output bit rate of 7.4 kbps.

Comparisons of MOS and delay for IS-641, ITU LD-CELP and GSM EFR are given in Tables 8.2 and 8.3.

Table 8.1 GSM enhanced full-rate vocoder (12.2 kbps).

	Bits per 20 ms
LP Filter Coefficients	38
Adaptive Excitation	30
Fixed or Algebraic Excitation (4 subframes/frame each with 35 bits)	$4 \times 35 = 140$
Gains (4 subframes/frame each with 9 bits)	$4 \times 9 = 36$
Total	244 or 12.2 kbps

Table 8.2 Comparison of MOS (5 is the best and 1 is the lowest).

Condition	Original	IS-641	LD-CELP	GSM EFR
Clean Speech	4.34	4.09	4.23	4.26
15 dB Babble	3.75	3.49	3.81	3.70
20 dB Car Noise	3.72	3.61	3.64	3.75
15 dB Office Noise	3.70	3.40	3.61	3.58
15 dB Music	3.99	3.82	3.98	3.99

Table 8.3 Comparisons of delay (ms).

Delay Cause	IS-641	GSM EFR	LD-CELP
Look-ahead	5	0	0
Frame Size	20	20	0.625
Processing	16	16	0.5
Bit Stream Buffer	0	0	19.375
Transmission	26.6	6.6	6.6
Total Delay	67.6	42.6	27.1

Results show an advantage for ITU LD-CELP and GSM EFR when compared to IS-641 for every condition. The disadvantage of the LD-CELP is its higher bit rate. This means fewer bits are available for error protection. In weaker channel conditions, GSM EFR will have an advantage over LD-CELP due to its lower bit rate, fewer sensitive bits to protect, and faster recovery from frame erasures. IS-641 also has the same advantages, and on an even weaker channel its performance would surpass that of GSM EFR.

ITU LD-CELP has a distinct advantage as far as total delay is concerned. The advantages of IS-641 over LD-CELP are in complexity and bit rate. LD-CELP is a low-delay coder and can produce better clear channel quality than IS-641 for a variety of conditions. GSM EFR has clear channel quality performance on a par with LD-CELP, has a small delay advantage compared to IS-641, and is a good candidate as an upgrade to IS-641 for strong channel conditions.

8.3.2   Adaptive Multiple Rate Codec

ETSI adaptive multiple rate (AMR) speech codec design (used for 3G UMTS) incorporates multiple codecs for use in full-rate or half-rate mode that are determined by channel quality. AMR operates in eight submodes that incorporate bit-exact versions of both 12.2 kbps US1/GSM EFR and 7.4 kbps IS-641 full-rate speech coders. The AMR can increase voice capacity by 150% compared to the GSM. Tables 8.4 and 8.5 provide details of AMR speech codecs.

The AMR codec allows dynamic management of voice quality and error control to provide good voice quality even under adverse radio conditions. AMR is not only used in GSM, but also in EDGE and UMTS networks [15]. It is designed to work with both GSM full-rate (one user per each of eight time slots in each radio channel) and GSM half-rate (two users per time slot). AMR defines multiple

Table 8.4 ETSI AMR speech coder full-rate (22.8 kbps) submodes.

Submode	Speech codec source rate (kbps)	Channel coding rate (kbps)
1	12.2 (US1/GSM EFR)	10.6
2	10.2	12.6
3	7.95	14.85
4	7.4 (IS-641 ACELP)	15.4
5	6.7	16.1
6	5.9	16.9
7	5.15	17.65
8	4.75	18.05

Table 8.5 ETSI AMR half-rate (11.4 kbps) submodes.

Submodes	Speech codec source rate (kbps)	Channel coding rate (kbps)
1	7.95	3.45
2	7.4 (IS-641 ACELP)	4.0
3	6.7	4.7
4	5.9	5.5
5	5.15	6.25
6	4.75	6.65

voice encoding rates, each with a different level of error control (see Table 8.4). The AMR codec dynamically responds to radio conditions by using the most effective mode of operation at each moment of time. Compared to the GSM EFR codec, AMR can operate under much worse radio conditions, such as with a heavily loaded network. AMR offers the following benefits:

- Greater spectral efficiency, hence higher capacity from higher frequency reuse with frequency hopping
- Better voice quality throughout the cell, particularly at cell edges and deep inside buildings, and increased overall coverage
- The potential of operating with toll-quality voice in half-rate mode, which reduces network costs

The 12.2 kbps rate of the AMR is the same as GSM EFR codec, and the 7.4 kbps rate is the same as the IS-136 TDMA codec. The gross bit rate of the channel (one time slot) is 22.8 kbps, which is divided into voice information and error control. As an example, 7.95 kbps mode means that more than half of the bit rate (i.e., $22.8 - 7.95 = 14.85$ kbps) can be allocated to channel coding. By reducing the AMR rate, resistance to errors increases further. *Figure 8.2* shows the resulting effect from voice quality (MOS) versus the *S/I* ratio.

Dynamic capability means AMR can compensate for the higher error rates arising from techniques such as tighter frequency reuse and higher fractional network loading, which inherently force the mobile to operate at lower *S/I*.

The six submodes of AMR half-rates are given in Table 8.5. Since the gross bit rate in a half-rate channel is only 11.4 kbps, a much smaller number of bits is available for channel coding, thus requiring a better *S/I*. But with a better radio signal, AMR can enable half-rate operation, which translates to more users in the same number of radio channels. AMR half-rate mode is further enhanced in EDGE radio networks where more bits per time slot are available.

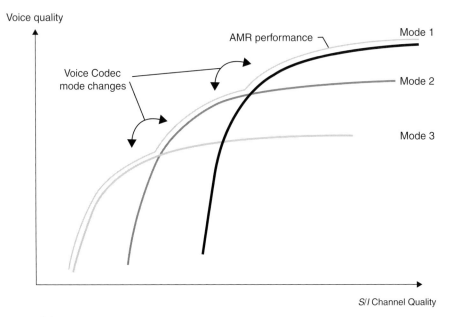

Figure 8.2(a)

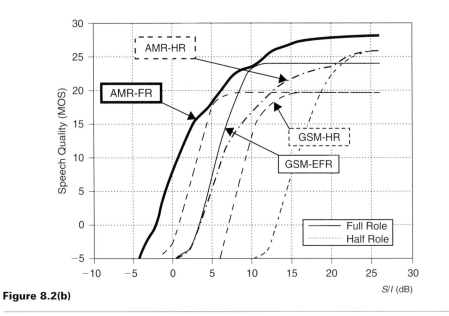

Figure 8.2(b)

Figure 8.2   MOS versus *S/I* for AMR codec.

The AMR codec operates on speech frames of 20 ms corresponding to 160 samples at the sampling frequency of 8000 samples per second. Depending on the air interface loading and quality of speech connections, the bit rate of the AMR speech connection can be controlled by the radio access network. During high loading, such as during busy hours, it is possible to use lower AMR bit rates to offer higher capacity while providing slightly lower speech quality. Also, if the mobile is running out of the cell coverage area and using maximum transmission power, a lower AMR bit rate can be used to extend the cell coverage area. With the AMR codec it is possible to achieve a trade-off between the network's capacity, coverage and speech quality according to the operator's requirements.

Example 8.1
The coverage gain in dB for the AMR is given as:

$$10\log\left[\frac{\text{DPDCH (kbps)} + \text{DPCCH}}{\text{DPDCH}[(\text{AMR bit rate (kbps)})] + \text{DPCCH}}\right] \text{dB}$$

Assuming the power difference between the dedicated physical control channel (DPCCH) and dedicated physical data channel (DPDCH) of the WCDMA to be -3.0 dB for 12.2 kbps AMR speech, calculate the gain in the link budget in dB by reducing the AMR bit rate from 12.2 to 7.95 kbps, and by reducing the AMR bit rate from 12.2 to 4.75 kbps.

Solution
(a)

$$\text{Coverage gain} = 10\log\left[\frac{12.2 + 12.2 \times 10^{-3/10}}{7.95 + 12.2 \times 10^{-3/10}}\right] = 10\log\left[\frac{18.315}{14.064}\right] = 1.15 \text{ dB}$$

(b)

$$\text{Coverage gain} = 10\log\left[\frac{12.2 + 12.2 \times 10^{-3/10}}{4.75 + 12.2 \times 10^{-3/10}}\right] = 10\log\left[\frac{18.315}{10.865}\right] = 2.27 \text{ dB}$$

8.4 CELP Speech Codec

Code excited linear predictive (CELP) codec dynamically selects one of the four data rates every 20 ms, depending on the speech activity. The four rates are 8 kbps (full-rate), 4 kbps (half-rate), 2 kbps (quarter-rate), and 1 kbps (eighth-rate). Typically, active speech is coded at the 8 kbps rate, while silence and background noise are coded at the lower rates. MOS testing has shown that QCELP provides speech quality equivalent to that of 8 kbps VSELP, while maintaining an

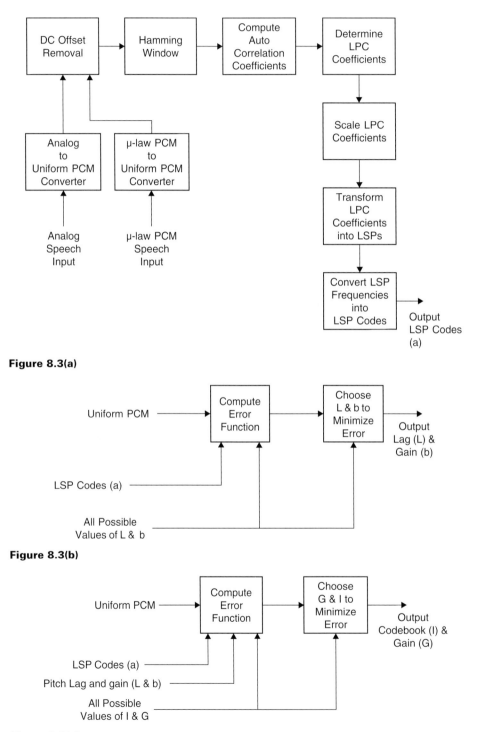

Figure 8.3(a)

Figure 8.3(b)

Figure 8.3(c)

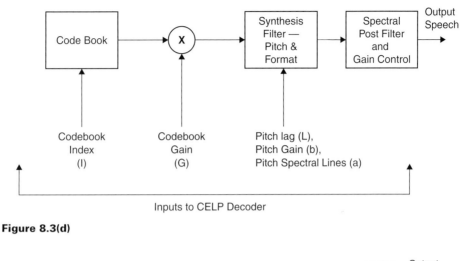

Figure 8.3(d)

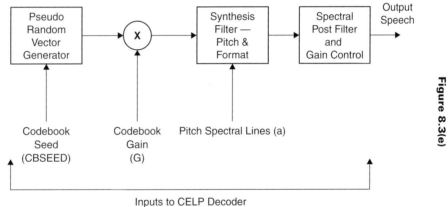

Figure 8.3(e)

Figure 8.3   Block diagram for CELP encoder/decoder.

Table 8.6 Bit allocation for CELP at 8 kbps (full rate).

LPC				40			
Pitch		10		10		10		10
Code-book Parameter	10	10	10	10	10	10	10	10

average data rate under 4 kbps in a typical conversation. Figure 8.3 shows block diagrams of the encoder/decoder. The bit allocation for each data rate is given in Tables 8.6, 8.7, 8.8, and 8.9, respectively. A 10th order LPC filter is used. Its coefficients are encoded using linear spectral pair (LSP) frequencies due to the good

Table 8.7 Bit allocation for CELP at 4 kbps (half rate).

LPC		20	
Pitch	10		10
Code-book Parameter	10	10	10	10

Table 8.8 Bit allocation for CELP at 2 kbps (1/4 rate).

LPC		10	
Pitch		10	
Code-book Parameter	10		10

Table 8.9 Bit allocation for CELP at 1 kbps (1/8 rate).

LPC	10
Pitch	0
Code-book Parameter	6

quantization, interpolation, and stability properties of LSPs. LSP is a representation of digital filter coefficients in pseudo-frequency domain.

8.5 Enhanced Variable Rate Codec

The CDMA Development Group's (CDG) enhanced variable rate codec (EVRC) takes advantage of signal processing hardware and software techniques to provide 13 kbps voice quality at 8 kbps data rate, thereby maximizing both quality and system capacity [13,14]. MOS data has shown that the 8 kbps EVRC (see Figure 8.4) compares favorably to 16 kbps LD-CELP and ADPCM, the industry standards for comparison. More important, the tests have shown that EVRC maintains superior quality over 13 kbps CELP as frame error rates (FERs) rise.

One of the important and unique aspects of CDMA wireless systems is that while there are limits to the number of mobile calls that can be handled by a given carrier at a time, this capacity limit is not a fixed number. In CDMA, cell coverage depends on the way the system is designed and implemented. System capacity, voice quality, and coverage are all interrelated, enabling a service provider to trade any one off against the other two.

To maximize the number of simultaneous calls that can be handled at any given time in the allocated frequency spectrum, digital wireless systems utilize

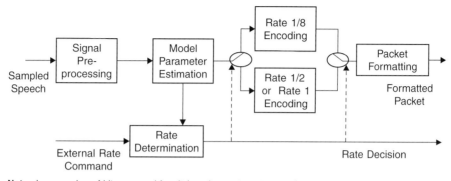

Note: Less number of bits are used for pitch and more to protect against errors

Figure 8.4 Enhanced variable rate codec (EVRC).

speech compression between the mobile and base stations. A lower speech transmission rate enables a higher number of simultaneous calls that a system can handle at any given time within the allocated carrier frequency spectrum. The variable rate vocoder in CDMA uses the smallest number of bits to represent each call without sacrificing voice quality.

The basis of the standard EVRC algorithm is the Relaxed Code-Excited Linear Predictive (RCELP) coding. RCELP coding is a generalization of the CELP speech coding algorithm. It is particularly well suited for variable rate operation and robustness in the CDMA environment. CELP uses 20 ms speech frames for coding and decoding. In each 20 ms time interval, the encoder processes 160 samples of speech. With variable rate coders, the encoder examines the contents of each speech frame to determine the necessary coding rate. Depending on the voice waveform (volume, pitch, rate, and so on), the coder represents speech at one of three bit rates: 8, 4, or 1 kbps. As a result, the average bit rate is less than 8 kbps. This differs from a half-rate or multirate coder, where the bit rate is determined once for each call. In addition, when no voice is detected, the vocoder drops its encoding rate and the effective bit rate goes to 1 kbps, further reducing the interference energy produced.

CELP codecs use three sets of bits to represent speech: linear predictor filter coefficients, pitch parameters, and excitation waveform. For each 20 ms speech frame, the CELP algorithm examines the data and generates 10 linear prediction coding filter coefficients. With EVRC, the coefficients are represented by a vector, which is a set of the most likely usable coefficients. Increased or decreased bit precision is applied as necessary.

CELP speech coders also perform long-term pitch analysis to generate a 7-bit pitch period and a 3-bit pitch gain. The analysis is based on a mathematical model of the human vocal track. At the different rates, the coder performs this pitch analysis on either four 5-ms subframes, two 10-ms subframes or one 20-ms frame. The result is a variable number of bits per frame representing the pitch information. EVRC makes two pitch measurements and uses wrapped pitch delays, removing the requirement for a large number of bits and increased computations for fractional pitch delays.

The excitation waveform for every frame or subframe is selected from a codebook consisting of a large number of candidate waveform vectors. The codebook vector chosen to excite the speech coder filters minimizes the weighted error between the original and synthesized speech. The technique used by EVRC to reduce the number of bits required for linear predictor coefficients and pitch synthesis enables the algebraic codebook to generate excitation. As a result, EVRC has higher voice quality.

Unlike conventional CELP encoders, EVRC does not attempt to match the original speech signal exactly. Instead, EVRC matches a time-wrapped version of the residual that conforms to a simplified pitch contour. The contour is obtained by estimating the pitch delay in each frame and linearly interpolating the pitch from frame to frame. While this adds to computational complexity, the result is higher voice quality per bit transmitted.

The simplified pitch representation also leaves more bits available in each packet for the stochastic excitation and channel impairment protection than would be possible if a traditional fractional pitch approach were used. The result is enhanced error performance without degraded speech quality at the small cost of added processing requirements.

EVRC also enhances call quality by suppressing background noise. The IS-127 [9] standard recommends a noise suppressor algorithm, but allows system designers to define their own. This is an important factor in choosing a processing platform, making programmable DSPs a desirable choice.

The EVRC algorithm offers a significant performance improvement over the IS-96A speech codec. Table 8.10 shows the performance of the CDMA Development Group's 13 kbps (CDG-13 kbps) speech codec along with IS-96A and EVRC codecs. The CDG-13 kbps offers high voice quality but results in a decrease in channel capacity of about 40% over the 8 kbps codec (the processing gain is reduced from 128 to 85.2, resulting in a 40% reduction in system capacity). Table 8.11 provides the bit allocations by packet type.

Table 8.10 Comparisons of CDG-13 kbps, IS-96A and EVRC vocoders in MOS.

Frame error rate (FER)%	CDG-13 kbps	IS-96A	EVRC
0	4.00	3.29	3.95
1	3.95	3.17	3.83
2	3.88	2.77	3.66
3	3.67	2.55	3.50

Table 8.11 Bit allocations by packet type in EVRC.

Field	Packet Type			
	Rate 1	Rate 1/2	Rate 1/8	Blank
Spectral Transition Indicator	1			
LSP	28	22	8	
Pitch Delay	7	7		
Delta Delay	5			
ACB Gain	9	9		
FCB Shape	105	30		
FCB Gain	15	12		
Frame Energy			8	
Unused	1			
Total Encoded Bits	171	80	16	
Mixed Mode Bit (MM)	1			
Frame Quality Indicator (CRC) (F)	12	8		
Encoder Tail Bits (T)	8	8	8	8
Total Bits	192	96	24	8
Rate (kb/s)	9.6	4.8	1.2	0.4

LSP (Line Spectral Pair) = A representation of digital filter coefficient in a pseudo frequency domain. This representation has good quantization and interpolation properties.
ACB = Adaptive Codebook
FCB = Fixed Codebook

8.6 Channel Coding

High-speed digital communication systems development demands the optimization of:

- Data transmission rate
- Data reliability
- Transmission energy
- Bandwidth
- System complexity
- Cost

Error correcting codes help to meet the requirements cost effectively with a reduced cost. A higher error correction code performance offers more flexibility to a designer in determining the required transmission energy, bandwidth, and system complexity. Signal-to-noise ratio (SNR) improvement of a communication channel depends on the error correction code used and the channel's characteristics.

Error control coding is the process of adding redundant information to a message to be transmitted that can then be used at the receiving end to detect and possibly corrects errors in the transmission. Since the redundant information adds overhead to the transmission, the type of coding must be chosen based upon how many errors the system is expected to see, and whether the capability to request retransmission of data is available. There are two basic error control coding classifications: *automatic repeat request* (ARQ) and *forward error correction* (FEC). ARQ is a detection-only type coding, in which transmission errors can be detected by the receiver but not corrected. The receiver must ask for any data received and request that detected errors be retransmitted. FEC allows not only detection of errors at the receiving end but correction of errors as well. We primarily focus on the FEC techniques and present the error correction codes (i.e., block, convolutional, and turbo) that are generally used for FEC.

Reed-Solomon (RS), Viterbi (V) Convolutional, and concatenated Reed-Solomon Viterbi (RSV) are the most common error correction codes implemented today. At a bit error rate (BER) of 10^{-6}, these codes are at least 2.5 to 3.0 dB short of the Shannon limit in an additive white Gaussian noise (AWGN) channel. Turbo codes have been shown to perform within 1 dB of the Shannon limit at a BER of 10^{-6}. Turbo codes break a complex decoding problem into simple steps, in which each step is repeated until a solution is reached.

8.6.1   Reed-Solomon (RS) Codes

RS coding is a type of FEC. It has been widely used because of its relatively large error correction capability when weighed against its minimal added overhead. RS codes are also easily scaled up or down in error correction capability to match the error rates expected in a given system. It provides a robust error control method for many common types of data transfer mediums, particularly those that are one-way or noisy and sure to produce errors.

In *block codes* a sequence of K information symbols is encoded in a block of N symbols, $N > K$, to be transmitted over the channel. For a data source that delivers the information bits at the rate B bps, every T seconds the encoder receives a sequence of $K = BT$ bits which defines a message. After K information bits have entered the encoder, the encoder generates a sequence of coded symbols of length N to be transmitted over the channel. In this transmitted sequence or codeword, N must be greater or equal to K in order to guarantee a unique relationship between each codeword and each of the possible 2^K messages. Such a code which maps a block of K information symbols into a block of N coded symbols is called an (N,K) block code. The *code rate* is $r = K/N$ bits/symbol, N is called the *block length*.

RS codes are an example of a block coding technique. The data stream to be transmitted is broken up into blocks and redundant data is then added to each

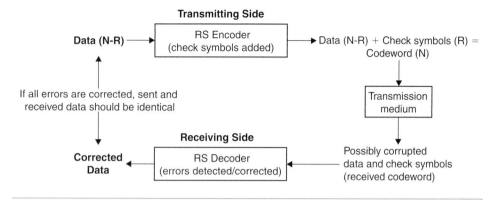

Figure 8.5 RS system level block diagram.

block. The size of these blocks and the amount of redundant data added to each block is either specified for a particular application or can be user-defined for a closed system. Within these blocks, the data is further subdivided into a number of symbols, which are generally from 6 to 10 bits in size. The redundant data then consists of additional symbols being added to the end of the transmission. The system-level block diagram for an RS codec is shown in Figure 8.5.

The original data, which is a block consisting of N-R symbols, is run through an RS encoder and R check symbols are added to form a code word of length N. Since RS can be done on any message length and can add any number of check symbols, a particular RS code is expressed as RS (N, N-R) code. N is the total number of symbols per code word; R is the number of check symbols per code word, and N-R is the number of actual information symbols per code word.

RS encoding consists of the generation of check symbols from the original data. The process is based upon finite field arithmetic. The variables to generate a particular RS code include field polynomial and generator polynomial starting roots. The field polynomial is used to determine the order of the elements in the finite field.

Another system-level characteristic of RS coding is whether the implementation is systematic or nonsystematic. A systematic implementation produces a code word that contains the unaltered original input data stream in the first R symbols of the code word. In contrast, in a nonsystematic implementation, the input data stream is altered during the encoding process. Most specifications require systematic coding.

The simplified schematic representation for a systematic RS encoder is shown in Figure 8.6. The input data stream is immediately clocked back out of the function into the check symbol generation circuitry. The fact that the input data stream is clocked out immediately without being altered means that the implementation is systematic. A series of finite fields adds and multiplies results in each register

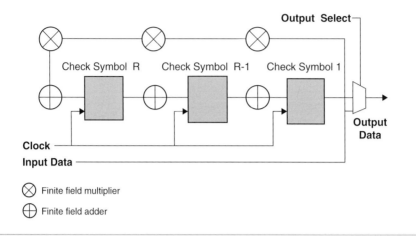

Finite field multiplier

Finite field adder

Figure 8.6 Systematic RS encoder schematic representation.

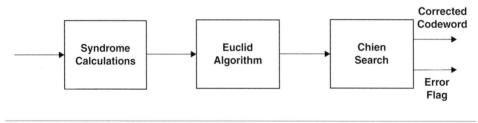

Figure 8.7 RS decoder block diagram.

containing one check symbol after the entire input stream has been entered. At that point, the output select is switched over to the check symbol registers, and the check symbols are shifted out at the end of the original message.

The size of the encoder is most heavily affected by the number of check symbols required for the target RS code. The total message length, as well as the field polynomial and first root value, do not have any appreciable effect on the device performance.

A typical RS decode algorithm consists of several major blocks. The first of these blocks is the syndrome calculation, where the incoming symbols are divided into the generator polynomial, which is known from the parameters of the decoder. The check symbols, which form the remainder in the encoder section, will cause the syndrome calculation to be zero in the case of no errors. If there are errors, the resulting polynomial is passed to the Euclid algorithm, where the factors of the remainder are found (see Figure 8.7). The result is evaluated for each of the incoming symbols over many iterations, and any errors are found and corrected. The corrected code word is the output from the decoder. If there are more errors in the code word than can be corrected by the RS code used, then the received code word is output with no changes and a flag is set, stating that the error correction has failed for that code word.

The error correction capability of a given RS code is a function of the number of check bits appended to the message. In general, it may be assumed that correcting an error requires one check symbol to find the location of the error, and a second check symbol to correct the error. In general then, a given RS code can correct R/2 symbol errors, where R is the number of check symbols in the given RS code. Since RS codes are generally described as an RS (N, N-R) value, the number of errors correctable by this code is [N- (N-R)]/2. This error control capability can be enhanced by use of erasures, a technique that helps to determine the location of an error without using one of the check symbols. An RS implementation supporting erasures would then be able to correct up to R errors.

Since RS codes work on symbols (most commonly equal to one 8-bit byte) as opposed to individual data bits, the number of correctable errors refers to symbol errors. This means that a symbol with all of the bits corrupted is no different than a symbol with only one of its bits corrupted, and error control capability refers to the number of corrupted symbols that can be corrected. RS codes are more suitable to correct consecutive bits. RS codes are generally combined with other coding methods such as Viterbi, which is more suited to correcting evenly distributed errors.

The effective throughput of an RS decoder is a combination of the number of clock cycles required to locate and correct errors after the code word has been received and the speed at which the design can be clocked. Knowing the latency and clock speed allows the user to determine how many symbols per second may be processed by the decoder. In the RS code, there are two RS decoder choices: a high-speed decoder and a low-speed decoder. The trade-off is that the low-speed decoder is usually approximately 20% smaller in device utilization. Note that both decoders operate at the same clock rate, but the low-speed decoder has a longer latency period, resulting in a slower effective symbol rate. As the number of check symbols decreases, the complexity of the decoder decreases, resulting in a smaller design and an increase in performance.

In a real-life RS coding implementation, functions that tend to reside on either side of the RS encoder or decoder are often implemented in programmable logic. One function that often resides after an RS encoder is an interleaver. The task of an interleaver is to scramble the symbols in several RS code words before transmission, effectively spreading any burst error that occurs during transmission over several code words. Spreading this burst error over several code words increases the chance of each code word being able to correct all of its induced errors.

8.6.2 Convolutional Code

Convolutional codes are suitable when the information symbols to be transmitted arrive serially in long sequences rather than in blocks. In convolutional codes, long sequences of information symbols are encoded continuously in serial form. To provide the extra bits required for error control, an output rate greater than the message bit is achieved by connecting two or more mod-2 adders to the shift registers (see Figure 8.8). Each message bit influences a span of $n(L + 1)$ successive output bits. The quantity $n(L + 1)$ is called the constraint length measured in terms of encoded output bits, where L is the encoder's memory measured in terms of input message. L represents the number of shift-registers in the encoder.

One at a time the symbols enter into the encoder, which has some finite memory capacity. The information symbols are sequentially shifted through an L-stage shift register, and following each shift some number, v, of coded symbols are generated and transmitted. These v coded symbols are obtained by parity checking, that is, by modulo-2 addition of the contents of various stages of the shift register according to the specific code. The length $K = L + 1$ is called the *constraint length* of the code, and the *code rate* is $r = 1/v$ bit per transmitted symbol.

A binary convolutional code of rate $1/v$ bits per symbol can be generated by a linear finite-state machine consisting of an L-stage shift register, v modulo-2 adders connected to some of the shift registers, and a commutator that scans the output of the modulo-2 adders. The whole system is called a *convolutional encoder* (see Figure 8.8).

The error-correcting capability of a convolutional coding scheme increases as rate r decreases. However, the channel bandwidth and decoder complexity both increase with K. The advantage of lower code rates when using a convolutional code with coherent phase-shift-keying (PSK), is that the required E_b/N_0 is

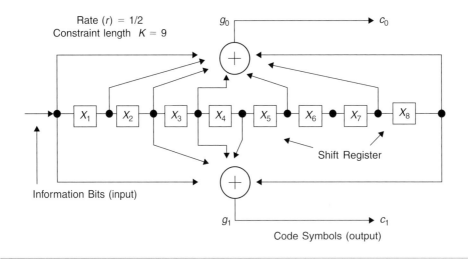

Figure 8.8   Convolutional encoder.

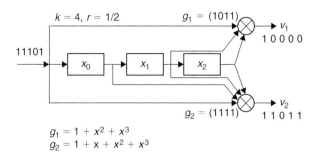

$$g_1 = 1 + x^2 + x^3$$
$$g_2 = 1 + x + x^2 + x^3$$

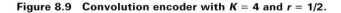

Figure 8.9 Convolution encoder with K = 4 and r = 1/2.

decreased, permitting the transmission of higher data rates for a given amount of power, or permitting reduced power for a given data rate. Simulation studies have indicated that for a fixed constraint length, a decrease in code rate from 1/2 to 1/3 results in reduction of the required E_b/N_0 of about 0.4 dB. However, the corresponding increase in decoder complexity is about 17%. For smaller values of code rate, the improvement in performance relative to the increased decoding complexity diminishes rapidly.

The major drawback of the Viterbi algorithm is that while error probability decreases exponentially with constraint length, the number of code states, and consequently decoder complexity, grows exponentially with constraint length.

Example 8.2

Figure 8.9 shows a convolution encoder with K = 4 and r = 1/2. The generator polynomials g_1 and g_2 are $1 + x^2 + x^3$ and $1 + x + x^2 + x^3$, respectively. If the input is 1 1 1 0 1 (first bit), calculate the output of the encoder.

Solution

Refer to Table 8.12.

Figures 8.10 and 8.11 indicate the implementations of convolutional coding in GSM for downlink and uplink. Figure 8.10 shows the downlink channel coding that is used with the GSM EFR speech coder and 8-PSK modulation. The bits are combined such that every two class IA bits are used with one class II bit to form the first 89 triads. The remaining 44 triads are formed by using three class IB bits.

For the uplink (Figure 8.11) the bits are combined in a somewhat similar manner. The 86 triads consist of two class IA bits and one class II bit, three triads contain two class IB bits along with one class II bit, and 35 triads consist of three class IB bits.

Table 8.12 Input /output state table of the encoder.

Input after each shift*	X0 (=x)	X1(=x²)*	X2(=x³)*	V1	V2
1	0	0	0	1	1
0	1	0	0	0	1
1	0	1	0	0	0
1	1	0	1	0	1
1	1	1	0	0	1

The bits in columns marked with an * are added using modulo-2 to produce V1 after each shift. The output of the encoder is 1 0 1 0 0 0 1 0 1 1.

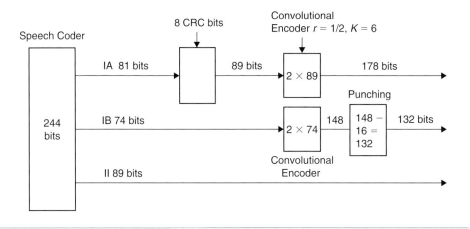

Figure 8.10   Channel coding (downlink) with GSM EFR speech coder.

The triads are reordered to provide additional interleaving gain, and then intra-slot interleaving is applied. There are three primary interleaving options: one-slot, two-slot, and three-slot. In the one-slot interleaving, there is no intra-slot interleaving and the current triads are simply transmitted in the current slot.

In the two-slot interleaving, certain triads (from the current triad vector) are transmitted in the current slot, and the remaining triads in the next slot. Thus, the current slot contains triads from the current and previous triad vector.

In the three-slot interleaving, the concept is extended to include another set of triads in each transmitted slot. Certain triads from the current triad vector are transmitted in the next slot, and the remaining triads are transmitted in the slot after the next slot. Thus, the current slot contains triads from the current triad vector, the previous triad vector, and one before the previous triad vector.

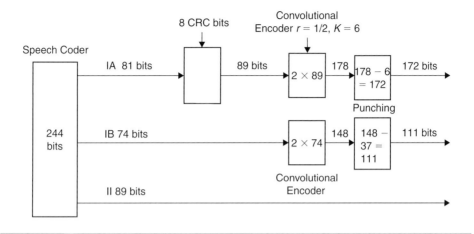

Figure 8.11   Channel coding (uplink) with GSM EFR speech coder.

8.6.3   Turbo Coding

A turbo decoder consists of two concatenated decoders, each providing soft information (see the next section) and so-called intrinsic information. Two main classes of algorithms available for turbo codes are *soft-output Viterbi algorithm* (SOVA) and iterative soft-input and soft-output decoding algorithms such as the symbol-by-symbol *maximum a posterior* algorithm, which is more complex but yields better performance. Since the capacity of a CDMA system is strongly dependent on the required E_b/N_0 value of the receiver, any improvement in the E_b/N_0 value due to coding translates directly into capacity increase.

The inner receiver's (in serial arrangement) task is to provide the best estimate for the outer receiver. It deals with the following signal impairments:

- The presence of multipath components
- The presence of multi-user interference (both inter- and intracell)
- The fading of each transmission path
- The near-far effect due to the relative position of all mobiles and base station
- The time-varying nature of these impairments

These issues are tackled with a combination of the following techniques:

- Channel estimation and tracking
- A maximum ratio combination-based (coherent) Rake receiver, to take advantage of multipath characteristics
- Multi-user detection (MUD) schemes, such as interference cancellation (IC) or decorrelating receivers

- Fast power control based on SNR estimation
- Antenna arrays (in the base station) to provide another form of diversity (space diversity)

Turbo codes may use serial (concatenated) and/or parallel recursive convolution codes. Recursive means the output not only depends on the input sequence but also depends on the previous output. A turbo encoder uses two convolutional encoders separated by an N-bit random interleaver or permuter, together with an optional puncturing mechanism. The interleaver is used to spread coded symbols from the other encoder or from the input. The input to an encoder can be from an input sequence and/or from a coded output from another encoder. Instead of cascading the encoders in the usual serial fashion, they can be arranged in parallel concatenation (see Figure 8.12).

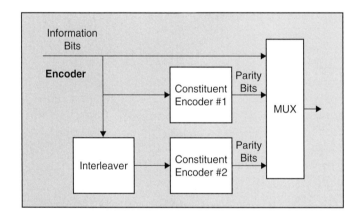

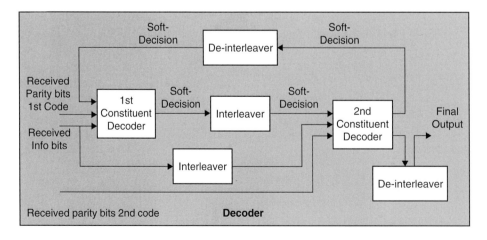

Figure 8.12   Turbo encoding and decoding.

The function of the permuter is to take each incoming block of N data bits and rearrange them in a pseudo-random fashion prior to encoding by the second encoder. Unlike the classical interleaver (e.g., block or convolution interleaver), which arranges the bits in some systematic manner, it is important that the permuter sort the bits in a manner that lacks any apparent order. It is also important that N be selected quite large (N > 1000). The role of the turbo code puncture is identical to that of its convolutional code counterpart, to periodically delete selected bits to reduce coding overhead.

Each decoding process yields a soft output decision for the next decoder. The key component is the soft input, soft output (SISO) decoder. To achieve the benefits of turbo code, several iterations are provided. Thus, the turbo codes are processing intensive and are applied to less delay-sensitive applications such as data.

Turbo codes have been implemented in both software and hardware as a single integrated circuit. Turbo codes don't suffer from the error at low BERs that have been attributed to other codes. Turbo codes are capable of providing high coding gain, even with high code rates.

Turbo codes provide about 1.5 to 3 dB bit energy-to-noise improvement over the current R-S and RSV error correction codes. A 1.5 to 3.0 dB performance gain allows the system developer to reduce transmitter power and/or bandwidth and still maintain the same BER performance. Conversely, a 1.5 to 3 dB improvement can be used to improve overall system performance if transmitter power is not reduced. The use of turbo codes can allow the system engineer to reduce antenna size, lowering system cost. When used to provide improved BER performance, turbo codes can also lead to a clearer image transmission, improved audio, greater range, or better data integrity. A turbo code is applicable where digital data is transmitted over a noisy channel because it spreads error uniformly over the interleaved duration. Turbo codes are better than convolutional codes due to their strength in nonlinear coding/decoding and feedback property. Turbo codes can support the data rates achieved with other error correction codes while offering improved correction capability.

Figure 8.13 provides a comparison of a turbo code having constraint length $K = 4$ decoded with 4 iterations with a convolutional code of constraint length $K = 9$ decoded with a Viterbi decoder. Figure 8.13 shows the performance of a rate 1/3 turbo code compared to the corresponding rate 1/3 convolutional code. The results indicate that the turbo code out-performs the corresponding convolutional code with the same decoding complexity for data rates larger than 9.6 kbps; the improvement in performance increases with the data rate for a fixed frame length of 20 ms. The performance improvement is due to the fact that the number of bits in the 20 ms frames increases with the data rate and the performance of a turbo code improves with the number of bits in the frame. With a large number of bits in a frame, the interleaver separating the codes can randomize the errors more effectively.

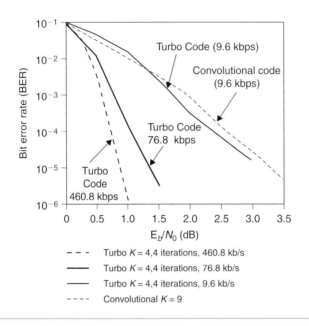

Figure 8.13   Performance of turbo code compared with convolutional code.

8.6.4   Soft and Hard Decision Decoding

The decoding of a block code can be performed with hard or soft decision input, and the decoder may output hard or soft decision data. In hard decision decoding, each received channel bit is assigned a value of 1 or 0 at the demodulator depending whether the received noisy data is higher or lower than a threshold. The decoder then uses the redundancy added by the encoder to determine if there are errors, and, if possible, to correct the errors. The desired output of the decoder is a corrected code word.

A soft decision decoder receives not only the binary value of 1 or 0, but also a confidence value associated with the given bit. If the demodulator is certain that the bit is a 1, it places very high confidence on it. If it is less certain, it places a lower confidence value. A soft input decoder can output either hard decision data or soft decision data. For example, a Viterbi decoder receives soft information from the demodulator and outputs hard decision data. The decoder can use the soft information to determine if a given bit is a solid 1 or solid 0, and it outputs this hard decision.

A SISO decoder both receives soft decision data and generates soft decision output. For each bit in the code word, the SISO decoder examines the confidence of the other bits in the code word, and, using the redundancy of the code, generates an updated soft output for the given bit.

RS error correction codes are the "standard" algorithm for FEC. RS codes are block codes that are very efficient for error correction implemented either in hardware or software. RS codes are hard decision codes. RS codes followed with concatenated Viterbi codes (RSV) offer an improvement over the stand-alone RS codes in terms of BER performance.

The concept of SISO decoders has been applied to turbo codes. A turbo code feeds demodulated soft decision data into a SISO decoder. The output of this decoder is then fed into the same (or a different) SISO decoder. The output of this decoder is then fed again. This iterative process continues until a confident solution is reached. The concept of feeding the output back into the input is similar to a turbo charger of an engine, therefore the name turbo code is used.

For a turbo code to be effective, the given data must be encoded with two (or more) different codes. Then, when decoding, each of the codes will modify the confidence of each bit. With each iteration, all of the codes modify the confidence of the data, so each code sees slightly different data for each iteration. Each code pushes the confidence of a given bit higher or lower, and consequently changes the hard decision value of the bits in error. Eventually, the data will settle on an arrangement in which all codes are pushing the confidence of all bits higher. The hard decision values at this point are closer to the transmitted data.

8.6.5 Bit-Interleaving and De-Interleaving

Transmission errors occur randomly in bursts. Burst errors happen when the signal undergoes deep fades. The ability of FEC is limited in correcting a long string of errors, bit-interleaving is used between the encoder and modulator. At the receiver, the de-interleaver is employed between the demodulator and FEC decoder.

The convolutional codecs perform poorly on bursts of errors. Bit-interleaving is used to randomize the errors so that the convolutional codec can correct them. The purpose of bit-interleaving is to avoid loss of consecutive information bits. Interleaving is performed by storing the data in a table containing rows and columns at the transmitter. The data is written in rows and transmitted in a vertical direction (according to columns). At the receiver, the data is written and read in the opposite manner.

GSM blocks of full-rate speech are interleaved over 8 bursts; i.e., 456 bits of one block are divided into 8 sub-blocks, each containing 57 bits. Each sub-block is carried by a different burst and in a different TDMA frame. Thus a burst contains contributions from two successive speech blocks, j and $j + 1$. To avoid proximity relations between successive bits, bits from block j use even positions in the burst and bits from block $j + 1$ use odd positions. This ensures sufficient redundancy within the interleaving process of the GSM signal structures to allow for one frame in the five to be lost without significant loss in voice quality.

De-interleaving in the receiver is the reverse operation of interleaving. The major drawback of de-interleaving and interleaving is that they introduce delay.

The delay amounts to the transmission time from the first burst to the last burst in the block and is equal to 8 TDMA frames, i.e., about 37 ms.

Example 8.3

Using a bit string of 0 0 0 0 0 1 1 1 0 0 0 1 0 0 0 1 (first bit) demonstrate operations of 4 × 4 bit interleaver/de-interleaver in converting a burst error into bit errors.

Solution

Input to the interleaver:

$$0\,0\,0\,0\,0\,1\,1\,1\,0\,0\,0\,1\,0\,0\,0\,1 \to \begin{bmatrix} 1 & 0 & 0 & 0 \\ 1 & 0 & 0 & 0 \\ 1 & 1 & 1 & 0 \\ 0 & 0 & 0 & 0 \end{bmatrix} \text{write the data in rows}$$

Output of the interleaver: 0 0 0 0 0 1 0 0 0 1 0 0 0 1 1 1

Input to the de-interleaver: $0\,0\,0\,0\,0\,1\,0\,0\,\underline{\textit{0}}\,\underline{\textit{1}}\,0\,0\,0\,1\,1\,1 \to \begin{bmatrix} 1 & 1 & 1 & 0 \\ 0 & \underline{\textit{0}} & \underline{\textit{1}} & \underline{\textit{0}} \\ \underline{\textit{0}} & 0 & 1 & 0 \\ 0 & 0 & 0 & 0 \end{bmatrix}$

Burst error is indicated by italic bold-face (under-score).

The output of the de-interleaver: $0\,0\,\underline{\textit{0}}\,0\,0\,1\,\underline{\textit{1}}\,1\,0\,0\,\underline{\textit{0}}\,1\,0\,\underline{\textit{0}}\,0\,1$

8.7   Summary

In this chapter, we presented speech codec attributes and briefly discussed different coding schemes. We provided details of the enhanced variable rate codec (EVRC) which takes advantage of a signal processing hardware and software technique to provide 13 kbps voice quality at 8 kbps data rate by maximizing both quality and system capacity. MOS data indicates that 8 kbps EVRC compares favorably to 16 kbps LD-CELP and ADPCM, the industry standards for comparison. The tests also show that EVRC maintains superior quality over 13 kbps CELP as frame error rates increase.

We focused on three types of channel coding schemes: RS codes, convolutional codes, and turbo codes. The error correcting capability of a convolutional coding scheme increases as rate decreases and constraint length increases. However, the number of code states and, consequently, decoder complexity, grows with the constraint length. Turbo codes use serial and/or parallel recursive convolutional codes. They appear to approach the Shannon limit with large enough iterations of decoding. Turbo codes spread error uniformly over the interleaved duration and, therefore, they are very useful for fading channels. Turbo codes provide an additional coding gain of 1 to 2 dB over convolutional codes.

Problems

8.1   Define speech coding methods used in digital communication.

8.2   What are the main features of a hybrid coding scheme?

8.3   List the main attributes of a speech codec.

8.4   What is a CELP codec?

8.5   List steps used in speech processing in a vocoder.

8.6   What is an AMR codec?

8.7   How is capacity improvement in a GSM system achieved with the AMR codec compared to the FER codec?

8.8   What are the information rates in EVRC codec for rate set 1 in the CDMA IS-95 B?

8.9   Discuss various forward error correcting schemes used in digital communication. Why are convolutional schemes used in wireless communications?

8.10   Calculate the output of the convolution encoder shown in Figure 8.8, if the input is 1 0 1 0 0 1 1 (first bit).

8.11   Why is bit-interleaving used in wireless communications?

8.12   Using a bit string of 1 0 1 1 0 0 0 0 0 1 1 1 0 0 0 1 0 0 0 1 0 1 0 1 0 demonstrate operations of 5 × 5 bit interleaver/de-interleaver in converting burst error into bit errors.

References

1. Atal, B. S. Predictive Coding of Speech Signals at Low Bit Rates. *IEEE Trans., Comm.* vol. 30, no. 4, 1982, pp. 600–614.

2. Atal, B. S., and Schroeder, M. R. Stochastic Coding of Speech at Very Low Bit Rate. *Proc. International Conf. Comm.* Amsterdam, 1984, pp. 1610–1613.

3. Bahl, L. R., Cocke, J., Jelinek, F., and Raviv, J. Optimal Decoding of Linear Codes for Minimizing Symbol Error Rate. *IEEE Trans. Info. Theory.* vol. IT-20, no. 2.

4. Chen, J., Cox, R., Lin, Y., Jayant, N., and Melchner, M. Coder for the CCITT 16 kb/s Speech Coder Standard. *IEEE Journal of Selected Areas of Communications* 6, 1988, pp. 353–363.

5. Furuskar, A., et al. System Performance of EDGE, a Proposal for Enhanced Data Rates in Existing Digital Cellular System. *IEEE VTC 98*, pp. 1284–1289.

6. Garg, V. K., and Wilkes, J. E. *Principles & Applications of GSM*. Upper Saddle River, NJ: Prentice Hall, 1999.

7. Hess, W. *Pitch Determination of Speech Signals*. Berlin: Springer Verlag, 1983.

8. Jarvinen, K., et al. GSM Enhanced Full Rate Speech Codec. *IEEE GLOBECOM '97*.

9. Kleijn, W. B., Ramachandran, R. P., and Kroon, P. Generalized Analysis-by-Synthesis Coding and Its Application to Pitch Prediction. *International Conf., Acoust. Speech Signal Processing*. San Francisco, 1992, pp. 1337–1340.

10. Kroon, P., and Deprettere, E. F. A Class of Analysis-by-Synthesis Prediction Coders for High Quality Speech Coding at Rates Between 4.8 and 16 kbps. *IEEE Journal of Selected Areas of Communications* 6, 1988, pp. 353–363.

11. TIA IS-96A, "Speech Service Option Standard for Wideband Spread Spectrum Digital Cellular System."

12. TIA IS-733, "13 kbps Speech Coder."

13. TIA IS-127, "Enhanced Variable Rate Codec (EVRC) 8.5 kbps Speech Coder."

14. TR45.5/98.04.03.03. "The cdma2000 ITU-R RTT Candidate Submission." April, 1998.

15. Van Nobelen, R., et al. An Adaptive Radio Link Protocol with Enhanced Data Rates for GSM Evolution. *IEEE Personal Communications*. vol. 6, no. 1, February. 1999, pp. 54–64.

CHAPTER 9

Modulation Schemes

9.1 Introduction

The digital signals that are developed in the process of transmitting voice, data, video, and signaling information are generated at low data rates. These data rates, typically 1 to 50 kbps, are so low in frequency that their transmissions directly from the transmitter to the receiver would require antennas that are thousands of meters long. Furthermore, the signals from one transmitter would interfere with signals from other transmitters if they all use the same frequency band. Therefore, baseband signals (see Chapter 4) are modulated onto a radio frequency carrier for transmission from the transmitter to the receiver [2,3]. The radio environment at $800-2000\,\text{MHz}$ (used for mobile communications) is hostile. We must therefore select modulation schemes that are robust. In addition to the modulation schemes we must also choose encoding algorithms that improve the performance of the system [9,10].

We introduced modulation in Chapter 4. In this chapter we focus primarily on three modulation schemes, minimum shift keying (MSK), Gaussian minimum shift keying (GMSK), and π/4-Differential Quadrature Phase Shift Keying (π/4-DQPSK) that have been used in different cellular systems [8]. GMSK is the modulation scheme used in GSM900, GSM1800, GSM1900, and digital enhanced cordless telecommunications (DECT). MSK was introduced as a first step toward GMSK. Personal wireless telecommunications (PWT) and PWT-E (enhanced), the variations of DECT for the licensed and unlicensed 1900-MHz band, respectively, in North America, use π/4-DQPSK. Since each of these schemes is descended from phase shift keying (PSK), we first discuss PSK and then show its relationship to others.

We also examine quadrature amplitude modulation (QAM) and M-*ary* frequency shift keying (MFSK) schemes [11,13].

9.2 Introduction to Modulation

When we want to send signals over any distance, baseband signaling is not sufficient [6]. We must therefore modulate the signals onto an RF carrier. When we transmit the digital bit stream, we convert the bit stream into the analog signal, $a(t)\cos(\omega t + \theta)$. This signal has amplitude $a(t)$, frequency $\omega/(2\pi)$, and phase θ. We can change any of these three characteristics to formulate the modulation scheme. The basic forms of the three modulation methods used for transmitting digital signals are [11]:

- Amplitude shift keying (ASK)
- Frequency shift keying (FSK)
- Phase shift keying (PSK)

When ω and θ remain unchanged, we have ASK. In this case, the transmitted carrier wave takes two amplitude values during the duration of the pulse. When $a(t)$ and θ remain unchanged, we have binary (or M-*ary*) FSK. In this case the carrier frequency is shifted up or down by a fixed value corresponding to a binary 1 or 0. When $a(t)$ and ω remain unchanged, we have binary (or M-*ary*) PSK. In binary PSK, the information is contained in the phase. During the transmission of a binary 1 the carrier phase is zero, and the carrier phase is changed to a value of π during the transmission of a binary 0. Hybrid schemes exist where two characteristics are changed with each symbol transmitted. The most common method is to fix ω and change $a(t)$ and θ. This method is known as quadrature amplitude modulation (QAM). Each modulation method results in a different transmitter and receiver design, different occupied bandwidth, and different error rate [4]. Since all signals have a theoretical bandwidth that is infinite, all modulation schemes must be band limited. Band limiting introduces detection errors, and the filter bandwidths must be selected to achieve an optimization between bandwidth and error rate [12].

The baseband outputs of the data transmitters are a series of binary data that cannot be sent directly over a radio link. The communications designer must choose radio signals that represent the binary data and allow the receiver to decode the data with minimal errors. For the simplest binary signaling system, we choose two signals denoted by $s_0(t)$ and $s_1(t)$ to represent the binary values of 0 and 1, respectively. Since no channel is perfect, the receiver will also have additive white Gaussian noise (AWGN), $n(t)$. The data receiver (Figure 9.1) will then process the signal and noise through a filter, $h(t)$, and, at the end of the signaling interval, T, make a determination of whether the transmitter sent a 0 or a 1.

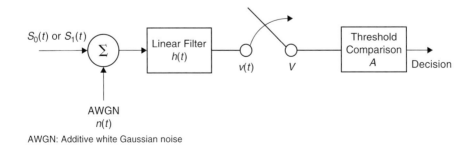

Figure 9.1 Receiver structure to detect binary signal with AWGN.

The energies of $s_0(t)$ and $s_1(t)$ in a T interval are assumed to be finite and denoted as E_0 and E_1, respectively. For simplicity we assume that the noise has a probability density function of amplitude that is Gaussian and that the noise spectral density is flat with frequency with the double-sided Power Spectral Density of $N_0/2$, where N_0 is the noise density.

When $s_0(t)$ is present at the filter as an input, its output at time $t = T$ is

$$V = S_0 + N, \text{ with } s_0(t) \text{ present} \tag{9.1}$$

where:
    S_0 = output signal component at $t = T$ for the input $s_0(t)$, and
    N = the output noise component

Similarly, when $s_1(t)$ is present at the filter as an input, its output at $t = T$ is

$$V = S_1 + N, \text{ with } s_1(t) \text{ present} \tag{9.2}$$

Since noise is Gaussian with zero mean (implied by its constant PSD), N is also Gaussian. The variance of the noise, σ^2, can be determined as:

$$\sigma^2 = \int_{-\infty}^{\infty} \left(|H(f)|^2 \cdot \frac{N_0}{2} \right) df \tag{9.3}$$

$$\sigma^2 = N_0 \int_0^{\infty} |H(f)|^2 \, df \tag{9.4a}$$

$$\sigma^2 = N_0 \cdot B_N \tag{9.4b}$$

where:
    $H(f)$ is the transfer function of filter, and

$B_N = \int_0^{\infty} |H(f)|^2 \, df$ is the noise-equivalent bandwidth or simply the noise bandwidth

of the receiver filter function $H(f)$

Given that $s_0(t)$ is present at the receiver input, the probability density function (PDF) of v is

$$p\left(v|s_0\right) = \frac{1}{\sqrt{2\pi} \cdot \sigma} e^{-\left[\frac{(v - S_0)^2}{2\sigma^2} \right]} \tag{9.5}$$

Similarly, when $s_1(t)$ is present at the receiver input, the conditional probability function of v is

$$p\left(v|s_0\right) = \frac{1}{\sqrt{2\pi} \cdot \sigma} e^{-\left[\frac{(v - S_1)^2}{2\sigma^2} \right]} \tag{9.6}$$

From Figure 9.2, given that $s_0(t)$ is present, the probability of error is

$$P\left(e|s_0\right) = \int_A^\infty p\left(v|s_0\right)dv \tag{9.7}$$

If $s_1(t)$ is present, it is

$$P\left(e|s_1\right) = \int_{-\infty}^A p\left(v|s_1\right)dv \tag{9.8}$$

If the a priori probability that $s_0(t)$ was sent is p, and the a priori probability that $s_1(t)$ was sent is $q = 1 - p$, then the average probability of error is

$$P_e = pP\left(e|s_0\right) + qP\left(e|s_1\right) \tag{9.9}$$

$$p_e = p\int_A^\infty \frac{1}{\sqrt{2\pi}\cdot\sigma}e^{-\left[\frac{(v-S_0)^2}{2\sigma^2}\right]}dv + q\int_{-\infty}^A \frac{1}{\sqrt{2\pi}\cdot\sigma}e^{-\left[\frac{(v-S_1)^2}{2\sigma^2}\right]}dv \tag{9.10}$$

If we simplify Equation 9.10, differentiate the result with respect to A, and then set the derivative equal to zero, we can determine the optimum choice for the threshold value, A, to minimize the probability of error P_e [13]:

$$A = A_{\text{opt}} = \frac{\sigma^2}{S_1 - S_0}\ln\frac{p}{q} + \frac{S_0 + S_1}{2} \tag{9.11}$$

In most systems, the values of 0 and 1 are equally likely; if they are not, then the designer usually redesigns the encoding method to ensure that they are equally likely. Thus, $p = q$, and

$$A = A_{\text{opt}} = \frac{S_0 + S_1}{2} \tag{9.12}$$

For the optimum value of A, with $p = q$, the probability of error is

$$P_e = \frac{1}{2}\text{erfc}\left[\frac{S_1 - S_0}{2\sqrt{2}\cdot\sigma}\right] = Q\left[\frac{S_1 - S_0}{2\sigma}\right] \tag{9.13}$$

where erfc is the complementary error function $= 1 - \text{erf}(u) = 2Q(\sqrt{2u})$. The error function, $\text{erf}(u)$ is defined as

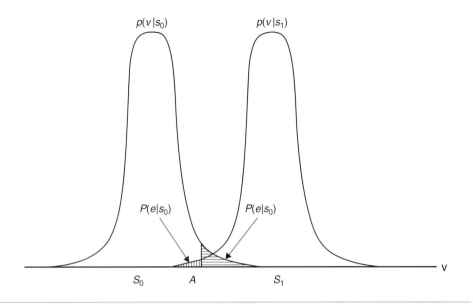

Figure 9.2   Conditional probability density function of the filter output at time _t_ = _T_.

$$\text{erf}(u) = \frac{2}{\sqrt{\pi}} \int_0^u e^{-t^2} dt \tag{9.14}$$

and

$$Q(u) = \frac{1}{\sqrt{2\pi}} \int_u^\infty e^{-\frac{x^2}{2}} dx \tag{9.15}$$

or

$$Q(u) = \frac{e^{-\frac{u^2}{2}}}{\sqrt{2\pi} \cdot u}, \, u \gg 1 \text{ Gaussian integral} \tag{9.16}$$

Next we want to find the filter that provides the minimum probability of error, as expressed by Equation 9.13. At time t_0, the sample value consists of a signal-related component $g_0(t_0)$ and noise component $n_0(t_0)$. This filter is known as a *matched filter* and has a transfer function $H_0(f)$ optimized to provide the maximum signal-to-noise ratio (SNR) at its output at time t_0. Schwartz shows that this filter must be the conjugate match of the signal, $s(t)$. Since we have two signals, $s_0(t)$ and $s_1(t)$, we will need two filters in our receiver design.

If we transmit a signal, $s(t)$, then it has the Fourier transform, $S(\omega)$, which is a complex function. The optimum or matched filter must have a frequency response, $H(\omega)$, where

$$H(\omega) = S^*(\omega)e^{-j\omega t_0} \tag{9.17}$$

where:

S^* is the complex conjugate of the Fourier transform of the signal

In general this filter is not realizable, since analysis will show that it must have output before there is input. However, we can design filters that approximate the ideal filter.

The SNR is defined as

$$\xi^2 = \frac{g^2_0(t_0)}{\sigma^2} \tag{9.18}$$

It can also be shown that the maximum value for SNR ξ^2 is twice the energy of the input signal (E_g) divided by the single-sided input noise spectral density, regardless of the input signal shape

$$\xi^2_{max} = \frac{Eg}{(N_0)/2} \tag{9.19}$$

For a binary system, Equation 9.19 becomes

$$\xi^2_{max} = \frac{1}{N_0} \int_0^T [s_1(t) - s_0(t)]^2 \, dt \tag{9.20}$$

Since the signals are zero outside the range (0, T), the probability of error corresponding to the optimum receiver filter becomes

$$P_e = \frac{1}{2}\mathrm{erfc}(\sqrt{z}) = Q(\sqrt{2z}) \tag{9.21}$$

where:

$$z = \frac{1}{4N_0} \int_0^T [s_1(t) - s_0(t)]^2 dt$$

If the transmitted pulses are allowed to take on any of M transmitted levels with equal probability, then the information rate per transmitted pulse is $\log_2 M$ bits. For a constant information rate, the bandwidth of the transmitted signal can be reduced by the same factor. With M-*ary* transmission, we will show that the error rates are higher, but if we have sufficient SNR then higher error rates will not matter. Thus, we are using excess SNR to code the signal and reduce its bandwidth.

When we add additional levels to a baseband system, we are reducing the distance between detection levels in the receiver output. Thus, the error rate of a multilevel baseband system can be determined by calculating the appropriate reduction in error distance. If the maximum amplitude is A, the error distance d_e between equally spaced levels at the detector is

$$d_e = \frac{A}{M - 1} \tag{9.22}$$

where:
   M = number of levels

Setting the error distance A of a binary system to that defined in Equation 9.22 provides the error probability of a multilevel system.

$$P_e = \frac{1}{\log_2 M}\left[\frac{M - 1}{M}\right]\text{erfc}\left[\frac{A}{(M - 1)(\sqrt{2} \cdot \sigma)}\right] \tag{9.23}$$

where:
   The factor $(M - 1)/M$ reflects that the interior signal levels are vulnerable to both positive and negative noise.

   The factor $1/\log_2 M$ arises because the multilevel system is assumed to be coded, so symbol errors produce single bit errors ($\log_2 M$ is the number of bits per symbol), and the probability of multiple bit error is assumed to be small and can be neglected.

   Equation 9.23 relates error probability to peak signal power A^2. To determine P_e with respect to average power, the average power of an M-level system is determined by averaging the power associated with various pulse amplitude levels.

$$[A^2]_{\text{avg}} = \frac{2}{M}\left[\left(\frac{A}{M - 1}\right)^2 + \left(\frac{3A}{M - 1}\right)^2 + \ldots + A^2\right] \tag{9.24}$$

$$[A^2]_{\text{avg}} = \frac{2A^2}{M(M - 1)^2}\sum_{j=1}^{M/2}(2j - 1)^2 \tag{9.25}$$

where:
   the levels $\frac{A}{M - 1}\left[\pm 1, \pm 3, \pm 5, \ldots, \pm(M - 1)\right]$ are assumed to be equally likely.

   If T is the signaling interval for a two-level system, the signaling interval T_M for an M-level system providing the same data rate is determined as

$$T_M = T\log_2 M \tag{9.26}$$

For a raised cosine filter, the noise bandwidth is $B_N = \dfrac{1}{2T_M}$.
From Equation 9.4, we get

$$\sigma^2 = \frac{N_0}{2T_M} \tag{9.27}$$

$$\sigma = \frac{1}{\sqrt{2}} \left[\frac{N_0}{T_M} \right]^{1/2} \tag{9.28}$$

Substituting Equation 9.28 into Equation 9.23, we get

$$P_e = \left[\frac{1}{\log_2 M} \right] \left[\frac{M-1}{M} \right] \text{erfc} \left[\frac{A}{(M-1)\left(\dfrac{N_0}{T_M}\right)^{1/2}} \right] \tag{9.29}$$

The energy per symbol

$$E_s = E_b \log_2 M = A^2 T_M$$

where:
E_b is the bit energy

$$\therefore A^2 = \frac{E_b \log_2 M}{T_M} \tag{9.30}$$

Substituting for A from Equation 9.30 into Equation 9.29, we get

$$P_e = \left[\frac{1}{\log_2 M} \right] \left[\frac{M-1}{M} \right] \text{erfc} \left[\left(\frac{E_b}{N_0}\right)^{1/2} \cdot \frac{(\log_2 M)^{1/2}}{M-1} \right] \tag{9.31}$$

$$\text{SNR} = \frac{\text{Signal power}}{\text{Noise power}} = \frac{E_b(\log_2 M)(1/T_M)}{N_0 \cdot \left(\dfrac{T_M}{2}\right)} \tag{9.32}$$

$$\therefore \text{SNR} = 2\log_2 M \left(\frac{E_b}{N_0} \right) \tag{9.33}$$

Another variation of baseband signaling is antipodal baseband signaling (APBBS) in which two signals of opposite polarities are sent. If $s_0(t) = -A$ and $s_1(t) = A$ for $0 \le t \le T$, then $s_1(t) - s_0(t) = 2A$.

We then calculate the value of z from Equation 9.21 as

$$z = \frac{1}{4N_0} \int_0^T (2A)^2 \, dt = \frac{A^2 T}{N_0} = \frac{E_b}{N_0} \tag{9.34}$$

where E_b is the energy in either $s_0(t)$ or $s_1(t)$, or the bit energy

$$P_e = \frac{1}{2}\, \mathrm{erfc}\left[\sqrt{\frac{E_b}{N_0}}\right] = Q\left[\sqrt{\frac{2E_b}{N_0}}\right] \tag{9.35}$$

APBBS is used to modulate some signal; we will compare its SNR with other modulation schemes.

9.3   Phase Shift Keying

For binary phase shift keying (BPSK) [5,7], we transmit two different signals. If the baseband signal is a binary 0, we transmit:

$$A\cos(\omega t + \pi) = -A\cos(\omega t) \tag{9.36}$$

and for binary 1, we transmit:

$$A\cos(\omega t) \tag{9.37}$$

BPSK can be considered a form of ASK where each non-return to zero (NRZ) data bit of value 0 is mapped into $a\ -1$ and each NRZ 1 is mapped into $a\ +1$. The resulting signal is passed through a filter to limit its bandwidth and then multiplied by the carrier signal $\cos\omega t$ (see Figure 9.3). The signal constellation for BPSK is shown in Figure 9.4.

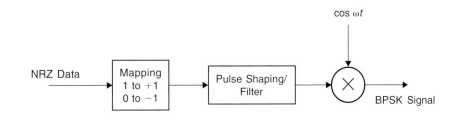

Figure 9.3   BPSK modulator.

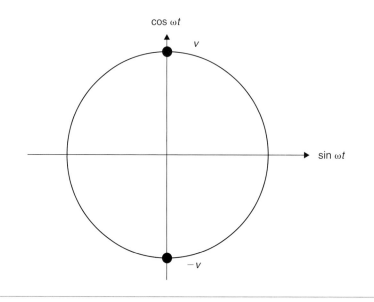

Figure 9.4   Signal constellation for BPSK.

We can also define PSK where there are M phases rather than two phases. In M-*ary* PSK, every b (where $M = 2^b$) bits of the binary bit stream are coded as a signal that is transmitted as $A \sin(\omega t + \theta_j)$, $j = 1, \ldots, M$.

The error distance of a PSK system with M phases is $A \sin(\pi/M)$ where A is the signal amplitude at the detector. A detector error occurs if noise of the proper polarity is present at the output of either of the two phase detectors. The probability of error is

$$P_e = \left(\frac{1}{\log_2 M}\right) \operatorname{erfc}\left[\sin\left(\frac{\pi}{M}\right)(\log_2 M)^{1/2}\left(\frac{E_b}{N_0}\right)^{1/2}\right] \tag{9.38}$$

The SNR is given as:

$$\mathrm{SNR} = \log_2 M\left(\frac{E_b}{N_0}\right) \quad \text{(for } M \geq 2\text{)} \tag{9.39}$$

Each symbol has a length of T. Therefore

$$E_b = \frac{A^2 T}{\log_2 M} \tag{9.40}$$

The rms noise σ is:

$$N_0 = \sigma^2(2T) \tag{9.41}$$

for noise in Nyquist bandwidth.

The bandwidth efficiency of the M-*ary* PSK is given as $\dfrac{R}{B_w} = \dfrac{\log_2 M}{2}$ where R is the data rate and B_w is the bandwidth. Next we examine several variations of PSK.

Example 9.1

Find E_b/N_0 in dB to provide $P_e = 10^{-6}$ for BPSK and coherent FSK.

Solution

For BPSK

$$P_e \approx \frac{e^{-E_b/N_0}}{2\sqrt{\pi\left(\dfrac{E_b}{N_0}\right)}}$$

$$10^{-6} = \frac{e^{-\xi}}{2\sqrt{\pi\xi}}$$

where:

$$\xi = E_b/N_0$$

$$\therefore \xi = 11.32 = 10.54\,\text{dB}$$

FSK requires $3\,\text{dB}$ more in terms of E_b/N_0 to give the same P_e as BPSK, i.e., $13.54\,\text{dB}$.

Example 9.2

The BPSK modulation is used in a channel that adds white noise with single-sided PSD $N_0 = 10^{-10}\,\text{W/HZ}$. Calculate the amplitude A of the carrier signal to give $P_e = 10^{-6}$ for a data rate of $100\,\text{kbps}$.

Solution

From Example 9.1, $E_b/N_0 = 10.54\,\text{dB} = 11.32$ for $P_e = 10^{-6}$

$$\frac{E_b}{N_0} = \frac{A^2 T}{2N_0} = \frac{A^2}{2N_0 R}$$

Quadrature

$$\therefore A = \left[2N_0 \left(\frac{E_b}{N_0} \right) R \right]^{1/2}$$

$$A = [2 \times 10^{-10} \times 11.32 \times 10^5]^{1/2} = 1.505 \, \text{mv}$$

9.3.1 Quadrature Phase Shift Keying (QPSK), Offset-Quadrature Phase Shift Keying (OQPSK) and M-PSK Modulation [5,7,11]

For QPSK, $M = 4$, we substitute $M = 4$ in Equation 9.38 to get

$$P_e = \left(\frac{1}{\log_2 4} \right) \text{erfc} \left[\sin \left(\frac{\pi}{4} \right) (\log_2 4)^{1/2} \left(\frac{E_b}{N_0} \right)^{1/2} \right]$$

$$P_e = \frac{1}{2} \text{erfc} \sqrt{\frac{E_b}{N_0}} = Q \left[\sqrt{\frac{2E_b}{N_0}} \right] \tag{9.42}$$

The input binary bit stream $\left[b_k \right]$, $b_k = \pm 1$; $k = 0, 1, 2, \ldots$, arrives at the modulator at a rate $1/T_b$ bps and is separated into two data streams, $a_I(t)$ and $a_Q(t)$, containing odd and even bits respectively (see Figure 9.5). The modulated QPSK signal $s(t)$ is given as:

$$s(t) = \frac{1}{\sqrt{2}} a_I(t) \cos \left(2\pi f_c t + \frac{\pi}{4} \right) + \frac{1}{\sqrt{2}} a_Q(t) \sin \left(2\pi f_c t + \frac{\pi}{4} \right) \tag{9.43}$$

$$s(t) = A \cos \left(2\pi f_c t + \frac{\pi}{4} + \theta(t) \right) \tag{9.44}$$

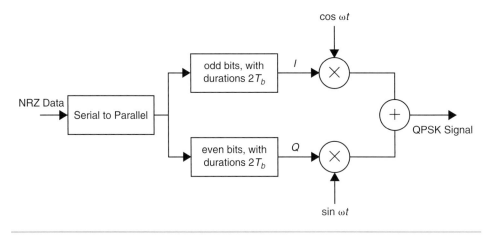

Figure 9.5   QPSK modulator.

where:

$$A = \sqrt{(1/2)(a_I^2 + a_Q^2)} = 1$$

$$\theta(t) = -a\tan\frac{a_Q(t)}{a_I(t)}$$

In the QPSK, the odd and even pulse streams are transmitted at the rate of $(1/2)T_b$ bps and are synchronously aligned. In QPSK, due to the coincident alignment of $a_I(t)$ and $a_Q(t)$, the carrier phase can change only once every $2T_b$. The carrier phase during any $2T_b$ interval can be any one of the four phases as shown in Figure 9.6, depending on the values of $a_I(t)$ and $a_Q(t)$ during that interval. During the next interval, if neither pulse stream changes sign, the carrier phase remains the same. If only one of the pulse streams changes sign, a phase shift of $\pm\pi/2$ occurs. A change in both streams results in a carrier phase shift of π. The 180° phase shifts cause the signal envelope to go through zero momentarily (see Figure 9.7).

In offset QPSK (OQPSK), the timing of the pulse stream $a_I(t)$ and $a_Q(t)$ is shifted such that the alignment of the two pulse streams is offset by T_b. In OQPSK, the pulse stream $a_I(t)$ and $a_Q(t)$ are staggered and thus do not change states simultaneously. The possibility of the carrier changing phase by π degree is

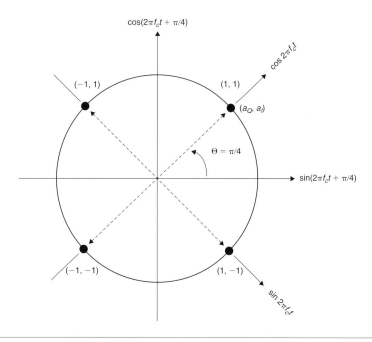

Figure 9.6 $\pi/4$-QPSK constellation.

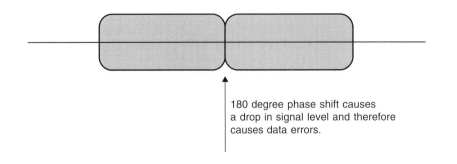

180 degree phase shift causes
a drop in signal level and therefore
causes data errors.

Figure 9.7   Signal amplitude of QPSK during 180° phase change.

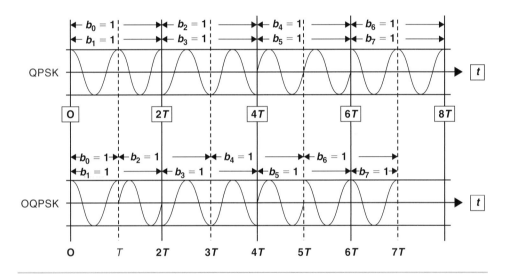

Figure 9.8   A comparison of QPSK and OQPSK modulation schemes.

eliminated, since only one component can make a transition at one time. Changes are limited to $0°$ and $\pm\pi/2$ every T_b seconds (see Figure 9.8). Since the phase transitions of $180°$ are avoided in OQPSK, the signal envelope does not pass through zero as it does in QPSK. The high-frequency components associated with the collapse of the signal envelope are not reinforced. Thus, out-of-band interference is avoided.

In theory, QPSK and OQPSK systems can improve the spectral efficiency of mobile communications. They do, however, require a coherent detector. Coherent detection requires an accurate carrier phase reference. In general, known pilot symbols need to be inserted periodically in the data stream to obtain acceptable performance with a coherent detector on a fading channel. These symbols increase both the bandwidth and the required SNR. In a multipath fading environment, the use of a coherent detector is difficult and often results in poor performance over noncoherently based systems. The coherent detection problem can be overcome by using a differential detector, but then QPSK is subject to intersymbol interference which results in poor system performance. The spectrum of OQPSK is given as:

$$\text{psd}(f) = T_b \left[\frac{\sin \pi f_c T_b}{\pi f_c T_b} \right] \tag{9.45}$$

Further improvement is possible if the OQPSK format is modified to avoid discontinuous phase transitions. This was the motivation in designing *continuous phase modulation* (CPM) schemes. Minimum shift keying (MSK) is one such scheme. A comparison of PSD for QPSK and MSK is shown in Figure 9.9.

Since we often encounter Rayleigh fading (see Chapter 3) in the outdoor environment, the BER performance of PSK under flat Rayleigh fading conditions is given by

$$P_e = \frac{1}{2} \left[1 - \frac{1}{1 + 1/(E_b/N_0)_{\text{avg}}} \right] \tag{9.46}$$

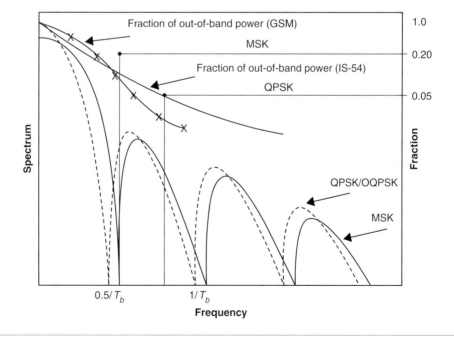

Figure 9.9 Comparison of PSD and QPSK and MSK.

The symbol error probability for M-PSK under AWGN conditions is given as

$$P_s \approx 2Q\left[\sin(\pi/M) \cdot \sqrt{\frac{2bE_b}{N_0}}\right] \tag{9.47}$$

where:

    b = number of bits per symbol

Using Gray code, a symbol error is likely to result in only one out of b bit errors.

$$\therefore P_b = \frac{2}{b} \cdot Q\left[\sin(\pi/M) \cdot \sqrt{\frac{2bE_b}{N_0}}\right] \tag{9.48}$$

For 8-PSK, $M = 8$ and $b = 3$:

$$P_b = \frac{2}{3}Q\left[\sin\frac{\pi}{8} \cdot \sqrt{\frac{6E_b}{N_0}}\right] = \frac{2}{3}Q\left[0.937\sqrt{\frac{E_b}{N_0}}\right] \tag{9.49}$$

For 16-PSK, $M = 16$ and $b = 4$:

$$P_b = \frac{Q}{2}\left[\sin\frac{\pi}{16} \cdot \sqrt{\frac{8E_b}{N_0}}\right] = \frac{Q}{2}\left[0.552\sqrt{\frac{E_b}{N_0}}\right] \tag{9.50}$$

In M-PSK modulation, the amplitude of the transmitted signal remains constant, thereby yielding a circular constellation.

9.3.2 π/4-DQPSK Modulation

We can design a PSK system to be inherently differential and thus solve detection problems. π/4-DQPSK [11] is a compromise modulation method because the phase is restricted to fluctuate between $\pm\pi/4$ and $\pm(3\pi)/4$ rather than the $\pm\pi/2$ phase changes for OQPSK. It has spectral efficiency about 20% higher than the GMSK modulation used for GSM and DECT.

π/4-DQPSK is essentially a π/4-QPSK with differential encoding of symbol phases. The differential encoding mitigates loss of data due to phase slips. However, differential encoding results in the loss of a pair of symbols when channel errors occur. This can be translated to an approximate 3-dB loss in E_b/N_0 relative to coherent π/4-QPSK.

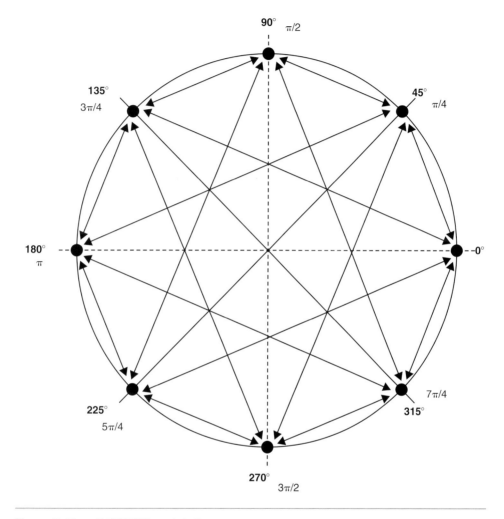

Figure 9.10   π/4-DQPSK modulation.

A π/4-DQPSK signal constellation (Figure 9.10) consists of symbols corresponding to eight phases. Consider that these eight phase points are formed by superimposing two QPSK signal constellations, offset by 45° relative to each other. During each symbol period, a phase angle from only one of the two QPSK constellations is transmitted. The two constellations are used alternately to transmit every pair of bits (di-bits). Thus, successive symbols have a relative phase difference that is one of the four phases as shown in Table 9.1.

Figure 9.10 shows the π/4-DQPSK signal constellation. When the phase angle of π/4-QPSK symbols are differentially encoded, the resulting modulation is π/4-DQPSK. This can be done either by differentially encoding the source bits

Table 9.1 Phase transitions of π/4-DQPSK.

Symbol	π/4-DQPSK phase transition
00	45°
01	135°
10	−45°
11	−135°

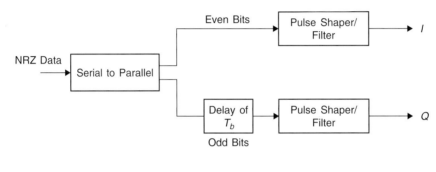

Note: Odd bits are delay by T_b

Figure 9.11 OQPSK encoding.

and mapping them onto absolute phase angles or, alternately, by directly mapping the pairs of input bits onto relative phase angles $[\pm\pi/4, (\pm 3\pi)/4]$ as shown in Figure 9.10. The binary stream $b_M(t)$ entering the modulator is converted by a serial to parallel converter into two binary streams $b_0(t)$ and $b_e(t)$. The bits are differentially encoded (see Figure 9.11).

The in-phase and quadrature components corresponding to the kth symbol are given as:

$$I_k = I_{k-1}\cos(\Delta\phi_k) - Q_{k-1}\sin(\Delta\phi_k) \qquad (9.51)$$

$$Q_k = I_{k-1}\sin(\Delta\phi_k) + Q_{k-1}\cos(\Delta\phi_k) \qquad (9.52)$$

where:

I_k and Q_k are the in-phase and quadrature components of the π/4-DQPSK signal corresponding to the kth symbol, and the amplitudes of I_k and Q_k are $\pm 1, 0, \pm 1/\sqrt{2}$

Since the absolute phase of $(k-1)$th symbol is ϕ_{k-1}, the in-phase and quadrature components can be expressed as:

$$I_k = \cos\phi_{k-1}\cos(\Delta\phi_k) - \sin\phi_{k-1}\sin(\Delta\phi_k) = \cos(\phi_{k-1} + \Delta\phi_k) \qquad (9.53)$$

$$Q_k = \cos\phi_{k-1}\sin(\Delta\phi_k) + \sin\phi_{k-1}\cos(\Delta\phi_k) = \sin(\phi_{k-1} + \Delta\phi_k) \qquad (9.54)$$

These component signals (I_k, Q_k) are then passed through a baseband filter having a raised cosine frequency response as:

$$|H(f)| = \begin{array}{c} 1 \\[6pt] \sqrt{\dfrac{1}{2}\left\{1 - \sin\left[\dfrac{\pi T_s}{\alpha}\left(|f| - \dfrac{1}{2T_s}\right)\right]\right\}} \\[6pt] 0 \end{array} \qquad \left[\begin{array}{c} 0 \le |f| \le \dfrac{1-\alpha}{2T_s} \\[6pt] \dfrac{1-\alpha}{2T_s} \le |f| \le \dfrac{1+\alpha}{2T_s} \\[6pt] |f| \ge \dfrac{1+\alpha}{2T_s} \end{array}\right. \qquad (9.55)$$

where:

   α = the roll-off factor
   T_s = the symbol duration

If $g(t)$ is the response to pulse I_k and Q_k at filter input, then the resultant transmitted signal is given as:

$$s(t) = \sum_k g(t - kT_s)\big(\cos\phi_k \cdot \cos\omega t\big) - \sum_k g(t - kT_s)\big(\sin\phi_k \cdot \sin\omega t\big) \quad (9.56)$$

$$s(t) = \sum_k g(t - kT_s) \cdot \cos\big(\omega t + \phi_k\big) \qquad (9.57)$$

where:

   $2\pi\omega$ = the carrier frequency of the transmission

The component ϕ_k results from differential encoding (i.e., $\phi_k = \phi_{k-1} + \Delta\phi_k$).

Depending on detection method (coherent or differential detection), the error performance of $\pi/4$-DQPSK can either be the same as or 3 dB worse than QPSK.

Example 9.3

The bit stream is 00100111010011101001010100 and the transmitted signal is $A\cos\big(\omega t + \phi_k\big)$. Calculate ϕ_k for the $\pi/4$-DQPSK modulation method.

Table 9.2 Phase angle for π/4-DQPSK.

$b_0 b_e$	$\Delta\phi_k$(degree)	ϕ_k (degree)
		0
00	45	45
10	−45	0
01	135	135
11	−135	0
01	135	135
00	45	180
11	−135	45
10	−45	0
10	−45	−45
01	135	90
01	135	225
00	45	270

Solution

See Table 9.2.

9.3.3 MSK and GMSK Modulation

The MSK is the constant envelope modulation [5]. It is derived from OQPSK by replacing the rectangular pulse in amplitude with a half-cycle sinusoidal pulse. Let us consider a data stream $\{a_k\}$, $k = 0,1,2$, where $a_k = \pm1$ at a rate of $R = 1/T_b$, and T_b is the bit duration. The in-phase and quadrature bit streams are:

$$a_I(t) = a_0 \quad a_2 \quad a_4 \quad \cdots$$

and

$$a_Q(t) = a_1 \quad a_3 \quad a_5 \quad \cdots$$

The rate of $a_I(t)$ and $a_Q(t)$ is $(1/2T_b)$ bit per second. The in-phase, $a_I(t)$ and quadrature $a_Q(t)$ signals are delayed by interval T_b from each other. The MSK signal is defined as:

$$s(t) = a_I(t)\left|\cos\left(\frac{\pi(t - 2nT_b)}{2T_b}\right)\right|\cos 2\pi f_c t + a_Q(t)\left|\sin\left(\frac{\pi(t - 2nT_b)}{2T_b}\right)\right|\sin(2\pi f_c t) \quad (9.58)$$

$$s(t) = \cos\left[2\pi f_c t + b_k(t) \cdot \frac{\pi(t - 2nT_b)}{2T_b} + \phi_k\right] \quad (9.59)$$

where:

$n = 0, 1, 3, \ldots$

$b_k = 1$ for $a_I \cdot a_Q = -1$

$b_k = -1$ for $a_I \cdot a_Q = 1$

$\phi_k = 0$ for $a_I = 1$

$\phi_k = \pi$ for $a_I = -1$

Note that, since the I and Q signals are delayed by a 1-bit interval, the cosine and sine pulse shapes in Equation 9.58 are actually both in the shape of a sine pulse.

MSK has the following properties:

- For a modulation bit rate of R, the high frequency, $f_H = f + 0.25R$ when $b_k = 1$, and the low frequency, $f_L = f - 0.25R$ when $b_k = -1$.
- The difference between the high frequency and the low frequency is $\Delta f = f_H - f_L = 0.5R = \dfrac{1}{2T_b}$, where T_b is the bit interval of the NRZ signal.
- The signal has a constant envelope.

The error probability for an ideal MSK system is

$$P_e = \frac{1}{2} \operatorname{erfc} \sqrt{\frac{E_b}{N_0}} = Q\left[\sqrt{\frac{2E_b}{N_0}}\right] \qquad (9.60)$$

This is the same as for QPSK/OQPSK.

MSK modulation makes the phase change linear and limited to $\pm\pi/2$ over a bit interval T_b. This enables MSK to provide a significant improvement over QPSK. Because of the effect of the linear phase change, the PSD has low side lobes that help to control adjacent-channel interference. However, the main lobe becomes wider than QPSK. Thus, it becomes difficult to satisfy the CCIR-recommended value of -60 dB side lobe power levels. The PSD for MSK is given as:

$$\text{PSD}\,(f_c) = \frac{16T_b}{\pi^2}\left[\frac{\cos 2\pi f_c T_b}{1 - 16f_c^2 T_b^2}\right] \qquad (9.61)$$

MSK has a narrower bandwidth than BPSK, and is significantly better in terms of side lobe level than any unfiltered linear scheme, but it still retains side lobes at a level that would give unacceptable adjacent channel interference in most practical systems. Any attempt to filter these side lobes out, post-modulation, would regenerate envelope variations. Hence, we require a mean to improve the spectrum of MSK further. This can be done by filtering the data signal before modulation. A wide range of filter responses is possible. The Gaussian filter, which leads to GMSK, is the most popular. The bandwidth (B) of the Gaussian filter is quantified in the time-bandwidth product, BT_b. The values of 0.3 to 0.5 are used for BT_b.

The Gaussian response is interesting, in that the frequency response has the same shape as the impulse response. The standard deviation σ of the impulse response is related to the filter bandwidth B by

$$B = \frac{\log_e 2}{\pi \sigma} = \frac{0.2206}{\sigma} \tag{9.62}$$

where the impulse response is

$$h(t) = \frac{1}{\sigma \sqrt{2\pi}} e^{(-t^2)/(2\sigma^2)}$$

The price of the resulting improvement in bandwidth efficiency is a degradation in power efficiency. However, if the time-bandwidth product is not too small (i.e., 0.3 or more), the effect is not excessive (less than 1 dB).

In the case of MSK, BER performance for coherent detection is the same as that for deferentially encoded BPSK with coherent detection. BER for GMSK is degraded due to inter-symbol interference (ISI) by the promulgation Gaussian filter. The BER performance of GMSK with coherent detection under AWGN conditions is given by

$$P_e \approx \text{erfc}\left(\sqrt{2\beta \frac{E_b}{N_0}} \right) \tag{9.63}$$

where β is a degradation factor due to premodulation filter, β = 1 corresponds to the performance for MSK.

The BER under flat Rayleigh fading conditions is given as

$$P_e = 1 - \frac{1}{\sqrt{1 + \frac{1}{\beta(E_b/N_0)_{avg}}}} \tag{9.64}$$

where $(E_b/N_0)_{avg}$ represents an average value

Table 9.3 shows relationship between the bandwidth of the premodulation Gaussian filter normalized by bit rate (BT_b), the 99.99% bandwidth normalized by a bit duration, and β.

Example 9.4

The GMSK modulation is used in the GSM system with a channel bandwidth of 200 kHz and a data rate of 270.8 kbps. Calculate (a) the frequency shift between binary 1 and binary 0, (b) the transmitted frequencies if the carrier frequency is 900 MHz, and (c) the bandwidth efficiency in bps/Hz.

Table 9.3 BT_b versus β for GMSK modulation.

BT_b	99.99% bandwidth	β
0.20	1.22	0.76
0.25	1.37	0.84
0.30	1.41	0.89
0.40	1.80	0.94
0.50	2.08	0.97

Solution

(a) The frequency shift is

$$\Delta f = f_H - f_L = 0.5R = 0.5 \times 270.83\,\text{kHz} = 135.415\,\text{kHz}$$

(b) The maximum and minimum frequency are:

$$f_H = f_c + 0.25R = 900\,\text{MHz} + 67.707\,\text{kHz} = 900.0677\,\text{MHz},$$

and

$$f_L = f_c - 0.25R = 900\,\text{MHz} - 67.707\,\text{kHz} = 899.9323\,\text{MHz}$$

(c)

$$\text{Bandwidth efficiency} = \frac{R}{B_w} = \frac{270.83}{200} = 1.35\,\text{bps/Hz}$$

Example 9.5

Determine the 3-dB bandwidth for a Gaussian low-pass filter that is used to generate 0.30 GMSK with a channel data rate of 270 kbps. What is the 99.99% power bandwidth in the RF channel? What is the bit error probability for GMSK if $E_b/N_0 = 6$ dB?

Solution

$$T_b = \frac{1}{R_b} = \frac{1}{270 \times 10^3} = 3.7\,\mu\text{s}$$

$$BT_b = 0.3$$

$$B = \frac{0.3}{T_b} = \frac{0.3}{3.7 \times 10^{-6}} = 81.08\,\text{kHz}$$

The 3-dB bandwidth is 81.08 kHz. To determine the 99.99% power bandwidth, we use Table 9.3 to find that 1.41 R_b is the required value. The occupied RF spectrum for a 99.99% power bandwidth will be RF bandwidth = 1.41 R_b = 1.41 × 270 × 10³ = 380.7 kHz, β = 0.89. For E_b/N_0 = 6 dB (3.9811), we get

$$P_e = \mathrm{erfc}\left[\sqrt{2\beta \cdot \frac{E_b}{N_0}}\right] = \mathrm{erfc}\,(2.662) \approx 11.5 \times 10^{-5}$$

9.4 Quadrature Amplitude Modulation

By allowing the amplitude to vary with phase, a modulation scheme referred to quadrature amplitude modulation (QAM) is obtained [1]. Figure 9.12 shows a constellation diagram of 16-QAM. The constellation consists of a square lattice of signal points.

The distance d between constellation points can be expressed as

$$d = a(2i - K + 1) + a(2j - K + 1)\,i, j = 0, 1, 2, \ldots K-1$$

$$K = \sqrt{M} \qquad M = 4, 16, 64, 256, \ldots$$

Average symbol error probability under AWGN conditions is given as:

$$P_s = \overline{n_n}Q\!\left(\frac{a}{\sigma}\right) \tag{9.65}$$

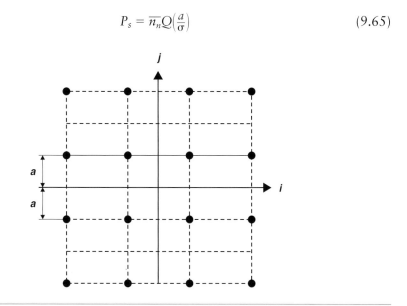

Figure 9.12   16-QAM constellation.

where:

$\bar{n}_n$ = average number of immediate neighbors

$\sigma^2 = N_0/T_s$

$\overline{A^2}$ = the average power

$\bar{n}_n$ and $\overline{A^2}$ are given as:

$$\bar{n}_n = \frac{4(K-2)^2 + 12(K-2) + 8}{M}$$

and

$$\overline{A^2} = \frac{4Ka^2\sum_{p=0}^{K/2-1}(2p+1)^2}{M}$$

For 16-QAM, $M = 16$ and $K = 4$:

$$\bar{n}_n = \frac{4(4-2)^2 + 12(4-2) + 8}{16} = 3$$

$$\overline{A^2} = \frac{16a^2\sum_{p=0}^{1}(2p+1)^2}{16} = \frac{16a^2(1+9)}{16} = 10a^2$$

$$a = \sqrt{\frac{\overline{A^2}}{10}}$$

$$P_s = 3Q\left(\sqrt{\frac{\overline{A^2}}{10\sigma^2}}\right) = 3Q\left(\sqrt{\frac{\overline{A^2}T_s}{10N_0}}\right) = 3Q\left(\sqrt{\frac{2E_s}{10N_0}}\right)$$

Since $E_s = 4E_b$:

$$P_e = \frac{3}{4}Q\left(\sqrt{\frac{2E_s}{10N_0}}\right) = \frac{3}{4}Q\left(\sqrt{\frac{E_s}{5N_0}}\right) = \frac{3}{4}Q\left(\sqrt{\frac{4E_b}{5N_0}}\right) \tag{9.66}$$

The disadvantage of 16-QAM compared to 16-PSK is that the signal amplitude has an inherent variation, regardless of filtering. This may cause problems in nonlinear amplifiers and on fading channels.

Example 9.6

A modulator transmits symbols at a rate of 19,200 symbols per second. Each symbol has 64 different possible states. What is the bit rate?

Solution

   Number of bits per symbol: $b = \log_2 M = \log_2 64 = \log_2 2^6 = 6$
   Bit rate of the modulator $= 6 \times 19{,}200 = 115.2\,\text{kbps}$

Example 9.7

A data stream with data rate $R_b = 144$ Mbps is transmitted on an RF channel with a bandwidth of 36 MHz. Assuming Nyquist filtering and Gaussian channel, determine the modulation scheme that should be used. If the probability of bit error is 3×10^{-5}, find the required E_b/N_0.

Solution

   Required spectral efficiency $= \dfrac{144}{36} = 4\,\text{bps/Hz}$

   Since the channel is bandwidth limited

$$\frac{R_b}{B_w} = 4 = \log_2 M$$

$$\therefore M = 2^4 = 16$$

16-QAM (see Equation 9.66) should be used as it is more efficient than 16-PSK (see Equation 9.50).

   For a rectangular constellation (see Figure 9.12), with a Gaussian channel and matched filter reception, the bit error probability is given as:

$$P_b = \frac{3}{4}\,Q\left[\sqrt{\frac{4}{5}\frac{E_b}{N_0}}\right] = 3 \times 10^{-5}$$

$$\therefore \frac{E_b}{N_0} = 19.5 = 12.9\,\text{dB}$$

Example 9.8

Compare the performance of 16-PSK with 16-QAM for a BER probability of 10^{-8}.

Solution

   16-PSK (Equation 9.50):

$$10^{-8} = \frac{1}{2}Q\left(0.552\,\sqrt{\frac{E_b}{N_0}}\right)$$

$$\therefore \frac{E_b}{N_0} = 20\,\text{dB}$$

16-QAM (Equation 9.66):

$$10^{-8} = \frac{3}{4} Q\left(\sqrt{\frac{4}{5}\frac{E_b}{N_0}}\right)$$

$$\therefore \frac{E_b}{N_0} \sim 16\,\text{dB}$$

Thus, 16-QAM has an advantage of about 4 dB compared to 16-PSK.

9.5   M-*ary* Frequency Shift Keying

The PSK and various forms of QPSK do not have a continuous phase, but they provide good bandwidth efficiency to allow more users for a given channel bandwidth [11]. If there is an interest in using low-cost amplifiers that are essentially nonlinear, we need constant-envelope modulation schemes. The frequency shift keying (FSK) modulation is one such schemes. In binary FSK, the carrier frequency is increased or decreased by a fixed value corresponding to a binary 1 or 0. Thus, two separate signals are transmitted. The FSK signals can be detected coherently or noncoherently. The bit error probability P_e for a coherent binary FSK in the AWGN channel is given as:

$$P_e = \frac{1}{2}\,\text{erfc}\left[\sqrt{\frac{E_b}{2N_0}}\right] \tag{9.67}$$

This is 3 dB worse than the coherent BPSK. For noncoherent systems the P_e is:

$$P_e = \frac{1}{2}\,e^{-E_b/(2N_0)} \tag{9.68}$$

As the frequency hopping spread spectrum (FHSS) uses orthogonal frequency division multiplexing (OFDM) with MFSK, we provide a brief description of MFSK. In MFSK, signal $s_i(t)$ at frequency f_i is given as

$$s_i(t) = \sqrt{\frac{2E_s}{T_s}} \cdot \cos\left(2\pi f_i t\right) \qquad 1 \leq i \leq M \tag{9.69}$$

where:
  $f_i = f_c + (2i - 1 - M)\,\Delta f$
  in which
    f_c = carrier frequency
    Δf = frequency step
    M = number of different signal elements = 2^b
    b = number of bits per signal element

E_s = energy per symbol = $E_b \log_2 M$
E_b = energy per bit
T_s = symbol duration = bT_b
T_b = bit duration
R_b = data rate
B_w = bandwidth
N_0 = noise density

- Total bandwidth required, $B_w = 2M\Delta f$
- Minimum frequency separation required = $2\Delta f = 1/T_s$

Average probability of symbol error P_s for the coherent orthogonal MFSK signal is given as:

$$P_s \leq (M-1)Q\left(\sqrt{\frac{E_b \log_2 M}{N_0}}\right) \qquad M > 2 \qquad (9.70)$$

Average probability of symbol error P_s for the noncoherent orthogonal MFSK signal is given as:

$$P_s = \frac{1}{M}\left((e^{-E_s/N_0})\sum_{i=1}^{M}(-1)^i \cdot \binom{M}{i} \cdot e^{E_s/(iN_0)}\right) \qquad M > 2 \qquad (9.71)$$

where:

$$\binom{M}{i} = \frac{M!}{i! \cdot (M-i)!}$$

The upper-bound of P_s for the orthogonal and noncoherent orthogonal MFSK signal is

$$P_s < \frac{M-1}{2} e^{-E_s/(2N_0)} \qquad M > 2 \qquad (9.72)$$

The relationship between bit error probability P_e and P_s for the orthogonal MFSK signal set is given as

$$\frac{P_e}{P_s} = \frac{2^{k-1}}{2^k - 1} = \frac{M/2}{(M-1)} \qquad M > 2 \qquad (9.73)$$

$$\lim_{k \to \infty}\left(\frac{P_e}{P_s}\right) = \frac{1}{2} \qquad (9.74)$$

The channel bandwidth efficiency of a coherent MFSK is

$$\frac{R_b}{B_w} = \frac{2 \log_2 M}{(M + 3)} \qquad M > 2 \tag{9.75}$$

The channel bandwidth efficiency of a noncoherent MFSK is

$$\frac{R_b}{B_w} = \frac{2 \log_2 M}{M} \qquad M > 2 \tag{9.76}$$

The bandwidth efficiency of an MFSK decreases with increasing M, MFSK signals are bandwidth inefficient. However, since all the M signals are orthogonal there is no crowding in the signal space and hence the power efficiency increases with M. MFSK can use nonlinear amplifiers with no performance degradation.

Example 9.9

If $M = 8$, $f_c = 250\,\text{kHz}$, $\Delta f = 25\,\text{kHz}$, what is the total bandwidth required? What is the bandwidth efficiency? What is the E_b/N_0 required for symbol bit error probability $= 10^{-6}$ of a coherent MFSK? How many bits per symbol are carried?

Solution

Total bandwidth required $= 2M\Delta f = 2 \times 8 \times 25\,\text{kHz} = 400\,\text{kHz}$

$$\eta_b = \frac{R_b}{B_w} = \frac{2 \log_2 M}{(M + 3)} = \frac{2 \times 3}{11} = 0.5455$$

$$10^{-6} = (M - 1)\, Q(z) = 7Q(z)$$

$$z \approx 5.08$$

$$\sqrt{\frac{E_b \log_2 M}{N_0}} = 5.08$$

$$\therefore \frac{E_b}{N_0} = \frac{(5.08)^2}{\log_2(8)} = 9.346\,\text{dB}$$

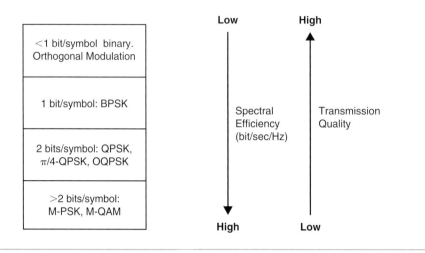

Figure 9.13   Modulation method selection.

9.6   Modulation Selection

Proper selection of a modulation scheme in mobile communications is needed. Several factors dictate the selection. In general a modulation method should have the following characteristics:

- spectrally efficient (i.e., higher bps/Hz);
- applicable to cellular/PCS in all environments (urban, suburban, rural);
- easy to implement;
- low adjacent-channel interference.
- good BER performance

    As shown in Figure 9.13, transmission quality and spectral efficiency of a modulation method are interrelated. Table 9.4 lists the modulation schemes and their spectral efficiencies used in different mobile systems.

9.7   Synchronization

The demodulation of a signal requires that the receiver be synchronized with the transmitter signal as it is received at the input of the receiver [11]. The synchronization must be for:

- *Carrier synchronization*. The receiver is on the same frequency as the transmitted signal, adjusted for the effects of Doppler shifts.
- *Bit synchronization*. The receiver is aligned with the beginning and end of each bit interval.
- *Word synchronization (also known as frame synchronization)*. The receiver is aligned with the beginning and end of each word in the transmitted signal.

Table 9.4 Modulation methods in mobile communication.

System	Modulation scheme	Information rate (kbps)	Channel spacing (kHz)	Spectral efficiency(bps/Hz)
IS-136	π/4-DQPSK	48.6	30	1.62
PDC	π/4-DQPSK	42.0	25	1.68
GSM	GMSK	270.8	200	1.35
CT-2	GMSK	72	100	0.72
DECT	GMSK ($BT_b = 0.5$)	1572	1728	0.67
CDMA	QPSK	1228.8	1230	0.99
UWC-136	16-QAM	384	1600	4.17

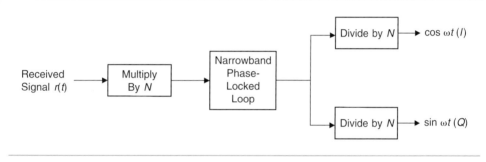

Figure 9.14   Carrier recovery for PSK.

If the synchronization in the receiver is not precise for any of these operations, then the BER of the receiver will not be the same as described by equations in previous sections of this chapter. The design of a receiver is an area for which standards are traditionally not specified. It is an art that enables one manufacturer to offer better performance in its equipment compared to a competitor. The methods of achieving the synchronization discussed are the traditional methods. A particular receiver may or may not use any of these methods, and proprietary methods are often used by many manufacturers.

For PSK, the carrier signal changes phase every bit interval (see Figure 9.14). If we multiply the received signal by an integer, N, we can convert all of the phase changes in the multiplied signal to multiples of 360°. The new signal then has no phase changes and we can recover it using a narrowband phase-locked loop (PLL). After the PLL recovers the multiplied carrier signal, it is divided by N to recover the carrier at the proper-frequency. By a suitable choice of digital dividing circuits, it is possible to get a precise 90° difference in the output of two dividers and thus generate both the cos ωt and sin ωt signals needed by the receiver. There are also some down-converter integrated circuits that have a precise phase shift network contained within them. The carrier recovery is typically performed at some lower intermediate frequency rather than directly at the received frequency. For BPSK we would need an N of 2, but an N of 4 would be used to enable the sine and cosine terms to be generated. For QPSK and its derivatives, an N of 4 is necessary; for $\pi/4$-DQPSK an N of 8 would be needed.

After we recover the carrier, we must reestablish the carrier phase to determine the values of the received bits. Somewhere in the transmitted signal must be a known bit pattern that we can use to determine the carrier phase. The bit pattern can be alternating 0s and 1s that we use to determine bit timing or it could be some other known pattern.

The advantage of differential keying (e.g., $\pi/4$-DQPSK) is that it is not important to know the absolute carrier phase. Only the change in carrier phase from one symbol to the next is important.

MSK is a form of frequency modulation; therefore, a different method of carrier recovery is required. In Figure 9.15, the MSK signal has frequency f and deviation $\Delta f = 1/2T_b$. We first multiply the signal by 2, thus doubling the deviation and generating strong frequency components at $2f + 2\Delta f$ and $2f - 2\Delta f$. We use two PLLs to recover these two signals.

$$s_1(t) = \cos(2\pi ft + \pi\Delta ft) \tag{9.77}$$

$$s_2(t) = \cos(2\pi ft - \pi\Delta ft) \tag{9.78}$$

We then take the sum and difference of $s_1(t)$ and $s_2(t)$ to generate the desired $I(t)$ and $Q(t)$ signals.

$$I(t) = s_1(t) + s_2(t) = 2\cos 2\pi ft \cos \pi \Delta ft \tag{9.79}$$

$$Q(t) = s_1(t) - s_2(t) = 2\sin 2\pi ft \sin \pi ft \tag{9.80}$$

The identical circuit can also be used for carrier recovery for a GMSK system.

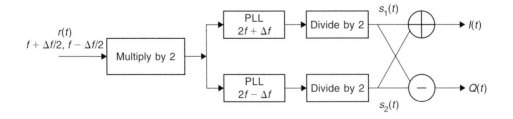

Figure 9.15   Carrier recovery for MSK.

The next step is to recover data timing or bit synchronization (Figure 9.16). Most communications systems transmit a sequence of 1s and 0s in an alternating pattern to enable the receiver to maintain bit synchronization. A PLL operating at bit timing is used to maintain timing. Once the PLL is synchronized on the received 101010 … pattern, it will remain synchronized on any other patterns except for long sequences of all 0s or all 1s.

MSK uses an additional circuit to achieve bit timing. The $s_1(t)$ and $s_2(t)$ signals are multiplied together and low-pass filtered (Figure 9.17).

$$s_1(t)s_2(t) = \cos(2\pi ft + \pi\Delta ft) \times \cos(2\pi ft - \pi\Delta ft)$$
$$= 0.5\cos 4\pi ft + 0.5\cos 2\pi\Delta ft \tag{9.81}$$

$$\text{low} - \text{pass filtered}\left[s_1(t)s_2(t)\right] = 0.5\cos 2\pi\Delta ft = 0.5\cos\frac{\pi t}{T_b} \tag{9.82}$$

The output of a low-pass filter is a clock signal at one-half of the transmitted bit rate. The one-half bit rate clock is the correct rate for demodulation of the signal since the I and Q signals are at one-half of the bit rate.

Frame synchronization is determined by the receiver correlating the received signal with a known bit pattern. The receiver performs an autocorrelation function

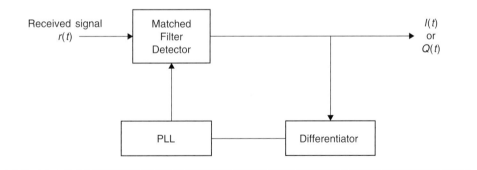

Figure 9.16   Generalized data timing recover circuit.

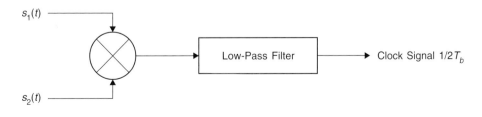

Figure 9.17   MSK data timing recovery circuit.

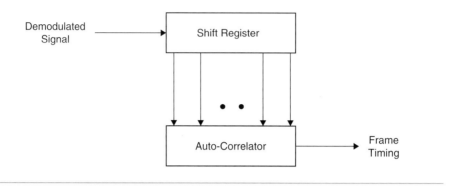

Figure 9.18   Generalized framing recovery circuit.

to determine when the bit pattern is received and then outputs a framing pulse (Figure 9.18).

9.8 Equalization

The received signal in a mobile radio environment travels from the transmitter to the receiver over many paths [7]. The signal fades in and out and undergoes distortion because of the multipath nature of the channel. For a transmitted signal $s(t) = a(t)\cos[\omega t + \theta(t)]$, we can represent the received signal, $r(t)$, as:

$$
\begin{aligned}
r(t) = \sum_{i=0}^{n} & x_i(t - \tau_i)a(t - \tau_i)\cos\left[\omega(t - \tau_i) + \theta(t - \tau_i)\right] \\
& + y_i(t - \tau_i)a(t - \tau_i)\sin\left[\omega(t - \tau_i) + \theta(t - \tau_i)\right]
\end{aligned}
\tag{9.83}
$$

We know that the received signal has Rayleigh fading statistics. But what are the characteristics of the $x_i(t)$ and $y_i(t)$ term in Equation 9.83? If the transmitter signal is narrow enough compared to the multipath structure of the channel, then the individual fading components, $x_i(t)$ and $y_i(t)$ will also have Rayleigh statistics. If a particular path is dominated by a reflection off a mountain, hill, building, or similar structure, then the statistics of that path may be Rician rather than Rayleigh. If the τ_i (delay spread) is small compared to the bit interval, then a little distortion of the received signal occurs. If the τ_i is greater than the bit interval, then the transmissions from one bit will interfere with transmissions of another bit, resulting in inter-symbol interference (ISI).

Spread spectrum systems use wide bandwidth signals and attempt to recover the signals in each of the paths and add them together in a diversity (Rake) receiver. For our discussion here, however, we are transmitting narrowband signals, and the multipath signals are an interference to the desired signal. We need a receiver that removes the effects of the multipath signal or cancels the undesired multipath.

Another way to describe the multipath channel is to describe the channel as having an impulse response $h(t)$. The received signal is then written as:

$$r(t) = \int_{-\infty}^{\infty} s(t)h(t - \tau)d\tau \qquad (9.84)$$

We can recover $s(t)$ if we can determine a transfer function $h^{-1}(t)$ (the inverse of $h(t)$). One reason it is difficult to perform the inverse function is that it is time varying. The circuit that performs the inverse function is called an *equalizer* (see Figure 9.19).

Generally, we are interested in minimizing the ISI at the same time we do our detection (the sample time in a sample-and-hold circuit). Thus, we can model the equalizer as a series of equal time delays (rather than random as is the general case) with the shortest delay interval being a bit interval. We then construct a receiver that determines $r_{eq}(t)$, the equalized signal:

$$r_{eq}(t) = \sum_{i=0}^{n} \alpha_i(t - \tau_n)r(t - \tau_n) \qquad (9.85)$$

We use the equalized signal $r_{eq}(t)$ as the input to our detector to determine the value of the *i*th transmitted bit. We must adjust the values of α_i to achieve some measure of receiver performance. A typical measure is to minimize the mean square error between the value of the detected bit at the output of the adder and the output of the detector. Other measures for equalizers are possible.

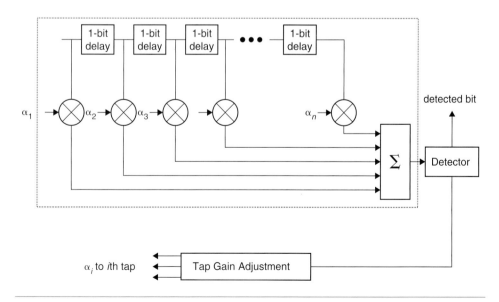

Figure 9.19 Block diagram of an equalizer.

9.9   Summary

In this chapter we studied the modulation and demodulation process applicable to cellular systems. Since baseband signals can be transmitted only over short distances with wires and require very long antennas to transmit them without wires, the baseband signals are modulated onto radio frequency carriers. We studied phase modulation and its derivatives $\pi/4$-QPSK, $\pi/4$-DQPSK, and OQPSK. We discussed MSK, a form of FM, as a first step toward GMSK which is used for DECT and GSM. Based on the literature, the error performance for GMSK also has the same-shaped curve as ASK with the proper definition of a correcting factor β. For GSM, where $BT_b = 0.3$, $\beta \sim 0.9$; for DECT, where $BT_b = 0.5$, $\beta \sim 0.97$.

We also studied a hybrid of amplitude and phase modulation, called QAM. We discussed the signal constellation and the bit error rate. Finally, we concluded the chapter by providing brief descriptions of the synchronization and equalization processes.

Problems

9.1   A digital signaling system is required to operate at 9.6 kbps. If a signal element encodes a 4-bit word, what is the minimum bandwidth of the channel?

9.2   Given a channel with intended capacity of 20 Mbps, the bandwidth of the channel is 3 MHz. What is the signal-to-noise ratio required to achieve this capacity?

9.3   The receiver in a communications system has a received signal power of -134 dBm, a received noise power spectral density of -174 dBm/Hz, and a bandwidth of 2000 Hz. What is the maximum rate of error-free information for the system?

9.4   Find the minimum required bandwidth for the baseband transmission of an 8-level PAM pulse sequence having a data rate, R_b, of 9600 bps. The system characteristics consist of a raised-cosine spectrum with 50% excess bandwidth.

9.5   Assuming that $\phi_0 = 0^0$, a bit stream 101100 is sent using a $\pi/4$-DPQSK modulation scheme. The leftmost bits are first applied to the transmitter. Determine phase ϕ_k, and the values of I_k and Q_k during transmission.

9.6   Find the BER probability, P_e, of an M-PSK system for a 14.4 kbps signal. Amplitude is 15 mW, and noise density, N_0, is 10^{-9} W/Hz.

9.7   A noncoherent orthogonal M-FSK system carries 3 bits per symbol. The system is designed for $E_b/N_0 = 6$ dB. What is the maximum bit error rate probability? Find the bandwidth efficiency of the system.

9.8 A noncoherent orthogonal M-FSK system carries 4 bits per symbol. The system is designed to have a maximum probability of symbol-error of 10^{-6}. What is the required E_b/N_0? What is the bandwidth efficiency?

References

1. Burr, Alister. *Modulation and Coding for Wireless Communications*. Prentice Hall, 2001.

2. Garg, V. K., and Wilkes, J. E. *Principles & Applications of GSM*. Upper Saddle River, NJ: Prentice Hall, 1999.

3. Glover, I. A., and Grant, P. M. *Digital Communications*. Prentice Hall, 1998.

4. Hamming, R.W. Error detecting and correcting codes. *Bell System Technical Journal*, 29:147−160, 1950.

5. Lathi, B. P. *Modern Digital and Analog Communications Systems*, 2nd Edition. Holt, Reinhart and Winston, 1989.

6. Nyquist, H. Certain topics in telegraph transmission theory. *AIEE Transactions*, 47: 617–644, 1928.

7. Proakis, J. G. *Digital Communications*, 3rd Edition. McGraw-Hill, 1995.

8. Rappaport, T. S. *Wireless Communications: Principles and Practice*, 2nd Edition. Prentice Hall, 2000.

9. Shannon, C. E. A mathematical theory of communication. Part 1. *Bell System Technical Journal*, 27:379, 1948.

10. Shannon, C. E. A mathematical theory of communication, Part 2. *Bell System Technical Journal*, 27:623, 1948.

11. Skalar, B. *Digital Communication—Fundamental and Applications*. Englewood Cliffs, NJ: Prentice Hall, 1988.

12. Sweeney, P. *Error Control Coding: An Introduction*. Prentice Hall, 1991.

13. Ziemer, R. E., and Peterson, R. L. *Introduction to Digital Communication*. New York: Macmillan Publishing Co., 1992.

CHAPTER 10

Antennas, Diversity, and Link Analysis

10.1 Introduction

The antenna system used for any radio communication platform is one of the critical and least understood parts of the system. The antenna system is the interface between the radio system and the external environment. Wireless communication systems require antennas at the transmitter and receiver to operate properly. The design and deployment of antennas can make or break a wireless system, and many poorly performing systems can be traced to improperly installed or placed antennas.

The antenna system can consist of a single antenna at the base station and one at the mobile station. Primarily the antenna is employed by the base station and the mobile for establishing and maintaining the communication link. There are many types of antennas available, all of which perform specific functions depending on the application at hand. The type of antenna used by a system operator can be colinear, folded dipole, or Yagi, to mention a few [2,8–9].

In this chapter we first define and discuss the roles of antennas. We then examine the interrelationships between antenna beam width, directivity, and gain. After completing our discussion of antennas we explore the concepts of diversity reception where multiple signals are combined to improve the signal-to-noise ratio (SNR) of the system [3,10–13,16]. We describe the noise that a receiver sees and use the concept to perform an analysis of the link between the base station and the mobile station and calculate the SNR at the receiver. Throughout the chapter, we present numerical examples to improve the understanding of each topic.

10.2 Antenna System

The radio communication system requires a reliable communication between a fixed base station and a mobile station. The goal of the system designer is to have the same performance in both the transmitting and receiving directions. This is not always possible since the base station typically has a higher output power than the mobile station. Also, the mobile station antenna is typically on the street at a height of 1 to 2 m compared to the base station antenna height of 50 to 60 m.

This results in the mobile receiver having a high noise level due to interference caused by the ignitions of nearby vehicles or other objects, whereas the base station, with its higher antenna, usually sees a quieter radio environment. These factors can combine to favor one particular direction over the other. The wireless system designer must carefully consider all these factors or the range of the system may be limited by poor performance in one direction.

The types of antennas, their gain and coverage patterns, the available power to drive them, the use of simple or multiple antenna configurations, and the polarization (the polarization of a radio wave is the orientation of its electric field vector) are the major factors that can be controlled by the system designer. The system designer has no control over the topography between the base station and mobile station antennas, the speed and direction of the mobile station, and the location of antenna(s) on the mobile station. Each of these factors significantly affects system performance. Sometimes the placement of the mobile antenna can severely limit system performance. While the mobile antenna is usually installed by a knowledgeable technician, the owner of the vehicle may force a nonoptimum placement of the antenna. Furthermore, cellular antennas are vertically polarized physically, but many vehicle antennas are no longer vertical after the vehicle is sent through a car wash.

Along with the type of antenna there is the relative pattern of the antenna, indicating in which direction the energy emitted or received will be directed. There are two primary classifications of antennas: omnidirectional and sectorized (directional). The omnidirectional antennas are used when the desire is to have a 360° radiation pattern. The sectorized antennas are used when a more refined pattern is needed. The directional pattern is usually required to facilitate system growth through frequency reuse.

The choice of antenna directly impacts the performance of both the base station and the overall network. The antenna selection must consider a number of design issues. Some involve the antenna gain, the antenna pattern, the interface or matching to the transmitter, the receiver used for the site, the bandwidth and frequency range over which the signals will travel, and the power handling capabilities.

10.3   Antenna Gain

The task of a transmitting antenna is to convert the electrical energy travelling along a transmission path into electromagnetic waves in space. The antennas are passive devices, the power radiated by the transmitting antenna cannot be greater than the power entering from the transmitter. It is always less because of losses. Antenna gain in one direction results from a concentration of power in that direction and is accompanied by a loss in other directions. Antenna gain [7] is the most important parameter in the design of an antenna system. A high gain is achieved by increasing the aperture area, A, of the antenna. Antennas obey reciprocity; the transmit gain and receive gain are the same, and the antenna can be analyzed by examining it as either a receive or transmit antenna. The amount of power captured by an antenna is given as:

$$P = pA \tag{10.1}$$

where:
   p = power density (power per unit area)
   A = aperture area

Antenna gain can be defined either with respect to an *isotropic* antenna or with respect to a *half-wave dipole* and is usually analyzed as a transmit antenna. An isotropic antenna is an idealized system that radiates equally in all directions. The half-wave dipole antenna is a simple, practical antenna which is in common use.

The gain of an antenna in a given direction is the ratio of power density produced by it in that direction divided by the power density that would be produced by an isotropic antenna. The term *dBi* is used to refer to the antenna gain with respect to the isotropic antenna. The term *dBd* is used to refer to the antenna gain with respect to a half-wave dipole ($0\,\text{dBd} = 2.1\,\text{dBi}$). While most analyses of system performance use a half-wave dipole as the reference, many times antenna gain figures are quoted in dBi to give a falsely inflated gain figure. The system designer must carefully read data sheets on antennas to use the correct gain figure. As a rule of thumb, if the gain is not quoted in either dBd or dBi, the gain is in dBi, with the dBi left out to inflate the gain figures.

For an isotropic antenna in free space, the received power density is given as

$$p_R = \frac{P_T}{4\pi d^2} \tag{10.2a}$$

where:
   P_T = transmitter power
   p_R = receiver power density
    d = distance between transmitter and receiver

When a directional transmitting antenna with power gain factor, G_T, is used, the power density at the receiver antenna is G_T times Equation 10.2a, i.e.,

$$p_R = G_T \cdot \frac{P_T}{4\pi d^2} \tag{10.2b}$$

The amount of power captured by the receiver is p_R times the aperture area, A_R, of the receiving antenna. The aperture area is related to the gain of the receiving antenna by

$$G_R = \frac{4\pi A_R}{\lambda^2} \tag{10.3}$$

where:

$\lambda = \frac{c}{f}$

f = the transmission frequency in Hz

$c = 3 \times 10^8$ m/s is the free-space speed of propagation for electromagnetic waves

A_R = the effective area of aperture, which is less than the physical area by an efficiency factor ρ_R; typical value for ρ_R ranges from 60 to 80%

The total received power, P_R is given as:

$$P_R = A_R p_R \tag{10.4}$$

Substituting the value of P_R and A_R from Equations 10.2b and 10.3 into Equation 10.4, together with the transmitting antenna gain G_T, we get

$$P_R = \left[\frac{\lambda}{4\pi d}\right]^2 P_T G_T G_R \tag{10.5}$$

Equation 10.5 includes the power loss only from the spreading of the transmitted wave. If other losses are also present, such as atmospheric absorption or ohm losses of the waveguides leading to antennas, Equation 10.5 can be modified as [4]:

$$\frac{P_R}{P_T} = \left[\frac{\lambda}{4\pi d}\right]^2 \cdot \frac{G_T G_R}{L_0} \tag{10.6}$$

$$\frac{P_R}{P_T} = \frac{G_T G_R}{L_0 L_p} \tag{10.7}$$

where:

$L_p = \left[\frac{4\pi d}{\lambda}\right]^2$ denotes the loss associated with propagation of electromagnetic waves from the transmitter to the receiver as discussed in Chapter 3.

L_p depends on carrier frequency and separation distance d. This loss is always present. L_0 is the loss factor for additional losses.

When we express Equation 10.7 in terms of decibels, we get

$$P_R = 20\log\left[\frac{\lambda}{4\pi d}\right] + P_T + G_T + G_R - L_0 \text{ dB} \qquad (10.8)$$

The product $P_T G_T$ is called the *effective isotropic radiated power (EIRP)* and term $20\log\left[\dfrac{\lambda}{4\pi d}\right]$ refers to *free space path loss* (L_p) in dB. Another term, *effective radiated power (ERP)*, is also used. It is the power input multiplied by the antenna gain measured with respect to a half-wave dipole antenna. The EIRP is related to ERP as

$$\text{EIRP} = \text{ERP} + 2.14 \text{ dB} \qquad (10.9)$$

In the free space, the path between two antennas has no obstruction (see Figure 10.1) and there is no object where reflection can occur. Thus, the received signal is composed of only one component. When the two antennas are located on the earth, then there are multiple paths from the transmitter to the receiver. The effect of multiple paths is to change the path loss between two points. The simplest case occurs when the antenna heights h_T and h_R are small compared to their separation distance d and the reflecting earth surface is assumed to be flat (see Chapter 3). The received signal can then be represented by a field that is approximated by a combination of a direct wave and a reflected wave as shown in Figure 10.2. In this case the received power, P_R, and transmitted power, P_T, are related as (see Chapter 3 and Appendix B for derivation):

$$\frac{P_R}{P_T} = \left[\frac{h_T h_R}{d^2}\right]^2 \cdot \frac{G_T G_R}{L_0} \qquad (10.10)$$

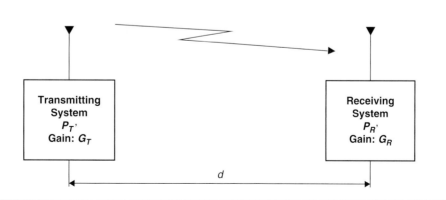

Figure 10.1   Free-space path-loss model.

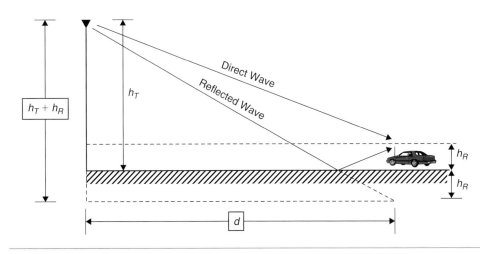

Figure 10.2   Path-loss model with reflection.

Expressing Equation 10.9 in decibels, we get

$$P_R = 20\log\left[\frac{h_T h_R}{d^2}\right] + P_T + G_T + G_R - L_0 \text{ dB} \qquad (10.11)$$

comparing it with Equation 10.8 we note that Equation 10.10 is independent of transmitting frequency.

Example 10.1

We consider a communication system in which the distance between the transmitter and receiver is 10,000 m. The transmitter EIRP is 30 dBW (G_T = 20 dBi; P_T = 10 dBW). The transmitting frequency is 1.5 GHz (λ = 0.2 m). The receiver antenna gain is 3 dBi; and total system losses are 6 dB. Assuming the receiver noise figure, N_f = 5 dB and bandwidth, B_w = 1.25 MHz, calculate the received signal power at the receiver antenna and the SNR of the received signal. Neglect any feed line losses between the antenna and receiver.

Solution

$$L_P: \text{free-space loss} = 20\log\left[\frac{0.2}{4\pi \times 10,000}\right] = -116 \text{ dB}$$

$$P_R = -116 + 30 + 3 - 6 = -89 \text{ dBW} = -59 \text{ dBm}$$

$$\text{Noise power, } P_N = N_f B_W N_0$$

$$\text{Noise Density, } N_0 = kT = \log(1.38 \times 10^{-23} \times 290) = -204\,\text{dBW} = -174\,\text{dBm}$$

$$P_N = 5 - 174 + 10\log(1.25 \times 10^6) = -108\,\text{dBm}$$

$$\text{SNR} = \frac{P_R}{P_N} = -59 - (-108) = 49\,\text{dB}$$

Example 10.2

Using the data in Example 10.1 and antenna heights at the receiver and transmitter units to be 40 m and 2 m, respectively, calculate the received signal power at the receiver antenna and the SNR of the received signal.

Solution

$$\text{Path loss} = 20\log\left[\frac{40 \times 2}{10{,}000^2}\right] = -122\,\text{dB}$$

$$P_R = -122 + 30 + 3 - 6 = -95\,\text{dBW} = -65\,\text{dBm}$$

$$P_N = -108\,\text{dBm (from Example 10.1)}$$

$$\therefore \text{SNR} = \frac{P_R}{P_N} = -65 - (-108) = 43\,\text{dB}$$

10.4 Performance Criteria of Antenna Systems

The performance of an antenna system [7,19] is not restricted to its gain and physical attributes. There are many other parameters that must be considered in evaluating antenna performance. The parameters that define the performance of an antenna system are:

- Antenna pattern
- Main and side lobe
- Radiation efficiency (η)
- Antenna bandwidth
- Horizontal beam width
- Vertical beam width
- Gain (G)
- Directivity (D)
- Antenna polarization
- Input impedance

- Front-to-back ratio (R_{FB})
- Front-to-side ratio (R_{FS})
- Power dissipation
- Intermodulation
- Construction
- Cost

The *antenna pattern* chosen should match the coverage requirements for the base station. An antenna has one *major lobe* and a number of *side lobes* (see Figure 10.3). The side lobes are important because they create potential problems by generating interference. Ideally there should be no side lobes for the antenna pattern. For down tilting, the side lobes are important because they can create secondary sources of interference.

The *radiation efficiency* is a ratio of total power radiated by an antenna to net power accepted by an antenna from the transmitter. The *bandwidth* defines the operating range of the frequencies for the antenna. It is the angular separation between two directions in which radiation interest is identical. The *half power point* for the beam width is the angular separation where there is 3 dB reduction off the main lobe. Normally, the wider the beam width, the lower the gain of the antenna. The *gain* is the ratio of the radiation intensity in a given direction to that of an isotropically radiated signal. The *directivity* is the gain calculated assuming a lossless antenna. Real antennas have losses, gain is the directivity multiplied by the efficiency of the antenna.

Polarization is important for an antenna because wireless mobile systems use vertical polarization. A vertical antenna is easiest to mount on a vehicle; therefore vertical polarization has been standardized. In general, horizontal or vertical polarization will work equally well.

Most cables used as feed line from the transmitter/receiver to the antenna are either 50 or 72/75 ohms. If the *input impedance* of the antenna is far removed from either of these values, it will be difficult to get an antenna to accept the power delivered to it and its radian efficiency will be low.

The *front to back ratio* is a ratio in respect to how much energy is directed in the exact opposite direction of the main lobe of the antenna (see Figure 10.3). The *front to side ratio* is a ratio in respect to how much energy is directed in the side lobes of the main lobe of the antenna (see Figure 10.3).

The *power dissipation* is a measure of the total power the antenna can accept at its input terminals. The antenna chosen should be able to handle the maximum envisioned power load without any problem. The amount of *intermodulation* which the antenna introduces to the network in the presence of strong signals as referenced from manufacturer should be considered in antenna selection. The *construction* attributes are associated with physical dimensions, mounting requirements, material used, wind loading, and connectors.

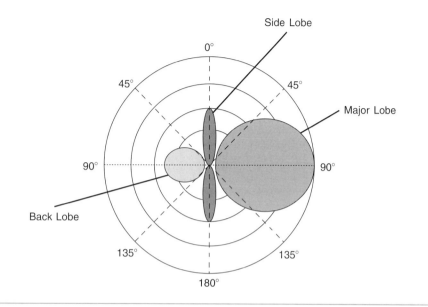

Figure 10.3   Radiation pattern of an antenna.

10.5   Relationship between Directivity, Gain, and Beam Width of an Antenna

Real antennas are not isotropic radiators but have a pattern of more and less power in different directions. The antenna engineer considers the pattern of the power radiated in the horizontal and vertical directions. The shape of the pattern describes the directionality of the antenna. The direction for maximum power is referred to as the primary beam or *main lobe*, whereas secondary beams are called the minor lobes (back and side lobes), see Figure 10.3 [7]. The pattern of the antenna has two desired effects: concentration of the power in a desired direction to improve the signal strength at the receiver and weakening the power in an undesired direction to reduce interference from or to other receivers. Thus, the minor lobes give undesired radiation or reception. Since the major lobe propagates and receives the most energy, this lobe is called the *front lobe*. Lobes adjacent to the front lobe are the *side lobes* and the lobe exactly opposite to the front lobe is called the *back lobe*. The front-to-back ratio of an antenna is defined as:

$$R_{FB} = 10 \log \frac{p_{mF}}{p_{mB}} \tag{10.12}$$

where:

p_{mF} = maximum power density of the front lobe
p_{mB} = maximum power density of the back lobe

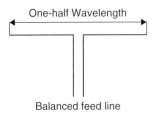

Figure 10.4   Half-wave dipole antenna.

The front-to-side ratio of an antenna is given as:

$$R_{FS} = 10\log \frac{p_{mF}}{p_{mS}} \qquad (10.13)$$

where:

p_{mS} = maximum power density of the side lobe

The half-wave dipole antenna has two parts (see Figure 10.4). The half-wave length is handy for impedance matching. Since the dipole antenna is the simplest one to build, it is often used as the reference to describe the gain of other antennas. The average power density for the dipole vertical antenna is given as:

$$p_{avg} = \frac{3P_T}{8\pi d^2}(\sin\theta)^2 \qquad (10.14)$$

In Equation 10.14 it should be noticed that power density at any point depends on the direction θ and distance d from the dipole. Since there is no dependence on ϕ, the antenna pattern is directional in x-z and y-z planes, but omnidirectional in the x-y plane.

A dipole mounted in a vertical direction provides an omnidirectional pattern that is useful for the base station antenna of cellular systems. While dipoles are often used as simple antennas, in real systems other types of antennas are used to provide higher gain than a dipole antenna.

10.5.1   The Relationship between Directivity and Gain

The directivity of an antenna is the ratio of the maximum power density from the antenna and power density from an isotropic antenna

$$D = \frac{|p|_{max}}{\left[\dfrac{P_T}{4\pi d^2}\right]} \qquad (10.15)$$

where:

$|p|_{max}$ = maximum power density from the antenna
P_T = transmitted power of an isotropic antenna

The gain of an antenna is the ratio of maximum power density from the antenna and input power density if the antenna is isotropic.

$$G = \frac{|p|_{max}}{\left[\dfrac{P_I}{4\pi d^2}\right]} = \frac{|p|_{max}}{\left[\dfrac{P_T}{4\pi d^2 \cdot \eta}\right]}$$ (10.16a)

Using Equation 10.15 in Equation 10.16a, we get:

$$G = \eta \times D$$ (10.16b)

where:

η = efficiency of the antenna
P_I = input power of an isotropic antenna

10.5.2   Relation between Gain and Beam Width

The receiver gain G_R can be related to its half-power beam width as [7]:

$$G_R = \frac{4\pi}{\theta_{HP}\phi_{HP}}$$ (10.17)

where:

θ_{HP} and ϕ_{HP} are the half-power beam widths in the θ and ϕ planes

The factor 4π is the solid angle subtended by a sphere in steradians (square radians)

$$4\pi \text{ steradians} = 4\pi \times \left[\frac{180}{\pi}\right]^2 = 41{,}250 \text{ degree}^2 = \text{solid angle in a sphere}$$

$$G_R = \frac{41{,}250}{\theta_{HP}\phi_{HP}}$$ (10.18)

For an ideal gain antenna, where power density is uniform inside the 3-dB beam width (for both θ and ϕ) and zero outside the 3-dB beam width, the gain G_R can be expressed as:

$$G_R = \frac{41{,}250}{\theta\phi}$$ (10.19)

where:

θ and ϕ are in degrees

For a real antenna with side lobes, the gain should be calculated using Equation 10.20, which includes the effect of side lobes.

$$G_R = \frac{32,400}{\theta\phi} \qquad (10.20)$$

If an antenna is designed with a circular pattern in one direction, i.e., a linear element is used, the approximate gain can be obtained for the vertical 3-dB beam width from Equation 10.19 as:

$$G_R \approx \frac{41,250}{360 \times \theta} = \frac{114.6}{\theta} \qquad (10.21)$$

The corresponding gain equation for a linear element including side lobes is given as [7]:

$$G_R \approx \frac{101.5}{\theta} \qquad (10.22)$$

10.5.3 Helical Antennas

Helical antennas provide circular polarized waves in the same direction as that of the helix. A helical antenna can be used to receive circular polarized waves and also receive plane polarized waves with the polarization in any direction. Helical antennas are often used with VHF and UHF satellite transmissions. The gain of a helical antenna is proportional to the number of turns (N). An approximate expression for the gain with respect to an isotropic antenna is given as [7]:

$$G = \frac{15NS(\pi D_H)^2}{\lambda^3} \qquad (10.23)$$

where:

G = antenna gain in dB

N = number of turns in the helix, $N > 3$

S = turn spacing in meters, $S \approx \frac{\lambda}{4}$

D_H = diameter of helix in meters, $D_H = \frac{\lambda}{\pi}$

λ = wave length in meters.

The radian pattern for this antenna type has one major lobe and several minor lobes. For the major lobe, the 3 dB beam width (in degrees) is approximately:

$$\theta = \frac{52\lambda}{\pi D_H} \cdot \sqrt{\frac{\lambda}{N \cdot S}} \qquad (10.24)$$

Example 10.3

We consider an antenna in which 12 W of power results in 3 W of power being dissipated as resistive losses in the antenna and the rest radiated by the antenna. If the directivity of the antenna is 7 dB (i.e., 5), what is its gain?

Solution

$$\eta = \frac{12 - 3}{12} = 0.75$$

$$G = \eta D = 0.75 \times 5 = 3.75 \ (5.74 \, \text{dB})$$

Example 10.4

What is the 3-dB beam width of a linear element antenna with a gain of 12 dBi?

Solution

$$\theta = \frac{101.5}{(10)^{12/10}} = 6.4°$$

Example 10.5

Consider a helical antenna, which has 12 turns and is designed for a frequency of 1.8 GHz.

  a. calculate the optimum diameter (D_H), spacing (S) for the antenna and total length of the antenna,
  b. calculate the antenna gain, and
  c. calculate the beam width of the antenna.

Solution
   a.

$$\lambda = \frac{c}{f} = \frac{300 \times 10^6}{1800 \times 10^6} = 0.1667 \, \text{m}$$

$$D_H = \frac{\lambda}{\pi} = \frac{0.1667}{\pi} = 0.053\,\text{m} = 53\,\text{mm}$$

$$S = \frac{\lambda}{4} = \frac{0.1667}{4} = 0.0417\,\text{m} = 41.7\,\text{mm}$$

$$L = NS = 12 \times 41.7 = 500\,\text{mm}$$

b.

$$G = \frac{15NS(D_H \cdot \pi)^2}{\lambda^3} = \frac{15 \times 12 \times 0.0417 \times (\pi \times 0.053)^2}{(0.1667)^3} = 44.92 = 16.5\,\text{dBi}$$

c.

$$\theta = \frac{52\lambda}{\pi D_H} \cdot \sqrt{\frac{\lambda}{NS}} = \frac{52 \times 0.1667}{\pi \times 0.053} \cdot \sqrt{\frac{0.1667}{12 \times 0.0417}} \approx 30\,\text{degrees}$$

10.6 Diversity

In Chapter 3, we pointed out that a radio channel is subjected to fading, time-dispersion, and other degradations. Diversity techniques are employed to overcome these impairments and improve signal quality [6,13,15,20]. The basic concept of diversity is that the receiver has more than one version of the transmitted signal available, and each version of transmitted signal is received through a distinct channel. When several versions of the signal, carrying the same information, are received over multiple channels that exhibit independent fading with comparable strengths, the chances that all the independently faded signal components experience the same fading simultaneously are greatly reduced. Suppose the probability of having a loss of communications due to fading on one channel is p and this probability is independent on all M channels. The probability of losing communications on all channels simultaneously is then p^M. Thus, a 10% chance of losing the signal for one channel is reduced to $0.1^3 = 0.001 = 0.1\%$ with three independently fading channels [5,17].

Typically, the diversity receiver is used in the base station instead of the mobile station, because the cost of the diversity combiner can be high, especially if multiple receivers are necessary. Also, the power output of the mobile station is limited by the battery. Handset transmitters usually lower power than mobile-mounted transmitters to preserve battery life and reduce radiation into the human body. The base station, however, can increase its power output or antenna height to improve the coverage to a mobile station.

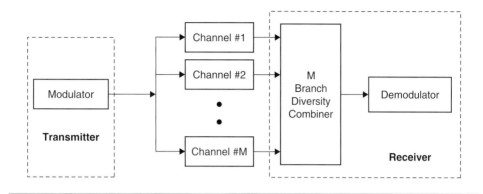

Figure 10.5 Diversity channel model.

Each of the channels, plus the corresponding receiver circuit, is called a *branch* and the outputs of the channels are processed and routed to the demodulator by a *diversity combiner* (see Figure 10.5).

Two criteria are required to achieve a high degree of improvement from a diversity system. First, the fading in individual branches should have low cross-correlation. Second, the mean power available from each branch should be almost equal. If the cross-correlation is too high, then fades in each branch will occur simultaneously. On the other hand, if the branches have low correlation but have very different mean powers, then the signal in a weaker branch may not be useful even though it has less fades than the other branches.

10.6.1 Types of Diversity
The following methods are used to obtain uncorrelated signals for combining:

1. **Space diversity:** Two antennas separated physically by a short distance *d* can provide two signals with low correlation between their fades. The separation *d* in general varies with antenna height *h* and with frequency. The higher the frequency, the closer the two antennas can be to each other. Typically, a separation of a few wavelengths is enough to obtain uncorrelated signals. Taking into account the shadowing effect (see Chapter 3), usually a separation of at least 10 carrier wavelengths is required between two adjacent antennas. This diversity does not require extra system capacity; however, the cost is the extra antennas needed.

2. **Frequency diversity:** Signals received on two frequencies, separated by coherence bandwidth (see Chapter 3) are uncorrelated. To use frequency diversity in an urban or suburban environment for cellular and personal communications services (PCS) frequencies, the frequency separation must be 300 kHz or more. This diversity improves link transmission quality at the cost of extra frequency bandwidths.

3. **Time diversity:** If the identical signals are transmitted in different time slots, the received signals will be uncorrelated, provided the time difference between time slots is more than the channel *coherence time* (see Chapter 3). This system will work for an environment where the fading occurs independent of the movement of the receiver. In a mobile radio environment, the mobile unit may be at a standstill at any location that has a weak local mean or is caught in a fade. Although fading still occurs even when the mobile is still, the time-delayed signals are correlated and time diversity will not reduce the fades. In addition to extra system capacity (in terms of transmission time) due to the redundant transmission, this diversity introduces a significant signal processing delay, especially when the channel coherence time is large. In practice, time diversity is more frequently used through bit interleaving, forward-error-correction, and automatic retransmission request (ARQ).

4. **Polarization diversity:** The horizontal and vertical polarization components transmitted by two polarized antennas at the base station and received by two polarized antennas at the mobile station can provide two uncorrelated fading signals. Polarization diversity results in 3 dB power reduction at the transmitting site since the power must be split into two different polarized antennas.

5. **Angle diversity:** When the operating frequency is ≥ 10 GHz, the scattering of signals from transmitter to receiver generates received signals from different directions that are uncorrelated with each other. Thus, two or more directional antennas can be pointed in different directions at the receiving site and provide signals for a combiner. This scheme is more effective at the mobile station than at the base station since the scattering is from local buildings and vegetation and is more pronounced at street level than at the height of base station antennas. Angle diversity can be viewed as a special case of space diversity since it also requires multiple antennas.

6. **Path diversity:** In code division multiple access (CDMA) systems, the use of direct sequence spread spectrum modulation allows the desired signal to be transmitted over a frequency bandwidth much larger than the channel coherence bandwidth. The spread spectrum signal can resolve in multipath signal components provided the path delays are separated by at least one chip period. A Rake receiver can separate the received signal components from different propagation paths by using code correlation and can then combine them constructively. In CDMA, exploiting the path diversity reduces the transmitted power needed and increases the system capacity by reducing interference.

10.7 Combining Methods

The goal of a combiner is to improve the noise performance of the system [1,11,14]. After obtaining the uncorrelated signals, we need to consider the method of processing these signals to obtain the best results. The analysis of combiners is generally performed in terms of SNR. We will examine several different types (selection, maximal-ratio, and equal-gain) of combiners and compare their SNR improvements over no diversity.

10.7.1 Selection Combiner

In this case, the diversity combiner selects the branch that instantaneously has the highest SNR (see Figure 10.6). We assume that the signal received by each diversity branch is statistically independent of the signals in other branches and is Rayleigh distributed with equal mean signal power P_0. The probability density function of the signal envelope, on branch i, is given as (see Chapter 3)

$$p(r_i) = \frac{r_i}{P_0} e^{-r_i^2/(2P_0)} \qquad (10.25)$$

where:
$2P_0$ = mean-square signal power per branch = $<r_i>$
r_i^2 = instantaneous power in the ith branch

Let $\xi_i = r_i^2/(2N_i)$ and $\xi_0 = (2P_0)/(2N_i)$, where N_i is the noise power in the ith branch

$$\therefore \frac{\xi_i}{\xi_0} = \frac{r_i^2}{2P_0} \qquad (10.26)$$

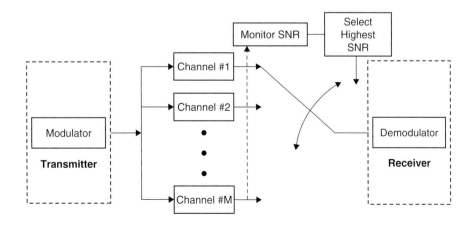

Figure 10.6 Diversity selection combiner.

The probability density function for ξ_i is given by

$$p(\xi_i) = \frac{1}{\xi_0} e^{-(\xi_i/\xi_0)} \tag{10.27}$$

Assuming that the signal in each branch has the same mean, the probability that the SNR on any branch is less than or equal to any given value ξ_g is

$$P[\xi_i \leq \xi_g] = \int_0^{\xi_g} p(\xi_i)d\xi_i = 1 - e^{-(\xi_g/\xi_0)} \tag{10.28}$$

Therefore, the probability that the SNRs in all branches are simultaneously less than or equal to ξ_g is given by

$$P_M(\xi_g) = P\left[\xi_1, \xi_2, ..., \xi_M \leq \xi_g\right] = \left[1 - e^{-(\xi_g/\xi_0)}\right]^M \tag{10.29}$$

The probability that at least one branch will exceed the given SNR value of ξ_g is given by:

$$P\,(\text{at least one branch} \geq \xi_g) = 1 - P_M(\xi_g) \tag{10.30}$$

The percentage of time the instantaneous output SNR ξ_M is below or equal to the given value, ξ_g, is equal to $P(\xi_M \leq \xi_g)$. The results of the selection combiner for $M = 1, 2,$ and 4 are plotted in Figure 10.7. Note that the largest gain occurs for the 2-branch diversity combiner.

By differentiating Equation 10.29 we get the probability density function

$$p_M(\xi_g) = M(\xi_g/\xi_0)\left[1 - e^{-(\xi_g/\xi_0)}\right]^{M-1} e^{-(\xi_g/\xi_0)} \tag{10.31}$$

The mean value of the SNR can be given as

$$\bar{\xi}_M = \int_0^\infty M\left(\frac{\xi_g}{\xi_0}\right)\left[1 - e^{-(\xi_g/\xi_0)}\right]^{M-1} e^{-(\xi_g/\xi_0)}d\xi_g \tag{10.32}$$

$$\therefore \frac{\bar{\xi}_M}{\xi_0} = \sum_{k=1}^M \frac{1}{k} \tag{10.33}$$

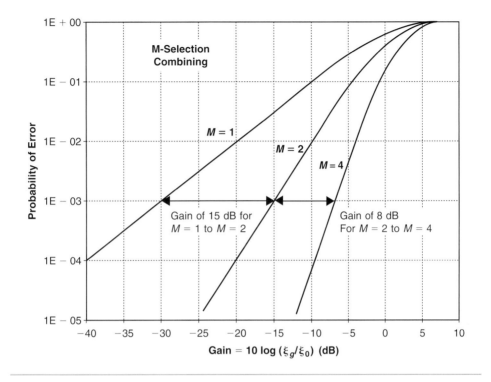

Figure 10.7 Probability for different values of *M*-selection combiners.

Example 10.6

We consider the 3-branch selection combiner in which each branch receives an independent Rayleigh fading signal. If the average SNR is 30 dB (1000), determine the probability that the SNR will drop below 10 dB (10). Compare the result with a case without any diversity.

Solution

$$\xi_g = 10 \, \text{dB and } \xi_0 = 30 \, \text{dB}$$

$$\therefore \frac{\xi_g}{\xi_0} = \frac{10}{1000} = 0.01$$

$$p_3(10 \, \text{dB}) = \left[1 - e^{-(\xi_g/\xi_0)} \right]^M = \left[1 - e^{-0.01} \right]^3 = 1 \times 10^{-6}$$

When diversity is not used, $M = 1$

$$p_1(10 \, \text{dB}) = \left[1 - e^{-(\xi_g/\xi_0)} \right]^M \left[1 - e^{-0.01} \right]^1 = 1 \times 10^{-2}$$

10.7.2   Switched Combiner

The disadvantage with selection combining is that the combiner must be able to monitor all M branches simultaneously. This requires M independent receivers which is expensive and complicated. An alternative is to use switched combining. In this case only one receiver is needed, and it is only switched between branches when the SNR on the current branch is lower than some predefined threshold value ξ_g. This is called a *switch and stay* combiner.

The performance of a switch combiner is less than that in selection combining, since unused branches may have SNRs higher than the current branch if the current SNR exceeds the threshold. The threshold therefore has to be carefully selected in relation to the mean power on each branch, which must also be estimated with sufficient accuracy.

10.7.3   Maximal Ratio Combiner

In maximal ratio combining, M signals are weighted proportionally to their signal-to-noise ratios and then summed (see Figure 10.8).

$$r_M = \sum_{i=1}^{M} a_i r_i(t) \tag{10.34}$$

where:

$\quad a_i$ = weight of ith branch
$\quad M$ = number of branches

Since noise in each branch is weighted according to noise power,

$$\overline{n_i^2(t)} = \sum_{j=1}^{M} \sum_{i=1}^{M} a_i a_j \overline{n_i(t) n_j(t)} \tag{10.35}$$

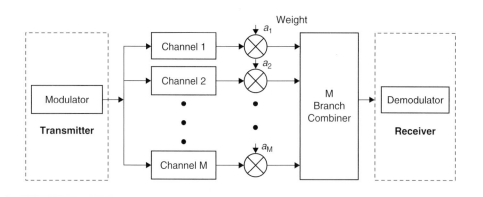

Figure 10.8   Maximal ratio combining.

$$N_T = \sum_{i=1}^{M} a_i^2 \overline{n_i^2(t)} = \sum_{i=1}^{M} |a_i|^2 N_i \tag{10.36}$$

where:

N_T = average noise power

$\overline{n_i^2(t)} = 2N_i$

The SNR at the output is given as:

$$\xi_M = \frac{1}{2} \cdot \frac{\left| \sum_{i=1}^{M} a_i r_i(t) \right|^2}{\sum_{i=1}^{M} |a_i|^2 N_i} \tag{10.37}$$

We want to maximize ξ_M. This can be done using the Schwartz inequality.

$$\left| \sum_{i=1}^{M} a_i r_i \right|^2 \leq \left[\sum_{i=1}^{M} |r_i^2| \right] \left[\sum_{i=1}^{M} |a_i|^2 \right] \tag{10.38}$$

If $a_i = r_i/(\sqrt{N_i})$, then using Equation 10.38 to define Equation 10.39 we get:

$$\xi_{M_{max}} = \frac{1}{2} \cdot \frac{\sum_{i=1}^{M} r_i^2 \sum_{i=1}^{M} \frac{r_i^2}{N_i}}{\sum_{i=1}^{M} r_i^2} \tag{10.39}$$

$$\therefore \xi_{M_{max}} = \frac{1}{2} \cdot \sum_{i=1}^{M} \frac{r_i^2}{N_i} = \sum_{i=1}^{M} \xi_i \tag{10.40}$$

Thus, the SNR at the combiner output equals the sum of the SNR of the branches.

$$\overline{\xi}_{M_{max}} = \sum_{i=1}^{M} \overline{\xi}_i = \sum_{i=1}^{M} \xi_0 = M\xi_0 \tag{10.41}$$

$$\therefore \frac{\overline{\xi}_M}{\xi_0} = M \tag{10.42}$$

The probability that $\xi_M \le \xi_g$ is given by:

$$p\left(\xi_M \le \xi_g\right) = 1 - e^{-\frac{\xi_g}{\xi_0}}\left[\sum_{k=1}^{M} \frac{\left(\frac{\xi_g}{\xi_0}\right)^{k-1}}{(k-1)!}\right] \tag{10.43}$$

$$p\left(\xi_M \ge \xi_g\right) = e^{-\frac{\xi_g}{\xi_0}}\left[\sum_{k=1}^{M} \frac{\left(\frac{\xi_g}{\xi_0}\right)^{k-1}}{(k-1)!}\right] \tag{10.44}$$

The plot of p for $M = 1$, 2, and 4 with maximal ratio combining is shown in Figure 10.9.

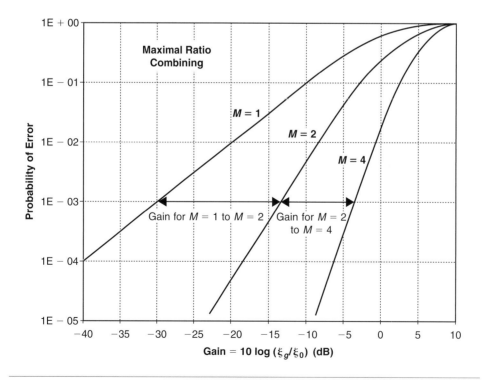

Figure 10.9   Probability for different values of maximal ratio combiner.

10.7.4 Equal Gain Combiner

Equal gain combining is similar to maximal ratio combining, but there is no attempt to weight the signal before addition; thus $a_i = 1$. The envelope of the output signal is given as:

$$r = \sum_{i=1}^{M} r_i(t) \tag{10.45}$$

and mean output SNR is given as:

$$\bar{\xi}_M = \frac{1}{2} \cdot \frac{\overline{\left[\sum_{i=1}^{M} r_i\right]^2}}{\sum_{i=1}^{M} \bar{N}_i} \tag{10.46}$$

Assuming that mean noise in each branch is the same (i.e., N); Equation 10.46 becomes

$$\bar{\xi}_M = \frac{1}{2NM} \overline{\left[\sum_{i=1}^{M} r_i\right]^2} = \frac{1}{2NM} \sum_{i,j=1}^{M} \overline{r_i r_j} \tag{10.47}$$

but $\overline{r_i^2} = 2P_0$; and $\bar{r}_i = \sqrt{(\pi P_0)/2}$.

The various branch signals are uncorrelated, $\overline{r_j r_i} = \overline{r_i r_j} = \bar{r}_i \bar{r}_j$, for i not equal to j. Therefore Equation 10.47 will become:

$$\bar{\xi}_M = \frac{1}{2NM}\left[2MP_0 + M(M-1) \cdot \frac{\pi P_0}{2}\right] = \xi_0\left[1 + (M-1)\frac{\pi}{4}\right] \tag{10.48}$$

$$\frac{\bar{\xi}_M}{\xi_0} = 1 + (M-1)\frac{\pi}{4} \tag{10.49}$$

For $M = 2$, the probability p can be written in the closed form as:

$$p(\xi_M \leq \xi_g) = 1 - e^{-\left(\frac{2\xi_g}{\xi_0}\right)} - \left(\sqrt{\pi\left(\frac{\xi_g}{\xi_0}\right)} \cdot e^{-\left(\frac{\xi_g}{\xi_0}\right)} \cdot \text{erf}\sqrt{\frac{\xi_g}{\xi_0}}\right) \tag{10.50}$$

For $M > 2$, the bit error probability should be obtained by a numerical integration technique.

Table 10.1 SNR ratio in dB.

M	Selection combiner	Maximal-ratio combiner	Equal-gain combiner
1	−20	−20	−20
2	−10	−8.5	−9.2
4	−4.0	−1.0	−2.0
6	−2.0	2.0	1.5

Table 10.2 SNR improvement in dB at 1% bit-error probability.

M	Selection combiner	Maximal-ratio combiner	Equal-gain combiner
2	10.0	11.5	10.8
4	16.0	19.0	18.0
6	18.0	22.0	21.5

Table 10.1 shows M versus SNR at 1% bit error probability for the selection, maximal ratio, and equal gain combiner. Table 10.2 shows SNR improvement for $M = 2$, 4, and 6 at 1% bit error probability for the selection, maximal ratio, and equal gain combiner. It can be observed that the selection diversity combiner has the poorest performance and the maximal ratio the best. The performance of the equal gain diversity combiner is sightly lower than that of the maximal ratio combiner. The implementation complexity for equal gain combining is significantly less than the maximal ratio combining because of the requirement of correct weighing factors. Performance of the three combining schemes is compared in Figure 10.10.

10.8 Rake Receiver

In 1958, Price and Green proposed a method of resolving multipath using wideband pseudo-random sequences modulated onto a transmitter using other modulation methods (AM or FM) [4]. The pseudo-random sequence (see Chapter 11) has the property that its time-shifted versions are almost uncorrelated. Thus, a signal that propagates from transmitter to receiver over multipath (hence different time delays) can be resolved into separately fading signals by cross-correlating the received signal with multiple time-shifted versions of the pseudo-random sequence. Figure 10.11 shows a block diagram of a typical system. In the receiver, the outputs are time shifted and therefore must be sent through a delay line before entering the diversity combiner. The receiver is called a Rake receiver since the block diagram looks like a garden rake.

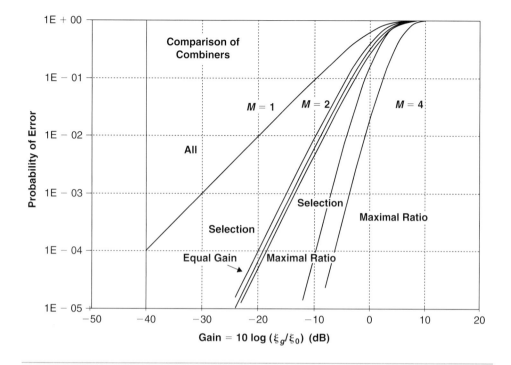

Figure 10.10   Performance comparison of various combining schemes.

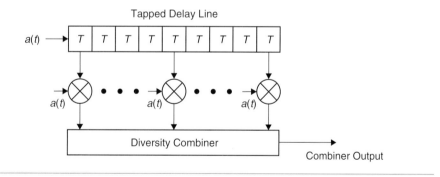

Figure 10.11   Rake receiver.

When CDMA systems were designed for cellular systems, the inherent wide bandwidth signals with their orthogonal Walsh functions (see Chapter 11) were a natural for implementing a Rake receiver. The Rake receiver reduces the effects of fading and provides spectral efficiency improvement of CDMA over other cellular systems.

In CDMA systems, the bandwidth (1.25 to 15 MHz) is wider than the coherence bandwidth of the cellular or PCS channel. Thus, when the multipaths are resolved in the receiver, the signals from each tap on the delay line are uncorrelated with each other. The receiver can combine them using any of the combining techniques described in the previous section. Thus, the CDMA system uses the multipath characteristics of the communication channel to its advantage to improve the operation of the system.

The performance of the Rake receiver will be governed by the combining scheme used. In the IS-95 and WCDMA, the maximal ratio combining with 3 to 4 branches is used. An important factor in the receiver design is to obtain sychronization of the signals in the receiver. Since in CDMA, adjacent cells are also on the same frequency with different time delays on the Walsh codes, the entire CDMA system is tightly synchronized.

Example 10.7

A Rake receiver is used in the wideband CDMA (WCDMA) system (spreading rate 3.84 Mcps) to reduce the multipath effects in the channel. What is the minimum delay difference to successfully resolve the multipath components and operate the Rake receiver?

Solution

In order to resolve multipath components, the chip duration should be equal to or greater than τ, where τ is defined as:

$$\tau = \frac{\text{delay distance}}{\text{speed of electromagnetic wave}} = \frac{d_d}{300 \times 10^6} \text{ s}$$

$$\text{Chip duration of WCDMA system} = \frac{1}{3.84 \times 10^6} = 0.26 \,\mu s$$

$$\frac{d_d}{300 \times 10^6} = 0.26 \times 10^{-6}$$

$$\therefore d_d = 78 \,\text{m}$$

10.9 Link Budgets

We discussed link margin in Chapter 3. In this section, we present a systematic procedure for developing a link budget of a wireless system [18]. A link budget is the calculation of the amount of power received at a given receiver based on the output power from the transmitter. The link budget considers all of the gains and losses that a radio wave experiences along the path from transmitter to receiver. We need to perform the calculation in both directions: from the mobile station to the base station (uplink) and from the base station to the mobile station (downlink). We determine the maximum allowable path loss in each direction and use the lesser of the two to calculate the coverage for the cell and service in question.

The link budget should include a margin to allow fading of the signal. In other words, we design the system such that service will be supported even if the signal fades significantly. The greater the fade margin, the greater the reliability

Table 10.3 Uplink budget for speech (outdoor pedestrian) service at 12.2 kbps.

Transmitter (Mobile Station)		Note
Mobile transmit power (dBm)	21	
Antenna gain (dBi)	0	
Body losses (dB)	3.0	
Equivalent isotropic radiated power (EIRP)	18	
Receiver (Base Station)		
Thermal noise density (dBm/Hz)	−174	$10\log(1.38 \times 10^{-23} \times 290) =$ $-204\,dBW = -174\,dBm$
Receiver noise figure (dB)	5.0	
Receiver noise power (dBm), calculated for 3.84 Mcps (see Chapter 11)	−103.2	$-174 + 5 + 10\log(3.84 \times 10^6) =$ $-103.2\,dBm$
Interference margin (dB)	4	Depends on cell loading
Total noise + interference (dBm)	−99.2	
Processing gain (dB) (see Chapter 11)	25	$10\log(3.84 \times 10^6/12.2 \times 10^3)$
Required E_b/N_0 (dB)	4	Depend on service type
Effective receiver sensitivity (dBm)	−120.2	Total noise + interference − Processing gain + Required E_b/N_0
Base station antenna gain (dBi)	18	
Base station feeder and connector losses (dB)	2	
Fast fading margin (dB)	4	Enables room for closed loop power control
Log normal fading margin (dB)	7.5	Enable for greater cell-area reliability
Building penetration loss (dB)	0	
Soft handoff gain (dB)	2	
Max. allowable path loss (dB)	**144.7**	**18 − (−120.2) + (18 − 2 − 4 −7.5 + 2) = 144.7**

Table 10.4 Downlink budget for speech (outdoor pedestrian) service at 12.2 kbps.

Transmitter (Base Station)		Note
BS transmit power (dBm)	40	10 W
BS antenna gain (dBi)	18	
BS feeder and connector losses (dB)	2.0	
Equivalent isotropic radiated power (EIRP) (dBm)	56	
Receiver (Mobile Station)		
Thermal noise density (dBm/Hz)	−174	$10\log(1.38 \times 10^{-23} \times 290) =$ −204 dBW = −174 dBm
Receiver noise figure (dB)	5.0	
Receiver noise power (dBm), calculated for 3.84 Mcps	−103.2	$-174 + 5 + 10\log(3.84 \times 10^{6}) =$ −103.2 dBm
Interference margin (dB)	0	Depends on cell loading
Total noise + interference (dBm)	−103.2	
Processing gain (dB)	25	$10\log(3.84 \times 10^{6}/12.2 \times 10^{3})$
Required E_b/N_0 (dB)	4	Service-dependent
Effective receiver sensitivity (dBm)	−124.2	Total noise + interference − Processing gain + Required E_b/N_0
Mobile station antenna gain (dBi)	0	
Body losses (dB)	3.0	
Fast fading margin (dB)	4	Enables room for closed-loop power control
Lognormal fading margin (dB)	7.5	Enable for greater cell-area reliability
Building penetration loss (dB)	0	
Soft handoff gain (dB)	2	
Max. allowable path loss (dB)	**167.7**	**−(−124.2) + 56 + (−3 − 4 − 7.5 + 2) = 167.7**

of the service. Tables 10.3 and 10.4 provide examples of the uplink and downlink link budget for the WCDMA speech service at 12.2 kbps (outdoor pedestrian). The maximum allowable path loss is 144.7 dB based on uplink budget. The cell coverage must be based on the uplink path loss.

10.10 Summary

In this chapter we discussed the basics of antenna technology. We presented performance criteria of an antenna system and developed relations between the antenna gain, directivity, and beam width. We then moved to discuss diversity, an extremely powerful method for improving the quality of communication systems. It is possible to achieve gains equivalent to power savings in excess of 10 dB. These gains are obtained at the expense of extra hardware, particularly in terms of extra antennas and receivers, which must be balanced against the benefits. The key requirements for achieving the maximum benefit are that the multiple branches of the system should have substantially equal mean power and near-zero cross-correlation of the fading signal. We concluded the chapter by presenting link budget calculations for a WCDMA system.

Problems

10.1 Define gain, beam width, and radiation efficiency of an antenna.

10.2 What is diversity? How is it provided in a communication system?

10.3 Among the selection, equal-gain, and maximal ratio combining, which scheme is the best and why?

10.4 In Problem 9.1, calculate the SNR ratio when a frequency of 6 GHz is used for the system. At this frequency, use 10 dB for the receiver noise figure.

10.5 Repeat Problem 9.2 for 3 and 6 GHz and use 10 dB for the receiver noise figure. What can you say about system design needs as frequency is raised?

10.6 If the 3-dB beam width of a linear element antenna is 8°, find its gain.

10.7 A sector antenna has vertical and horizontal beam widths of 50° and 60°, respectively. Calculate its gain (1) ignoring the side lobes and (2) accounting the side lobes.

10.8 Calculate the time separation required for two signals to achieve a high degree of time diversity in a classical Rayleigh channel at 900 MHz with a mobile speed of 20 km/hour.

10.9 Given a 2-branch selection combining system operated with independent Rayleigh fading, calculate the diversity gain for a probability of 10^{-6}.

10.10 Repeat Problem 10.9 for the 2-branch maximal ratio combining system.

References

1. Brennan, D. J. Linear Diversity Combining Techniques. Proceedings of IRE, June 1959, 1075–1101.

2. Bryson, W. B. Antenna System for 800 MHz. IEEE Vehicular Technology Conference, San Diego, May 1982, p. 287.

3. Halpern, S. W. The Theory of Operation of an Equal-Gain Predetection Regenerative Diversity Combiner With Rayleigh Fading Channels. *IEEE Transaction on Communications Technology*, COM-22(8), August 1974, 1099–1106.

4. Jakes, W. C. *Microwave Mobile Communications*. New York: John Wiley & Sons, 1974.

5. Kahn, L. R. Radio Squarer. Proceedings of the IRE, 42, November 1954, 1704.

6. Kozono, S., et al. Base Station Polarization Diversity Reception for Mobile Radio. *IEEE Transactions, Vehicular Technology*, 33(4), 301–306, 1984.

7. Kraus, J. D. *Antennas*. New York: McGraw-Hill, 1988.

8. Lee, W. C .Y. Antenna Spacing Requirements for a Mobile Radio Base-Station Diversity. *Bell System Technical Journal*, 50, no. 6 (July-August 1971).

9. Lee, W. C. Y. *Mobile Communication Engineering*. New York: McGraw-Hill, 1982.

10. Lee, W. C. Y. Mobile Radio Performance for a Two-Branch Equal-Gain Combining Receiver with Correlated Signals at the Land Site. *IEEE Transaction EA*, VT-27, November 1978, 239–243.

11. Mahrotra, A. *Cellular Radio Performance Engineering*. Boston: Artech House, 1994.

12. Mark, J. W., and Zhuang, Weihua. *Wireless Communications and Networking*. Upper Saddle River, NJ: Prentice Hall, 2003.

13. Parsons, J. D., et al. Diversity Techniques for Mobile Radio Reception. *Radio and Electronic Engineer*, 45, no. 7 (July 1975): 357–367.

14. Parsons, J. D., and Gardiner, J. G. *Mobile Communications Systems*. New York: Halsted Press, 1989.

15. Price, R., and Green, P. E., Jr. A Communications Technique for Multipath Channels. Proceeding of the IRE. March 1958, 555–570.

16. Saunders, S. R. *Antennas and Propagation for Wireless Communication Systems*. New York: John Wiley & Sons, 1999.

17. Schwartz, M., et al. *Communication System Techniques*. New York: McGraw-Hill, 1966.

18. Smith, C., and Collins, D. *3G Wireless Networks*. New York: McGraw-Hill, 2002.

19. Tilston, W. V. On Evaluating the Performance of Communication Antennas. *IEEE Communications Society Magazine* (September 1981).

20. Vaughan, R. G. Polarization Diversity in Mobile Communications. *IEEE Transactions Vehicular Technology*, 39(3), 177–186, 1990.

CHAPTER 11

Spread Spectrum (SS) and CDMA Systems

11.1 Introduction

We introduced spread spectrum techniques in Chapter 6. In this chapter, we present details of direct-sequence spread spectrum (DSSS) and frequency-hop spread spectrum (FHSS) systems [1,2,4]. We show how signal spreading and despreading is achieved with binary phase shift keying (BPSK) and quadrature phase shift keying (QPSK) modulation in the DSSS [11,12]. We then address multipath issues in wireless communications and show how code division multiple access (CDMA) takes advantage of multipath in improving system performance with a Rake receiver [6,13]. We conclude the chapter by presenting a summary of the challenges in implementing a CDMA system and providing some highlights of the Telecommunication Industries Association (TIA) IS-95 CDMA system. Those who are not familiar with spreading codes should refer to Appendix D.

11.2 Concept of Spread Spectrum

In a wideband spread-spectrum (SS) system, the transmitted signal is spread over a frequency band that is much larger, in fact, than the maximum bandwidth required to transmit the information bearing (baseband) signal [3]. An SS system takes a baseband signal with a bandwidth of only a few kilohertz (kHz), and spreads it over a band that may be many megahertz (MHz) wide. In SS systems, an advantage in signal-to-noise ratio (SNR) is achieved by the modulation and demodulation process.

The SS signal is generated from a data-modulated carrier. The data-modulated carrier is modulated a second time by using a wideband spreading signal. An SS signal has advantages in the areas of security, resistance to narrowband jamming, resistance to multipath fading, and supporting multiple-access techniques.

The spreading modulation may be phase modulation or a rapid change of the carrier frequency, or it may be a combination of these two schemes. When spectrum spreading is performed by phase modulation, we call the resultant signal a *direct-sequence spread spectrum* (DSSS) signal (see Figure 11.1) [15]. When spectrum spreading is achieved by a rapid change of the carrier frequency, we refer to the resultant signal as a *frequency-hop spread spectrum* (FHSS) signal [14]. When both direct-sequence and frequency-hop techniques are employed, the

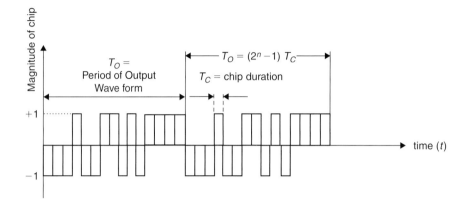

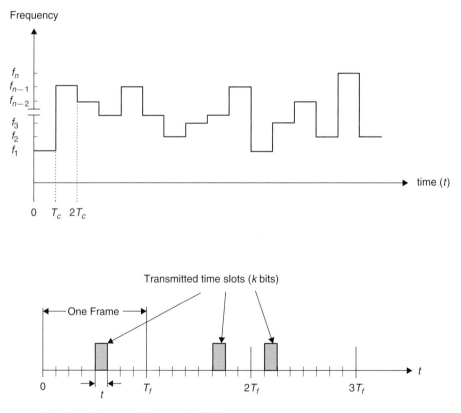

M = time slots in each frame; $t = T_f/M$
Note: M = 8 in this example

Figure 11.1   Spread spectrum techniques.

resultant signal is called a hybrid DS-FH SS signal. Another way to also generate an SS signal is the *time-hop spread spectrum* (THSS) signal. In this case, the transmission time is divided into intervals called frames. Each frame is further divided into time slots. During each frame, one and only one time slot is modulated with a message (details of THSS are not given in this chapter). The DSSS is the averaging technique to reduce interference whereas FHSS and THSS are the avoiding techniques to minimize interference.

The spreading signal is selected to have properties to facilitate demodulation of the transmitted signal by the intended receiver, and to make demodulation by an unintended receiver as difficult as possible. These same properties also make it possible for the intended receiver to differentiate between the communication signal and jamming. If the bandwidth of the spreading signal is large relative to the data bandwidth, the spread-spectrum transmission bandwidth is dominated by the spreading signal and is independent of the data signal bandwidth.

Consider the channel capacity as given by the Shannon equation [10]:

$$C = B_w \log_2(1 + S/N) \tag{11.1}$$

where:
  B_w = channel bandwidth in Hertz (Hz)
  C = channel capacity in bits per second (bps)
  S = signal power
  N = noise power

Equation 11.1 provides the relationship between the theoretical ability of a communication channel to transmit information without errors for a given signal-to-noise ratio and a given bandwidth of the channel. The channel capacity can be increased by increasing the channel bandwidth, the transmitted power, or a combination of both.

Shannon modeled the channel at baseband. However, Equation 11.1 is applicable to a radio frequency (RF) channel by assuming that the intermediate frequency (IF) filter has an ideal (flat) bandpass response with a bandwidth that is at least $2B_w$. This bound assumes that channel noise is additive white Gaussian noise (AWGN). AWGN is often used in the modeling of an RF channel. This assumption is justified since the total noise is generated by random electron fluctuations. The central limit theorem provides us with the assumption that the output of an IF filter has a Gaussian distribution and is frequency independent. For most communications systems that are limited by thermal noise, this assumption is valid. For interference-limited systems, this assumption is not valid and the results may be different. The Shannon equation does not provide a method to achieve the bound. Approaching the bound requires complex channel coding and modulation schemes. In many cases, achieving an implementation that provides performance near this bound is impractical due to the resulting complexity. Spread spectrum systems can be engineered to operate at much lower SNRs since the channel bandwidth can be traded for the SNR to achieve good performance at a very low SNR.

We rewrite Equation 11.1 as:

$$\frac{C}{B_w} = 1.44 \log_e (1 + S/N) \tag{11.2}$$

Since

$$\log_e(1 + S/N) = \frac{S}{N} - \frac{1}{2}\left(\frac{S}{N}\right)^2 + \frac{1}{3}\left(\frac{S}{N}\right)^3 - \frac{1}{4}\left(\frac{S}{N}\right)^4 + \cdots \tag{11.3}$$

Assuming that the S/N ratio is small (e.g., $S/N \le 0.1$), we can neglect the higher-order terms and rewrite Equation 11.2 as:

$$B_w \approx \frac{C}{1.44} \times \frac{1}{(S/N)} \tag{11.4}$$

For any given S/N ratio we can have a low information error rate by increasing the bandwidth used to transmit the information. As an example, if we wish a system to operate on an RF link in which the data information rate is 20 kilobits per second (kbps) and the S/N ratio is 0.01, we should use a bandwidth of

$$B_w = \frac{20 \times 10^3}{1.44 \times 0.01} = 1.38 \times 10^6 \, \text{Hz or } 1.38 \, \text{MHz} \tag{11.5}$$

Information can be modulated into the SS signal by several methods. The most common method is to multiply the information with the spread-spectrum code (refer to Appendix D) before it is used to modulate the carrier frequency (see Figure 11.2). This technique applies to any SS system that uses a spreading code to determine RF bandwidth. If the signal that is being sent is analog (voice, for example), the signal must be digitized before being modulated by the spreading code.

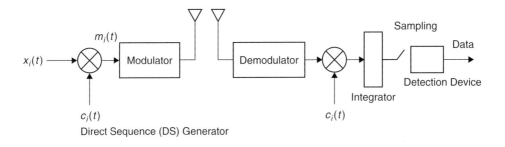

Figure 11.2 Direct sequence spread spectrum system.

11.3   System Processing Gain

In the DSSS system the baseband signal is spread over a large bandwidth by a spreading code. One of the major advantages of a DSSS system is the robustness to interference [8,9]. The system processing gain, G_p, quantifies the degree of interference rejection. The system processing gain is the ratio of spreading rate, R_c, to the information rate, R_b, and is given as:

$$G_p = R_c/R_b \tag{11.6}$$

where:

R_b = information bit rate
R_c = spreading rate

Typical processing gains of a DSSS system lie between 10 and 60 dB. With a DSSS system, the total noise level is determined both by thermal noise and by interference. For a given user, the interference is processed as noise. The input and output S/N ratios are related as:

$$(S/N)_o = G_p(S/N)_i \tag{11.7}$$

We express input $(S/N)_i$ ratio as:

$$\left(\frac{S}{N}\right)_i = \left(\frac{E_b \times R_b}{N_0 \times B_w}\right)_i = \left(\frac{E_b}{N_0}\right)_i \times \frac{1}{G_p} \tag{11.8}$$

where:

E_b = bit energy
N_0 = noise power spectral density

Using Equation 11.8 we rewrite Equation 11.7 as:

$$\left(\frac{S}{N}\right)_o = G_p \cdot \left(\frac{S}{N}\right)_i = \left(\frac{E_b}{N_0}\right)_i \tag{11.9}$$

but

$$\left(\frac{S}{N}\right)_o = \left(\frac{E_b \times R_b}{N_0 \times B_w}\right)_o = \left(\frac{E_b}{N_0}\right)_o \cdot \frac{1}{G_p} \tag{11.10}$$

Therefore using Equation 11.9 and Equation 11.10 we get,

$$\left(\frac{E_b}{N_0}\right)_o = G_p \cdot \left(\frac{E_b}{N_0}\right)_i \tag{11.11}$$

Example 11.1

A DSSS system has a 1.2288 megachips per second (Mcps) code clock rate and a 9.6 kbps information rate. Calculate the processing gain. How much improvement in information rate is achieved if the code generation rate is changed to 5 Mcps and the processing gain to 256?

Solution

$$\frac{1.2288 \times 10^6}{9.6 \times 10^3} = 128 = 10 \log 128 = 21 \, dB$$

$$R_b = \frac{R_c}{G_p}$$

$$R_b = \frac{5 \times 10^6}{256} = 19.53 \, kbps$$

Improvement in information rate = $19.53 - 9.6 = 9.93 \, kbps$

In a DSSS system the information signal, $x_i(t)$, is multiplied by a wideband code signal, $c_i(t)$, which is the output signal of the direct sequence (DS) generator (see Figure 11.2) [7,9]. The signal $x_i(t) \times c_i(t) \equiv m_i(t)$ is modulated and transmitted. This signal occupies a bandwidth far in excess of the minimum bandwidth required to transmit the information signal $x_i(t)$. We can observe from Figure 11.3 that the combined signal waveform has more high frequency changes in the data information since $1/T_c \gg 1/T_b$, where T_b is the bit interval of the information stream and T_c is the bit interval of the DSSS stream. T_c is called a *chip interval*. In a DSSS system, the pulse shape is a sequence of short rectangular pulses called *chips*. The chips are a pseudo-random sequence of 1's and -1's known at the receiver and often having a repetition period equal to the symbol period.

$$G_p = \frac{T_b}{T_c}$$

Since $R_b = 1/T_b$ and $B_w = 1/T_c$,

$$G_p = \frac{T_b}{T_c} = \frac{B_w}{R_b}$$

When $x_i(t)$ and $c_i(t)$ have the same rate, the product $m_i(t)$ contains all the information of $x_i(t)$ and has the same rate as $c_i(t)$. The spectrum of the signal remains unchanged, and the incoming bit stream is said to be encrypted or *scrambled*. However, when $c_i(t)$ has a higher rate than $x_i(t)$, $m_i(t)$ contains all the information of $x_i(t)$, has a higher bit rate compared to $x_i(t)$, and is said to have had its spectrum spread (refer to Figure 11.3).

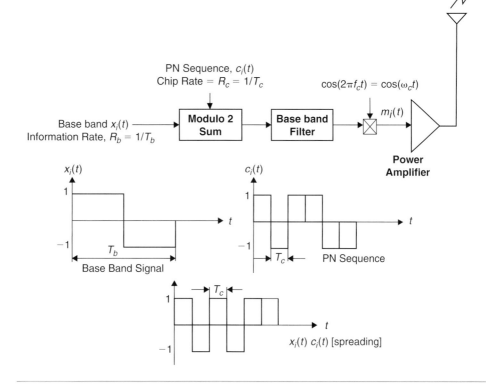

Figure 11.3   Direct sequence (DS) spreading.

Let us consider the downlink CDMA operation, in which the base station generates a data stream for mobile 1, 2, and 3 ... $x_i(t)$, and multiplies the data stream by an appropriate DS code, $c_i(t)$. Next, we add the coded data streams (see Figures 11.4, 11.5, and 11.6):

$$z(t) = \sum_{j=1}^{3} m_j \tag{11.12}$$

We modulate the resultant baseband spread spectrum signal, Equation 11.12, by a carrier (frequency $\omega_c = 2\pi f_c$) to obtain

$$z_i(t) = A_i \left\{ \sum_{j=1}^{3} c_j(t)x_j(t) \right\} \cos \omega_c t \tag{11.13}$$

The spread signal at the receiver for mobile, i, is $z_i(t)$ + noise, where $z_i(t)$ is transmitted over the allocated bandwidth.

$$z_i(t) = A_i \left[\sum_{j=1}^{3} c_j(t - \tau_j)x_j(t - \tau_j) \right] \cos(\omega_c t + \phi_i) \qquad (11.14)$$

Equation 11.14 contains the desired user signal and other users' signals. After multiplication by a coherent carrier phase ϕ_i estimated by the receiver, a locally generated DS sequence that is the exact replica of the desired code transmitted multiplies the incoming signal—despreading. We assume that this DS sequence is in perfect synchronization (receiver estimates delay τ_j) with the transmitted signal so that:

$$c_j(t - \tau_j)c_j(t - \tau_j) = 1 \qquad (11.15)$$

The multiplier output yields the desired data signal $x_i(t - \tau_j)$ plus interfering terms due to other users. Ideally, the integrator, an *integrate-and-dump* over T_b, should produce a cross-correlation between the desired signal and interferers that is 0. Hence, the output for mobile i is the transmitted data stream, $x_i(t - \tau_j)$.

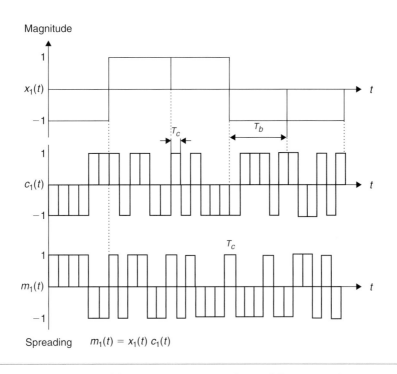

Figure 11.4   Downlink DSSS operation—signals for mobile 1, 2, and 3.

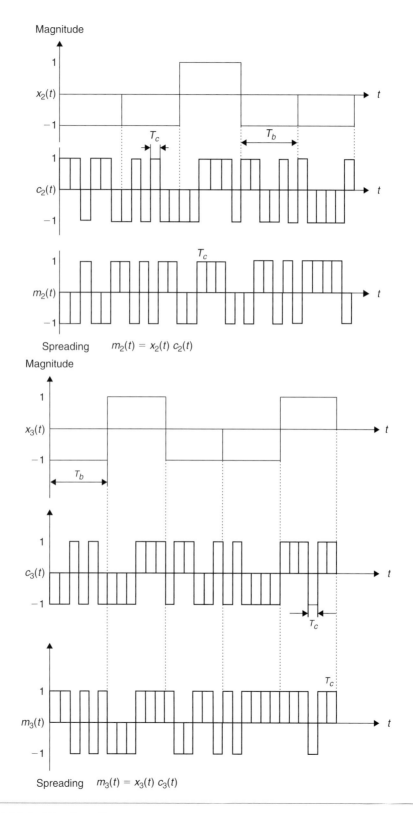

Spreading $m_2(t) = x_2(t) \, c_2(t)$

Spreading $m_3(t) = x_3(t) \, c_3(t)$

Figure 11.4   (Continued).   325

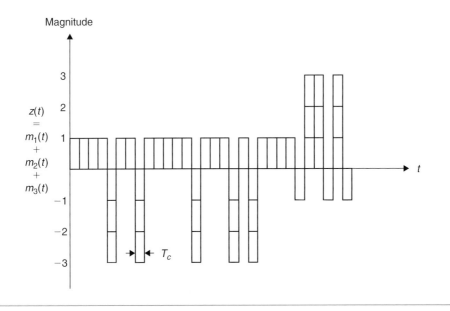

Figure 11.5   Resultant demodulated signal at a mobile.

Example 11.2

We consider signals transmitted from a base station to mobile 1, 2, and 3. The data streams $x_1(t)$, $x_2(t)$, and $x_3(t)$ are multiplied with codes $c_1(t)$, $c_2(t)$, and $c_3(t)$, respectively (see Figure 11.4). The resultant demodulated signal $z(t)$, sum of $m_1(t)$, $m_2(t)$, and $m_3(t)$ is given in Figure 11.5. Show that the transmitted signals to mobiles 1, 2, and 3 are recovered at the mobile receivers by despreading the resultant signal $z(t)$. Neglect propagation delay.

Solution

From Tables 11.1, 11.2, and 11.3, we observe that the transmitted signals for mobile 1, 2, and 3 are recovered by despreading at the receivers.

Table 11.1 $z(t)c_1(t)$ [see Figure 11.6(a)].

	T_b	$2T_b$	$3T_b$	$4T_b$	$5T_b$
Value of integration at end of bit period	−6	4	6	−2	−12
Bit value (if value of integration is <0, then bit value equals 1)	1	0	0	1	1

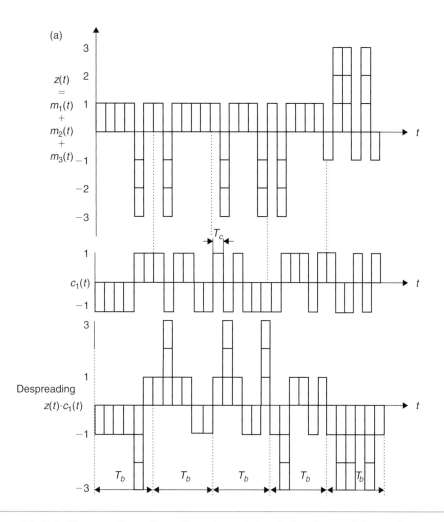

Figure 11.6(a)   Despreading of resultant demodulated signal at mobile receivers.

Table 11.2 $z(t)c_2(t)$ [see Figure 11.6(b)].

	T_b	$2T_b$	$3T_b$	$4T_b$	$5T_b$
Value of integration at end of bit period	-2	-6	8	-4	-8
Bit value	1	1	0	1	1

Table 11.3 $z(t)c_3(t)$ [see Figure 11.6(c)].

	T_b	$2T_b$	$3T_b$	$4T_b$	$5T_b$
Value of integration at end of bit period	-6	4	-8	-8	8
Bit value	1	0	1	1	0

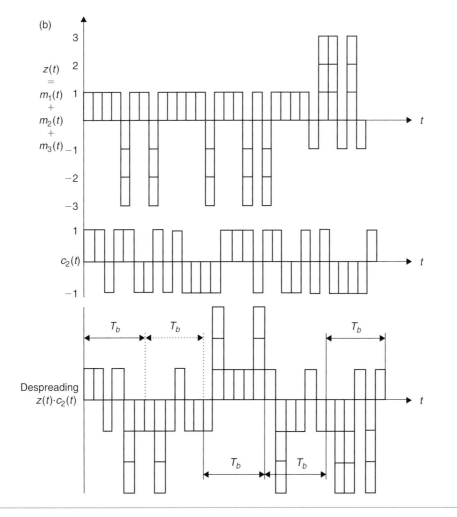

Figure 11.6(b) **(Continued).**

11.4 Requirements of Direct-Sequence Spread Spectrum

In the DSSS system, the entire bandwidth of the RF carrier is made available to each user. The DSSS system satisfies the following requirements [16]:

- The spreading signal has a bandwidth much larger than the minimum bandwidth required to transmit the desired information which, for a digital system, is baseband data.
- The spreading of the information is performed by using a spreading signal, called the *code signal* (see Appendix D). The code signal is independent of the data and is of a much higher chip rate than the data signal.

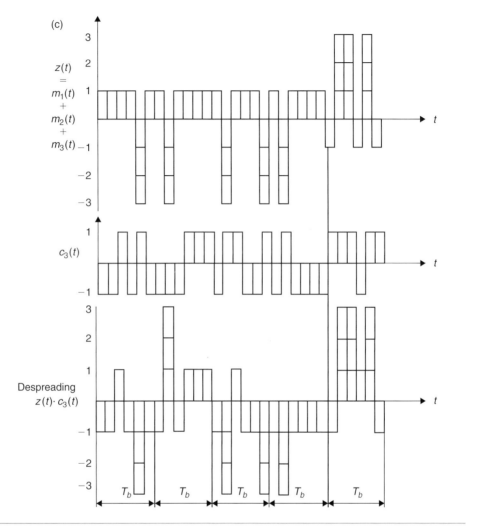

Figure 11.6(c)   (Continued).

- At the intended receiver, despreading is accomplished by cross-correlation of the received spread signal with a synchronized replica of the same code signal used to spread the data.

11.5   Frequency-Hopping Spread Spectrum Systems

In FHSS systems, the binary pseudo-random noise (PN) code generator drives the frequency synthesizer to hop to one of the many available frequencies chosen by the PN sequence generator. When the hopping rate is higher than the symbol rate, we have a *fast frequency-hop* system. If the hopping rate is lower than the symbol rate, i.e., there are several symbols transmitted per frequency hop, we have a *slow frequency-hop* system.

In an FHSS system b bits are used as an integer index for selecting the hop frequency. The term *chip* is also used in FHSS, but its meaning is different from the DSSS meaning of the word. It depends on whether the system is a slow frequency hop (FH) or a fast FH. In FH/*M-ary* frequency shift keying (MFSK) a system chip is the *tone of shortest duration*. The chip rate, R_c, is the maximum of R_s and R_h, where R_s is the symbol rate and R_h is the hop rate. Thus, for slow FHSS systems, the chip rate is equal to the symbol rate (i.e., $R_c = R_s$), whereas for fast FHSS systems, the chip rate is equal to the hop rate (i.e., $R_c = R_h$).

In a slow FHSS system with MFSK, the selected M frequencies must be an integer number of symbol rates apart. The spacing is required to maintain orthogonality between the frequencies and to allow reliable noncoherent detection. This implies that the minimum bandwidth of an MFSK signal should be about MR_s.

The modulation scheme commonly used with a fast FHSS system is MFSK, where $b = \log_2 M$ bits are used to determine which one of M frequencies should be used. The position of the *M-ary* signal set is shifted pseudo-randomly by the frequency synthesizer over a hopping bandwidth B_w. The frequency synthesizer produces a transmission tone based on the simultaneous instructions of the PN code and the data. At each frequency hop-time a PN generator feeds the frequency synthesizer a frequency word (a sequence of b chips) which decides one

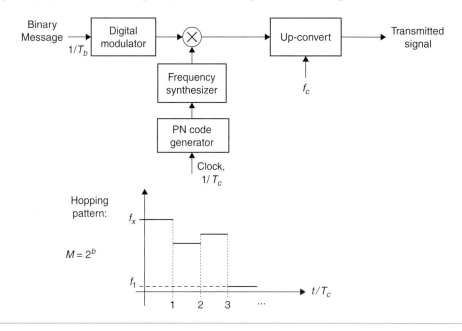

Figure 11.7   FHSS system.

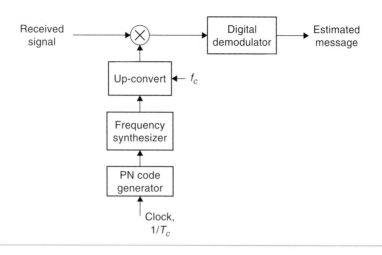

Figure 11.7 (Continued).

of 2^b symbol-set positions (see Figure 11.7). The frequency hopping bandwidth, B_w, and the minimum frequency spacing between consecutive hop positions, Δf, determine the minimum number of chips necessary in the frequency word. The processing gain, G_p, of an FHSS system is given as:

$$G_p = \frac{\text{Hopping Bandwidth}}{\text{Minimum Frequency Spacing}} = \frac{M \cdot \Delta f}{\Delta f} = M \qquad (11.16)$$

In the fast frequency-hop systems, there are L frequency hops during a symbol interval (T_s) (i.e., $T_s = LT_c$ or $R_c = LR_s$.)

$$\therefore \text{Hopping Bandwidth} = (KM\Delta f)L \qquad (11.17)$$

where:
  K = factor for frequency multiplication
  $M = 2^b$ is the number of frequencies produced by frequency synthesizer
  b = bit in a symbol
  L = frequency hops per symbol

$$\therefore G_p = \frac{KM\Delta fL}{\Delta f} = MKL \qquad (11.18)$$

The processing gain of a fast frequency hop system is dependent upon the number of frequencies used (M), the number of hops per symbol (L), and the frequency multiplication factor (K).

Example 11.3

In an FHSS system, a hopping bandwidth of 100 MHz and a frequency spacing of 10 kHz is used. What is the minimum number of PN chips that are required for each frequency symbol?

Solution

$$\text{Number of frequency tones in hopping bandwidth} = \frac{100 \times 10^6}{10^3} = 10^4$$

$$\text{Minimum number of chips} = \left\lceil \log_2(10^4) \right\rceil = 13 \text{ chips}$$

Example 11.4

A communication system transmits at 120 kbps and uses 32-FSK (Frequency Shift Keying). A hop rate of 2000 hops per second is used over an available spectrum of 10 MHz. Assuming a negligible synthesizer switching time between hops, calculate (a) data symbol transmitted per hop, and (b) the number of nonoverlapping hop frequencies.

Solution

For 32-FSK, we have $32 = 2^b$ i.e., $b = 5$ bits per symbol

$$\text{Symbol rate} = \frac{120 \text{ kbps}}{5} = 24 \text{ kbps}$$

Since the symbol rate is higher than the hop rate, the system is a slow FHSS system.

$$\text{Number of symbols per hop} = \frac{24,000}{2,000} = 12$$

Minimum bandwidth B_w of an M-*ary* FSK $\sim MRs$:

$$B_w = 12 \times 24 = 288 \text{ kHz}$$

Number of nonoverlapping hop frequencies:

$$n_{FH} = \frac{10 \text{ MHz}}{288 \text{ kHz}} \approx 35$$

Example 11.5

Consider an FHSS system in which the input data rate is 200 bits per second. The modulation scheme of 32-*ary* FSK is used to generate the modulation symbol. The frequency hopping rate is 200 hops per second. Calculate: (a) minimum separation between frequency tones; (b) number of frequency tones produced by a frequency synthesizer; (c) processing gain; and (d) hopping bandwidth. Assume a frequency multiplication factor $K = 1$.

Solution

With the 32-FSK modulation scheme there are 5 chips per symbol:

$$\text{Symbol rate: } R_s = \frac{200}{5} = 40 \text{ symbols/sec}$$

The hop rate is higher than symbol rate, the system is a fast FHSS system.

$$\text{Symbol duration} = \frac{1\text{s}}{40} = 25\,\text{ms}$$

$$L = \frac{200}{40} = 5 \text{ hops/symbol}$$

$$\text{Chip duration} = \frac{25}{5} = 5\,\text{ms}$$

$$\text{Minimum separation between tones} = \frac{1}{5 \times 10^{-3}} = 200\,\text{Hz}$$

$$M = 2^5 = 32 \text{ frequency tones}$$

$$\text{Frequency hopping bandwidth} = KM\Delta fL = 1 \times 32 \times 200 \times 5 = 64\,\text{kHz}$$

$$G_p = \text{MKL} = 32 \times 1 \times 5 = 160$$

11.6 Operational Advantages of SS Modulation

The following are the operational advantages of SS modulation:

- **Low probability of intercept:** Low probability of intercept implies that a third party cannot easily eavesdrop on the conversation or has to utilize expensive means to accomplish this. A standard communications receiver selects a demodulation circuitry, such as an amplitude demodulator, a frequency demodulator, or a phase demodulator, depending on the modulation scheme used at the transmitter. In an SS system, the receiver demodulates the transmitted energy through some correlation process and effectively combines various components within a wider bandwidth. A single-channel receiver will thus detect a small portion of the transmitted signal which will be too weak for normal detection, and even if it could be magnified to a detectable level would be incomplete and thus not possible to understand.

- **Low probability of position fix:** Conventional radio transmitters are easily pinpointed by simple and inexpensive direction finders. The spectral-spreading concepts make this task much more demanding as greater processing power and integration time will be required within each resolution.

- **Low probability of signal exploitation:** This refers to the possibilities which exist within a communication scenario to exploit the communication link by some manipulation of the waveforms used to carry the message. These could be:
  - destruction of synchronization messages;
  - destruction or alteration of the message contents;
  - invisible or concealed addition of data bits.

- **High resistance to jamming and interference:** SS systems are inherently more robust against jamming and interference than systems not using spreading techniques. This property of SS modulation is used in military systems where the main design objective is to develop a system which is able to deliver a message through a very hostile and impenetrable medium. SS modulation techniques provide an additional factor which is not seen in conventional systems. This is due to the fact that a code is used in the spreading process and unless the jammer/interferer manages to get hold of this code the impact of the jamming/interference is reduced by a significant amount. There are two very important aspects of SS with respect to jamming and interference. The first relates to the actual protection provided by an SS code. The SS receiver provides a post-detection signal-to-noise and signal-to-interference improvement. This means that if the SS signal can be received with sufficient clarity and strength and without interference signals, intelligent jammers might redirect the same waveform with more power and cause problems for the detection process in the data link. This is a practical limitation since a limited number of spreading codes are used in an SS modem and thus the transmission tends to repeat the same codes a large number of times. The vulnerability lies in the fact that the code might be revealed if it sticks out clearly only once—for instance if it is received at a very short distance from the transmitter. This points at the very important requirement of SS systems, that the spectral-spreading scheme should be sufficiently agile and use frequent change of code structures. The next important aspect of SS modulation with respect to jamming and interference is that SS modulation provides a practical way of coping with frequency-band congestion.

- **High time resolution/reduction of multipath effects:** Multipath effects are one of the unavoidable effects in radio communication. Multipath implies that the signal reaching the receiver antenna has travelled by two or more paths. Because these routes inevitably are of different lengths, the time delays of signals that have come along the respective paths are different and the signal will fade in or out with small displacements of the transmitter and receiver (displacements of the order of half a wavelength). Methods exist whereby the signal is coded such that the signals reaching the receiver via different paths add up in phase at the receiver. Adaptive methods capitalizing on the wider bandwidth of SS waveforms make it possible to use radio communication under extremely severe multipath conditions. Example of such are wireless local area networks used inside rooms or buildings where the propagation conditions are very poor.

- **Cryptographic capabilities:** The coding aspects of SS modulation have implications for possible security function in a communication link. The spreading codes can be chosen such that they serve the dual purpose of spreading the frequency spectrum of the transmitted signal and making it difficult to decipher the message. This requires that the code provide the required spectral signature (usually reasonably flat), has good anti-cryptographic property, and is sufficiently redundant. As these criteria place rather tough requirements on the coding strategy, a more common approach is to implement SS coding and scramble the code for cryptography independently and usually sequentially.

11.7 Coherent Binary Phase-Shift Keying DSSS

The simplest form of a DSSS communication system uses coherent binary phase-shift keying (BPSK) for both data modulation and spreading modulation. But the most common form uses BPSK for data modulation and quadrature phase-shift keying (QPSK) for spreading modulation.

We first consider the simplest case. The ith mobile station is assigned a spreading code signal $c_i(t)$ (a periodic spreading code sequence with chips of width T_c). Each mobile station has its own such code signal. Information bits are transmitted by superimposing the data bits onto the code signal. If the ith mobile station transmits the binary data waveform $x_i(t)[x_j(t) = \pm 1]$, it forms the binary sequence.

$$m_i(t) = x_i(t)c_i(t) \tag{11.19}$$

Equation 11.19 represents the modulo-2 addition of $c_i(t)$ and $x_i(t)$ as a multiplication because the binary 0 and 1 represent values of 1 and -1 into the modulator.

The transmitted signal from the ith mobile is

$$s_i(t) = x_i(t)c_i(t)\sqrt{2P}\cos(\omega_c t + \phi) \tag{11.20}$$

where:

$x_i(t) =$ baseband signal for the ith mobile
$c_i(t) =$ spreading code for the ith mobile
$\omega_c =$ carrier frequency
$P =$ signal power
$\phi =$ data phase modulation

If T_b is the bit period of $s_i(t)$, then T_b may correspond to either a full period for $c_i(t)$, or to a fraction of a period. If T_b is less than one code period, then the data bits are modulating the polarity of a portion of a code period. The code $c_i(t)$ serves as a subcarrier for the source data. Since each mobile station uses the entire channel bandwidth and since Equation 11.19 has a code chip rate of $1/T_c$ chips per second, each BPSK carrier uses an RF bandwidth of

$$B_w = \frac{1}{T_c} \tag{11.21}$$

The available channel RF bandwidth determines the minimum chip width, and the code period determines its relation to the bit time. The number of code chips per bit is given by:

$$G_p = \frac{T_b}{T_c} = \frac{B_w}{R_b} \tag{11.22}$$

The ratio B_w/R_b is the CDMA processing gain, G_p, or simply, the spreading ratio of code modulation. This shows how much the RF bandwidth must be spread relative to the bit rate, R_b, to accommodate a given spreading code length. Each mobile station (MS) uses the same RF carrier frequency and RF bandwidth, but with its own spreading code $c_i(t)$.

The signal in Equation 11.20 is transmitted using a distortionless path with transmission. The signal is received together with some type of interference and/ or Gaussian noise. Demodulation is performed in part by remodulating with the spreading code appropriately delayed as shown in Figure 11.8. This correlation of the received signal with the delayed spreading waveform is the despreading. This is a critical function in all spread spectrum systems. The signal component of the output of the despreading is:

$$x_i(t - \tau_d) \sqrt{2P} c_i(t - \tau_d) \times c_i(t - \hat{\tau}_d)\cos(\omega_c(t - \tau_d) + \phi) \tag{11.23}$$

where:

$\hat{\tau}_d$ = receiver's best estimate of the transmission delay

Since $c_i(t) = \pm 1$, the product $c_i(t - \tau_d) \times c_i(t - \hat{\tau}_d)$ will be unity if $\tau_d = \hat{\tau}_d$, that is, if the code at the receiver is synchronized with the spreading code at the transmitter. When correctly synchronized, the signal component of the output of the receiver despreading is equal to $\sqrt{2P} x_i(t - \tau_d) \cos(\omega_c(t - \tau_d) + \phi)$, which can be demodulated using a conventional coherent phase modulator.

The bit error probability, P_e, associated with the coherent BPSK spread spectrum signal is the same as with the BPSK signal and is given as:

$$P_e = \frac{1}{2}\operatorname{erfc}\left[\sqrt{\left(\frac{E_b}{N_0}\right)_o}\right] = Q\left[\sqrt{2\left(\frac{E_b}{N_0}\right)_o}\right] = Q\left[\sqrt{2\left\{Gp \cdot \left(\frac{E_b}{N_0}\right)_i\right\}}\right] \tag{11.24}$$

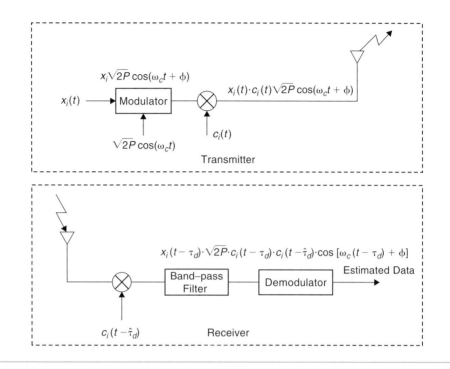

Figure 11.8   DSSS system with BPSK.

where:

$$Q(u) \approx \frac{e^{-u^2/2}}{(\sqrt{2\pi}u)}, \; u \gg 1 \text{ is known as } Q \text{ function (see Appendix C)}$$

$$E_b = P/R_b = \text{energy per bit}$$

11.8   Quadrature Phase-Shift Keying DSSS

Sometimes it is advantageous to transmit simultaneously on two carriers which are in phase quadrature. The main reason for this is to save spectrum because, for the same total transmitted power, we can achieve the same bit error probability, P_e, using one-half the transmission bandwidth. The quadrature modulations are more difficult to detect in low probability of detection applications. Also, the quadrature modulations are less sensitive to some types of jamming. We refer to Figure 11.9 and write:

$$s(t) = \sqrt{P}c_I(t)\cos[\omega t + \phi] + \sqrt{P}c_Q(t)\sin[\omega t + \phi] \quad (11.25)$$

$$s(t) = a_I(t) + a_Q(t) \quad (11.26)$$

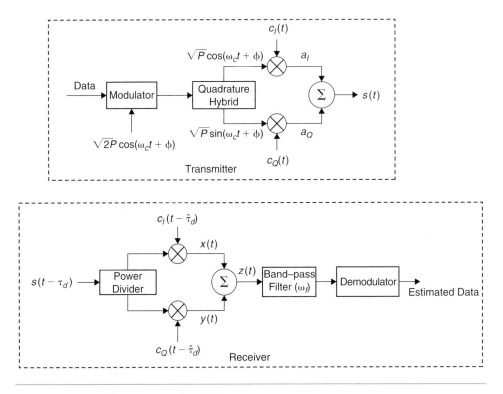

Figure 11.9   DSSS system with QPSK.

where:

$c_I(t)$ and $c_Q(t)$ are the in-phase and quadrature spreading codes and $a_I(t)$ and $a_Q(t)$ are orthogonal. This condition is satisfied in the present case since $c_I(t)$ and $c_Q(t)$ are independent code waveforms.

The receiver for the transmitted signal is shown in Figure 11.9. The bandpass filter is centered at frequency ω_f and has a bandwidth sufficiently wide to pass the data-modulated carrier without distortion.

$$
\begin{aligned}
x(t) = \sqrt{P/2}\,c_I(t - \tau_d)c_I(t - \hat{\tau}_d)\cos[\omega_c t + \phi] + (\sqrt{P/2})c_Q \\
\times\ (t - \tau_d)c_I(t - \hat{\tau}_d)\sin[\omega_c t + \phi]
\end{aligned}
\tag{11.27}
$$

$$
\begin{aligned}
y(t) = \sqrt{P/2}\,c_I(t - \tau_d)c_Q(t - \hat{\tau}_d)\sin(\omega_c t + \phi) + (\sqrt{P/2})c_Q \\
\times\ (t - \tau_d)c_Q(t - \hat{\tau}_d)\cos[\omega_c t + \phi]
\end{aligned}
\tag{11.28}
$$

If the receiver-generated replicas of spreading codes are correctly phased then

$$
c_I(t - \tau_d)c_I(t - \hat{\tau}_d) = c_Q(t - \tau_d)c_Q(t - \hat{\tau}_d) = 1
\tag{11.29}
$$

$$z(t) = x(t) + y(t) = \sqrt{2P}\cos[\omega_c t + \phi] \tag{11.30}$$

The signal $z(t)$ is the input to a conventional phase demodulator where data is recovered.

When the spreading codes are staggered one-half chip interval with respect to each other, the QPSK is called offset-QPSK (OQPSK). In OQPSK, the phase changes every one-half chip interval, but it does not change more than $\pm 90°$. This limited phase change improves the uniformity of the signal envelope compared to BPSK and QPSK, since zero-crossings of the carrier envelope are avoided. Neither QPSK nor OQPSK modulation can be removed with a single stage of square-law detection. Two such detectors and the associated loss of signal-to-noise ratio are required. QPSK and OQPSK offer some low probability of detection advantages over the BPSK method.

11.9 Bit Scrambling

Referring to Table 11.4, we consider the following activities at a given transmitter location (see Figure 11.10):

1. An arbitrary data sequence $s_i(t)$ is generated by a digital source.
2. An arbitrary code sequence $c_i(t)$ is produced by a DS generator.
3. Two sequences are modulo-2 added and transmitted to a distant receiver.
4. At the distant location, the resulting sequence (assuming no propagation delay) is picked up by the receiver (see Figure 11.10).

Table 11.4 Operations with modulo-2 addition.

Transmitter	1	$s_i(t)$	1	1	0	1	0	0	1	1	1	1
	2	$c_i(t)$	1	0	0	1	1	1	0	1	0	0
	3	$s_i(t) \oplus c_i(t)$	0	1	0	0	1	1	1	0	1	1
Receiver	4	$s_i(t) \oplus c_i(t)$	0	1	0	0	1	1	1	0	1	1
	5	$c_i(t)$	1	0	0	1	1	1	0	1	0	0
	6	$s_i(t) \oplus c_i(t) \oplus c_i(t) = s_i(t)$	1	1	0	1	0	0	1	1	1	1

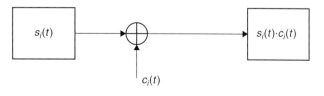

Figure 11.10 Mobile transmitter.

5. The code $c_i(t)$ used at the transmitter is also available at the receiver.

6. The original data sequence is recovered by modulo-2 adding the received sequence with the locally available code $c_i(t)$.

Next, referring to Table 11.5, we consider the following set of activities at the given transmitter location.

1. An arbitrary data sequence $s_i(t)$ is generated by a digital source. In this case, we use $+1$s and -1s to represent 0s and 1s.

2. An arbitrary code sequence $c_i(t)$ is produced by a DS generator.

3. We multiply $s_i(t)$ and $c_i(t)$. The output of the multiplier is transmitted to a distant receiver.

4. At the distant location, the resulting sequence (again assuming no propagation delay) is picked up by the receiver (Figure 11.11).

5. The code $c_i(t)$ used at the transmitting location is also available at the receiver.

6. The original data sequence is recovered by multiplying the received sequence by the locally available code, $c_i(t)$.

From Tables 11.4 and 11.5 we conclude that modulo-2 addition using 1s and 0s binary data is equivalent to multiplication using -1 and 1 binary data as long as we remain consistent in mapping 0s to $+1$s and 1s to -1s as shown in Table 11.5. (For circuit implementation, modulo-2 addition is preferred since exclusive OR

Table 11.5 Operations without modulo-2 addition.

Transmitter	1	$s_i(t)$		-1	-1	1	-1	1	1	-1	-1	-1	-1
	2	$c_i(t)$		-1	1	1	-1	-1	-1	1	-1	1	1
	3	$s_i(t) \cdot c_i(t)$		1	-1	1	1	-1	-1	-1	1	-1	-1
Receiver	4	$s_i(t) \cdot c_i(t)$		1	-1	1	1	-1	-1	-1	1	-1	-1
	5	$c_i(t)$		-1	1	1	-1	-1	-1	1	-1	1	1
	6	$s_i(t) \cdot c_i(t) \cdot c_i(t) = s_i(t)$		-1	-1	1	-1	1	1	-1	-1	-1	-1

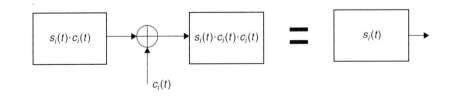

Figure 11.11   Mobile transmitter.

gates are cheaper than multiplication circuits. However, for modeling purposes, the multiplication method is usually easier to formulate and understand than the modulo-2 approach.)

We notice that for the output of the receiver to be identical to the original data, the following relationship must be satisfied:

$$s_i(t)c_i(t)c_i(t) = s_i(t) \tag{11.31}$$

In other words, $c_i(t)c_i(t)$ must be equal to 1. Note that $c_i(t)$ is a binary sequence made up of 1s and -1s, therefore

$$\text{If } c_i(t) = 1, c_i(t)c_i(t) = 1 \tag{11.32}$$

$$\text{If } c_i(t) = -1, c_i(t)c_i(t) = 1 \tag{11.33}$$

In these discussions we assume that there is no propagation delay and no other processing delay incurred between the transmitter and receiver input. Thus the code copy used at the receiver is perfectly lined up with the initial code used at the transmitter. The two codes are said to be in phase or in synchronization. In practice, however, propagation delay and other processing delays (τ_i) occur between the transmitter and the receiver. Therefore, the receiver may be time-shifted relative to the initial code at the transmitter. The two codes are no longer in synchronization. As a result, the output of the receiver will no longer be identical to the original data, $s_i(t)$.

In order to recover the original data, $s_i(t)$, we must "tune" the receiver code sequence to that of the incoming code from the transmitter. In other words, we must time-shift the receiver code in order to line it up with the incoming code. It should be noted that by synchronizing or "tuning" the receiver code to the phase of the incoming code $c_i(t - \tau_i)$, the original data (shifted by propagation delay) can now be recovered at the output of the receiver.

In these discussions, the data sequence and code sequence are assumed to have the same length (one code bit for each data bit) and are used for encrypting the data bits. The process is referred to as *bit scrambling*.

11.10 Requirements of Spreading Codes

To spread the data sequence, the code sequence must be much faster than the data sequence, and exhibit some random properties.

By multiplying the data sequence with the faster code sequence, the resulting product yields a sequence with more transitions than the original data. Using suitable random-like codes, the resulting sequence will have the same rate as the code sequence. It is desirable to use a set of orthogonal codes (see Appendix D) to provide good isolation between users. However, in practice, the codes used are not perfectly orthogonal, but they exhibit good isolation characteristics, i.e., they have low cross-correlation.

11.11   Multipath Path Signal Propagation and Rake Receiver

In the absence of a direct line-of-sight signal from the base station (BS) to the mobile station (or the received signal at the mobile station from the base station), the signal is made up of the sum of many signals, each travelling over a separate path. Since these path lengths are not the same, the information carried on the radio link experiences a spread in delay as it travels between the base station and the mobile. In addition, to delay spread, the same multipath environment causes severe local variations in signal strength as these multipath signals are added constructively and destructively at the receiving antenna. This effect is called *Rayleigh fading* (see Chapter 3). The movement of a mobile causes each received signal to be shifted in frequency as a function of the relative direction and speed of the mobile. This effect is called *Doppler shift* (see Chapter 3).

Multipath is treated as causing delayed versions of the signal to add to the system noise when the differential delay exceeds the chip time, T_c. Substantial performance improvement can occur by detecting each additional path separately, thereby enabling the signals to be combined coherently. A receiver can be implemented to resolve each individual path such that the paths can be combined to produce an overall gain. This type of receiver is known as a *Rake receiver*.

In the Rake receiver for user # 1 (refer to Example 11.2) baseband demodulated signal $Z(t)$ is the sum of N signals which arrives on N different paths (see Figure 11.12). We consider path 2, the multiplication of $Z(t)$ by $c_1(t - \Delta_2)$. The integration starts at time Δ_2 and ends at T_b, to yield the peak response for the path 2. (The output of the integrator is the value of the correlation function of $c_1(t)$ for a particular delay. For path 2, this delay is zero, whereas for the other paths the delay exceeds the time duration of a chip.) The contributions from other paths average out to 0, since the differential delays exceed the chip duration, T_c. The response from each path is summed to produce the stronger signal. We illustrate this concept with Example 11.6.

Example 11.6

We consider the downlink in Example 11.2 where the demodulated signal at the mobile is $z(t) = m_1(t) + m_2(t) + m_3(t)$ (see Figure 11.13(a)) for mobile 1, 2, and 3. We assume two equal-strength paths and write the demodulated signal as $Z(t) = z(t) + z(t - 2T_c)$ (see Figure 11.13(c)). The differential delay between these two paths is taken as $2T_c$ for simplicity. We will show how mobile 1 will

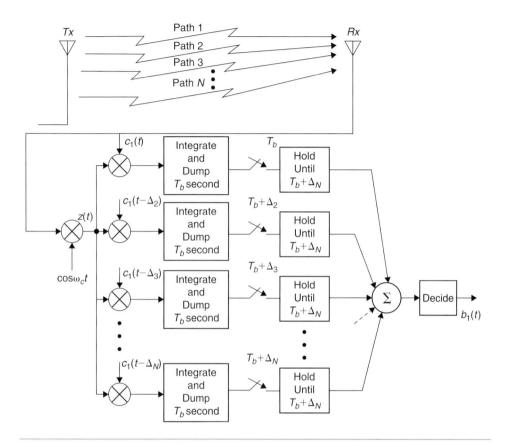

Figure 11.12   Simplified Rake receiver for user 1 (equal gain combining).

detect its information using a two-path Rake receiver. The results are shown in Figures 11.13(a), 11.13(b), 11.13(c), and 11.14.

Solution

Individual path outputs (Tables 11.6 and 11.7) yields an error in a particular bit position. The Rake combining strengthens the signal and removes the error, as shown in Table 11.8.

Table 11.6 $Z(t) \cdot c_1(t)$.

	T_b	$2T_b$	$3T_b$	$4T_b$	$5T_b$
Value of integration at end of bit period	-4	-4	8	-8	-12
Detected bit value	1	1	0	1	1
Actual bit value	1	0	0	1	1

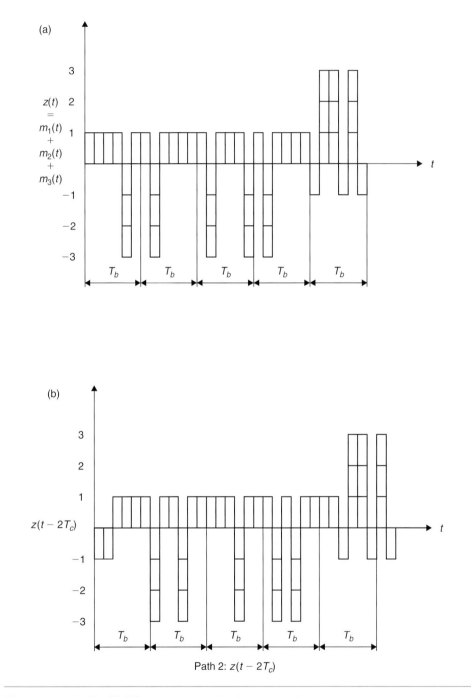

Figure 11.13(a, b) Mobile #1 sequence locked on path 1.

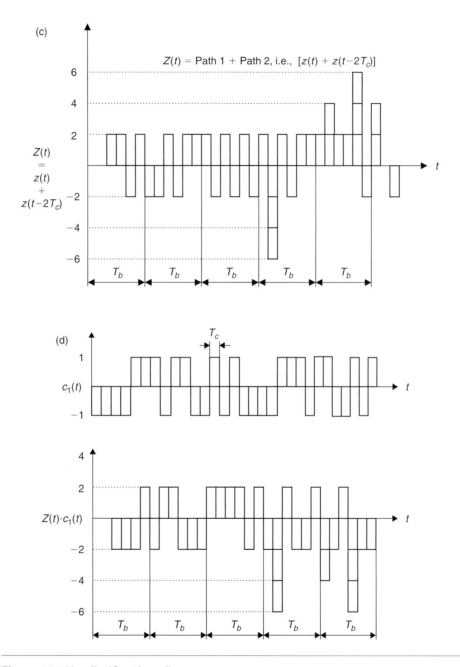

Figure 11.13(c, d)   (Continued).

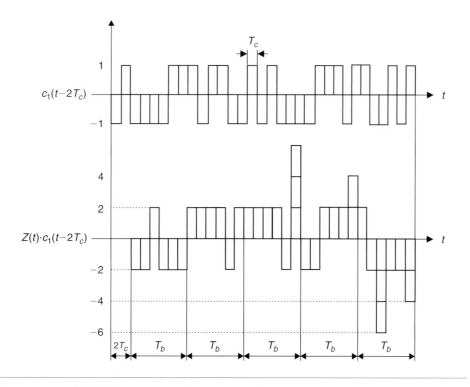

Figure 11.14   Mobile #1 sequence locked on path 2.

Table 11.7 $Z(t) \cdot c_1(t - 2T_c)$.

	T_b	$2T_b$	$3T_b$	$4T_b$	$5T_b$
Value of integration at end of bit period	−8	8	12	4	−12
Detected bit value	1	0	0	0	1
Actual bit value	1	0	0	1	1

Table 11.8 Sum of path 1 and path 2 integrator.

	T_b	$2T_b$	$3T_b$	$4T_b$	$5T_b$
Path 1: Integrator output	−4	−4	8	−8	−12
Path 2: Integrator output	−8	8	12	4	−12
Sum of integrator outputs (Rake receiver output)	−12	4	20	−4	−24
Detected bit value	1	0	0	1	1
Mobile 1 bits	1	0	0	1	1

11.12 Critical Challenges of CDMA

Code division multiple access (CDMA) is based on DSSS. CDMA is more complex than other multiple access technologies and as such poses several critical challenges.

- All users in a given cell transmit at the same time in the same frequency band. Can they be made not to interfere with each other?

- Will a user who is near the base station saturate the base station altogether so that it cannot receive users who are farther away (known as near-far problem)?

- CDMA uses a reuse factor of one. This means that the same frequency is used in adjacent cells. Can the codes provide sufficient separation for this to work well in most situations?

- CDMA uses soft handoffs where a moving user can receive and combine signals from two or more base stations at the same time. What is the impact on base station traffic handling ability?

11.13 TIA IS-95 CDMA System

Qualcomm proposed the CDMA radio system for digital cellular phone applications. It was optimized under existing U.S. mobile cellular system constraints of the advanced mobile phone system (AMPS). The CDMA system uses the same frequency in all cells and all sectors. The system design has been standardized by the TIA as IS-95 and many equipment vendors sell CDMA equipment that meet the standard.

The IS-95 CDMA system operates in the same frequency band as the AMPS using frequency division duplex (FDD) with 25 MHz in each direction.[*] The uplink (mobile to base station) and downlink (base station to mobile) bands use frequencies from 869 to 894 MHz and from 824 to 849 MHz, respectively. The mobile station supports CDMA operations on the AMPS channel numbers 1013 through 1023, 1 through 311, 356 through 644, 689 through 694, and 739 through 777, inclusive. The CDMA channels are defined in terms of an RF frequency and a code sequence. Sixty-four Walsh codes (see Appendix D) are used to identify the forward channels, whereas unique long PN code offsets are used for the identification of the reverse channels. The modulation and coding features of the IS-95 CDMA system are listed in Table 11.9.

Modulation and coding details for the forward and reverse channels differ. Pilot signals are transmitted by each cell to assist the mobile radio to acquire and track the cell site downlink signals. The strong coding helps these radios to operate effectively at an E_b/N_0 ratio of a 5 to 7 dB range.

The CDMA system (IS-95) uses power control and voice activation to minimize mutual interference. Voice activation is provided by using a variable rate vocoder (see Chapter 8) which for Rate set 1 codec operates at a maximum rate of 8 kbps to a

---

[*] The frequency spectrum for the A-System cellular service provider is split such that the spectrum is not divisible by 1.25 MHz. Thus the A-System cellular provider cannot partition the spectrum into ten 1.25 CDMA channels. This restriction is not imposed for the B-System, however.

Table 11.9 Modulation and coding feature of IS-95 CDMA system.

Modulation	Quadrature phase-shift keying (QPSK)
Chip rate	1.2288 Mcps
Nominal data rate	9600 bps, full rate with Rate Set 1
Filtered bandwidth	1.25 MHz
Coding	Convolution with Viterbi decoding
Interleaving	With 20-msec span

minimum rate of 1 kbps, depending on the level of voice activity. With the decreased data rate, the power control circuit reduces the transmitter power to achieve the same bit error rate. A precise power control, along with voice activation circuit, is critical to avoid excessive transmitter signal power that is responsible for contributing the overall interference in the system. The Rate set 2 coding algorithms at 13 kbps are also supported.

A bit-interleaver with 20 msec span is used with error-control coding to overcome multipath fading and shadowing (see Chapter 3). The time span used is the same as the time frame of voice compression algorithm.

A Rake receiver used in the CDMA radio takes advantage of a multipath delay greater than 1 μs, which is common in cellular/personal communication service networks in urban and suburban environments.

11.13.1   Downlink (Forward) (BS to MS)

The downlink channels include one pilot channel, one synchronization (synch) channel, and 62 other channels including up to 7 paging channels. (If multiple carriers are implemented, paging channels and synch channels do not need to be duplicated). The information on each channel is modulated by the appropriate Walsh code and then modulated by a quadrature pair of PN sequences at a fixed

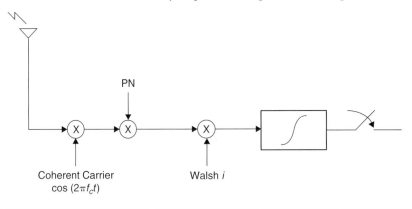

Figure 11.15   Application of PN sequence and Walsh code in CDMA.

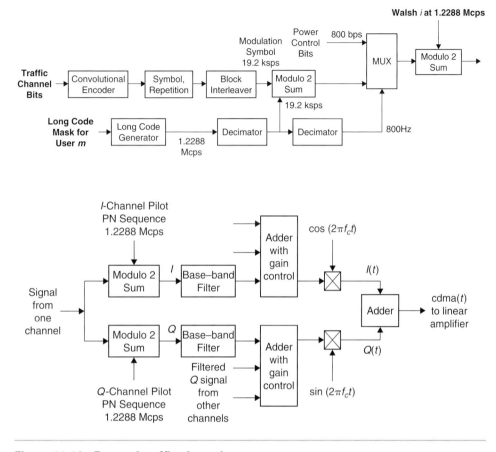

Figure 11.16   Forward traffic channel.

chip rate of 1.2288 Mcps (see Figure 11.15). The pilot channel is always assigned to code channel number zero. If the synch channel is present, it is given the code channel number 32. Whenever paging channels are present, they are assigned the code channel numbers 1 through 7 (inclusive) in sequence. The remaining code channels are used by forward traffic channels (see Figure 11.16).

The synch channel operates at a fixed data rate of 1200 bps and is convolutionally encoded to 2400 bps, repeated to 4800 bps, and interleaved.

The forward traffic channels are grouped into sets. Rate set 1 has four rates: 9600, 4800, 2400, and 1200 bps. Rate set 2 contains four rates: 14,400, 7200, 3600, and 1800 bps. All radio systems support Rate set 1 on the forward traffic channels. Rate set 2 is optionally supported on the forward traffic channels. When a radio system supports a rate set, all four rates of the set are supported.

Speech is encoded using a variable rate vocoder (see Chapter 8) to generate forward traffic channel data depending on voice activity. Since frame duration is fixed at 20 ms, the number of bits per frame varies according to the traffic rate. Half rate convolutional encoding is used, which doubles the traffic rate to give rates from 2400 to 19,200 bits per second. Interleaving is performed over 20 ms. A long PN code of $2^{42} - 1$ ($= 4.4 \times 10^{12}$) is generated using the user's electronic serial number (ESN) embedded in the mobile station long code mask (with voice privacy, the mobile station long code mask does not use the ESN). The scrambled data is multiplexed with power control information which steals bits from the scrambled data. The multiplexed signal on the traffic channel remains at 19,200 bps and is modulated at 1.2288 Mcps by the Walsh code, W_i, assigned to the ith user traffic channel. The signal is spread at 1.2288 Mcps by quadrature pseudo-random binary sequence signals, and the resulting quadrature signals are then weighted. The power level of the traffic channel depends on its data transmission rate.

The paging channel data is processed in a similar manner to the traffic channel data. However, there is no variation in the power level on a per frame basis. The paging channels provide the mobile stations with system information and instructions, in addition to acknowledging messages following access requests on the mobile stations' access channels. The 42-bit mask is used to generate the long code. The paging channels operate at a data rate of 9600 or 4800 bps.

All 64 channels are combined to give single I and Q channels. The signals are applied to quadrature modulators and resulting signals are summed to form a QPSK signal, which is linearly amplified.

The pilot CDMA signal transmitted by a base station provides a reference for all mobile stations. It is used in the demodulation process. The pilot signal level for all base stations is much higher (about 4 to 6 dB) than the traffic channel. The pilot signals are quadrature pseudo-random binary sequence signals with a period of 32,768 chips. Since the chip rate is 1.2288 Mcps, the pilot pseudo-random binary sequence corresponds to a period of 26.66 ms, which is equivalent to 75 pilot channel code repetitions every 2 seconds. The pilot signals from all base stations use the same pseudo-random binary sequence, but each base station is identified by a unique time offset of its pseudo-random binary sequence (short code). These offsets are in increments of 64 chips providing 512 unique offset codes. These large numbers of offsets ensure that unique base station identification can be obtained, even in a dense microcellular environment.

A mobile station processes the pilot channel to find the strongest multipath signal components. The processed pilot signal provides an accurate estimation of time delay, phase, and magnitude of the multipath components. These components are tracked in the presence of fast fading, and coherent reception with combining is used. The chip rate on the pilot channel and on all frequency carriers is locked to precise system time by using the global positioning system (GPS). Once the mobile station identifies the strongest pilot offset by processing the multipath components from the pilot channel correlator, it examines the signal on its synch channel which is locked to the pseudo-random binary sequence signal on the pilot channel. Since the synch channel is time aligned with its base station's pilot channel, the mobile station finds the information pertinent to this particular base station. The synch channel message contains time-of-day and long code synchronization to ensure that long code generators at the base station and mobile station are aligned and identical. The mobile station now attempts to access the paging channel and listens for system information. The mobile station enters the idle state when it has completed acquisition and synchronization. It listens to the assigned paging channel and is able to receive and initiate calls.

11.13.2 Uplink (Reverse) (MS to BS)

The uplink channel is separated from the downlink channel by 45 MHz at cellular frequencies and 80 MHz at PCS frequencies (1.8 to 1.9 GHz). The uplink uses the same 32,768 chip code as is used on the downlink. The two types of uplink channels are the access channel and reverse traffic channels (see Figure 11.17). The access channel enables the mobile station to communicate nontraffic information, such as originating calls and responding to paging. The access rate is fixed at 4800 bps. All mobile stations accessing a radio system share the same frequency assignment. Each access channel is identified by a distinct access channel long code sequence having an access number, a paging channel number associated with the access channel, and other system data. Each mobile station uses a different PN code; therefore, the radio system can correctly decode the information from an individual mobile station. Data transmitted on the reverse traffic channel is grouped into 20 ms frames. All data on the reverse traffic channel is convolutionally encoded, symbol repeated, block interleaved, and modulated by Walsh symbols transmitted for each six-bit symbol block. The symbols are from the set of the 64 mutually orthogonal waveforms.

The reverse traffic channel for Rate set 1 may use either 9600, 4800, 2400, or 1200 bps data rates for transmission. The transmission varies proportionally with the data rate, being 100% at 9600 bps to 12.5% at 1200 bps. An optional second rate set is also supported in the PCS version of CDMA and new versions of cellular CDMA. The actual burst transmission rate is fixed at 28.8 ksps. Since six code symbols are modulated as one of 64 modulation symbols for transmission, the modulation symbol transmission rate is fixed at 4800 modulation symbols per second. This results in a fixed Walsh chip rate of 307.2 kcps. The rate of spreading PN sequence is fixed at 1.2288 Mcps, so that each Walsh chip is spread by 4 PN chips. Table 11.10 provides the signal rates and their relationship for the various transmission rates on the reverse traffic channel.

Following orthogonal spreading, the reverse traffic channel and access channel are spread in quadrature. Zero-offset I and Q PN sequences are used for spreading. These sequences are periodic (short code) with 32,768 PN chips in length and are based on characteristic polynomials $g_I(x)$ and $g_Q(x)$.

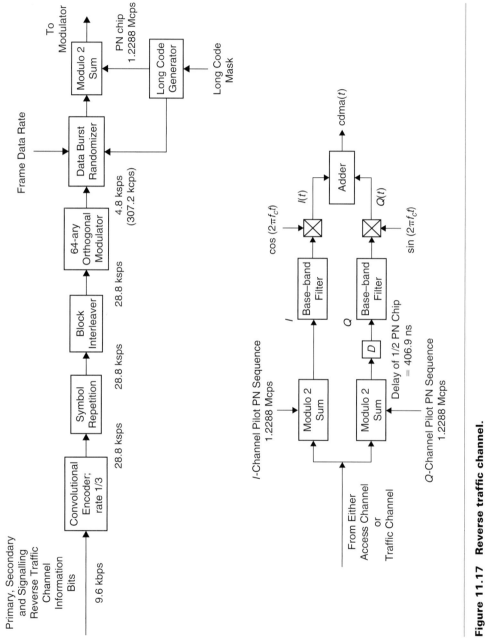

Figure 11.17 Reverse traffic channel.

Table 11.10 CDMA reverse traffic channel modulation parameters (rate set 1).

Parameter	9600 bps	4800 bps	2400 bps	1200 bps	Units
PN chip rate	1.2288	1.2288	1.2288	1.2288	Mcps
Code rate	1/3	1/3	1/3	1/3	bits per code symbol
Transmitting duty cycle	100	50	25	12.5	%
Code symbol rate	3 × 9600 = 28,800	28,800	28,800	28,800	symbol per second
Modulation	6	6	6	6	code symbol per mod. symbol
Modulation symbol rate	28,800/6 = 4800	4800	4800	4800	symbol per second
Walsh chip rate	64 × 4800 = 307.2	307.2	307.2	307.2	kcps
Mod. symbol duration	1/4800 = 208.33	208.33	208.33	208.33	μs
PN chips/ code symbol	12,288/288 = 42.67	42.67	42.67	42.67	PN chip per code symbol
PN chips/ mod. symbol	1,228,800/ 4800 = 256	256	256	256	PN chip per mod. symbol
PN chips/ Walsh chip	4	4	4	4	PN chips per Walsh chip

The maximum-length linear feedback register sequences $I(n)$ and $Q(n)$, based on these polynomials, have a period of $2^{15}-1$ and are generated by using the following recursions:

$$I(n) = I(n-15) \oplus I(n-8) \oplus I(n-7) \oplus I(n-6) \oplus I(n-2) \quad (11.34)$$

based on $g_I(x)$ as the characteristic polynomial, and

$$Q(n) = q(n-15) \oplus q(n-12) \oplus q(n-11) \oplus q(n-10)$$
$$\oplus \; q(n-9) \oplus q(n-5) \oplus q(n-4) \oplus q(n-3) \quad (11.35)$$

based on $q_Q(x)$ as the characteristic polynomial, where $I(n)$ and $Q(n)$ are binary numbers (0 and 1) and the additions are modulo-2.

To obtain the I and Q sequences, a 0 is inserted in $I(n)$ and $Q(n)$ after 14 consecutive 0 outputs (this occurs only once in each period). Therefore, the short PN sequences have one run of 15 consecutive 0 outputs instead of 14.

The chip rate for the short PN sequence is 1.2288 Mcps and its period is 26.666 ms. There are exactly 75 repetitions in every 2 seconds. The spreading modulation is OQPSK (see Figure 11.18). The data spread by Q PN sequence is

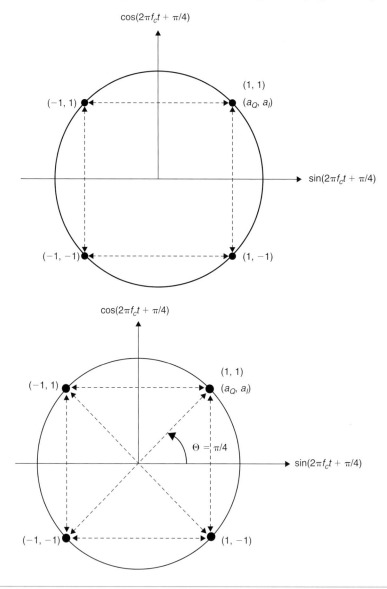

Figure 11.18   Signal constellation and phase transition of OQPSK and QPSK.

delayed by half a chip time (406.901 ns) with respect to the data spread by I PN sequence. Table 11.11 describes the characteristics of OQPSK.

Table 11.12 defines the signal rates and their relationship on the access channel. Each base station transmits a pilot signal of constant power on the same frequency. The received power level of the pilot signal enables the mobile station to adjust its transmitted power such that the base station will receive the signal at the requisite power level. The base station measures the mobile station's received power and informs the mobile station to make the necessary adjustments to its transmitter power. One command every 1.25 ms allows to adjust the transmitted power from the mobile station in steps of ± 1 dB. The base station uses frame errors reported by the mobile station to increase or decrease its transmitted power.

In summary, an IS-95 CDMA system operates with a low E_b/N_0 ratio, exploits voice activity and uses the same frequency in all sectors of all cells. Each sector has

**Table 11.11 Reverse CDMA channel
I and Q mapping.**

I	Q	Phase
0	0	$\pi/4$
1	0	$(3\pi)/4$
1	1	$-(3\pi)/4$
0	1	$\pi/4$

Table 11.12 CDMA access channel modulation parameters.

Parameter	4800 bps	Units
PN chip rate	1.2288	Mcps
Code rate	1/3	bits/code symbol
Code symbol repetition	2	symbols/code symbol
Transmit duty cycle	100	%
Code symbol rate	28,800	sps
Modulation	6	code symbol/mod. symbol
Modulation symbol rate	4800	sps
Walsh chip rate	307.2	kcps
Mod. symbol duration	208.33	μs
PN chips/code symbol	42.67	PN chip/code symbol
PN chips/mod. symbol	256	PN chip/mod. symbol
PN chips/Walsh chip	4	PN chips/Walsh chip

up to 64 CDMA channels on each carrier frequency. It is a synchronized system with Rake receivers to provide path diversity at the mobile station and at the cell site.

11.14   Power Control in CDMA

A proper power control on both the uplink and downlink has several advantages:

- System capacity is improved or optimized.
- Mobile battery life is extended.
- Radio path impairments are properly compensated for.
- Quality of service (QoS) at various bit rates can be maintained.

The reverse link (uplink) uses a combination of *open loop* and *closed loop* power control to command the mobile station to make power adjustments (see Figure 11.19).

The mobile station and the base station receiver measure the received power and use the measurements to maintain a power level for adequate performance. The mobile unit measurement is part of the open loop power control while the base station measurement is part of the closed loop power control. In the closed loop mode, the mobile station transmitter power is controlled by a signal from the base station site. Each base station demodulator measures the received SNR for that mobile station and sends a power command either to increase or decrease mobile station power. The measure-command-react cycle is performed at a rate of 800 times per second for each mobile station in IS-95.

The power adjustment command is combined with the mobile's open loop estimate and the result is used to adjust the transmitter gain. This solves the near-far interference problem, reduces interference to other mobiles using the same

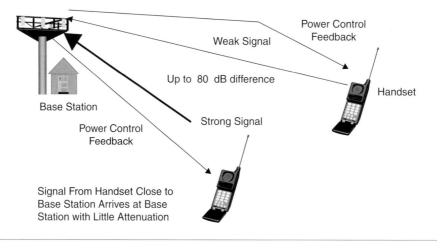

Figure 11.19   Power control in CDMA.

CDMA radio channel, helps to overcome fading, and conserves battery power in portable and mobile units.

On the uplink, the objective of the mobile station is to produce a nominal received power signal at the base station receiver. Regardless of the mobile's position or propagation loss, each mobile should be received at the base station with almost the same power level. If the mobile's signal arrives at the base station with a lower power level than the required power level, its error rate performance will be high. On the other hand, if the mobile's signal is too high, it will interfere with other users with the same CDMA radio channel causing performance degradation unless the traffic load is decreased.

Similarly, a combination of open loop and closed loop power control is used on the forward link (downlink) to keep SNR at the mobile almost constant. Forward link power control mitigates the corner problem. Mobiles at the edges of cells normally require more power than those close to the center of the base station for two reasons: more transmission loss and more interference from adjacent base stations. This is known as the *corner problem*. Forward link power control minimizes interference to mobiles in the same base station (in multipath environments) as well as mobiles in other base stations. Using the downlink power control, the base station transmits the minimum required power, hence, minimizes the interference to mobiles in the surrounding base stations.

The outer loop power control is the finer power control over the closed loop power control. It adjusts the target signal-to-interference ratio (SIR) in the base station according to the needs of the individual radio links and aims at a constant quality, which is usually defined as a certain target bit error rate (BER) or frame error ratio (FER). The required SIR depends on the mobile speed and multipath profile. The outer loop power control is typically implemented by having the base station to each uplink user data frame with frame quality indicator, such as a cyclic redundancy check (CRC) result, obtained during decoding of the particular user data frame.

11.14.1   Open Loop Power Control

In the open loop power control, the mobile uses the received signal to estimate the transmission loss from the mobile unit and the base station.

$$T(r) = 10 \log \frac{p_r(r)}{p_t} \tag{11.36}$$

$$L(r) = T(r) - G_t - G_r \tag{11.37}$$

where:

$$p_t = \text{transmitter's power output}$$
$$p_r(r) = \text{received power at distance } r \text{ from the transmitter}$$
$$G_t = \text{antenna gain of transmitter}$$

G_r = antenna gain of receiver
$L(r)$ = path loss at distance r
$T(r)$ = reverse link transmission power at distance r

This estimate is used to allow a rapid response to a sudden improvement in the channel while disallowing a rapid response to a sudden degradation in the channel. This is important because if the channel for one mobile unit suddenly improves, then the signal received at the base station from this mobile unit suddenly increases in power and causes additional interference to all other mobiles sharing the same CDMA channel. The approach to solve this problem is to tolerate a temporary degradation in one mobile unit performance in order to prevent degradation of all mobile units. The signal-to-noise ratio for the mobile at distance r is given as:

$$\text{SNR}(r) = \frac{p_t^m(r)T(r)}{(N_0 B_w)_{\text{cell}} + (M/f_r - 1) \cdot v_f \cdot p_t^m(r)T(r)} \tag{11.38}$$

where:

$p_t^m(r)$ = mobile power amplifier output
$T(r)$ = reverse link transmission power at distance r
M = number of users per cell
v_f = channel activity factor
f_r = frequency reuse efficiency
$(N_0 B_w)_{\text{cell}}$ = cell thermal noise

The mobile estimates the $T(r)$ from its measurements of the total received power and the prior knowledge of the base station power amplifier output. The total received power can be expressed as:

$$p_r^m(r) = p_t^c \cdot T(r) + (N_0 B_w)_{\text{cell}} + I_{oc} B_w = (1 + \xi_1 + \xi_2)[p_t^c \cdot L(r)] \text{ dB} \tag{11.39}$$

where:

P_t^c = base station power amplifier output (dB)
$P_r^m(r)$ = mobile received total power (dB)
I_{oc} = spectral density of interference from other cells (dBm/Hz)
ξ_1 = ratio of power from other cells to power from the mobile's home cell
ξ_2 = ratio of thermal noise to power from mobile's home cell
B_w = bandwidth (Hz)

To minimize the power transmitted by each mobile, $(\text{SNR})_{\text{min}}$ for all reverse channels should be maintained.

$$\text{SNR}_{\text{min}} = \frac{p_t^m(r) \cdot L(r)}{(N_0 B_w) + (M/f_r - 1)[(v_f \cdot p_t^m) \cdot L(r)]} \tag{11.40}$$

$$\text{SNR}_{\min} = \frac{p_t^m(r) \cdot \dfrac{p_r^m(r)}{(1 + \xi_1 + \xi_2) \cdot p_t^c}}{(N_0 B_w) + ((M/f_r) - 1) \dfrac{[v_f \cdot p_t^m(r)] \cdot p_r^m(r)}{(1 + \xi_1 + \xi_2) \cdot p_t^c}} \tag{11.41}$$

There are two extreme cases:

- The mobile is close to its base station in this case, the total received power is mainly due to its base station and

$$L(r) \approx (p_r^m(r))/p_t^c$$

- The mobile is close to the edge of its base station; in this case the total received power is the contribution of more than one base station. Worst case conditions are when the mobile is equidistant from three adjacent base stations. Assuming that all base stations transmit the same power, then

$$(1 + \xi_1 + \xi_2) \leq 3$$

The quantity

$$(N_0 B_w) + ((M/f_r) - 1) v_f \cdot p_t^m(r) \cdot \frac{p_r^m(r)}{(1 + \xi_1 + \xi_{2!}) \cdot p_t^c}$$

depends only on base station loading

$$1 + \frac{((M/f_r) - 1) \cdot v_f \cdot p_t^m(r)}{(N_0 B_w)} \cdot \frac{p_r^m(r)}{(1 + \xi_1 + \xi_2) \cdot p_t^c} = \frac{1}{1 - \rho}$$

where:

$$\rho = \text{cell loading} = M/M_{\max}$$
$$M = \text{number of users in the cell}$$
$$M_{\max} = \text{maximum capacity of the cell}$$

$$\text{SNR}_{\min} = p_t^m(r) + p_r^m(r) - p_t^c - 10\log(1 + \xi_1 + \xi_2) - (N_0 B_w)_{\text{cell}}$$
$$+ 10\log(1 - \rho)\,\text{dB} \tag{11.42}$$

In order to maintain a constant SNR at the base station, the mobile should transmit

$$p_t^m(r) = (\text{SNR})_{\min} + p_t^c + (N_0 B_w)_{\text{cell}} - 10 \log(1 - \rho) + 10 \log(1 + \xi_1 + \xi_2)$$
$$- p_r^m(r) \; \text{dB} \tag{11.43}$$

$$p_t^m(r) = \left(\frac{E_b}{N_t}\right) - 10 \log G_p + p_t^c + (N_0 B_w)_{\text{cell}} - 10 \log(1 - \rho) + 10 \log(1 + \xi_1 + \xi_2)$$
$$- p_r^m(r) \tag{11.44}$$

$$p_t^m(r) = \overline{K} - p_r^m(r) \; \text{dB} \tag{11.45}$$

where:

$$\overline{K} = \left(\frac{E_b}{N_t}\right) - 10 \log G_p + p_t^c + (N_0 B_w)_{\text{cell}} - 10 \log(1 - \rho) + 10 \log(1 + \xi_1 + \xi_2)$$

As an example, let us consider IS-95, in which: $G_p = 128$, $E_b/N_t = 7\,\text{dB}$, $p_t^c = 25$ watts, $N_0 = -174\,\text{dBm/Hz}$, $B_w = 1.23\,\text{MHz}$, $\rho = 0.5$, cell noise figure $N_f = 4\,\text{dB}$, $1 + \xi_1 + \xi_2 = 2$, then $(\overline{K}) = -73\,\text{dB}$ (*Note*: this value is used in Example 11.7).

$$p_t^m(r) = -73 - p_r^m(r) \; \text{dB} \tag{11.46}$$

The constant $(\overline{K})$ is the part of the broadcast message that is sent to the mobile by the base station on the paging channel. The speed at which the open loop power control tracks the changes in the channel depends on the time constant of the automatic gain control (AGC) filter.

Example 11.7

An IS-95 CDMA mobile measures the signal strength from the base stations as $-97\,\text{dBm}$, what should the mobile transmitter power be set to as a first approximation? After the connection is made with this power level, the base station requires the mobile station to change its power to $+18\,\text{dBm}$. How long would it take for the mobile station to make this change?

Solution

$$p_t^m(r) = \overline{K} - p_r^m(r) = -73 - p_r^m(r) = -73 - (-97) = 24\,\text{dBm}$$

Power reduction $= 24 - 18 = 6\,\text{dBm}$

Mobile requires 6 decrements each at $1.25\,\text{ms}$ $(1/800\,\text{sec})$

$\therefore$ Time required $= 6 \times 1.25 = 7.5\,\text{ms}$

11.15 Softer and Soft Handoff

During soft handoff, a mobile station is in the overlapping cell coverage area of two sectors belonging to two different base stations. The communications between mobile station and base station occur concurrently via two air interface channels from each base station separately. Both channels (signals) are received at the mobile station by maximal combining Rake processing (see Figure 11.20). Soft handoff occurs in about 20–40% of calls.

Soft handoffs are an integral part of CDMA design. The determination of which pilots will be used in the soft handoff process has a direct impact on the quality of the call and the capacity of the system. Therefore, setting soft handoff parameters is a key element in the system design for CDMA.

In the uplink direction, soft handoff differs significantly from softer handoff: the code channel of the mobile station is received from both base stations, but the received data is routed to the base station controller (BSC) for combining. This is done so that the same frame reliability indicator as provided for outer loop power

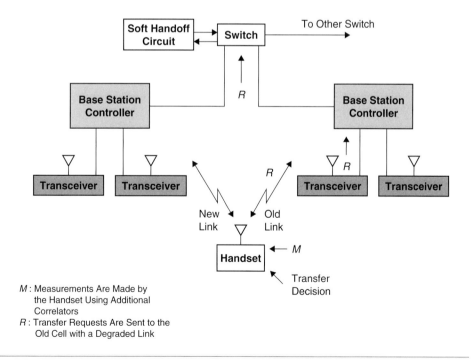

Figure 11.20 Soft handoff in CDMA.

control is used to select the better frame between two possible candidates within the BSC.

A brief description of each type of pilot set is given below:

- The *active set* is the set of pilots associated with downlink traffic channels assigned to the mobile units. The active set can contain more than one pilot because a total of three carriers, each with its own pilot, could be involved in a soft handoff process.
- The *candidate set* consists of the pilots that the mobile unit has reported are of a sufficient signal strength to be used. The mobile unit also promotes the neighbor set and remaining set pilots that meet the criteria to the candidate set.
- The *neighbor set* is the list of the pilots that are not currently on the active or candidate pilot lists. The neighbor set is identified by the base station via the neighbor list and neighbor list update messages.
- The *remaining set* contains all possible pilots in the system that can possibly be used by the mobile unit. However, the remaining set pilots that the subscriber unit looks for must be a multiple of Pilot_Inc.

The parameters used to control the movement of a pilot from a neighbor to a candidate, to active, and then back to neighbor set are given below:

1. Pilot strength exceeds **T_ADD** and the mobile unit sends a pilot strength measurement message (PSMM) and transfers the pilot to the candidate set.
2. The pilot strength drops below **T_DROP** and the mobile unit begins the handoff drop time (**T_TDROP**).
3. **T_COMP** is used into decision matrix for adding and removing pilots from the neighbor, candidate, and active set.

For more details about power control and soft handoff in CDMA, refer to [5].

Example 11.8

Given the current active set contains pilots P1, P2, P3, and P4. At a particular time, the mobile station measures the signal strength of P1, P2, P3, and P4 as $-95\,\text{dBm}$, $-100\,\text{dBm}$, $-101\,\text{dBm}$, and $-105\,\text{dBm}$, respectively. Pilot P5 is in the candidate set. The mobile station measures P5 signal strength as $-102\,\text{dBm}$ at the same time. Determine a possible pair of the SOFT-SLOPE and ADD-INTERCEPT values that will trigger the mobile station to send a PSMM to the base station. The mobile station is IS-95B compliant and uses the following criterion to add the pilot from the candidate set to the active set.

$$10\log(P_{cj}) \geq \text{Max}$$

$$\left\{ \left((\text{SOFT-SLOPE}) \times 10\log\left[\sum_{i=1}^{N_A} P_{ai}\right] + (\text{ADD-INTERCEPT}) \right), T_{\text{ADD}} \right\}$$

where:

P_{cj} = received power of the jth pilot in the candidate set
P_{ai} = received power of the ith pilot in the active set
N_a = number of pilots in the active set
T_{ADD} = threshold for adding a pilot from the candidate set into the active set

Also, if the mobile station is IS-95 compliant, find the value of T-COMP that could trigger the mobile station to generate a PSMM. Assume mobile receiver sensitivity, i.e., noise power $= -107\,\text{dBm}$ and $T_{\text{ADD}} = -13\,\text{dB}$.

Solution

(a)

$P_{a1} = \text{P1} - \text{Receiver Sensitivity} = -95 - (-107) = 12\,\text{dBm}$
$P_{a2} = \text{P2} - \text{Receiver Sensitivity} = -100 - (-107) = 7\,\text{dBm}$
$P_{a3} = \text{P3} - \text{Receiver Sensitivity} = -101 - (-107) = 6\,\text{dBm}$
$P_{a4} = \text{P4} - \text{Receiver Sensitivity} = -105 - (-107) = 2\,\text{dBm}$
$P_{c5} = -102 - (-107) = 5\,\text{dBm}$

$$10\log\left[10^{1.2} + 10^{0.7} + 10^{0.6} + 10^{0.2}\right] = 14.22$$

$$\therefore (\text{Max}\ \{14.22 \times (\text{SOFT-SLOPE}) + (\text{ADD-INTERCEPT}), -13\} \leq 5)$$

Thus, we have the following two relations:

$$14.22 \times \text{SOFT-SLOPE} + \text{ADD-INTERCEPT} > -13$$

and

$$14.22 \times \text{SOFT-SLOPE} + \text{ADD-INTERCEPT} \leq 5$$

Solving these equations, we get SOFT-SLOPE = 0.5 and ADD-INTERCEPT $= -4$

(b)

For an IS-95-compliant mobile station

$$(P_{cj} - P_{ai}) \geq 0.5 \times \text{T-COMP}$$

Since P1 > P2 > P3 > P4, we replace P4

$$[-102 - (-105)] \geq 0.5 \times \text{T-COMP}$$

$$\text{T-COMP} \leq 6\,\text{dB} \leq 4$$

11.16 Summary

In this chapter we first discussed the concept of spread spectrum systems and provided the main features of the direct sequence spread spectrum and frequency hop spread spectrum systems. A key component of spread spectrum performance is the calculation of the processing gain of the system, which is the relationship between the input and output signal-to-noise ratio of a spread spectrum receiver.

Spread spectrum systems trade bandwidth for processing gain, and code division systems use a variety of orthogonal or almost orthogonal codes to allow multiple users in the same bandwidth. Thus, CDMA systems can have a higher capacity than either analog or time division multiple access (TDMA) digital systems. However, because of practical constraints on CDMA systems, it is not possible to achieve the Shannon bound in system design. The upper bound of the capacity of a CDMA system is limited by the processing gain of the system, receiver modulation performance, power control accuracy, interference from other cells, voice activity, cell sectorization, and the ability to maintain accurate synchronization of the system.

We concluded the chapter by discussing the high level features of the IS-95 system and listing the challenges in implementing a CDMA system. Two important aspects of the CDMA system, power control and softer and soft handoff, were also discussed.

Problems

11.1 Define open loop, closed loop, and outer loop power control.

11.2 What are the softer and soft handoffs?

11.3 Differentiate between scrambling and spreading.

11.4 What are the requirements for a DSSS?

11.5 What are the basic differences between the DSSS and FHSS systems?

11.6 Which parameters are used to control the movements of the pilot in a CDMA system?

11.7 How many times are power adjustments performed in an IS-95 CDMA system?

11.8 Determine $(S/N)_o$ for a spread spectrum system having bandwidth, $B_w = 1.2288\,\text{MHz}$, information rate, $R = 9.6\,\text{kb/s}$ and $(S/N)_i = 15\,\text{dB}$. Relate your result to E_b/N_0.

11.9 Use the data in Example 11.6 and show that the bit pattern at mobile 2 and 3 can be determined from the demodulated signal using Path 1 and Path 2.

11.10 Consider a case where eight chips per bit are used to generate the Walsh codes (see Appendix D). Mobile stations A, B, C, and D are assigned W_1, W_2, W_3, and W_4, respectively. The stations use the Walsh code to send a "1" binary bit and use negative Walsh to send a "0" binary bit. Assume all stations are synchronized in time; therefore, chip sequences begin at the same instant. When two or more stations transmit simultaneously, their bipolar signals add linearly. Consider the following four different cases when one or more stations transmit and show that the receiver recovers the bit stream of stations B and C. (*Note:* dash (−) means no transmission by the station).

Station (A, B, C, D)	Transmitting Stations
− − 1 0	C + D
1 1 1 −	A + B + C
1 1 − −	A + B
1 1 0 0	A + B + C + D

11.11 An IS-95 CDMA mobile measures the signal strength from the base station as −100 dBm, what should the mobile transmitter power be set to as a first approximation?

11.12 Given the current active set contains pilots P1, P2, P3, and P4. At a particular time, the mobile station measures the signal strength of P1, P2, P3, and P4 as −98 dBm, −101 dBm, −103 dBm, and −104 dBm, respectively. Pilot P5 is in the candidate set. The mobile station measures P5 signal strength as −103 dBm at the same time. Determine a possible value of the SOFT-SLOPE, that will cause the mobile station to send a PSMM report to the base station, if the ADD-INTERCEPT value is −4. The mobile station is IS-95B compliant and uses the following criterion to add the pilot from the candidate set to the active set.

$$10 \log (P_{cj}) \geq \text{Max}$$

$$\left\{ \left(\left(\text{(SOFT-SLOPE)} \times 10 \log \left[\sum_{i=1}^{N_A} P_{ai} \right] + \text{(ADD-INTERCEPT)} \right), T_{\text{ADD}} \right\} \right.$$

where:

P_{cj} = received power of the jth pilot in the candidate set

P_{ai} = received power of the ith pilot in the active set

N_a = number of pilots in active set

T_{ADD} = threshold for adding a pilot from the candidate set into the active set

Assume mobile receiver sensitivity i.e., noise power = −106 dBm and T_{ADD} = −14 dB.

11.13 Also, if the mobile station is IS-95 compliant in Problem 11.12, find the value of T-COMP that could cause the mobile station to generate a PSMM report.

References

1. Bhargava, V., Haccoum, D., Matyas, R., and Nuspl, P. *Digital Communications by Satellite*. New York: John Wiley & Sons, 1981.

2. Dixon, R. C. *Spread Spectrum Systems*, Second Edition. New York: John Wiley & Sons, 1984.

3. Feher, K. *Wireless Digital Communications Modulation and Spread Spectrum Applications*. Upper Saddle River, NJ: Prentice-Hall, 1995.

4. Garg, V. K., and Wilkes, J. E. *Wireless and Personal Communications Systems*. Prentice Hall, 1996.

5. Garg, V. K. *CDMA IS-95 and cdma2000*. Prentice Hall, 2000.

6. Holma, H., and Toskala, A. (editors). *WCDMA for UMTS*. New York: John Wiley & Sons, 2000.

7. Lee, W. C. Y. *Mobile Cellular Telecommunication Systems*. New York: McGraw-Hill, 1989.

8. Pahlwan, K., and Levesque, A. H. *Wireless Information Networks*. New York: John Wiley & Sons, 1995.

9. Smith, C., and Collins, D. *3G Wireless Networks*. New York: McGraw-Hill, 2002.

10. Shannon, C. E. Communications in the Presence of Noise. Proceedings of the IRE. no. 37, pp. 10–21, 1949.

11. Steele, R. *Mobile Radio Communications*. New York: IEEE Press, 1992.

12. Skalar, B. *Digital Communications—Fundamentals & Applications.* Englewood Cliffs, NJ: Prentice Hall 1988.

13. TIA/EIA IS-95. "Mobile Station—Base Station Compatibility Standard for Dual-mode Wideband Spread Spectrum Cellular System," PN-3422, 1994.

14. Torrien, D. *Principle of Secure Communication Systems.* Boston: Artech House, 1992.

15. Virterbi, A. J. *CDMA.* Reading, MA: Addison-Wesley Publishing Company, 1995.

16. Viterbi, A. J. and Padovani, Roberto. Implications of Mobile Cellular CDMA. *IEEE Communication Magazine*, vol. 30, no. 12, pp. 38–41, 1992.

CHAPTER 12

Mobility Management in Wireless Networks

12.1 Introduction

Mobility is important in mobile communication [1–3]. It can be categorized as *radio mobility* and *network mobility*. Radio mobility is mainly concerned with the handoff process, whereas network mobility mainly deals with mobile location management (i.e., location updating and paging). As mobiles travels across system boundaries, whether they are cell, location, or a mobile switching center (MSC) areas, the network must be able to locate the mobile subscriber and automatically route the call to him or her [4–6]. Security mechanisms, such as subscriber identity authentication and equipment validation (discussed in Chapter 13) provide a level of protection against fraud to subscribers and service providers [8,13].

The public land mobile network (PLMN) is an integrated service digital cellular network providing wireless access for mobile subscribers to other networks and network services, including other PLMNs. A PLMN area is divided into regions called *location areas* (LAs). Each LA is made up of one or more cell areas. A mobile station (MS) registers with the visitor location register (VLR) each time it enters a new LA. The mobile station is free to move inside a given LA without a registration (see Figure 12.1). Each cell has a unique identification number called a *cell global identification* (CGI), which contains *location area identification* (LAI) and *cell identification* (CI) (see Figure 12.2) [10–12]. The LAI consists of *mobile country code* (MCC), *mobile network code* (MNC), and *location area code* (LAC). The CI is a specific cell identity within a given location area [7].

Mobility management in wireless networks is the primary set of functions that are supported by the network to enable subscriber mobility. Mobility management enables the network to keep track of the subscriber's status and location to deliver calls to that subscriber. It also enables the network to authorize a subscriber for service in a given service area. The key component to mobility management is the subscribers' service profile. The service profile is a database record in the network that contains information about each subscriber. This information includes dynamic data such as current location and status of a subscriber as well as permanent data such as the service profile, international mobile subscriber identity (IMSI), and so on of a subscriber.

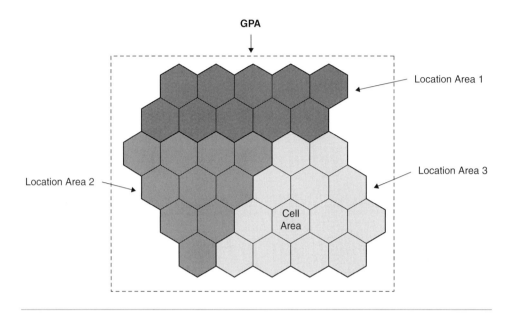

Figure 12.1   Wireless PLMN.

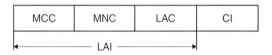

Figure 12.2   Cell global identification.

12.2   Mobility Management Functions

Mobility management generally deals with automatic roaming, authentication, and intersystem handoff.

Automatic roaming includes a set of network functions that allow a subscriber to obtain service outside the home service provider area. These functions are automatic and do not require special subscriber actions. The automatic roaming functions are divided into:

- Mobile station (MS) service qualification
- MS location management
- MS state management
- Home location register (HLR), and VLR fault recovery

The authentication process requires that end users of the system are authenticated, i.e., the identity of each subscriber is verified (see Chapter 13).

Handoff is one of the essential features that guarantees the subscriber mobility in a mobile network, where the subscriber can move around. Maintaining connection with a moving subscriber is possible with the help of the handoff function. The basic concept is simple: when the subscriber moves from the coverage area of one cell to another, a new connection with the target cell has to be set up and connection with the old cell may be released. Controlling the handoff mechanism is, however, quite a complicated issue in cellular systems.

12.3 Mobile Location Management

Location management schemes are based on subscribers' mobility and incoming call rate characteristics. Using the location update (LU) procedure the network keeps track of the mobile subscriber's location to direct the incoming call.

The paging process transmits paging messages to all those cells where mobile terminals can be located. There is a trade-off between the LU cost and paging cost. If the LU cost is high, the paging cost will be low, as paging is performed over a small area. On the other hand, if the LU cost is low, the paging cost will be high, as paging is performed over a wider area.

Location management uses either the periodic LU or the LU-on-LA-crossing. The VLR stores the LAI, and the HLR keeps the VLR identifier. In the periodic LU method, an MS periodically sends its identity to the network. The drawback of this method is its resource consumption. As an example, if the mobile subscriber does not move from an LA for several hours, resources are consumed unnecessarily.

In the LU-on-LA-crossing method, each base station (BS) periodically broadcasts the identity of its LA. The MS is required to regularly listen to network broadcast information and store the current LA identity if the received identity differs from the one stored in the MS. An LU procedure is automatically triggered by the MS. The advantage of this method is that it requires LUs only when the MS actually moves. A highly movable mobile generates a large number of LUs, a low mobility mobile triggers only a few.

A hybrid location management scheme combines the two methods. The MS generates its LU each time it detects an LA crossing. However, if no communication has occurred between the MS and the network for a specified duration of time, the MS generates an LU. This is a periodic LU that typically allows the network to recover subscriber location data in case of a database failure.

Generally LU procedures involve intra-VLR LU, inter-VLR LU using temporary mobile subscriber identity (TMSI), inter-VLR LU using IMSI, and the IMSI-attached procedure that is triggered when the mobile is powered on in the LA where it was powered off. To minimize the location management cost (i.e., LU plus paging traffic processing), the LA must be carefully designed by using a proper subscriber mobility model discussed next.

12.3.1 Mobility Model

A mobility model describes the occurrence of procedures such as LU and handoff. Several models have been proposed.

- **Fluid model:** This model assumes traffic flow to be like the flow of a fluid. The model suggests that the amount of traffic flowing out of an area is proportional to the population density of the area, the average velocity of movement, and the length of the area boundary. For an area, the average number of crossings per unit time is given as:

$$\lambda = \frac{v\rho L}{\pi} \qquad\qquad (12.1)$$

where:

   λ = number of crossings per unit of time
   ρ = average population density
   v = average movement velocity in the area
   L = perimeter of the area

One of the limitations of the fluid model is that it describes aggregate traffic and is difficult to apply to situations where individual movement patterns are desired. The fluid model is more applicable to areas with a large population because it uses the average population density and the average movement velocity of the area.

- **Markovian model:** This is also known as the random walk model. The Markovian model describes individual movement. In this model, a mobile subscriber either remains within the region or moves to an adjacent region according to a transition probability distribution. One of the limitations of this model is that there is no concept of trips.

- **Gravity model:** Variations of the gravity model have been employed in transportation research to model human movement behavior. Gravity models have been applied to regions of varying sizes, from city models to national and international models. The main difficulty in using the gravity model is that many parameters are required in the calculations; it is therefore difficult to model geography with many regions.

Next, we show an application of the simple mobility model based on the fluid flow concept for a wireless network in an urban area with small cells and high user density. We evaluate the impact of LUs on RF resource occupancy at the network level and determine the number of transactions processed by MSC/VLR [9]. The transaction is defined here as the message received or transmitted by the MSC/VLR. Assumptions used in our calculations are: (1) cells are hexagonal, (2) maximum blocking probability is 1%, (3) mobiles are uniformly distributed in the cell area, and (4) movements of the mobiles are uncorrelated; the directions of their movements are uniformly distributed over 0 to 2π.

The optimal number of cells (N_{opt}) per LA is given as:

$$N_{opt} = \sqrt{\frac{C_{page}}{C_{LU}} \cdot \frac{v}{\pi R}} \qquad (12.2)$$

where:

C_{page} = cost of paging (in terms of the number of paging messages required to find an MS)

C_{LU} = cost of LUs (in terms of the number of LU messages required to update the location of an MS)

R = cell radius

v = mean mobile velocity in LA

Using Equation 12.1 for number of LUs, λ^j_{LU} in an LA perimeter of the jth cell per hour will be:

$$\lambda^j_{LU} = \frac{v\rho L}{\pi} \qquad (12.3)$$

where:

$L = 6R\left[\dfrac{1}{3} + \dfrac{1}{2\sqrt{N_c} - 3}\right]$, length (km) of the cell exposed perimeter in an LA

R = cell radius (km)

N_c = number of cells per location area

ρ = mobile density in the cell (mobiles per km^2)

The resource occupancy R^j_{LU} in the jth cell due to MS LUs is given as:

$$R^j_{LU} = \lambda^j_{LU}\left[\sum_{k=1}^{3} p^{j,k}_{LU} \cdot t^k_{LU}\right] \qquad (12.4)$$

where:

$p^{j,k}_{LU}$ = percent of LUs in the kth case for the jth cell

t^k_{LU} = average duration of an LU in the kth case (k = 1: intra-VLR; k = 2: inter-VLR with TMSI; and k = 3: inter-VLR with IMSI)

λ^j_{LU} = number of transactions processed by MSC/VLR in an LA perimeter of the jth cell per hour

Table 12.1 Number and duration of transaction for different LU types.

Transaction type	Number of transaction/LU	Duration of a transaction
Intra-VLR LU	2	600 ms
Inter-VLR LU with TMSI	14	3500 ms
Inter-VLR LU with IMSI	16	4000 ms

The total number of transactions due to LUs generated in the $N_{LA} \times N_p$ location area perimeter cells (which we number from 1 to $N_{LA}N_p$) and processed per hour by the MSC/VLR is given as:

$$TN_{LU} = \lambda_{LU}^i \sum_{j=1}^{N_{LA} N_p} \left[\sum_{k=1}^{3} p_{LU}^{j,k} \cdot t_{LU}^k \right] \tag{12.5}$$

where:

N_{LA} = number of LAs managed by MSC/VLR

$N_p = 6\sqrt{\dfrac{N_c}{3}} - 3$ = number of cells located on the perimeter of an LA, and

Example 12.1

Using the following data for Groupe Special Mobile/Global System for Mobile Communications 1800, evaluate the impact of LUs on the radio resource and calculate the MSC/VLR transaction load using the fluid flow model.

- Density of mobiles in the cell = 10,000 mobiles/km^2
- Cell radius = 500 m
- Average moving velocity of a mobile = 10 km/hour
- Number of cells per LA = 10
- Number of LAs per MSC/VLR = 5
- Number of transactions and duration of each transaction to MSC/VLR per LU for different LU types are given in Table 12.1.

We consider two cases: (1) an optimistic situation in which generated LUs in a cell are only intra-VLR LUs, and (2) a pessimistic situation where generated LUs in a cell are inter-VLR LUs.

Solution

Case 1: In the first case, the jth cell located at the border of two LAs is related to the same MSC/VLR, only intra-VLR LUs are processed in the cell.

$$L = 6R\left[\frac{1}{3} + \frac{1}{2\sqrt{N_c} - 3}\right] = 6 \times \frac{500}{1000}\left[\frac{1}{3} + \frac{1}{2\sqrt{10} - 3}\right] = 1.902 \text{ km}$$

$$\lambda_{LU}^{j} = \frac{v\rho L}{\pi} = \frac{10 \times 10{,}000 \times 1.902}{\pi} = 60{,}543 \text{ LUs per hour}$$

$$R_{LU}^{j} = \lambda_{LU}^{j} \left[\sum_{k=1}^{3} p_{LU}^{j,k} \cdot t_{LU}^{k} \right] = \frac{60{,}543(1 \times 600/1000)}{3600} = 10.1 \text{ Erlangs}$$

This requires 18 channels at 1% blocking (refer to the Erlang-B table, Appendix A) or $18/8 = 2.25$ traffic channel (about 1/4 of an RF channel, assuming there are 8 traffic channels per RF channel).

Case 2: In this case the *j*th cell is located at the border of two LAs related to two different VLRs. In this case, only inter-VLR LUs will be processed in the cell. We assume 80% of LUs are with TMSI and 20% of LUs are with IMSI.

$$R_{LU}^{j} = \lambda_{LU}^{j} \left[\sum_{k=1}^{3} p_{LU}^{j,k} \cdot t_{LU}^{k} \right] = \frac{60{,}543}{3600} \left[0.8 \times \frac{3500}{1000} + 0.2 \times \frac{4000}{1000} \right] = 60.45 \text{ Erlangs}$$

This requires 75 channels at 1% blocking (refer to the Erlang-B table, Appendix A) or $75/8 = 9.38$ traffic channels (about 1.25 RF channels).

MSC/VLR Transaction Load

We assume that one LA is in the center of the region and the remaining four LAs are on the border of the region. We also assume that, in the perimeter cells at the border LAs, only intra-VLR LUs are generated. For half of the perimeter cells at the border LAs, only inter-VLR LUs are generated.

$$N_p = 6 \sqrt{\frac{N_c}{3}} - 3 = 6 \sqrt{\frac{10}{3}} - 3 = 7.9545 \approx 8 \text{ cells located on the perimeter}$$
of an LA.

Number of cells where inter-VLR LUs occur will be:

$$0.5 \times 4 \times N_p = 2 \times 8 = 16$$

Number of cells where intra-VLR LUs occur will be:

$$4 \times N_c - 16 = 4 \times 10 - 16 = 24$$

$$TN_{LU} = \lambda_{LU}^{i} \sum_{j=1}^{N_{LA} \, N_p} \left[\sum_{k=1}^{3} p_{LU}^{j,k} \cdot t_{LU}^{k} \right]$$

$$= 60{,}543[2 \times 24 + 16(0.8 \times 14 + 0.2 \times 16)]$$

$$= 16.85 \times 10^{6} \text{ transactions at peak hour}$$

Transactions at Peak Hour

The results show that, under heavy traffic conditions, the impact of LUs can be significant. In terms of radio channels used, we find that between 0.25 to 1.25 GSM RF carriers can be used for LA boundary crossings. Although this load cannot directly cause call blocking on the radio interface (in a GSM1800 network with 4×3 reuse cluster, the average number of RF carriers per cell with 3 operators is about 10), it is nevertheless not a negligible impact on traffic channel (TCH), consumption. In terms of processing at the MSC/VLR side, with a processing load of about 16.85×10^{6} transactions per hour, it is obvious that blocking can rapidly occur with the given scenario. The MSC/VLR resources dedicated to LU processing cannot be used to provide other services. A major concern of operators, where objectives are to provide users with rich and sophisticated services, is to have more processing resources at the MSC/VLR.

12.4 Mobile Registration

The detection of a subscriber in a new serving system is an example of a registration event. The registration is a subset of automatic roaming functions. It primarily consists of MS service qualification and MS location management. These functions enable a subscriber to register with the network; that is, to indicate to the network location register functional entities the location and status of the MS.

The registration-related functions generally must be performed before the network can provide service to the subscriber; however, there are exceptions, such as access to emergency services by dialing 911.

The process of registering a mobile subscriber is unique to wireless networks. The registration is not used for wireline systems because the subscriber location is fixed and does not change. The types of registration supported in a network are dependent on the protocol used for the air interface between the mobile station and radio system and on the algorithms implemented in the serving network. The air interface standards for advanced mobile phone systems (AMPS), time division multiple access (TDMA), and code division multiple access (CDMA) support different types of registration than does GSM.

Registration may be initiated by the MS or the network, or may be implied during MS access. Upon receiving the registration request from the MS, the radio system (base station) constructs the *Registration Update Request* message and sends it to the network (MSC). The Registration Update Request message contains the MS's identification and location information and may contain authorization parameters. The network (MSC) may respond with a request for authorization (optional) and finally with a *Registration Update Response* message.

The network (MSC) sends a Registration Update Response message to the BS when a registration procedure has been successfully completed. This message indicates whether the MS's registration has been accepted or rejected. The message may contain additional parameters to be sent to the MS. Upon receipt of this message, the BS sends an appropriate response to the MS. The three possible results in registration requests are: successful registration, unsuccessful registration, and cancellation of registration.

A mobile station registering on an access channel may perform any one of the following registration types:

- **Distance-based registration:** when the distance between the current cell and the cell where the mobile last registered exceeds a threshold.

- **Geographic-based registration:** whenever a mobile enters a new area of the same system. A service area may be segmented into smaller regions, known as location areas, which are groups of one or more cells. The MS identifies the current location area via parameters transmitted by the MS on the forward control channel. Location-based registration reduces the paging load of the system by allowing the network to page only in the location area(s) where a mobile station is registered.

- **Parameter change registration:** when specific operating parameters in the mobile are changed.

- **Periodic registration:** when the system sets parameters on the forward control channel to indicate that all or some of the mobile stations must register. The registration can be directed to a specific mobile or a class of mobile stations.

- **Power-down registration:** when the mobile is switched off. This allows the network to deregister a mobile immediately upon its power-down.

- **Power-up registration:** when power is applied to the mobile, and used to notify the network that the mobile is now active and ready to place or receive calls.

- **Timer-based registration:** when a timer expires in the mobile. This procedure allows the database in the network to be cleared if a registered MS does not register after a fixed time interval. The time interval can be varied by setting parameters on the control channel.

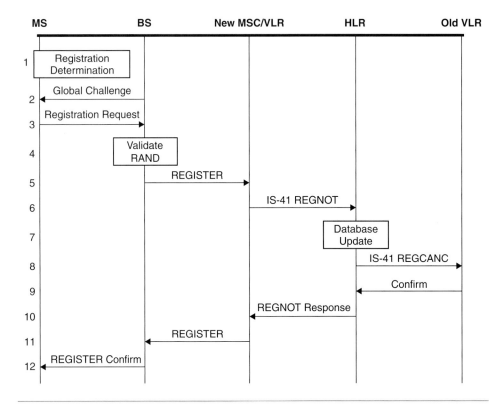

Figure 12.3   Call flows for MS registration of all mobiles listening to a control channel.

It should be noted that all registration types are not supported in a network.

The following are the call flows for the registration (ANSI-41 standard) of all MSs listening to a control channel (see Figure 12.3):

1. The MS determines that it must register with the system.
2. The MS listens on the control channel for the global challenge, random number (RAND).
3. The MS sends a message to the BS with IMSI, RAND, and other parameters, as needed, in the MS Registration Request.
4. The BS validates RAND.
5. The BS sends a REGISTER message to the new MSC/VLR.
6. If the MS is not currently registered to the serving VLR, the VLR sends a REGistration NOTification (REGNOT) message to the user's HLR containing the IMSI and other data as needed.

7. The MS's HLR receives the REGNOT message and updates its database accordingly (stores the location of the VLR that sent the REGNOT message).

8. The MS's HLR sends an IS-41 REGistration CANCel (REGCANC) message to the old VLR where the MS was previously registered so that the old VLR can cancel the MS's previous registration.

9. The old VLR returns a Confirmation message that includes the current value of the call count (CHCNT).

10. The user's HLR then returns a REGNOT Response message to the new VLR and passes along information that the VLR needs (e.g., user's profile, interexchange carrier ID, shared secret key for authentication, and current value of CHCNT). If the registration is a failure (due to invalid IMSI, service not permitted, nonpayment of bill, or other reason), then the REGNOT message will include a failure indication.

11. Upon receiving the successful REGNOT message from the user's HLR, the VLR assigns a TMSI. The MSC receives the message, retrieves the data and sends a REGISTER message to the BS.

12. The BS receives the REGISTER message and forwards it to the MS to Confirm Registration.

Note: For details of various messages refer to the ANSI-41 standard.

12.4.1 GSM Token-Based Registration

When an MS registers with the new network, it sends its TMSI and LAI. The LAI informs the system where to find the old VLR. The network then queries the old VLR for data and uses the data to authenticate the MS. The new VLR then communicates with the HLR to update the location of the MS. The HLR sends a registration cancellation message to the old VLR.

The operation of the token-based system is slightly different from the ANSI-41 system in that the segmentation of call processing between BS, MSC, and VLR is defined differently. Therefore the call flow does not, in most cases, distinguish where call processing is done. The call flows for token-based registration are shown in Figure 12.4. The following are the steps in the call flows:

1. The MS sends a Registration message to the visiting system with old TMSI and old LAI.

2. The visiting system queries the old VLR for data.

3. The old VLR returns security-related information (e.g., unused triplets and location of HLR).

4. The visiting system issues a challenge to the MS.

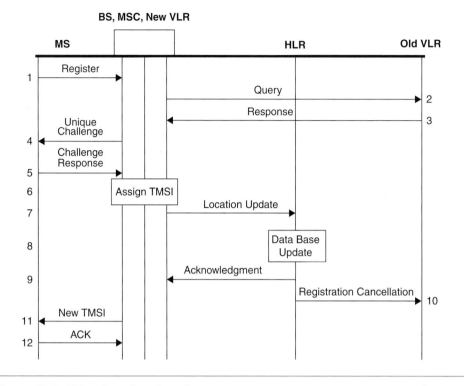

Figure 12.4 Token-based registration.

5. The MS responds to the challenge.

6. The visiting system assigns a new TMSI.

7. The visiting system sends a message to the HLR with location update information.

8. The HLR updates its location database with the new location of the MS.

9. The HLR acknowledges the message and may send additional security-related data (additional security triplets).

10. The HLR sends a Registration Cancellation message to the old system.

11. The visiting system sends an encrypted message to the MS with new TMSI.

12. The MS acknowledges the message.

Note: Steps 7–10 and 11–12 can occur in any order.

  If the old VLR is not reachable for any reason, then the network will request the MS to send its IMSI, and communications with the HLR will then occur.

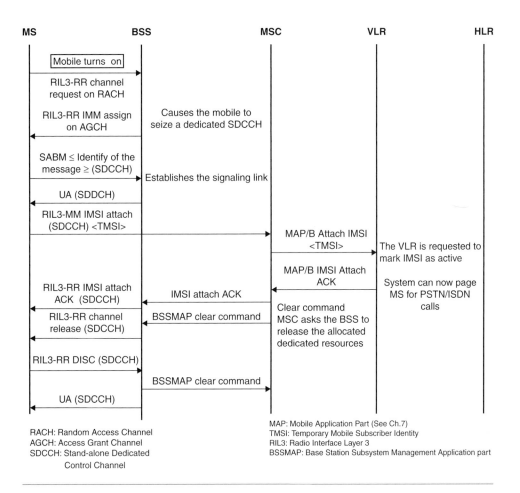

Figure 12.5 IMSI attach process in GSM.

12.4.2 IMSI Attach and IMSI Detach (Registration and Deregistration) in GSM

In GSM, mobile power-up implies *IMSI attach* which causes a mobile registration. On the other hand, mobile power-down implies *IMSI detach* causing mobile deregistration. Figures 12.5 and 12.6 show call flows for IMSI attach and detach (for details see GSM standards).

12.4.3 Paging in GSM

In this scenario we assume that the mobile is already registered with the system and has acquired a TMSI. It is also assumed that the mobile is located in its home

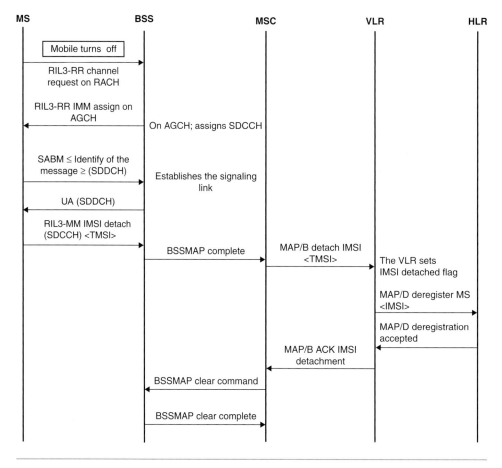

Figure 12.6   IMSI detach process in GSM.

network. A land subscriber dials the directory number of the mobile subscriber. Figure 12.7 shows the call flow for this scenario.

1. The public switched telephone network (PSTN) routes the call to the MSC assigned this directory number. The directory number in the Initial Address Message (IAM) is the MS ISDN Number (MSISDN).

2. The MSC sends the Send Routing Information Message to the HLR to provide the routing information for the MSISDN.

3. The HLR acknowledges the Send Routing Information Message to the MSC. This message contain the MS routing number (MSRN). If the MS is roaming within the serving area of this MSC, the MSRN returned by the HLR will most likely be same as MSISDN. In this scenario we assume that the mobile is not roaming.

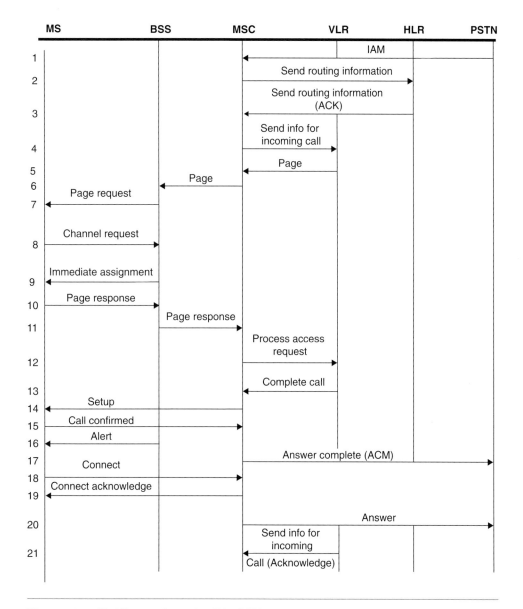

Figure 12.7   Mobile-terminated call in GSM.

4. The MSC informs its VLR about the incoming call using a Send Info for Incoming Call Message that includes MSRN.

5. The VLR responds to the MSC through a Page Message that specifies the LAI and TMSI of the MS. If the MS is barred from receiving the calls, the VLR informs the MSC that a call cannot be directed to the MS. The MSC

would connect the incoming call to an appropriate announcement (e.g., "The mobile phone that you have called is not permitted to receive calls").

6. The MSC uses the LAI provided by the VLR to determine which base station subsystems (BSSs) will page the MS. The MSC sends the Page Message to each of the BSSs to perform the paging of the MS.

7. Each BSS broadcasts the TMSI of the MS in the Page Request Message on the paging channel.

8. When the MS hears its TMSI broadcast on the paging channel, it responds to the BSS with a Channel Request Message over the common access channel, random access channel (RACH).

9. On receiving the Channel Request Message from the MS, the BSS allocates an SDCCH (Stand-alone Dedicated Control Channel) and sends the Immediate Assignment Message to the MS over the access grant channel (AGCH). It is over the SDCCH that the MS communicates with the BSS and the MSC until a traffic channel is assigned.

10. The MS sends a Page Response Message to the BSS over the SDCCH. The message contains the mobile TMSI and LAI.

11. The BSS forwards the Page Response Message to the MSC.

12. The MSC sends a Process Access Request Message to the VLR.

13. The VLR responds with a Complete Call Message.

14. The MSC then sends a Setup Message to the MS.

15. The MS responds with a Call Confirmed Message.

16. The MSC then sends an Alert Message to the MS.

17. The MSC sends an Address Complete Message (ACM) to the PSTN.

18. When the user answers, the MS sends a Connect Message to the MSC.

19. The MSC sends a Connect Acknowledge Message to the MS.

20. The MSC sends an Answer Message to the PSTN. The two parties can now talk.

21. The VLR closes the dialog with the MSC by sending a Send Info for Incoming Call (acknowledge) Message to the MSC.

12.5   Handoff

As the mobile moves from one cell area to another, an active call must undergo a switch from one channel to another. This process is called a *handover* or *handoff*. In the TDMA system, a handoff usually involves both a change of channel carrier frequency and time slot. It also may require a reconfiguration of the wireline facilities by dropping the connection to the serving (old) BSS and switching to a connection to the new (target) BSS as shown in Figure 12.8.

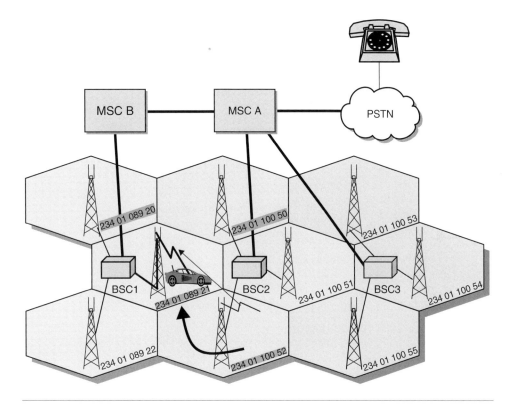

Figure 12.8 Inter-MSC, inter-BSS handoff.

In this example the mobile tunes from carrier frequency 6 on time slot 3 served by the old BSS to carrier frequency 9 on time slot 7 served by the new BSS. At the same time, the MSC switches the call from the old BSS to the new BSS. All this process may take less than 150 ms and go unnoticed by the subscriber.

Handoff is initiated because of a variety of reasons. Signal strength deterioration is the most common cause for handoff at the edge of a cell. Other reasons may include load balancing where the handoff is network initiated to relieve traffic congestion by shifting calls in a highly congested cell to a lightly loaded cell. The handoff could be synchronous or it may be asynchronous. Because the mobile does not have to resynchronize itself in the synchronous situation, handoff delay is much smaller (100 ms in synchronous versus 200 ms in asynchronous).

Handoffs are either initiated by the base station controller (BSC), based upon radio subsystem criteria such as RF level (RXLEV), signal quality (RXQUAL) or distance, or they are a result of network traffic loading. Appropriate decisions are made by a handoff algorithm. The measurements performed by the MS and base station transceiver (BTS) which are collated by the BSC include: (1) by MS—RX-LEV, RXQUAL (for serving downlink and adjacent BSs), and (2) by BTS—RXLEV, RXQUAL, Distance (uplink for serving BS).

Within the BSC these measurements are compared with a set of threshold values. If any one of the thresholds is exceeded, the BSC attempts to rectify the situation by means of power control. If the adjustment of power control cannot bring the parameters within the thresholds, a handoff is required.

There are two types of handoff: internal and external. If the serving and target base stations are located within the same BSS, the BSC for the BSS performs the handoff without the involvement of MSC. This type of handoff is referred to as *intra-BSS* handoff. However, if the serving and target base stations do not reside within the same BSS, an external handoff is performed. In this case, the MSC coordinates the handoff and performs the switching tasks between the serving and target base stations. The external handoffs can be either within the same MSC (*intra-MSC*) or between different MSCs (*inter-MSC*).

12.5.1 Handoff Techniques

Handoff techniques can be classified as *mobile-controlled handoff* (MCHO), *network-controlled handoff* (NCHO), and *mobile-assisted handoff* (MAHO). MCHO is the most popular technique for low-tier radio systems (indoor). It has been used by European digital enhancements of cordless telephony (DECT) and North American personal access communications system (PACS). In this case, the mobile continuously monitors the signal strength and signal quality from the serving base station and several handoff candidate base stations. When the handoff criteria are met, the mobile checks the best candidate base station for an available traffic channel and launches a handoff request. The combined control of automatic link transfer (ALT) and time slot transfer (TST) by the mobile is considered desirable because of the following reasons.

- Offload the handoff task from the network.
- Ensure robustness of the radio link by allowing reconnecting of calls even when radio channels suddenly become poor.
- Prevent simultaneous triggering of the two processes by control of both ALT and TST.

Network-controlled handoff is employed by a low-tier CT-2+ and a high-tier AMPS system. In this case, the base station monitors signal strength and signal quality from the mobile. When these fall below some thresholds, the network arranges for a handoff to another base station. The network asks all surrounding base stations to monitor the signal from the mobile and report the measurements back to the network. The network selects a new base station for handoff and informs both the mobile and the new base station.

Mobile-assisted handoff strategy has been employed by high-tier 2G (GSM, IS-95 CDMA, IS-136 TDMA) and 3G (WCDMA, cdma2000) systems. In this approach, the network provides the mobile with a list of base station frequencies (those of nearby base stations). The network asks the mobile to measure the signal strengths and signal quality (usually determined from bit error rate) from the surrounding base stations (as well as the serving base station) and report the measurements back to the serving base station so that the network can decide whether a handoff is required and to which base station. Since MAHO has been widely used for the high-tier cellular systems, we focus only on this technique in subsequent sections.

12.5.2 Handoff Types

Handoff can be categorized as hard handoff, soft handoff, and softer handoff. The hard handoff can be further divided into intrafrequency and interfrequency hard handoffs. During the handoff process, if the old connection is terminated before making the new connection, it is called a *hard handoff*. In the case of an *interfrequency* hard handoff, the carrier frequency of the new radio access is different from the old carrier frequency to which the MS was connected. On the other hand, if the new carrier frequency, to which the MS is accessed after the handoff procedure, is the same as the original carrier then it is referred to as an *intrafrequency* handoff. In the 2G TDMA systems, the majority of handoffs are intrafrequency hard handoffs.

Interfrequency handoffs may occur between two different radio access networks, for example, between GSM and Universal Mobile Telecommunications Services. In this case, it can also be called *intersystem* handoff. An intersystem handoff is always a type of interfrequency, since different frequencies are used in different systems.

The handoff is referred to as *soft handoff* if the new connection is established before the old connection is released. In the 3G systems, the majority of handoffs are intrafrequency soft handoffs. A soft handoff performed between two sectors belonging to different base stations but not necessarily to the same BSC is called a 2-way soft handoff. A soft handoff may be more than 2-way if the number of sectors involved in the handoff process is more than two.

In *softer handoff,* the BS transmits through one sector but receives from more than one sector. In this case the MS has active uplink radio connections with the network through more than one sector belonging to the same BS. When soft and softer handoffs occur simultaneously, the term *soft-softer handoff* is usually used.

12.5.3 Handoff Process and Algorithms

A basic handoff process consists of three main phases including measurements, decision, and execution phase (see Figure 12.9). The overall handoff process discussed here is related to MAHO strategy. The MS continuously measures the signal strength of the serving and the neighboring cells, and reports the results to the network. From a system performance standpoint, handoff measurement phase is an important task. The signal strength of the radio channel may vary significantly due to fading and signal path loss, resulting from the cell environment and user mobility. Also, an excess of measurement reports by MS or handoff execution by the network increases the overall signaling load, which is not desired.

The decision phase consists of an assessment of the overall quality of service (QoS), of the connection and comparing it with the requested QoS attributes and estimates from neighboring cells. Depending on the outcome of this comparison, the handoff procedure may or may not be triggered.

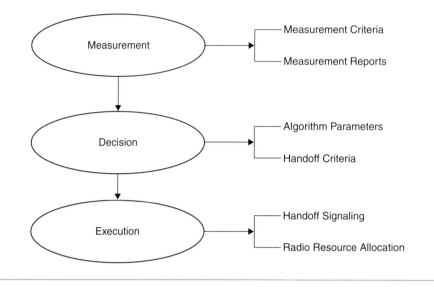

Figure 12.9   Handoff process.

The execution phase involves handoff signaling and radio resource allocation.

Radio signal strength (RSS) measurements from the serving cell and neighboring cells are primarily used in most of the networks. Alternatively or in conjunction, the path loss, signal-to-interference ratio (SIR), bit error rate (BER), and block error rate (BLER) have been used in certain voice and data networks. The following parameters are generally used in the handoff algorithm:

- *Upper threshold* is the level at which the signal strength of the connection is at the maximum acceptable level with respect to the required QoS.
- *Lower threshold* is the level at which the signal strength of the connection is at the minimum acceptable level to satisfy the required QoS. Thus, the signal strength of the connection must not fall below this level.
- *Handoff margin* is a predefined parameter that is set at the point where the signal strength of the neighboring cell has exceeded the signal strength of the serving cell by a certain amount and/or a certain time.

Some of the traditional handoff algorithms are as follows:

- *RSS type:* The BS with the largest signal strength is selected.
- *RSS plus threshold type:* A handoff is performed if the RSS of a new BS exceeds that of the serving BS and the signal strength of the serving BS is below the lower threshold value.
- *RSS plus handoff margin type:* A handoff is performed if the RSS of a new BS is more than that of the serving BS by a handoff margin.

12.5.4 Handoff Call Flows
In the following sections we discuss intra-MSC and inter-MSC handoff scenarios for GSM.

Intra-MSC Handoff
The MS constantly monitors the signal quality of the BSS-MS link. The BSS may also optionally forward its own measurements to the MS. When the link quality is poor, the MS will attempt to maintain the desired signal quality of the radio link by requesting a handoff. The following steps occur in the handoff process (see Figure 12.10).

1. The MS determines that a handoff is required; it sends the Measurement Report Message to the serving BSS. The message contains the RSS measurements.
2. The serving BSS sends a Handoff Request Message to the MSC. The message contains a rank-ordered list of the target BSSs that are qualified to receive the call.

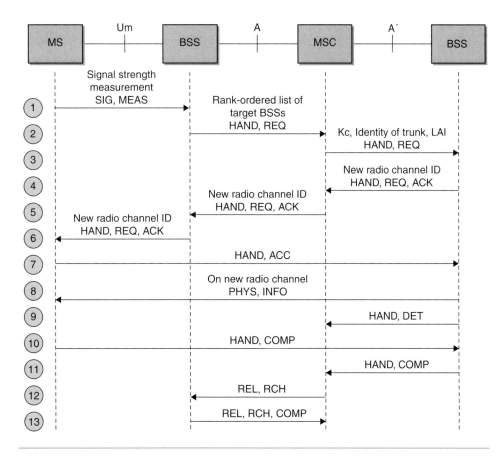

Figure 12.10   Call flow for intra-MSC handoff.

3. The MSC reviews the global cell identity (GCI) associated with the best candidate to determine if one of the BSSs that it controls is responsible for the cell area. In this scenario the MSC determines that the cell area associated with the target BSS is under its control. To perform an intra-MSC handoff, two resources are required: a trunk between the MSC and the target BSS and a radio channel in the new cell. The MSC reserves a trunk and sends a Handoff Request Message to the target BSS. The message includes the desired cell area for handoff, the identity of the MSC-BSS trunk that was reserved, and the mobile encryption key (K_i).

4. The target BSS selects and reserves the appropriate resources to support the handoff pending the connection execution. The target BSS sends a Handoff Request Acknowledgment to the MSC. The message contains the new radio channel identification.

5. The MSC sends the Handoff Command Message to the serving BSS. In this message the new radio channel identification supplied by the target BSS is included.

6. The serving BSS forwards the Handoff Command Message to the MS.

7. The MS retunes to the new radio channel and sends the Handoff Access Message to the target BSS on the new radio channel.

8. The target BSS sends the Physical Information Message to the MS.

9. The target BSS informs the MSC when it begins detecting the MS handing over with the Handoff Detected Message.

10. The target BSS and the MS exchange messages to synchronize/align the MS's transmission in the proper time slot. On completion, the MS sends the Handoff Completed Message to the target BSS.

11. At this point the MSC switches the voice path to the target BSS. Once the MS and target BSS synchronize their transmission and establish a new signaling connection, the target BSS sends the MSC the Handoff Completed Message to indicate that the handoff is successfully completed.

12. The MSC sends a Release Message to the serving BSS to release the old radio channel.

13. At this point the serving BSS releases all resources with the MS and sends the Release Complete Message to the MSC.

In hard handoff, the open interval gap starts when the MS retunes to the new radio channel and ends after synchronization without any loss in voice/data transmission in the BSS or MSC. It should be noted that GSM recommendations require that the open interval gap during a handoff should not exceed 150 ms for 90% handoffs.

Inter-MSC Handoff

In this scenario we assume that a call has already been established. The serving BSS is connected to the serving MSC and the target BSS to the target MSC (see Figure 12.11).

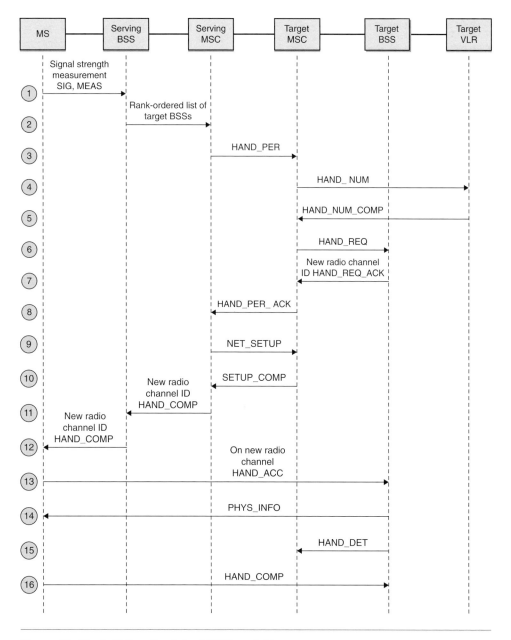

Figure 12.11   Call flow for inter-MSC handoff.

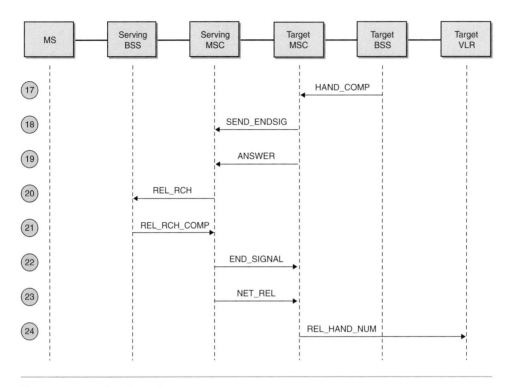

Figure 12.11   (Continued).

1. Same as step 1 in the intra-MSC handoff: handoff required

2. Same as step 2 in the intra-MSC handoff: rank-ordered list of target BSSs

3. When the call is handed over from the serving MSC to the target MSC via PSTN, the serving MSC sets up an inter-MSC voice connection by placing the call to the directory number that belongs to the target MSC. When the serving MSC places this call, the PSTN is unaware that the call is a handoff and follows the normal call routing procedures, delivering the call to the target MSC. The serving MSC sends a Prepare Handoff Message to the target MSC.

4. The target MSC sends a handoff directory number (HAND_NUM) message to its VLR to assign TMSI.

5. The target VLR sends the TMSI in the directory number completion (HAND_NUM_COMP) message.

6. Same as step 3 in the intra-MSC handoff: handoff request to target BSS

7. Same as step 4 in the intra-MSC handoff: target BSS acknowledges with new radio channel

8. The target MSC sends the handoff perform request acknowledgment (HAND_PER_ACK) message to the serving MSC indicating that it is ready for the handoff.

9. The serving MSC sends the network setup (NET_SETUP) message to the target MSC to set up the call.

10. The target MSC acknowledges this message with the network setup completion (SETUP_COMP) message to the serving MSC.

11. Same as step 5 in the intra-MSC handoff: serving BSS handoff

12. Same as step 6 in the intra-MSC handoff: mobile station handoff

13. Same as step 7 in the intra-MSC handoff: mobile retunes and handoff accepted

14. Same as step 8 in the intra-MSC handoff: channel information to mobile

15. Same as step 9 in the intra-MSC handoff: target BSS detects handoff

16. Same as step 10 in the intra-MSC handoff: synchronization of new radio channel

17. Same as step 11 in the intra-MSC handoff: voice path to target BSS established

18. The target MSC sends the send end signal (SEND_ENDSIG) message to the serving MSC as a reminder to inform the target MSC of the call release.

19. At this point the handoff has been completed, a new voice path is set up between the MS and target BSS. The target MSC sends an ANSWER message to the serving MSC.

20. Same as step 12 in the intra-MSC handoff: old radio channel release request

21. Same as step 13 in the intra-MSC handoff: old radio channel release complete

22. The serving MSC sends the message to the target MSC: call released

23. The serving MSC releases the network resources and sends the NET_REL message to the target MSC.

24. The target MSC sends the release handoff directory number (REL_HAND_NUM) message to the VLR to release the connection.

12.6   Summary

In this chapter we focused on mobility management in wireless wide area networks (WWANs). Mobility is concerned with radio mobility and network mobility. Radio mobility involves a handoff process, whereas network mobility deals with mobile paging and location update. We presented an application of the simple mobility model based on a fluid flow concept for a wireless network in an urban area with small cells and high user density. Next, we discussed mobile registration procedures and message flows of ANSI-41 standards for North American cellular systems and highlighted differences with the token-based registration scheme used in GSM. We concluded the chapter by explaining different types of handoff (hard, soft, and softer), handoff procedure, and message flows for the intra- and inter-MSC handoff.

Problems

12.1 Describe mobility models that are used in cellular systems to determine location updates and number of handoffs.

12.2 Repeat Example 12.1 using the following data and other data in the example:

- Density of mobile in the cell = 6000 mobiles/km^2
- Cell radius = 1 km
- Average moving velocity of a mobile = 20 km/hour
- Number of cells per LA = 8
- Number of LAs per MSC/VLR = 4

12.3 What is mobile registration? Why is it used?

12.4 Describe different types of registration schemes that can be used in a cellular system.

12.5 Describe the main features of the GSM token-based registration scheme.

12.6 What is handoff in a cellular system? Why is handoff used?

12.7 Describe various handoff techniques used in mobile networks.

12.8 What are the differences between hard and soft handoff methods?

12.9 Describe the steps used in the handoff process of a cellular system.

12.10 Develop a flow diagram for an intra-MSC handoff in the CDMA IS-95 network.

References

1. Akyildiz, I. F., et al. Mobility Management in Current and Future Communication Networks. *IEEE Networks*, July/August, 1998.

2. Bhagwat, P., Perkins, C., and Tripathi, S. Network Layer Mobility: An Architecture and Survey. *IEEE Personal Communications,* vol. 3, no. 3, June 1996, pp. 54–64.

3. Brown, T. X., and Mohan, S. Mobility Management for Personal Communications Systems. *IEEE Transactions Vehicular Technology*, vol. 46, no. 2, May 1997, pp. 269–278.

4. 3GPP, http://www.3gpp.org.

5. 3GPP2, http://www.3gpp2.org.

6. GSM Specification Series 8.01–8.60, "BSS-MSC Interface, BSC-BTS Interface."

7. GSM Specification Series 9.01–9.11, "Network Interworking, MAP."

8. Kaaranen, H., et al. *UMTS Networks: Architecture, Mobility and Services*. John Wiley & Sons, 2001.

9. Lam, D., Cox, D. C., and Widom, J. Teletraffic Modeling for Personal Communications Services. *IEEE Communications Magazine*, February 1997, pp. 79–87.

10. Lin, Y-B., and Chlamtac, I. *Wireless and Mobile Network Architecture*. John Wiley & Sons, 2001.

11. Markoulidakis, G., et al. Mobility Modeling in Third-Generation Mobile Telecommunications Systems. *IEEE Personal Communications*, vol. 4, August 1997, pp. 41–56.

12. McNair, J., Akyildiz, I. F., and Bender, M. An Intersystem Handoff Technique for IMT-2000 System. Tel Aviv, Israel: *Infocom00*. March 2000.

13. Mouly, M., and Pautet, M. *The GSM System for Mobile Communications*. Palaiseau, France, 1992.

CHAPTER 13

Security in Wireless Systems

13.1 Introduction

Although radio has existed for almost 100 years, most of the population uses wireline phones. Only over the last 30 years have large numbers of people used wireless or cordless phones. With this exposure, users of wireless phones and the news media have challenged two bedrocks of the telecommunications industry: privacy of conversation and billing accuracy.

The current concepts of privacy of communications and accuracy of billing are based on the telephone company's ability to route an individual pair of wires to each residence and office. Thus, when a call is placed on a pair of wires, the telephone company can correctly associate the call on a wire with the correct billing account [1–4]. Similarly, since there is a pair of wires from a home to the telephone company central office, no one can easily listen to the call. For most people, a wiretap is an abstract concept that only concerns someone who is involved in illegal activities.

Communications on shared media can be intercepted by any user of the media. When the media are shared, anyone with access to the media can listen to or transmit on the media. Thus, communications are no longer private. In shared media, the presence of a communication request does not uniquely identify the originator, as it does in a single pair of wires per subscriber. In addition, all users of the network can overhear any information that an originator sends to the network and can resend the information to place a fraudulent call. The participants of the phone call may not know that their privacy is compromised (see Figure 13.1). When the media are shared, privacy and authentication are lost unless some method is established to regain it. Cryptography provides the means to regain control over privacy and authentication [5].

In the past, there have been attempts to control privacy and authentication through noncryptographic means. These have failed thus far. The designers of the original cellular service in the United States implemented authentication of the mobile telephone using a number assignment module (NAM) and an electronic serial number (ESN). The NAM would be implemented in a programmable read only memory (PROM) for easy replacement when the phone number changed. The ESN would be implemented in a tamper-resistant module that could not be changed without damaging the cellular telephone. In practice, many manufacturers

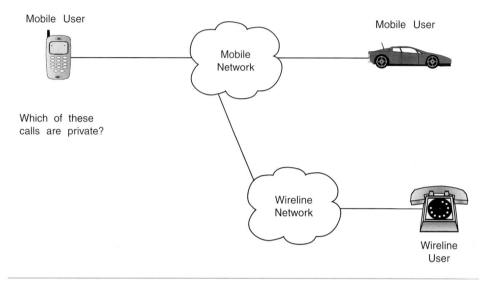

Figure 13.1   Mobile system privacy.

implement the NAM and the ESN in either battery-backed random access memory (RAM) or electrically erasable PROM (EEPROM). The manufacturer and the installer place the data in the phone via external programming.

Similarly, the designers assumed that privacy of cellular communications would occur because 900-MHz scanners would be difficult and too expensive to build. When those scanners became easily available, the U.S. Congress passed the Electronic Communications Privacy Act in 1986, and in 1992 the FCC banned the importation and manufacture of scanners covering cellular phone bands. In practice, the laws do not help since there are millions of scanners in existence today. Furthermore, cellular test equipment is easy to build or buy, and most cellular phones can be placed in a maintenance mode that allows them to monitor any channel. Any cellular phone can be easily converted to a cellular scanner.

To provide the proper privacy and authentication for a mobile station, a cryptographic system is essential. Some of the cryptographic requirements are in the air interface between the mobile station and base station. Other requirements are on databases stored in the network and on information shared between systems in the process of handoff to provide service for roaming units.

In this chapter we examine the requirements needed for privacy and authentication of wireless systems, and then we discuss how each of the cellular and personal communications services (PCS) systems supports these requirements. The chapter discusses four levels of voice privacy. We then identify requirements in the areas of privacy, theft resistance, radio system requirements, system lifetime, physical requirements as implemented in mobile stations, and law enforcement needs. We will examine different methods that are in use to meet these needs.

13.2 Security and Privacy Needs of a Wireless System

13.2.1 Purpose of Security

Most frauds result in a loss to the service provider. It is important to recognize that this loss may be in terms of:

- No direct financial loss, but results in lost customers and an increase in use of the system with no revenue.

- Direct financial loss, where money is paid out to others, such as other network carriers and operators of value-added networks such as a premium rate service line.

- Potential loss of business, where customers may move to another service provider because of the lack of security.

- Failure to meet legal and regulatory requirements, such as license conditions, or data protection legislation.

The objective of security for most wireless systems is to make the system as secure as the public switched telephone network. The use of radio as the transmission medium allows a number of potential threats from eavesdropping on the transmissions. It was soon apparent in the threat analysis that the weakest part of the system was the radio path, as this can be easily intercepted.

The technical features for security are only a small part of the security requirements; the greatest threat is from simpler attacks such as disclosure of the encryption keys, an insecure billing system, or corruption. A balance is required to ensure that these security processes meet these requirements. At some point in time judgment must be made of the cost and effectiveness of the security measure limitation.

13.2.2 Privacy Definitions

When most people think of privacy, they think of either of two levels [6,13]: none, and privacy that is used by military users.

However, as we describe here, there are four levels of privacy that need to be considered.

- **Level 0: None.** With no privacy enabled, anyone with a digital scanner could monitor a call.

- **Level 1: Equivalent to wireline.** As discussed earlier, most people think wireline communications are secure. Anyone in the industry knows that they are not, but the actions to tap a line often show the existence of the tap. With wireless communications, the tap can occur without anyone's knowledge. Therefore, the actions to tap a wireline call must be translated into a different requirement for a wireless system. With this level of security, the types of conversations that would be protected are the routine everyday conversations of most people. These types of communications would be personal discussions that most people would not want exposed to the general public—for example, details of a recent operation or other medical procedure, family financial matters, mail order using a credit card, family discussions, request for emergency services (911), discussions of vacation plans (thus revealing when a home will be vacant).

- **Level 2: Commercially secure.** This level would be useful for conversations in which the participants discuss proprietary information—for example, stock transactions, lawyer-client discussions, mergers and acquisitions, or contract negotiations. A cryptography system that allows industrial activities to be secure for about 10–25 years would be adequate. If one particular conversation was broken, the same effort would be needed to break other conversations.

- **Level 3: Military and government secure.** This is the level that an average person thinks of when cryptography is discussed. This would be used for the military activities of a country and nonmilitary government communications. The appropriate government agency would define requirements for this level.

13.2.3 Privacy Requirements

In this section we discuss the privacy needs of a wireless telephone user. Figure 13.2 is a high-level diagram of a wireless system that shows areas where intruders can compromise privacy. A user of a mobile system needs privacy in the following areas:

- **Privacy of call setup information.** During a call setup, the mobile station will communicate information to the network. Some of the information that a user or mobile station could send includes calling number, calling card number, or type of service requested. The system must send all this information in a secure fashion.

- **Privacy of speech.** The system must encrypt all spoken communications so that intruders cannot intercept the signals by listening on the airwaves.

- **Privacy of data.** The system must encrypt all user communications so that intruders cannot intercept the data by listening on the airwaves.

- **Privacy of user location.** A user should not transmit information that enables an eavesdropper to determine the user's location. The usual method to meet this requirement is to encrypt the user ID. Three levels of protection are often needed:

  1. Eavesdropping of radio link
  2. Unauthorized access by outsiders to the user location information stored in the network visitor location register (VLR) and home location register (HLR)
  3. Unauthorized access by insiders to the user location information stored in the network. This level is difficult to achieve, but not impossible

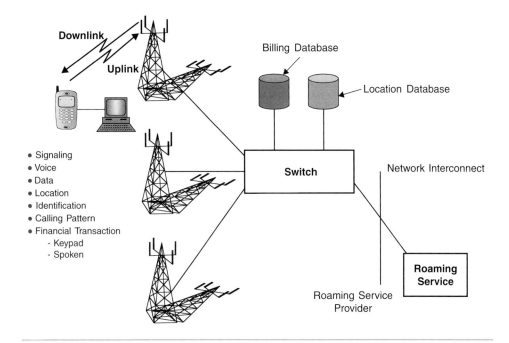

Figure 13.2   Privacy requirements.

- **Privacy of user identification.** When a user interacts with the network, the user ID is sent in a way that does not show user identification. This prevents analysis of user calling patterns based on user ID.

- **Privacy of calling patterns.** No information must be sent from a mobile that enables a listener of the radio interface to do traffic analysis on the mobile user. Typical traffic analysis information is:

  - Calling number
  - Frequency of use of the mobiles
  - Caller identity
  - Privacy of financial transactions

If the user transmits credit card information over any channel, the system must protect the data. Users may order items from mail order houses via a telephone that is wireless. Users may choose to voice their credit card numbers rather than dialing them via touch-tone phone.

Users may access bank voice response systems, where they send account data via tone signaling. Users may access calling card services of carriers and may speak or use tone signaling to send the card number.

All these communications need to be private. Since the user can send the information on any channel—voice, data, or call control—the system must encrypt all channels.

13.2.4 Theft Resistance Requirements

The system operator may or may not care if a call is placed from a stolen mobile station as long as the call is billed to the correct party. The owner of a mobile station will care if the unit is stolen.

The mobile terminal design should reduce theft of the mobile station by making reuse of a stolen mobile station difficult. Even if the mobile station is registered to a new legitimate account, the use of the stolen mobile station should be stopped. The mobile station design should also reduce theft of services by making reuse of a stolen mobile station unique information difficult. Requirements needed to accomplish the reduction in theft are:

- **Clone-resistant design**. In the current wireless systems, cloning of mobile stations is a serious problem; methods must be put in place to reduce or eliminate fraud from cloning. To achieve fraud reduction, mobile station unique information must not be compromised by any of the following means:

    1. Over the air: someone listening to a radio channel should not be able to determine information about the mobile station and then program it into a different mobile station;

    2. From the network: The databases in the network must be secure. No unauthorized person should be able to obtain information from those databases.

    3. From network interconnect: Systems will need to communicate with each other to verify the identity of roaming mobile stations. A system operator could perpetrate fraud by using the security information about roaming mobile stations to make clone mobile stations.

    4. The communication scheme used between systems to validate roaming mobile stations should be designed so that theft of information by a fraudulent system does not compromise the security of the mobile station.

    5. Thus, any information passed between systems for security checking of roaming mobile stations must have enough information to authenticate the roaming mobile station. It must also have insufficient information to clone the roaming mobile station.

    6. From users cloning their own mobile station: Users can perpetrate fraud on the system. Multiple users could use one account by cloning mobile stations. The requirements for reducing or eliminating this fraud are the same as those to reduce repair and installation fraud described below.

- **Installation and repair fraud.** Theft of service can occur when the service is installed or when a terminal is repaired. Multiple mobile stations can be programmed with the same information (cloning). The cryptographic system must be designed so that installation and repair cloning is reduced or eliminated.

- **Unique user ID.** More than one person may use a handset. It is necessary to identify the correct person for billing and other accounting information. Therefore, the user of the system must be uniquely identified in the system.

- **Unique mobile station ID.** When all security information is contained in a separate module (smart card), the identity of the user is separate from the identity of the mobile station. Stolen mobile stations can then be valuable for obtaining service without purchasing a new (full price) mobile station. Therefore, the mobile station should have unique information contained within it that reduces or eliminates the potential for stolen mobile stations to be registered with a new user.

13.2.5 Radio System Requirements

When a cryptographic system is designed, it must function in a hostile radio environment characterized by bit errors caused by:

- **Multipath fading and thermal noise.** The characteristics of the radio channel affect the choice of cryptographic algorithms. The radio signals will take multiple diverse routes from the mobile station to the base station. The effect of multiple diverse routes that can be severe and cause burst errors is fading. Although the system may be interference limited, there may be conditions when the limiting factor on performance is thermal noise. The choice of cryptographic modes must include both of these channel characteristics.

- **Interference.** The mobile systems may initially share a radio spectrum with other users. The modulation scheme and cryptographic system must be designed so that interference with shared users of the spectrum does not compromise the security of the system.

- **Jamming.** Although usually thought about only in the context of military communications, civilian systems can also be jammed. As wireless communication becomes ubiquitous, jamming of the service can also be a method of breaking the security of the system. Therefore, cryptographic systems must work in the face of jamming.

- **Support of handoff.** When the call handoff occurs to another radio port in the same or adjacent mobile system, the cryptographic system must maintain synchronization.

13.2.6   System Lifetime Requirements

It has been estimated that computing power doubles every 18 months. An algorithm that is secure today may be breakable in 5 to 10 years. Since any system being designed today must work for many years after design, a reasonable requirement is that the procedures must last at least 20 years. The algorithm must have provisions to be upgraded in the field.

13.2.7   Physical Requirements

Any cryptographic system used in a mobile station must work in the practical environment of a mass-produced consumer product. Therefore, the cryptographic system must meet the following requirements:

- **Mass production.** It can be produced in mass quantities (million of units per year).
- **Exported and/or imported.** The security algorithm must be capable of being exported and imported. Two problems are solved with export and import restrictions lifted:

  1. It can be manufactured anywhere in the world.
  2. It can be carried on trips outside the United States.

     As an alternative, if an import/export license for the algorithm cannot be obtained, the following restrictions must apply:

     - Either only U.S. manufacturing or two-stage manufacturing
     - All mobile stations must be made in the United States or all mobile stations made outside the United States will have final assembly in the United States
     - All mobile stations must be impounded on leaving the United States.

- **Basic handset requirements.** Any cryptographic system must have minimum impact on the following mobile station requirements:
  - Size
  - Weight
  - Power drain
  - Heat dissipation
  - Microprocessor speed
  - Reliability
  - Cost
- **Low-cost level 1 implementation.** Level 1 implementation would be expected as a baseline for most mobile systems. Therefore, level 1 implementation must

be low cost. Designers obtain low-cost solutions by implementations that can be done either in software or in low-cost hardware. Software solutions are attractive. Often mobile stations have spare read only memory (ROM), RAM, and central processing unit (CPU) cycles in microprocessors.

13.2.8 Law Enforcement Requirements

When a valid court order is obtained in the United States, current telephones (either wired or wireless) are relatively easy to tap by the law enforcement community. The same requirements described in this chapter to ensure privacy and authentication of wireless mobile communications make it more difficult to execute legitimate court wiretap orders.

The law enforcement community can wiretap mobile stations after properly obtaining court orders. When an order is obtained, there are several ways a mobile system operator can meet the needs of the order. Any method used must not compromise the security of the system. Figure 13.3 shows possible approaches to tapping the call. The tap can be done over the air or at a central switch.

This discussion assumes that only the radio portion of the link is encrypted and the call appears in the clear in the wired portion of the network. If end-to-end

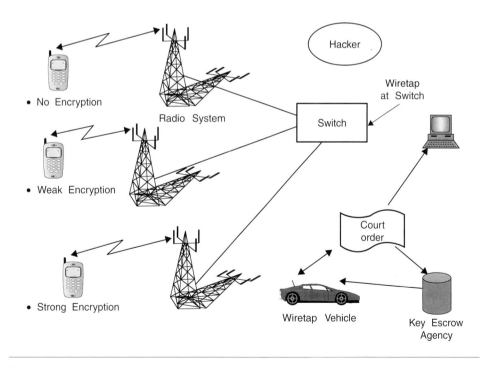

Figure 13.3   Law enforcement requirements.

encryption is used, other means must be considered to obtain the information since the call never appears clear except at the end points.

Over-the-Air Tap

When the tap is done over the air, a wiretap van is required. The van is driven to inside the cell where the call is placed. A centrally located base station (BS) receives interference from mobile stations in many cells or may not be able to receive a low-power mobile station at all.

In a large-cell mobile system, wiretap stations could be deployed in each cell, but in a small cell system, the number of tap points would be too high. Therefore, a wiretap van is needed and is driven to the correct cell where the call is placed.

After the van is driven to the correct cell, it needs to be close to the mobile station. A van might have an antenna that is a maximum of 6 to 10 feet high versus a BS antenna has a height of 25 to 100 feet or more. Thus, the van must be closer to the mobile station than a cell radius. A quick rule of thumb for the wiretap van is that if the mobile station is in line of sight, then the wiretap van can receive the mobile station transmission.

If a wiretap van is used, then the transmissions of the mobile station must be decrypted. The following are possibilities:

- No encryption: This approach makes tapping the easiest; if no encryption is used, anyone can listen to a call over the airwaves. Thus, law enforcement personnel can listen to and record a call, and so can anyone else.

- Breakable algorithms: If the algorithm is weak enough, law enforcement agencies can break the algorithm when permitted to do so by an appropriate court order. Unfortunately, given the proliferation of desktop/laptop personal computers, any algorithm that can be easily broken by the law enforcement community will also be quickly broken by anyone else.

- Strong encryption: Strong encryption makes it difficult, if not impossible, for the wiretap van to decrypt the transmission. One method to resolve this dilemma is to use a key escrow system where all cryptographic keys would be available from an appropriate key escrow agency. With a court order, the information could be obtained by law enforcement agencies so that they could listen to and record a call.

Wiretap at Switch

Since all mobile calls must be routed through a central switch, those calls that use radio-link-only encryption can be tapped at the central switch under a court order.

This is the preferred method for low-power wireless calls. This method leaves it to the user and system provider to have appropriate levels of security in the wireless portion of the call.

13.3 Required Features for a Secured Wireless Communications System

For wireless communications to be secure the following features must be available [8–12]:

- **User authentication** proves that the users are who they claim to be.

- **Data authentication** consists of data integrity and data origin authentication. With data integrity the recipient can be sure that the data has not changed. Data origin authentication proves to the recipient that the stated sender has originated the data.

- **Data confidentiality** means the data is encrypted so that it is not disclosed while in transit.

- **Non-repudiation** corresponds to a security service against denial by either party of creating or acknowledging a message.

- **Authorization** is the ability to determine whether an authenticated entity has the permission to execute an action.

- **Audit** is a history of events that can be used to determine whether anything has gone wrong and, if so, what it was, when it went wrong, and what caused it.

- **Access control** enables only authorized entities to access resources.

- **Availability** ensures that resources or communications are not prevented from access or transmission by malicious entities.

- **Defense against denial of service** is the attack corresponding to the security service of availability.

13.4 Methods of Providing Privacy and Security in Wireless Systems

North American and European cellular and PCS systems support a variety of air interface protocols. They include:

- The Advanced Mobile Phone System (AMPS)
- The IS-54/IS-136 TDMA protocol
- The IS-95 CDMA
- The cdma2000
- The Global System for Mobile communications (GSM)
- The Wideband CDMA (WCDMA) system

Across these protocols, there are four security models that have been used for cellular/ PCS phones in the United States and Europe.

1. MIN/ESN: The original AMPS used a 10-digit mobile identification number (MIN) and a 32-bit ESN. All data is sent in clear text. Data is shared between systems with bad (incorrect) MINs, ESNs, and MIN/ESN pairs. When a mobile station (MS) roams into a system, first the bad list is checked, and then a message is sent to the home system to validate the MIN/ESN pair. The intersystem communications are sent via Signal System 7 (SS7) using an ANSI IS-41 protocol.

As an improvement to this approach, some systems require that a user enter a PIN before placing the calls. The main advantage of the Personal identification number (PIN) is that it can be changed in the network when it is compromised, and the user can continue to have the same phone number. Cellular phones that are cloned must have their phone number (MIN) changed to stop the fraudulent use.

2. Shared secret data (SSD): The TDMA and CDMA systems in the United States use SSD stored in the network and the mobile phone. At service initiation time, a secret key is stored in the phone and the network. AMPS, IS-95 CDMA, IS-54/IS-136 TDMA, and cdma2000 all support SSD. The intersystem communications are sent via SS7 using an ANSI IS-41 protocol.

All mobile stations are assigned an ESN at the time of manufacturing. They are also assigned a 15-digit international mobile subscriber identity (IMSI), that is unique worldwide, an A-key, and other data at the time of service installation. When the MS is turned on, it must register with the system. When it registers, it sends its IMSI and other data to the network. The VLR in the visiting system then queries the HLR for the security data and service profile information. The VLR then assigns a temporary mobile subscriber identity (TMSI) to the MS. The MS uses the TMSI for all further access to that system. The TMSI provides anonymity of communications since only the MS and the network know the identity of the MS with a given TMSI. When the MS roams into a new system, some air interfaces use the TMSI to query the old VLR and then assigns a new TMSI; other air interfaces request that the MS send its IMSI and then assign a new TMSI.

Each time an MS places or receives a call, a call counter (CHCNT) is incremented. The counter is also used for clone detection since clones will not have a call history identical to the legitimate phone.

3. Security triplets (token based): GSM uses its own unique algorithm and does not share secrets between cellular or PCS systems. It uses a token-based authentication scheme. When an MS roams into a system, a message is sent to the home system asking for sets (3 to 5 typically) of triplets (unique challenge, response to the challenge, and a voice privacy key derived from the challenge). Each call that is placed or received uses one triplet. After all triplets are used up, the visited system must send a new message to the home system to get another set of triplets. The intersystem communications use the CCITT SS7 and GSM mobile application part (MAP) protocol.

Each system operator can choose its own authentication method. The MS and the HLR each support the same method and have common data. Each MS sends a registration request; then the network sends a unique challenge. The MS calculates the response to its challenge and sends a message back to the network. The VLR contains a list of triplets; the network compares a triplet with responses it receives from the MS. If the response matches, the MS is registered with the network. The just-used triplet is discarded.

4. Public key: The public key system is analogous to the lock and its combinations. Public key algorithm relies on two cryptographic keys, intimately related to each other but each not derivable from the other. Public key systems do not need communications to the home system to validate the MS. The intersystem communications are still needed to validate the account and get user profile information.

13.5 Wireless Security and Standards

The National Institute of Standards and Technology (NIST) expects that future IEEE 802.11 (and possibly other wireless technologies) products will offer advanced encryption standard (AES)-based data link level cryptographic services that are validated under the U.S. Federal Information Processing Standard (FIPS) 140-2 [7]. As these will mitigate most concerns about wireless eavesdropping or active wireless attacks, their use is strongly recommended when they become available.

- **IEEE 802.11—WLAN.** Data security using encryption is an optional functionality of medium access control (MAC). The functionality is called wired equivalent privacy (WEP). Encryption is only supplied between stations and not on an end-to-end basis. No key management is specified. Authentication is performed by assigning an Extended Service Set ID (ESSID) to each access point (AP) in the network and by using the ESSID in a challenge-response authentication scheme. WEP was shown to have severe security weaknesses. Wi-Fi protected access (WPA) was introduced by the Wi-Fi Alliance as an intermediate solution to WEP insecurities. WPA implemented a subset of IEEE 802.11i specifications which will be discussed in the following section.

- **European and North American Systems.** Almost all information being sent between an MS and the network is encrypted, and sensitive information is not transmitted over a radio channel.

13.6 IEEE 802.11 Security

The IEEE 802.11 Wi-Fi wireless local area network (WLAN) (see Chapter 18) standard addressed security with the WEP protocol, which proved relatively easy to crack and was shown to have major security weaknesses. IEEE 802.11i, also known as Wi-Fi protected access 2 (WPA2), is an improved security protocol for IEEE 802.11. IEEE 802.11i includes stronger encryption, authentication, and key management strategies that go a long way toward guaranteeing data and system security.

The new data-confidentiality protocols in 802.11i are the *temporal key integrity protocol* (TKIP) and *counter-mode/block chaining message authentication code protocol* (CCMP). 802.11i also uses an 802.1X key distribution system to control access to the network. Because 802.11 handles unicast and broadcast traffic differently, each traffic type has different security concerns. 802.11i uses a negotiation process to select the correct confidentiality protocol and key system for each traffic type. Other features introduced in 802.11i include key caching and preauthentication.

The TKIP is a data confidentiality protocol, which improves the security of products using WEP. Among WEP's numerous flaws are its lack of a message integrity code and its insecure data-confidentiality protocol. The message integrity code enables devices to authenticate that the packets are coming from the claimed source. This authentication is important in a wireless system where traffic can be easily injected. The TKIP uses a mixing function to defeat weak-key attacks. The mixing function creates a per frame key to avoid the WEP weaknesses.

The CCMP is a data-confidentiality protocol to handle packet authentication as well as encryption. For confidentiality CCMP uses AES in counter mode. For authentication and integrity, CCMP uses a cipher block chaining message authentication code (CBC-MAC). In 802.11i, CCMP uses a 128-bit key. The block size is 128 bits. The CBC-MAC size is 8 octets, and nonce size is 48 bits. There are two bytes of 802.11 overhead. The CBC-MAC, the nonce, and the 802.11 overhead make the CCMP packet 16 octets larger than an unencrypted 802.11 packet. Although slightly slower, the larger packet is not a bad exchange for increased security.

The CCMP protects some fields that are not encrypted. The additional parts of the 802.11 frame that are protected are known as additional authentication data (AAD). AAD includes the packet source and destination and protects against attackers replaying packets to different destinations.

The 802.1X provides a framework to authenticate and authorize devices connecting to the network. It prevents access to the network until such devices pass authentication. The 802.1X also provides a framework to transmit key information between authenticator and supplicant. For 802.11i, the access point takes the role of the authenticator and the client card the role of supplicant. The supplicant authenticates with the authentication server through the authenticator. In 802.1X, the authenticator enforces authentication. The remote authentication dial-in user service (RADIUS) protocol (see Section 13.9) is typically used between authenticator and authentication server. Once the authentication server concludes authentication with the supplicant, the authentication server informs the authenticator of the successful authentication and passes established keying material to the authenticator. At that point, the supplicant and authenticator share established key material through extensive authentication protocol over LANs (EAPOL)-key exchange. If all exchanges have been successful, the authenticator allows traffic to flow through the controlled port giving the client access to the network.

The 802.11i EAPOL-key exchange uses a number of keys and has a key hierarchy to divide initial key material into useful keys. The two key hierarchies are: pair-wise key hierarchy and group key hierarchy. In the 802.11i specification, these exchanges are referred to as the 4-way handshake and the group key handshake. The 4-way handshake does several things:

- Confirms the pairwise master key (PMK) between the suppliant and authenticator
- Establishes the temporal keys to be used by the data-confidentiality protocol
- Authenticates the security parameters that were negotiated
- Performs the first group key handshake
- Provides keying material to implement the group key handshake

Wireless clients often roam back and forth between access points. This has a negative effect on the system performance. Key caching reduces the load on the authentication server and reduces the time required to get connected to the network. The basic concept behind the key caching is for a client and access point to retain a security association when client roams away from the access point. When the client roams back to the access point, the security association can be restarted.

Preauthorization enables a client to establish a PMK security association to an access point with which the client has yet not been associated. Preauthorization provides a way to establish a PMK security association before a client associates. The advantage is that the client reduces the time that it is disconnected from the network. Preauthorization has limitations. Clients performing preauthorization will add load to the authorization server. Also, since preauthorization is done at the IEEE 802 layer, it does not work across IP subnets.

13.7 Security in North American Cellular/PCS Systems

The ANSI-41 authentication features are independent of the air-interface protocol used to access the network, and subscribers are never involved in the process. A successful outcome of authentication occurs when it can be shown that the MS and the network possess identical results of a calculation performed in both the MS and the network. The authentication center (AC) is the primary functional entity in the network responsible for performing this calculation (see Chapter 7), although the serving system (i.e., the VLR) may also be allocated certain responsibilities. The authentication calculations are based on a set of algorithms, collectively known as the *cellular authentication and voice encryption* (CAVE) algorithm.

The authentication process and algorithm are based on the following two secret numbers:

1. Authentication key (A-key) (64-bit)
2. Shared secret data (SSD) (128-bit)

The A-key is a 64-bit secret number that is the permanent key used by the authentication calculations in both the MS and the AC. The A-key is permanently installed into the MS and is securely stored at the AC in the network when a new subscription is obtained.

Once the A-key is installed in the MS, it should not be displayed or retrievable. The MS and the AC are the only functional entities ever aware of the A-key; it is never transmitted over the air or passed between systems. The primary function of the A-key is as a parameter used in calculation to generate the SSD.

The COUNT is a 6-bit parameter that is intended to provide additional security in case the A-key or SSD is compromised. The current value of the COUNT is maintained by both the MS and the authentication controller. The respective counts should generally be the same—they may not always match exactly due to radio transmission problems or system failures in the network. If the respective counts differ by a large enough range, or frequently do not match, the AC may assume that a fraudulent condition exists and take corrective action. Note that a COUNT mismatch detection does not conclusively indicate that the particular MS accessing the system is fraudulent—only that a clone may exist.

13.7.1 Shared Secret Data Update

The SSD is a 128-bit secret number that is essentially a temporary key used by authentication calculations in both the MS and the AC. The SSD may also be shared with the serving system via a number of ANSI-41 messages. The SSD is a semipermanent value. It can be modified by the network at any time, and the network can command the MS to generate a new value.

The SSD is obtained from calculations using the A-key, the ESN, and a random number shared between the MS and the network. SSD calculation results in two separate 64-bit values, SSD_A and SSD_B. SSD_A is the value used for the authentication process, whereas SSD_B is used for encryption algorithms for privacy and to encrypt and decrypt selected messages on the radio traffic channel. Figure 13.4 shows the SSD generation process. At any time, the network can order the MS to update the SSD by generating the new SSD with a new SSD random number for security purposes.

13.7.2 Global Challenge

For a global and unique challenge authentication process, the ANSI-41 standard is used [8,9]. In a global challenge the serving system presents a numeric authentication challenge to all mobile stations that are using a particular radio control

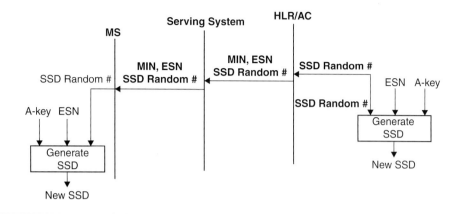

Figure 13.4 SSD generation.

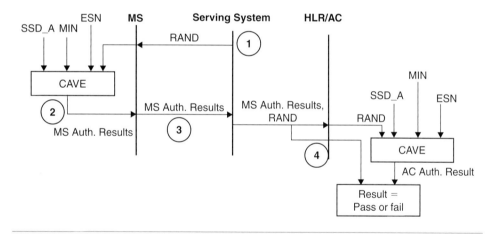

Figure 13.5 Global challenge authentication process (no SSD sharing with serving system).

channel. The ANSI-41 AC verifies that the numeric authentication response from an MS attempting to access the system is correct. This is called a *global challenge* because the challenge indicator and random number used for the challenge are broadcast on the radio control channel and are used by all mobile stations accessing that control channel.

The authentication process flow diagram (when SSD is not shared with the serving system) is given in Figure 13.5.

1. The serving system generates a random number (RAND) and sends it to the MS in the overhead message on the control channel.

2. MS calculates an authentication result using CAVE and transmits that result back to the serving system when it accesses the system for registration, call origination, or paging response purposes.

3. The serving system forwards the authentication result and the random number to AC.

4. The AC independently calculates an authentication result and compares it to the result received from the MS. If the results match, the MS is considered successfully authenticated. If the results do not match, the MS may be considered fraudulent and service may be denied.

If the SSD is shared, then the serving system performs the calculations.

13.7.3   Unique Challenge

In the ANSI-41 unique challenge, the authentication controller directs the serving system to present a numeric authentication challenge to a single MS that either is requesting service from the network or is already engaged in a call. The serving system presents the numeric authentication challenge to the MS and verifies that the numeric authentication response provided by the MS is correct. The unique challenge is so named because the challenge indicator and the random number used for the challenge are directed to a particular MS, whereas a global challenge is required by each MS. Figure 13.6 shows the basic unique challenge procedure for authentication when SSD is not shared.

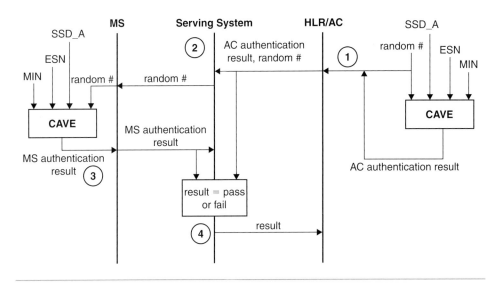

Figure 13.6   Basic unique-challenge authentication process when SSD is not shared.

1. The AC generates a random number and uses it to calculate an authentication result. The AC sends both the random number and authentication result to the serving system.

2. The serving system forwards the random number to the MS.

3. The MS calculates an authentication result and sends it to the serving system.

4. The serving system compares the result from the AC with the result from the MS. If the results match, the MS is considered to have successfully responded to the challenge. If they do not match, the MS may be considered fraudulent and service may be denied. Either way, the serving system reports the results to the AC.

If SSD is shared, the serving system may initiate the unique challenge process and would report a failure to the AC.

13.8 Security in GSM, GPRS, and UMTS

13.8.1 Security in GSM

GSM allows three-band phones to be used seamlessly in more than 160 countries. In GSM, security is implemented in three entities:

- **Subscriber identity module** (SIM) contains IMSI, TMSI, PIN, MSISDN, authentication key K_i (64-bit), ciphering key (K_c) generating algorithm A8, and authentication algorithm A3. SIM is a single chip computer containing the operating system (OS), the file system, and applications. SIM is protected by a PIN and owned by an operator. SIM applications can be written with a SIM tool kit.
- **GSM handset** contains ciphering algorithm A5.
- **Network** uses algorithms A3, A5, A8; K_i and IDs are stored in the authentication center.

Both A3 and A8 algorithms are implemented on the SIM. The operator can decide which algorithm to use. Implementation of an algorithm is independent of hardware manufacturers and network operators.

A5 is a stream cipher. It can be implemented very efficiently on hardware. Its design was never made public. A5 has several versions: A5/1 (most widely used today), A5/2 (weaker than A5/1; used in some countries), and A5/3 (newest version based on the Kasumi block cipher).

The authentication center contains a database of identification and authentication information for subscribers including IMSI, TMSI, location area identity (LAI), and authentication key (K_i). It is responsible for generating (RAND), response (RES), and ciphering key (K_c) which are stored in HLR/VLR for authentication and encryption processes. The distribution of security credentials and encryption algorithms provides additional security.

GSM uses information stored on the SIM card within the phone to provide encrypted communications and authentication. GSM encryption is only applied to communications between a mobile phone and the base station. The rest of the transmission over the normal fixed network or radio relay is unprotected, where it could easily be eavesdropped or modified. In some countries, the base station encryption facility is not activated at all, leaving the user completely unaware of the fact that the transmission is not secure.

GSM encryption is achieved by the use of a shared secret key. If this key is compromised it will be possible for the transmission to be eavesdropped and for the phone to be cloned (i.e., the identity of the phone can be copied). The shared secret key could easily be obtained by having physical access to the SIM, but this would require the attacker to get very close to the victim. However, it has been shown by research that the shared secret key can be obtained over the air from the SIM by transmitting particular authentication challenges and observing the responses.

If the base station can be compromised then the attacker will be able to eavesdrop on all the transmission being received. The attacker will also have access to the shared secret keys of all the mobile phones that use the base station, thus allowing the attacker to clone all of the phones.

Authentication in the GSM system is achieved by the base station sending out a challenge to the mobile station. The MS uses a key stored on its SIM to send back a response that is then verified. This only authenticates the MS, not the user.

A 64-bit key is divided to provide data confidentiality. It is not possible to encrypt all the data; for example, some of the routing information has to be sent in clear text.

GSM Token-based challenge

The security-related information consisting of triplets of RAND, signature response (SRES), and K_c are stored in the VLR. When a VLR has used a token to authenticate an MS, it either discards the token or marks it used. When a VLR needs to use a token, it uses a set of tokens that is not marked as used in preference to a set that is marked used.

When a VLR successfully requests a token from the HLR or an old VLR, it discards any tokens that are marked as used. When an HLR receives a request for tokens, it sends any sets that are not marked as used. Those sets shall then be deleted or marked as used. The system operator defines how many times a set may be reused before being discarded. When HLR has no tokens, it will query the authentication center for additional tokens.

The token-based challenge can be integrated into various call flows (e.g., registration, handoff). It is described separately here for clarity. Figures 13.7 and 13.8 show the call flows of token-based challenges.

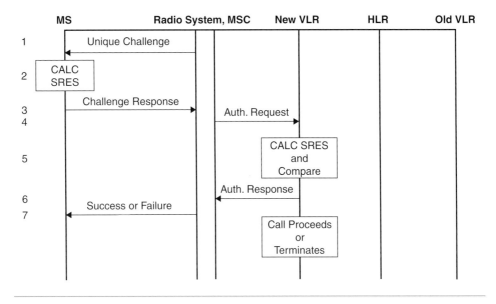

Figure 13.7 GSM token-based unique challenge.

1. The serving system sends a RAND to the MS.
2. The MS computes the SRES using RAND and the authentication key (K_i) in the encryption algorithm.
3. The MS transmits the SRES to the serving system.
4. The MSC sends a message to the VLR requesting authentication.
5. The VLR checks the SRES for validity.
6. The VLR returns the status to the MSC.
7. The MSC sends a message to the MS with a success or failure indication.

Both GSM and North American systems use the international mobile equipment identity (IMEI) stored in the equipment identity register (EIR) (see Chapter 7) to check malfunctions and fraudulent equipment. The EIR contains a valid list (list of valid mobiles), a suspect list (list of mobiles under observation), and a fraudulent list (list of mobiles for which service is barred) (see Figure 13.9 for call flow).

13.8.2 Security in GPRS

The general packet radio service(GPRS) allows packet data to be sent and received across a mobile network (GSM). GPRS can be considered an extension to the GSM network to provide 3G services. GPRS has been designed to allow users to connect to the Internet, and as such is an essential first step toward 3G networks

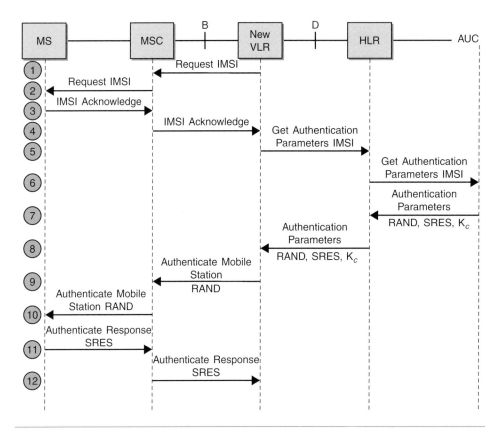

Figure 13.8   GSM token-based unique challenge with ciphering.

for all mobile operations. In GPRS, TMSI is replaced by P-TMSI and P-TMSI signature as alternative identities. The HLR GPRS register maps between internet protocol (IP) addresses and IMSI.

GPRS security functionality is equivalent to the existing GSM security. Authentication and encryption setting procedures are based on the same algorithms, keys, and criteria as in GSM systems.

GPRS provides identity confidentiality to make it difficult to identify the user. This is achieved by using a temporary identity where possible. When possible, confidentiality also protects dialed digits and addresses. As in GSM, the device is authenticated by a challenge-response mechanism. This only verifies that the smart card within the device contains the correct key.

GPRS does not provide end-to-end security so there is a point where the data is vulnerable to eavesdropping or attack. If this point can be protected, e.g., in a physically secure location, this is not a problem. However, if end-to-end security

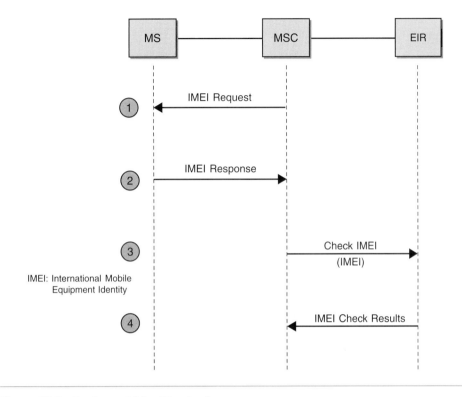

Figure 13.9   Equipment identity check.

is required, there are other standards that can be used over GPRS; such as the wireless application protocol (WAP) and Internet protocol security (IPSec).

In GPRS authentication is performed by serving GPRS support node (SGSN) instead of VLR. The encryption is not limited to radio part, but it is up to SGSN. An IP address is assigned after authentication and ciphering algorithm negotiation.

13.8.3   Security in UMTS

The security in universal mobile telecommunications services (UMTS) is built upon the security of GSM and GPRS. UMTS uses the security features from GSM that have proved to be needed and robust. UMTS security tries to ensure compatibility with GSM in order to ease interworking and handoff between GSM and UMTS. The security features in UMTS correct the problems with GSM by addressing its real and perceived security weaknesses. New security features are added as necessary for new services offered by UMTS and to take into account the changes in network architecture. In UMTS the SIM is called UMTS SIM (USIM).

UMTS uses public keys. In UMTS mutual authentication between the mobile and BS occurs; thus there is no fake BS attack. UMTS has increased key lengths and provides end-to-end security. The other security features of UMTS are listed below:

- Subscriber individual key K;
- Authentication center and USIM share
  - User-specific secret key K,
  - Message authentication functions f_1, f_2
  - Key generating functions f_3, f_4, f_5
- The authentication center has a random number generator.
- The authentication center has a scheme to generate fresh sequence numbers.
- USIM has a scheme to verify freshness of received sequence numbers.
- Authentication functions f_1, f_2 are:
  - MAC (XMAC);
  - RES (XRES).
- Key generating functions f_3, f_4, f_5 are:
  - f_3: ciphering key CK (128 bit);
  - f_4: integrity key IK (128 bit) and
  - f_5: anonymity key AK (128 bit).
- Key management is independent of equipment. Subscribers can change handsets without compromising security.
- Assure user and network that CK/IK have not been used before.
- For operator specific functions, UMTS provides an example called Milenage based on the Rijndael block cipher.
- Integrity function f_9 and ciphering function f_8 are based on the Kasumi block cipher.

13.9   Data Security

The primary goals in providing data security are confidentiality, integrity, and availability. Confidentiality deals with the protection of data from unauthorized disclosures of customers and proprietary information. Integrity is the assurance that data has not been altered or destroyed. Availability is to provide continuous operations of hardware and software so that parties involved can be assured of uninterrupted service.

In this section, we focus upon commonly used data security methods including firewalls, encryption, and authentication protocols.

13.9.1   Firewalls

Firewalls have been used to prevent intruders from securing Internet connection and making unauthorized access and denial of service attacks to the organization network. This could be for a router, gateway, or special purpose computer. The firewalls examine packet flowing into and out of the organization network and restrict access to the network. There are two types of firewalls: (1) packet filtering firewall, and (2) application-level gateway.

The packet filter examines the source and destination address of packets passing through the network and allows only the packets that have acceptable addresses. The packet filter also examines IP addresses and TCP (transmission control protocol) ports. The packet filter is unaware of applications and what an intruder is trying to do. It considers only the source of data packets and does not examine the actual data. As a result, malicious viruses can be installed on an authorized user computer, giving the intruder access to the network without authorized user knowledge.

The application-level gateway acts as an intermediate host computer between the outside client and the internal server. It forces everyone to login to the gateway and allows access only to authorized applications. The application-level gateway separates a private network from the rest of the Internet and hides individual computers on the network. This type of firewall screens the actual data. If the message is deemed safe, then it is sent to the intended receiver. These firewalls require more processing power than packet filters and can impact network performance.

13.9.2   Encryption

Encryption is one of the best methods to prevent unauthorized access of an intruder. Encryption is a process of distinguishing information by mathematical rules. The main components of an encryption system are: (1) plaintext (not encrypted message), (2) encryption algorithm (works like a locking mechanism to a safe), (3) key (works like the safe's combination), and (4) ciphertext (produced from plaintext message by encryption key).

Decryption is the process that is the reverse of encryption. It does not always use the same key or algorithm. Plaintext results in decryption. The following types of keys are used in encrypting data.

Secret Key (symmetric encryption)

Both sender and recipient share a knowledge of the same secret key. The scrambling technique is called encryption. The message is referred to as plaintext or clear text, and the encrypted version of it is called ciphertext. The encryption of a plaintext x into a ciphertext y using a secret key e_k is given as (see Figure 13.10):

$$y = e_k(x) \quad \text{Ciphertext}$$

The corresponding decryption yields

$$x = d_k(y) \quad \text{Plaintext}$$

where: d_k is the decryption key

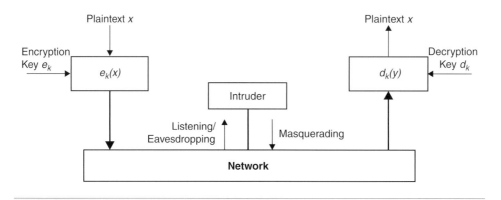

Figure 13.10   Encryption using secret key.

Ideally, the encryption scheme should be such that it cannot be broken at all. Because there are no practical methods of achieving such an unconditional security, encryption schemes are designed to be computationally secure. The encryption and decryption algorithms use the same key, and, hence, such algorithms are called *symmetric key algorithms*. The symmetric key algorithm is vulnerable to interception and key management is a challenge. The strength of this algorithm depends upon length of key. Longer keys are more difficult to break. If the length of a secret key is n bits, at least 2^{n-1} steps would be required to break the encryption.

The data encryption standard (DES) defined by US NIST performs encryption in hardware thereby speeding up the encryption and decryption operation. Additional features of DES are:

1. DES is a block cipher and works on a fixed-size block of data. The message is segmented into blocks of plaintext, each comprising 64 bits. A unique 56-bit key is used to encrypt each block of plaintext into a 64-bit block of ciphertext. The receiver uses the same key to perform the decryption operation on each 64-bit data block it receives, thereby reassembling the blocks into a complete message.

2. The larger the key, the more difficult it is for someone to decipher it. DES uses a 56-bit key and provides sufficient security for most commercial applications. Triple-DES is the extended version of DES which applies DES three times with two 56-bit keys.

International data encryption algorithm (IDEA) is a block cipher method similar to DES. It operates on 64-bit blocks of plaintext and uses a 128-bit key. The algorithm can be implemented either in hardware or software. It is three times faster than DES and is considered superior to DES.

The key sizes used in current wireless systems are not sufficiently large enough for good security. IS-136 uses a 64-bit A-key that is secure, but is still considered to be weak.

Public Key (or asymmetric encryption)

Public key encryption uses longer keys than does symmetric encryption. The key management problem is greatly reduced because the public key is publicized and the private key is never distributed. There is no need to exchange keys.

In a public key system, two keys are used, one for encrypting and one for decrypting. The two keys are mathematically related to each other but knowing one key does not divulge the other key. The two keys are called the "public key" and the "private key" of the user. The network also has a public key and a private key.

The sender uses a public key to encrypt the message. The recipient uses its private key to decrypt the message. Public key infrastructure (PKI) is a set of hardware, software, organizations, and policies to public key encryption work on the Internet. There are security firms that provide PKI and deploy encrypted channels to identify users and companies through the use of certificates—VeriSign Inc. Xcert offers products based on PKI.

Public Key Algorithms

Rivet-Shamir-Adleman (RSA) Algorithm The RSA algorithm [7] is based on public key cryptography. The pretty good privacy (PGP) version of RSA is a public domain implementation available for noncommercial use on the Internet in North America. It is often used to encrypt e-mail. Users make their public keys available by posting them on web pages. Anyone wishing to send an encrypted message to that person copies the public key from the web page into the PGP software and sends the encrypted message using the person's public key.

Two interrelated components of the RSA are (see Figure 13.11):

1. Public key and the private key
2. The encryption and decrypting algorithm

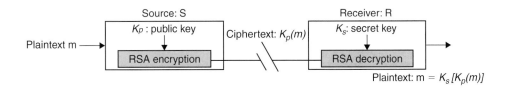

Figure 13.11 RSA algorithm operation.

Steps in the RSA algorithm are:

- Choose two large prime numbers, p and q (RSA labs recommend that the product of p and q be on the order of 768 bits for personal use and 1024 bits for corporate use).
- Compute $n = pq$ and $z = (p - 1) \times (q - 1)$.
- Choose a number, e, less than n, which has no common factors (other than 1) with z (in this case e and z are the prime numbers).
- Find a number d such that $ed - 1$ is exactly divisible by z.
- The public key available to the world is the pair of numbers (n, e); and the private key is the pair of numbers (n, d).

$$\text{Encrypted value} \qquad m^e \, mod(n) = C \qquad (13.1)$$

$$\text{Plaintext} \qquad m = C^d \, mod(n) \qquad (13.2)$$

Example 13.1

Using the prime numbers $p = 5$ and $q = 7$, generate public and private keys for the RSA algorithm.

Solution

- $n = pq = 5 \times 7 = 35$, $z = (p - 1) \cdot (q - 1) = 4 \times 6 = 24$
- choose $e = 5$, because 5 and 24 have no common factors except 1
- choose $d = 29$ since $ed - 1 = 5 \times 29 - 1 = 144$. This is exactly divisible by z (24).
- Public key (35, 5)
- Private key (35, 29)

If the sender sends a letter e that has a numeric representation of 5, show that the receiver gets the letter e. The calculations are shown below.

Sender:

Plaintext letter	m: numeric representation	m^e	ciphertext: $C = m^e$ mod n
e	5	$5^5 = 3125$	10

Receiver:

Ciphertext	c^d	$m = c^d$ mod n	plaintext letter
10	10^{29}	5	e

Diffie-Hellman (DH) Algorithm The Diffie-Hellman key exchange algorithm was proposed in 1976. It is a widely used method for key exchange and is based on cyclic groups. In practice, multiplicative groups of prime field Zp or the group of an elliptic curve are most often used. If the parameters are chosen carefully, the DH protocol is secure against passive (i.e., attacker can only eavesdrop) attacks. The DH key exchange is a cryptographic protocol that allows two parties that have no prior knowledge of each other to jointly establish a shared secret key over an insecure communications channel. This key can then be used to encrypt subsequent communications using a symmetric key cipher. The implementation of protocol uses the multiplicative groups of integers modulo p, where p is prime and g is primitive mod p.

The algorithm works as follows (see Figure 13.12):

1. Ron and Mike agree to use a prime number p and base g.
2. Mike chooses a secret integer $a \in \{2, 3, 4, ..., p - 1\}$ and sends Ron g^a mod p.
3. Ron chooses a secret integer $b \in \{2, 3, 4, ..., p - 1\}$ and sends Mike g^b mod p.
4. Ron computes $(g^a \bmod p)^b \bmod p = K$.
5. Mike computes $(g^b \bmod p)^a \bmod p = K$.
6. Mike and Ron use K as the secret key for encryption.

It should be noted that only a, b and $g^{ab} = g^{ba}$ are kept secret. All other values are sent in clear. Once Mike and Ron compute the shared secret key they can use it as an encryption key, known to them only, for sending messages across the same open communications channel.

Example 13.2

Determine the secret encrypting key, K, using the Diffie-Hellman key exchange algorithm, if two parties agree to use a prime number $p = 23$ and base $g = 5$. Party A selects its secret number $a = 6$ and party B chooses its secret number $b = 15$.

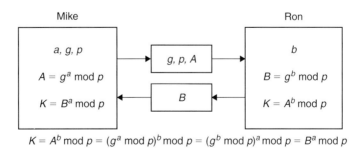

$$K = A^b \bmod p = (g^a \bmod p)^b \bmod p = (g^b \bmod p)^a \bmod p = B^a \bmod p$$

Figure 13.12   Diffie-Hellman key exchange algorithm.

Solution

1. Party A sends to party B $g^a \bmod p = 5^6 \bmod 23 = 8$.
2. Party B sends to party A $g^b \bmod p = 5^{15} \bmod 23 = 19$.
3. Party A computes $(g^b \bmod p)^a \bmod p = 19^6 \bmod 23 = 2$.
4. Party B computes $(g^a \bmod p)^b \bmod p = 8^{15} \bmod 23 = 2$.
5. Party A and B use $K = 2$ as the secret key for encryption.

One-time Key Method

The one-time key method is based on the generation of a new key every time data is transmitted. A single-use key is transmitted in a secure (encoded) mode and, once used, becomes invalid. In some implementations, the central system does not issue a key for a new connection until the user supplies the previously used key.

Elliptic Curve Cryptography (ECC)

The features of the ECC are discussed below:

- ECC is a public key encryption technique that is based on elliptic curve theory.
- ECC can be used in conjunction with most public key encryption methods, such as RSA and Diffie-Hellman.
- ECC can yield a level of security with a 164-bit key, while other systems require a 1024-bit key.
- Because ECC helps to establish equivalent security with lower computing power and battery resources, it is widely used for mobile applications.
- Many manufacturers (3COM, Cylink, Motorola, Pitney Bowes, Siemens, TRW, and VeriFone) have included support for ECC in their products.

Digital Signature

A digital signature provides a secure and authenticated message transmission (enabled by public key enabling (PKE)). It provides proof identifying the sender. The digital signature includes the name of the sender and other key contents (e.g., date, time, etc). The features of the digital signature method are discussed below:

- A digital signature can be used to ensure that users are who they claim to be.
- The signing agency signs a document, m, using a private decryption key, d_B, and computes a digital signature $d_B(m)$.
- The receiver uses the agency's public key, e_B, and applies it to the digital signature, $d_B(m)$, associated with the document, m, and computes $e_b[d_b(m)]$ to produce m.
- This algorithm is very fast, especially with hash functions.

- It is only used in message authentication codes when a secure channel is used to transmit unencrypted messages, but needs to verify their authenticity.
- It is also used in the secure channels of a secure socket layer (SSL).

13.9.3 Secure Socket Layer

SSL is a protocol that uses a session-level layer in the Internet to provide a secure channel. SSL is widely used on the web. In SSL, the server sends its public key and encryption technique to be used to the browser. The browser generates a key for the encryption technique and sends it to the server. Communications between server and browser are encrypted using the key generated by the browser.

The features of SSL are:

- Negotiate cipher suite which is a collection of encryption and authentication algorithms.
- Bootstrapped secure communication, which eliminates the need for third parties, and uses unencrypted communications for initial exchanges.
- Public-key crypto for secret keys and secret-key crypto for data.

13.9.4 IP Security Protocol (IPSec)

IPSec is a widely used protocol that can be employed with other application layer protocols (not just for web applications such as SSL). The operations of IPSec between A and B involve:

- A and B generate and exchange two random keys using Internet key exchange (IKE).
- A and B combine the two numbers to create an encryption key to be used between them.
- A and B negotiate the encryption technique to be used such as DES or 3DES.
- A and B then begin transmitting data using either the transport mode in which only the IP payload is encrypted or tunnel mode in which the entire IP packet is encrypted.

13.9.5 Authentication Protocols

Authentication of a user is used to ensure that only the authorized user is permitted into the network and into the specific resource inside the network. Several methods used for authentication are user profile, user account, user password, biometrics, and network authentication.

The user profile is assigned to each user account by the manager. The user profile determines the limits of a user in accessing the network (i.e., allowable login day and time of day, allowable physical locations, allowable number of incorrect login attempts). The user profile specifies access details such as data and network resources that a user can access and type of access (e.g., read, write, create, delete). The form of access to the network may be based on the password, card, or one-time password.

With a biometric-based form of access, the user can gain access based on finger, hand, or retina scanning by a biometric system. It is convenient and does not require remembering a password. Biometric-based methods are used in high security applications.

Network authentication requires a user to login to an authentication server, which checks user ID and password against a database, and issues a certificate. The certificate is used by the user for all transactions requiring authentications. Kerberos is one of many commonly used authentication protocols. Two other authentication protocols that have been used are remote authentication dial-in user service (RADIUS) and terminal access controller access control system+(TACACS+).

Kerberos is a secret-key network authentication protocol that uses a DES cryptographic algorithm for encryption and authentication. It was designed to authenticate requests for network resources. Kerberos is based on the concept of a trusted third party that performs secure verification of users and services. The primary use of Kerberos is to verify that users and the network services they use really are who and what they claim to be. To accomplish this, a trusted Kerberos server issues tickets to users. These tickets, which have a limited life span, are stored in a user's credential cache. The tickets are used in place of standard user name and password authentication mechanisms.

RADIUS is a distributed client/server system that secures the network against unauthorized access. In the Cisco implementation, RADIUS clients run on Cisco routers and send authentication requests to a server. The central server contains all user authentication and network service access information. RADIUS is the only security protocol supported by wireless authentication protocol.

TACACS+(improved TACAS) is a security application that provides centralized validation of users attempting to gain access to a router or network access server. TACACS+ services are maintained in a database on a TACACS+ daemon running on a UNIX, Windows NT, Window 2000 workstation. TACACS+ provides for separate and modular authentication, authorization, and accounting facilities.

A network administrator may allow remote users to have network access through public services based on remote-access solutions. The network must be designed to control who is allowed to connect to it, and what they are allowed to do once they get connected. The network administrator may find it necessary to configure an accounting system that tracks who logs in, when they log in, and what they do once they have logged in.

Authentication, authorization, and accounting (AAA) security services provide a framework for these kinds of access control and accounting functions. The user dials into an access server that is configured with challenge handshake authentication protocol (CHAP). The access server prompts the user for a name and password. The access server authenticates the user's identity by requiring the user name and password. This process of verification to gain access is called *authentication*. The user may now be able to execute commands on that server once it has been successfully authenticated.

The server uses a process for authorization to determine which commands and resources should be made available to that particular user. Authorization asks the question, what privileges does this user have? Finally, the number of login attempts, the specific commands entered, and other system events can be logged and time-stamped by the accounting process. Accounting can be used to trace a problem, such as a security breach, or it may be used to compile usage statistics or billing data. Accounting asks questions such as: what did this user do and when was it done? The following are some of the advantages in using AAA:

- AAA provides scalability. Typical AAA configurations rely on a server or group of servers to store user names and passwords. This means that local databases don't have to be built and updated on every router and access server in the network.
- AAA supports standardized security protocols—TACACS+, RADIUS, and Kerberos.
- AAA lets the administrator configure multiple backup systems. For example, an access server can be configured to consult a security server first and a local database second.
- AAA provides an architectural framework for configuring three different security features: authentication, authorization, and accounting.

13.10 Air Interface Support for Authentication Methods

The various air interfaces used for PCS and cellular systems in Europe and North America support one or more of the different authentication methods. Only the older AMPS supports MIN/ESN as the authentication method. All of the digital systems in North America, except for GSM1900, support SSD. GSM supports only token-based authentication. UMTS supports token-based authentication along with some advanced security features. cdma2000 supports SSD. Table 13.1 summarizes this information.

Table 13.1 Summary of authentication methods for PCS and cellular systems in Europe and North America.

	Type of Authentication				
Air interface	MIN/ESN	SSD	Token-based	Public key	Type of voice privacy supported
AMPS	X				None
CDMA IS-95		X			Strong
TDMA IS-136		X			Strong
GSM			X		Strong
cdma2000		X			Strong
UMTS			X		Strong

13.11   Summary of Security in Current Wireless Systems

Each of the security methods satisfies the security needs for a wireless system in different ways. The older AMPS has poor security. The digital systems using either SSD or tokens meet most of the security needs of the wireless systems except full anonymity. The public-key-based security system meets all the requirements, including anonymity, but is not yet fully implemented. Privacy of communications is maintained via encryption of signaling, voice, and data for the digital systems. The AMPS sends all data in the clear and has no privacy unless the user adds it to the system. The following is a summary of the support for security requirements for the PCS and cellular systems in North America and Europe (see Table 13.2).

Table 13.2 Summary of support for security requirements for PCS and cellular systems in Europe and North America.

Feature	MIN/ESN (AMPS)	SSD	Token-based	Public key
Privacy of Communication				
• Signaling	None	High: messages are encrypted	High: messages are encrypted	High: messages are encrypted
• Voice	None	High: voice is encrypted	High: voice is encrypted	High: voice is encrypted
• Data	None	High: data is encrypted	High: data is encrypted	High: data is encrypted
Billing Accuracy				
• Accuracy	None: phones can be cloned	High: if authentication is done	High: if authentication is done	High: if authentication is done
Privacy of User Information				
• Location	None	Moderate: using IMSI/ TMSI	Moderate: using IMSI/ TMSI	High: public key provides full anonymity
• User ID	None	Moderate: using IMSI/ TMSI	Moderate: using IMSI/ TMSI	High: public key provides full anonymity
• Calling Pattern	None	High: using TMSI and encryption	High: using TMSI and encryption	High: public key provides full anonymity
Theft Resistance of MS				
• Over the air	None	High	High	High
• From Network	Depends on system design	Depends on system design	Depends on system design	Depends on system design

(Continued)

Feature	MIN/ESN (AMPS)	SSD	Token-based	Public key
• From inter-connection	Depends on system design	Depends on system design	Depends on system design	Depends on system design
• Cloning	None	High	Medium	High
Handset Design	Algorithm run in microproces-sor of handset	Algorithm run in microproces-sor of handset	Algorithm run in microproces-sor of handset	Microprocessor speed may be fast enough for some algorithms
Law Enforce-ment Needs	Easily met on the air interface (if van is nearby to MS or at the switch)	Must wiretap at the switch	Must wiretap at the switch	Must wiretap at the switch

13.11.1 Billing Accuracy

Since AMPS phones can be cloned from data intercepted over the radio link, billing accuracy for AMPS is low to none. For other systems, when authentication is done, billing accuracy is high. If a system operator gives service before authentication or even if authentication failure occurs, then billing accuracy will be low.

13.11.2 Privacy of Information

Privacy of user information is high for the public key system, moderate for the SSD and token-based systems (since sometimes IMSI is sent in cleartext) and low for the AMPS.

13.11.3 Theft Resistance of MS

MS theft resistance is high over-the-air transmission for all systems except the AMPS. Since the token-based system in GSM doesn't support a call history count, it has a lower resistance to cloning than the SSD or public key systems. Earlier AMPS phones using MIN/ESN have no resistance to cloning, but now they support SSD. The resistance of stealing data from network interconnects or from operations systems (OS) in the network depends on the system design.

13.11.4 Handset Design

All of the authentication and privacy algorithms easily run in a standard 8-bit microprocessor used in mobile stations, except the public key systems.

13.11.5 Law Enforcement

The AMPS is relatively easy to tap at the air interface. The digital systems will require a network interface since privacy is maintained over the air interface.

The network requirements currently meet most of the needs of the law enforcement community doing legal wiretaps.

13.12 Summary

In this chapter, we discussed the requirements for strong privacy and authentication of wireless systems, and outlined how each of the cellular and PCS systems supports these requirements. Four levels of voice privacy were presented. We then identified requirements in the areas of privacy, theft resistance, radio system requirements, system lifetime, physical requirements as implemented in mobile stations, and law enforcement needs. We also examined different methods of authentication that are in use to satisfy these needs.

The chapter described the requirements that any cryptographic system should meet to be suitable for use in a ubiquitous wireless network. We also examined security models and described how they met security requirements.

Problems

13.1 Define the equivalent to wireline and commercially secure systems.

13.2 What are the privacy requirements of a wireless telephone user?

13.3 Define the methods that are used in wireless systems to provide privacy and security.

13.4 What are the main differences between global and unique challenge methods used in North American cellular systems?

13.5 Describe the token-based authentication scheme used in the GSM system.

13.6 Describe the public-key-based authentication scheme.

13.7 Describe how you would design a mobile station so that the security data stored in the terminal is tamper resistant.

13.8 Describe data encryption keys.

13.9 Describe the call flows that would be necessary for a shared secret key registration when the old VLR must first be queried for IMSI before the correct HLR can be queried.

13.10 Describe the call flows that would be necessary for a token-based registration in GSM when the old VLR cannot be queried for IMSI and the network must request that the mobile station send its IMSI before messages can be exchanged with the HLR.

References

1. D'Angelo, D. M., McNair, B., and Wilkes, J. E. Security in Electronic Messaging Systems. *AT&T Technical Journal,* 73, no. 3, May/June 1994.

2. JTC(AIR)/94.03.25-257R1. "Minimum Requirements for PCS Air Interface Privacy and Authentication."

3. Karygiannis, T., and Owens, L. *Wireless Network Security.* NIST Special Publication 800-48, November 2002.

4. Owens L., and Crowe, David. *Wireless Security Perspectives.* Calgary, Canada: Cellular Networking Perspectives Ltd., 1999–2001.

5. Paar, C. *Lectures Notes — Applied Cryptography and Data Security,* version 2.5, January 2005.

6. Report of the Joint Experts Meeting on Privacy and Authentication for PCS. Phoenix, Arizona, November 8–12, 1993.

7. Rivet, R. L., Shamir, A., and Adleman, L. A Method for Obtaining Digital Structures and Public-Key Crypto Systems. *Communications ACM* 21, no. 2, February 1978, 120–127.

8. Snyder, R. A., and Gallagher, M. D. *Wireless Telecommunications Networking with ANSI-41,* second edition. New York: McGraw-Hill, 2001.

9. TIA Interim Standard, IS-41 C, "Cellular Radio Telecommunication Intersystem Operations."

10. TR-46 P&A ad hoc/94.04.17.01R5, "TR-46 PCS Privacy and Authentication, Volume 1, Common Requirements," Version 6, November 1994.

11. TR-46 P&A ad hoc/94.04.17.02R4, "TR-46 PCS Privacy and Authentication, Volume 2, PCS1900 Based Requirements."

12. TR-46 P&A ad hoc/94.04.17.02R3, "TR-46 PCS Privacy and Authentication, Volume 3, Shared Secret Data Requirements."

13. Wilkes, J. E. Privacy and Authentication Needs of PCS. *IEEE Personal Communications,* 2, no. 4, August 1995.

CHAPTER 14

Mobile Network and Transport Layer

14.1 Introduction

In this chapter, we first introduce concepts of the Internet and briefly discuss IPv4 and IPv6. We then focus on mobile internet protocol (MIP) and session initiation protocol (SIP) designed to provide a variety of IP data services in mobile networks [1–9]. We conclude the chapter by discussing transmission control protocol (TCP) enhancements for mixed networks (wireline and wireless). The chapter is primarily intended for those readers who are not familiar with the Internet and TCP/IP. The chapter may be omitted by those who have sufficient exposure to TCP/IP and the Internet.

The Internet is a global network that supports a variety of interpersonal and interactive multimedia applications and operates in a *packet-switched* mode [15,20]. The Internet consists of a large number of different *access networks* that are interconnected by a *global internetwork*. Associated with each access network is the *Internet Service Provider* (ISP) network, intranet, enterprise network, local area network (LAN), and so on, that is referred to as a *gateway*. The global internetwork consists of an interconnected set of regional, national, and international networks all of which are joined together using high bit rate leased lines and *routers*. Each of the routers in the interconnection network has a number of access networks attached to it by means of access gateways. Each access network is an LAN with a single *netid* and the *address resolution protocol* (ARP) in each host is used to carry out routing within the access network. Hence, the ARP in each gateway acts as an agent (proxy ARP) on behalf of all hosts at the site/campus to relay packets to and from the interconnection network.

ARP and *reverse address resolution protocol* (RARP) are specialized protocols used only with certain types of network interfaces (such as Ethernet and token-ring) to convert between the addresses used by the IP layer and the addresses used by the network interfaces.

The interconnection network comprises four routers (R1–R4) that are interconnected by, say, leased lines (refer to Figure 14.1). Each leased line has a pair of numbers associated with it. The first number is used as a leased line identifier and the second is referred to as the cost of the line. Let us assume that the host

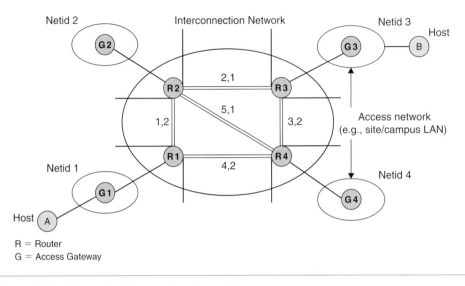

Figure 14.1   Example of internetwork topology.

A on *netid* 1 wants to send a datagram to host B on *netid* 3. On determining that the destination *netid* in the datagram header is for a different netid from its own, gateway G1 forwards the datagram to router R1 over the connecting line/link. On receipt of the datagram, R1 proceeds to forward it first to R3 over the interconnection network and then it is forwarded by R3 to G3. At that point, the IP in G3 knows how to route the datagram to host B using the *hostid* part of the destination IP address and the related media access control (MAC) address of the host B in its ARP cache. Thus, each IP address has two parts: *netid* that is centrally managed by the Internet information center and *hostid* that is allocated by the local administrator of the access network to which the host is attached.

14.2   Concept of the Transmission Control Protocol/Internet Protocol Suite in Internet

Assume a process associated with port 1 at host **A** wishes to send a message to another process associated with port 3 at host **B** (see Figure 14.2). The process at host **A** hands the message to the TCP with instructions to send it to host **B**, port 3. TCP hands the message to IP with instructions to send it to host **B**. Note that IP need not to be told the identity of the destination port. All it needs to know is that the data is intended for host **B**. Next, IP hands the message down to the network access layer (NAP) (e.g., Ethernet logic) with instructions to send it to router *J* that is the first hop on the way to **B**. To control this operation, control information as well as user data must be transmitted.

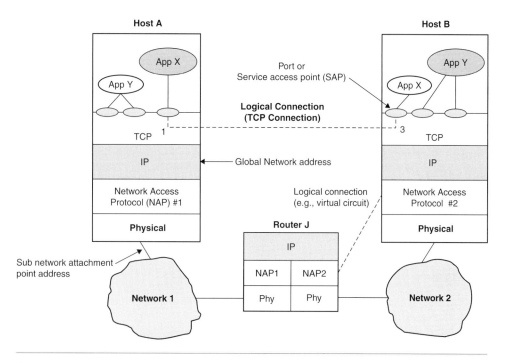

Figure 14.2 TCP/IP concept.

We assume the network interface card in all the hosts that are attached to an access network communicate with other hosts using the TCP/IP protocol stack. The various access networks have different operational parameters associated with them in terms of their bit rate, frame format, maximum frame size, and type of address that are used. Therefore, the routing and forwarding operations associated with gateway are performed at the network layer. In the TCP/IP protocol stack the network layer protocol is IP. In order to transfer packets of information from one host to another, the IP in the two hosts together with the IP in each Internet gateway and router are involved in performing the routing and other harmonization functions necessary. The IP in each host has a unique address assigned to it.

TCP/user datagram protocol (UDP) passes the block of information to its local IP together with the IP address of the intended recipient. The source IP adds the destination and source IP addresses to the head of the block, together with an indication of the source protocol (TCP or UDP), to form an IP datagram. The IP forwards the datagram to its local gateway.

The TCP/IP protocol suite allows computers of all sizes, supplied by many different computer vendors, running totally different operating systems, to communicate with each other. TCP/IP is a 4-layer system (PL, LL, NL, TL)

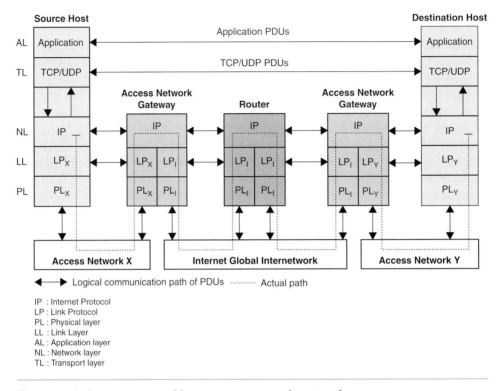

Figure 14.3   Internet networking components and protocols.

(see Figure 14.3). The *link layer* (LL) normally includes the device driver in the operating system and the corresponding network interface card in the computer. Together they handle all the hardware details of physically interfacing with cable (or whatever type media is being used).

The *network* or *internet layer* handles the movement of packets around the network. Internet protocol (IP), Internet control message protocol (ICMP), and Internet group management protocol (IGMP) are used at the network layer.

IP is the workhorse protocol of the TCP/IP protocol suite. All TCP, UDP, ICMP, and IGMP data get transmitted as IP datagrams [10,11,19,30]. IP provides an unreliable connectionless datagram delivery service. By unreliable we mean there are no guarantees that an IP datagram successfully gets to its destination. IP provides a best effort service when something goes wrong, such as a router temporarily running out of buffers. IP has a simple error handling algorithm: throw away the datagram and try to send an ICMP message back to source. The term connectionless means that the IP does not maintain any state information about successive datagrams. Each datagram is handled independently from all other datagrams. IP datagrams are delivered out of order.

ICMP is an adjunct to IP. It is used by an IP layer to exchange error messages and other vital information with the IP layer in another host or router. IGMP is used with multicasting: sending a UDP datagram to multiple hosts.

The *transport layer* (TL) provides a flow of data between two hosts, for the application layer above. In the TCP/IP protocol suite two different transport protocols, TCP and UDP, are used. TCP provides a reliable flow of data between two hosts. It is concerned with data segmentation passed to it from application into appropriately sized segments for the network layer below, acknowledging received packets, setting timeouts to make certain the other end acknowledges packets that are sent, and so on.

UDP provides a much simpler service to the application layer. It sends packets of data called datagrams from one host to the other, but there is no guarantee that the datagrams will reach the other end. Any desired reliability must be added by the application layer.

14.3 Network Layer in the Internet

The IP provides the basis for the interconnections of the Internet [16–18]. IP is a datagram protocol. The packets contain an IP header. The basic header, without options, is shown in Figure 14.4.

The *version* field contains the version of IP—IPv4 or IPv6. The *internet header length* (IHL) field specifies the actual length of the header in multiples of 32-bit words. The minimum length is 5. The maximum permissible length is 15. The *type of service* (TOS) field allows an application protocol/process to specify the relative priority of the application data and the preferred attributes associated with the path to be

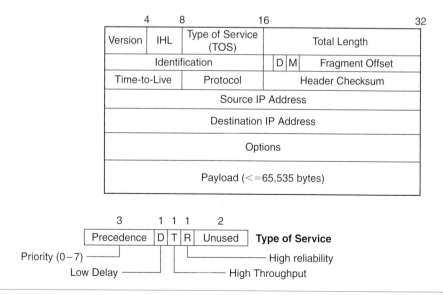

Figure 14.4 IP header.

followed. It is used by each gateway and router during the transmission and routing of the packet to transmit packets of higher priority first and to select a line/route that has the specified attributes should a choice be available. The *total length* field defines the total length of the initial datagram including the header and payload parts. When the contents of the initial datagram need to be transferred in multiple packets, then the value in this field is used by the destination host to reassemble the payload contained within each smaller packet—known as a fragment—into the original payload.

The *identification* field enables the destination host to relate each received packet fragment to the same original datagram. *Don't fragment* or *D-bit* is set by a source host and is examined by routers. A D-bit indicates that the packet should not be fragmented. *More fragment* or *M-bit* is used during the reassembly procedure associated with data transfers involving multiple smaller packets/fragments. It is set to 1 in all but the last packet/fragment in which it is set to 0. The *fragment offset* field is used to indicate the position of the first byte of the fragment contained within a smaller packet in relation to the original packet payload. All fragments except the last one are in multiples of 8 bytes.

The *time-to-live* field defines the maximum time for which a packet can be in transit across the Internet. The value is in seconds and is set by the IP in the source host. It is decremented by each gateway and router by a defined amount and should the value become zero, the packet is discarded. The *protocol* field is used to enable the destination IP to pass the payload within each received packet to the same (peer) protocol that sent the data. This can be an internal network layer protocol such as the ICMP or a higher-layer protocol such as TCP or UDP. The *header checksum* applies just to the header part of the datagram and is a safeguard against corrupted packets being routed to incorrect destinations.

The *source and destination* Internet addresses indicate the sending host and the intended recipient host for this datagram.

The *options* field is used in selected datagrams to carry additional information relating to security, source routing, loose source routing, route recording, stream identification, and time-stamp. The last field is the *payload*.

A symbolic address, or name, of the form user@domain can be used instead of an Internet address. It is translated into an Internet address by directory tables that are organized along the same hierarchy as the addressing. Typically, the domain is of the form machine.institution.type.country. The *type* is edu for educational institutions, *com* for companies, *gov* for governmental agencies, *org* for nonprofit organizations, and *mil* for military. The *country* field is omitted for the United States and is a two-letter country code for the other countries (e.g., fr for France). For instance, the author's address is *vgarg@uic.edu.*

With best-effort delivery service (optional quality of service (QoS)), IP packets may be lost, corrupted, delivered out-of-order, or duplicated. The upper layer entities should anticipate and recover on an end-to-end basis.

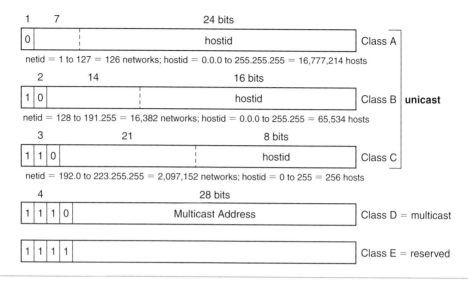

Figure 14.5 Internet addresses.

14.3.1 Internet Addresses

Three classes of Internet addresses (unicast) are used (see Figure 14.5):

- *Class A*—7 bits for *netid* and 24 bits for *hostid*, they are used with networks having a large number of hosts (2^{24})
- *Class B*—14 bits for *netid* and 16 bits for *hostid*, they are used with networks having a medium number of hosts (2^{16})
- *Class C*—21 bits for *netid* and 8 bits for *hostid*, they are used with networks having a small number of hosts (2^8)

It should be noted that the *netid* and *hostid* with all 0s or all 1s have special meaning.

- An address with *hostid* of all 0s is used to refer to the network in netid part rather than a host
- An address with a *netid* of all 0s implies the same network as the source network/netid
- An address of all 1s means broadcast the packet over the source network
- An address with a *hostid* of all 1s means broadcast the packet over the destination network in netid part

A class A address with a *netid* of all 1s is used for test purposes within the protocol stack of the source host. It is known as the loop-back address.

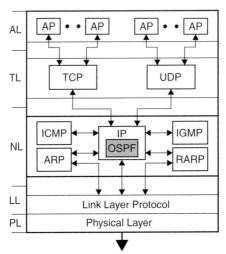

AP: Application Protocol/process
ARP: Address Resolution Protocol
RARP: Reverse ARP
ICMP: Internet Control Message Protocol
IGMP: Internet Group Message Protocol
OSPF: Open Shortest Path First
UDP: User Datagram Protocol
TCP: Transmission Control Protocol

Figure 14.6   Adjunct protocols.

14.3.2   IP Adjunct Protocols

Figure 14.6 shows the IP adjunct protocols [25–29].

- *Address resolution protocol (ARP) and reverse ARP (RARP)* are used by IP in hosts that are attached to a broadcast LAN (such as Ethernet or token ring) in order to determine the physical MAC address of a host or gateway given its IP address (ARP), and, in case of the RARP, the reverse function.

- *Open shortest path first (OSPF) protocol* is a routing protocol used in the global internetwork. Such protocols are present in each internetwork router. They are used to build up the contents of the routing table used to route packets across the global internetwork.

- *Internet control message protocol (ICMP)* is used by the IP in a host or gateway to exchange errors and other control messages with IP in another host or gateway.

- *Internet group message protocol (IGMP)* is used with multicasting to enable a host to send a copy of a datagram to the other hosts that are part of the same multicast group.

The ICMP forms an integral part of all IP implementations. It is used by hosts, routers, and gateways for a variety of functions, and especially by network management. The main functions associated with the ICMP are as follows:

- Error reporting
- Reachability testing
- Congestion control
- Route-change notification
- Performance measuring
- Subnet addressing

The standard way to send an IP packet over any point-to-point link is either dial-up modems (e.g., asynch framing), leased lines (e.g., bit synchronous framing), or ISDN, IS-99 CDMA (e.g., octet-synchronous framing). The *link control protocol* (LCP) runs during initial link establishment and negotiates link-level parameters (e.g., maximum frame size, etc.). The *IP control protocol* (IPCP) establishes the IP address of the client (the point-to-point (PPP) server, allocates a temporary address, or the client notifies the server of the fixed address) and negotiates for the use of TCP/IP header compression.

14.3.3   QoS Support in the Internet

QoS requirements include a defined minimum mean packet throughput rate and a maximum end-to-end packet transfer delay. To meet the varied set of QoS requirements, two schemes have been standardized:

- Integrated Services (IntServ)
- Differentiated Services (DiffServ)

Packets relating to different types of call/session are each allocated a different value in the precedence bits of the *type of service* (TOS) field of the IP packet header. This is used by routers within the Internet to differentiate between the packet flows relating to different types of calls.

Integrated Services (*IntServ*)

The *IntServ* solution defines three different classes of service:

- *Guaranteed*: It specifies maximum delay and jitter, and an assured level of bandwidth is guaranteed.
- *Controlled load* (also known as predictive): No firm guarantee is provided but the flow obtains a constant level of service equivalent to that obtained with the best-effort service at light loads.
- *Best-effort*: This is intended for text-based applications.

To cater for three different types of packet flows within each router, three separate output queues are used for each line—one for each class. Appropriate scheduling mechanisms are used to ensure that the QoS requirements of each class are met.

Differentiated Service (*DiffServ*)

Incoming packet flows relating to individual calls are classified by the router/gateway at the edge of the *DiffServ* compliant net/Internet into one of the defined service/traffic classes by examining selected fields in various headers in the packet. The *TOS* field in the IP packet header is replaced by a new field called the *differentiated service* (*DS*) field. Within the *DiffServ* network, a defined level of resources in terms of buffer space within each router and the bandwidth of each output line is allocated to each traffic class.

Internet Protocol version 6 (IPv6)

Today's Internet operates over the common network layer datagram protocol, Internet Protocol version 4 (IPv4). In the early 1990s, a new design of addressing scheme was initiated within the Internet Engineering Task Force (IETF) due to the recognized weaknesses of IPv4. The result was IPv6 (see Figure 14.7). The single most significant advantage IPv6 offers is increased destination and source addresses. IPv6 quadruples the number of network address bits from 32 bits in

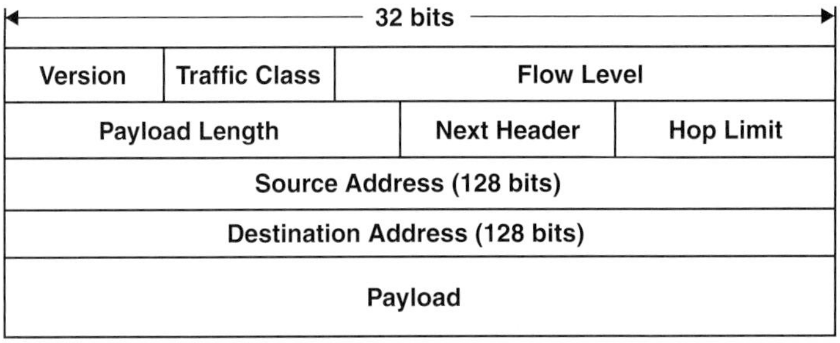

◄─────────────── 32 bits ───────────────►			
Version	Traffic Class	Flow Level	
Payload Length		Next Header	Hop Limit
Source Address (128 bits)			
Destination Address (128 bits)			
Payload			

- Version: 4-bit field to identify the IP version number
- Traffic class: 8-bit field is similar to Type of Service field in IPv4
- Flow level: This 20-bit field is used to identify a "flow" of datagrams
- Payload length: 16-bit is treated as an unsigned integer to give the number of bytes in the datagram
- Next header: This field identifies the protocol to which the contents of this datagram will be delivered (for example TCP or UDP). The field uses the same values as Protocol field in IPv4 header
- Hop limit: The contents of this field are decremented by one by each router that forwards the datagrams. If the hop limit count reaches zero, the datagram is discarded
- Source and destination address: 128-bit field

Figure 14.7   IPv6.

IPv4 to 128 bits, which provides more than enough globally unique IP addresses for every network device on the planet. This will lead to network simplification, first, through less need to maintain a routing state within the network and second, through reduced need for address translation; hence, it will improve the scalability of the Internet.

IPv6 will allow a return to a global end-to-end environment where the addressing rules of the network are transparent to applications. The current IP address space is unable to satisfy the potentially large increase in number of users or the geographical needs of Internet expansion, let alone the requirements of emerging applications such as Internet-enabled personal digital assistants (PDAs), personal area networks (PANs), Internet-connected transportation, integrated telephony services, and distributed gaming.

The use of globally unique IPv6 addresses simplifies the mechanisms used for reachability and end-to-end security for network devices, functionally crucial to the applications and services driving the demand for the addresses.

The lifetime of IPv4 has been extended using techniques such as address reuse with translation and temporary use allocations. Although these techniques appear to increase the address space and satisfy the traditional client/server setup, they fail to meet the requirements of new applications. The need for an always-on environment to be connectable precludes these IP address conversion, pooling, and temporary allocation techniques, and the "plug and play" required by consumer Internet applications further increases address requirements. The flexibility of the IPv6 address space provides the support for private addresses but should reduce the use of network address translation (NAT) because global addresses are widely available. IPv6 reintroduces end-to-end security that is not always readily available throughout an NAT-based network.

The success of IPv6 will depend ultimately on the innovative applications that run over IPv6. A key part of IPv6 design is its ability to integrate into and coexist with existing IP networks. It is expected that IPv4 and IPv6 hosts will need to coexist for a substantial time during the steady migration from IPv4 to IPv6, and the development of transition strategies, tools, and mechanisms has been part of the basic IPv6 design from the start. Selection of a deployment strategy will depend on current network environment, and factors such as the forecast of traffic for IPv6 and availability of IPv6 applications on end systems.

IPv6 does not allow for fragmentation and reassembly at an intermediate router; these operations can be performed only by the source and destination. If an IPv6 datagram received by a router is too large to be forwarded over the outgoing link, the router simply drops the datagram and sends a *packet too big* ICMP message back to sender. The checksum field in IPv4 was considered redundant and was removed because the transport layer and data link layer protocols perform checksum.

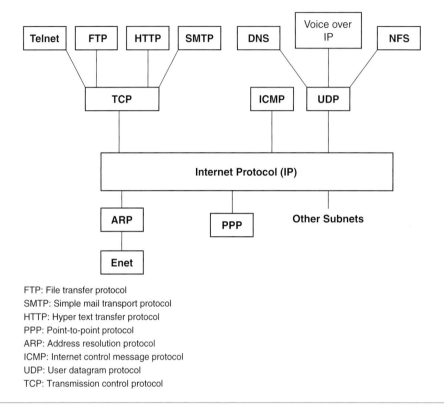

FTP: File transfer protocol
SMTP: Simple mail transport protocol
HTTP: Hyper text transfer protocol
PPP: Point-to-point protocol
ARP: Address resolution protocol
ICMP: Internet control message protocol
UDP: User datagram protocol
TCP: Transmission control protocol

Figure 14.8   TCP/IP protocol suite.

14.4   TCP/IP Suite

The TCP/IP suite (Figure 14.8) occupies the middle five layers of the 7-layer open system interconnection (OSI) model (see Figure 14.9) [30]. The TCP/IP layering scheme combines several of the OSI layers. From an implementation standpoint, the TCP/IP stack encapsulates the network layer (OSI layer 3) and transport layer (OSI layer 4). The physical layer, the data-link layer (OSI layer 1 and 2, respectively) and application layer (OSI layer 7) at the top can be considered non-TCP/IP-specific. TCP/IP can be adapted to many different physical media types.

IP is the basic protocol. This protocol operates at the network layer (layer 3) in the OSI model, and is responsible for encapsulating all upper layer transport and application protocols. The IP network layer incorporates the necessary elements for addressing and subnetting (dividing the network into subnets), which enables TCP/IP packets to be routed across the network to their destinations. At a parallel level, the ARP serves as a helper protocol, mapping physical layer addresses typically referred to as MAC-layer addresses to network layer (IP) addresses.

There are two transport layer protocols above IP: the UDP and TCP. These transport protocols provide delivery services. UDP is a connectionless

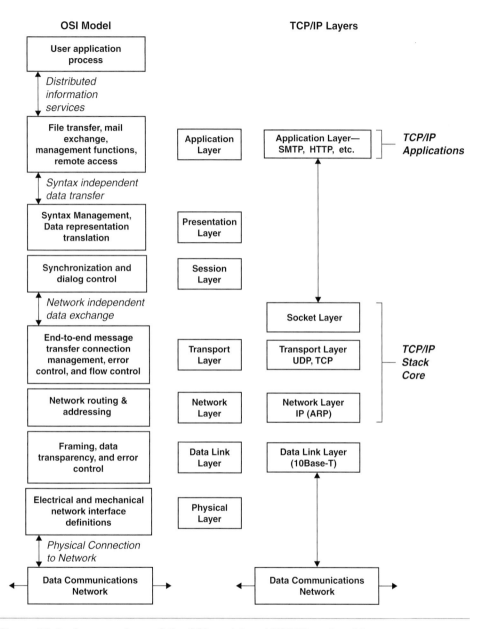

Figure 14.9   A comparison of the OSI model and TCP/IP protocol layers.

delivery transport protocol and used for message-based traffic where sessions are unnecessary. TCP is a connection-oriented protocol that employs sessions for ongoing data exchange. File transfer protocol (FTP) and Telnet are examples of applications that use TCP sessions for their transport. TCP also provides the reliability of having all packets acknowledged and sequenced. If data is dropped or arrives out-of-sequence, the stack's TCP layer will retransmit and resequence. UDP is an unreliable service, and has no such provisions. Applications such as the *simple mail transport protocol* (SMTP) and *hyper text transfer protocol* (HTTP) use transport protocols to encapsulate their information and/or connections. To enable similar applications to talk to one another, TCP/IP has what are called "well-known port numbers." These ports are used as sub-addresses within packets to identify exactly which service or protocol a packet is destined for on a particular host.

TCP/IP serves as a conduit to and from devices, enabling the sharing, monitoring, or controlling those devices. A TCP/IP stack can have a tremendous effect on a device's memory resources and CPU utilization. Interactions with other parts of the system may be highly undesirable and unpredictable. Problems in TCP/IP stacks can render a system inoperable.

14.5   Transmission Control Protocol

The TCP [31,32] is the connection-oriented transport layer protocol designed to operate on the top of the datagram network layer IP. The two widely used protocols are known under the collective name TCP/IP. TCP provides a reliable end-to-end byte stream transport. The segmentation and reassembly of the messages are handled by IP, not by TCP.

TCP uses the *selective repeat protocol* (SRP) with positive acknowledgments and time-out. Each byte sent is numbered and must be acknowledged. A number of bytes can be sent in the same packet, and the acknowledgment (ACK) then indicates the sequence number of the next byte expected by the receiver. ACK carrying sequence number m provides acknowledgment for all packets up to, and including, packets with sequence number $m - 1$. If a packet is lost, the receiver sends duplicate ACK for a subsequent correctly received packet.

The TCP header is at least 20 bytes, and has 16 error detection bits for the data and the header. The error detection bits are calculated by summing the 1's complements of the groups of 16 bits that make up the data and the header, and by taking the 1's complement of that sum. The number of data that can be sent before being acknowledged is the *window size* (W_{max}) which can be adjusted by either the sender or the receiver to control the flow based on the available buffers and the congestion. Initial sequence numbers are negotiated by means of a three-way handshake at the outset of connection. Connections are released by means of a three-way handshake.

TCP transmitter (Tx) uses an adaptive window based transmit strategy. Tx does not allow more than W_{max} unacknowledged packets outstanding at any given time. With the congestion window lower limit at time t equal to $X(t)$, packets up to $X(t)-1$ have been transmitted and acknowledged. Tx can send starting from

$X(t)$. $X(t)$ has a nondecreasing sample path. With congestion window width at time t equal to $W(t)$, this is the amount of packets Tx is allowed to send starting with $X(t)$. $W(t)$ can increase or decrease (because of window adaptation), but never exceed W_{max}. Transitions in $X(t)$ and $W(t)$ are triggered by receipt of ACK. Receiver (Rx) of an ACK increases $X(t)$ by an amount equal to the amount of data acknowledged. Changes in $W(t)$, however, depend on the version of TCP and the congestion control process.

Tx starts a timer each time a new packet is sent. If the timer reaches a *round trip time-out* (RTO) value before the packet is acknowledged, a time-out occurs. Retransmission is initiated on time-out. RTO value is derived from a *round trip timer* estimation procedure. RTO is sent only in multiples of a timer granularity.

The window adaptation procedure is as follows:

1. *Slow start phase.* At the beginning of the TCP connection, the sender enters the slow start phase, in which window size is increased by 1 maximum segment size (MSS) for every ACK received; thus, the TCP sender window grows exponentially in round trip timer.

   If $W < W_{th}$, $W \leftarrow W + 1$ for each ACK received;

   W_{th} is the slow-start threshold.

2. *Congestion avoidance phase.* When the window size reaches W_{th}, the TCP sender enters the congestion avoidance phase. TCP uses a sliding window-based flow control mechanism allowing the sender to advance the transmission window linearly by one segment upon reception of an ACK, which indicates that the last in-order packet was received successfully by the receiver.

   If $W \leftarrow W + 1/W$ for each ACK received

3. *Upon time-out.* When packet loss occurs at a congested link due to buffer overflow at the intermediate router, either the sender receives duplicate ACKs, or the sender's RTO timer expires. These events activate TCP's fast retransmit and recovery, by which the sender reduces the size of the congestion window to half and linearly increases the congestion window as in the congestion avoidance phase, resulting in a lower transmission rate to relieve the link congestion.

   $W + \leftarrow 1$ and $W_{th} \leftarrow W/2$

Assuming long running connections and large enough window sizes, the upper bound on throughput, R, of a TCP connection is given by:

$$R = \frac{0.93\text{MSS}}{\text{RTT}\sqrt{p}} \tag{14.1}$$

where:
   MSS = maximum segment size
   RTT = average end-to-end round trip time of the TCP connection
   p = packet loss probability for the path.

Equation 14.1 neglects retransmissions due to errors. If the error rate is more than 1%, these retransmissions have to be considered. This leads to the following formula:

$$R = \frac{\text{MSS}}{\text{RTT}\sqrt{1.33p} + \text{RTO} \cdot p \cdot (1 + 32p^2) \cdot \min(1, 3\sqrt{0.75p})} \qquad (14.2)$$

where:
   RTO = retransmission time-out ~ 5RTT

For a given object, the *latency* is defined as the time from when the client initiates a TCP connection until the time at which the client receives the requested object in its entirety. Using the following assumptions, we provide expressions for latency with a static and dynamic congestion window.

- The network is not congested.
- The amount of data that can be transmitted is dependent on the sender's congestion window size.
- Packets are neither lost nor corrupted.
- All protocol header overheads are negligible and ignored.
- The object to be transferred consists of an integer number of segments of size MSS.
- The only packets with non-negligible transmission times are packets that carry maximum-size TCP segments. Request messages, acknowledgments, and TCP connection establishment segments are small and have negligible transmission times.
- The initial threshold in the TCP congestion-control mechanism is a large value that is never attained by the congestion window.

Static Congestion Window

$$L = 2\text{RTT} + \frac{O}{R} + \{(K - 1)[S/R + \text{RTT} - (WS)/R]\}^* \qquad (14.3)$$

where:
   L = latency of the connection
   $\{x\}^* = \max(x, 0)$
   $K = O/(WS)$ round-up to the nearest integer
   W = congestion window size

O = Size of the object to be transmitted
R = Transmission rate of the link from the server to the client
Maximum Segment Size (MSS) = S
RTT = round-trip time

Dynamic Congestion Window

$$L = 2\text{RTT} + \frac{O}{R} + P\left[\text{RTT} + \frac{S}{R}\right] - (2^P - 1)\frac{S}{R} \qquad (14.4)$$

where:
  $P = \min\{Q, K - 1\}$
  in which
  $Q = \log_2[1 + \text{RTT}/(S/R)] + 1$

ISO defined five classes (0 to 4) of connection-oriented transport services (ISO 8073). We briefly describe class 4, which transmits packets with error recovery and in the correct order. This protocol is known as *Transport Protocol Class 4* (TP4) and is designed for unreliable networks. The basic steps in the TP4 connection are given below:

- **Connection establishment:** This is performed by means of a three-way handshake to agree on connection parameters, such as a credit value that specifies how many packets can be sent initially until the next credit arrives, connection number, the transport source and destination access points, and a maximum time-out before ACK.

- **Data transfer:** The data packets are numbered sequentially. This allows resequencing. ACKs may be done for blocks of packets. There is a provision for *expedited data* transport in which the data packets are sent and acknowledged one at a time. Expedited packets jump to the head of the queues. Flow is controlled by windows or by credits.

- **Clear connection:** Connections are released by an expedited packet indicating the connection termination. The buffers are then flushed out of the data packets corresponding to that connection.

In practice, TCP has been tuned for a traditional network consisting of wired links and stationary hosts. TCP assumes that congestion in the network is the primary cause of packet losses and unusual delay. TCP performs well over wired networks by adapting to end-to-end delays and congestion losses. TCP reacts to packet losses by dropping its transmission (congestion) window size before retransmitting packets, initiating congestion control or avoidance mechanisms. These measures result in a reduction in the load on the intermediate links, thereby controlling the congestion in the network. While slow start is one of the most useful mechanisms in wireline networks, it significantly reduces the efficiency of TCP when used together with mobile receivers or senders.

Example 14.1

We consider sending an object of size $O = 800$ kilobytes from a server to a client. If maximum segment size $(S) = 536$ bytes and RTT $= 100$ msec, and the transport protocol uses a static window with window size W, determine the minimum possible latency for a transmission rate of 1 Mbps. What is the minimum window size that achieves this latency?

Solution

$$L_{min} = 2RTT + \frac{O}{R} = 0.2 + \frac{800 \times 1000}{1 \times 10^6} = 1.0 \sec.$$

$$\text{For minimum latency } \frac{S}{R} + RTT - \frac{WS}{R} = 0$$

$$\therefore W = 1 + \frac{RTT}{\left(\frac{S}{R}\right)} = 1 + \frac{100 \times 10^{-3}}{\frac{536 \times 8}{10^6}} = 1 + 23.3 = 24.3 \text{ segments}$$

Example 14.2

Find the upper bound of the throughput for a TP connection if RTT $= 100$ msec, maximum segment size (MSS) $= 536$ bytes, and packet loss probability for the path is 1%. What is the throughput with retransmissions due to errors?

Solution

$$R = \frac{0.93 \times (536 \times 8)}{0.100 \times \sqrt{0.01}} = 0.3988 \text{ Mbps}$$

With retransmission due to error:

$$R = \frac{536 \times 8}{0.1\sqrt{0.0133} + 0.5 \times 0.01 \times [1 + 32(0.01)^2] \min(1, 3\sqrt{0.75 \times 0.01})} = 0.3349 \text{ Mbps}$$

14.5.1   TCP Enhancements for Wireless Networks

TCP was primarily designed for wired networks. Its parameters were selected to maximize its performance on wired networks where packet delays and losses are caused mainly by congestion. In wired networks, random bit error rate is negligible. In a wireless network, packet losses occur due to handoff or fading and can be random. When TCP responds to packet losses by invoking congestion control or an avoidance algorithm, a degraded end-to-end performance in wireless network results. A wireless environment violates many of the assumptions made by TCP. Several approaches have been suggested to improve end-to-end TCP performance over wireless links [12–14]. They can be classified into three categories.

1. *end-to-end TCP protocols*, where loss recovery is performed by the sender, such as explicit loss notification (ELN) option

2. *link-layer protocols* that provide local reliability using techniques such as forward error correction (FEC) and retransmission of lost packets in response to automatic repeat request (ARQ) messages

3. *split TCP connection protocol* that breaks the end-to-end TCP connection into two parts at the base station, one between the sender and the base station, and the other between the base station and the receiver.

All wireless networks face a high bit error rate. In heterogeneous networks, to explicitly differentiate the cause of packet loss is the primary goal of TCP design. Such efforts aim to find an explicit way to inform the sender of the cause of packet loss, be it congestion or random errors. Thus, the sender is able to make appropriate decisions on how to adjust the congestion window.

In the end-to-end case, the link-layer ARQ mechanism is used to improve the error rate seen by TCP. The IS-95 CDMA data stack uses this approach. In the link-layer case, network layer software is modified at the base station to monitor every passing packet in either direction. Cache packets at the base station are used and local retransmissions across wireless links are performed. In the split, the TCP mode wireless portion is separated from the fixed portion. With split TCP, TCP may get ACK even before the packet is successfully delivered to the receiver. It also involves software overhead. Several schemes that have been used with the goal to improve TCP's performance in wireless and mobile environment are:

- Indirect TCP (I-TCP) (see Figure 14.10)
- Snooping TCP (see Figure 14.11)
- Mobile TCP (M-TCP)
- Fast retransmit/fast recovery

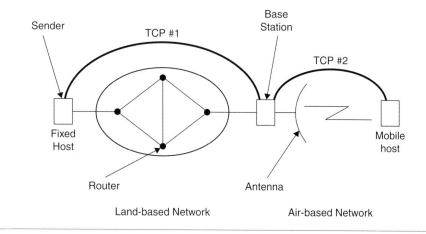

Figure 14.10 Split connection (indirect) TCP in wireless environment.

- Transmission/time-out freezing
- Selective retransmission

Table 14.1 gives comparisons of these schemes.

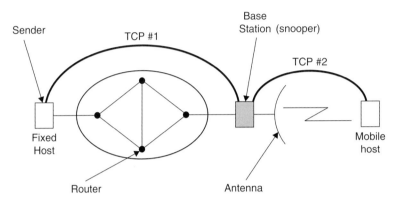

Note: Base station snoops and acts like TCP (generates ACKs, delete ACKs, resent segments)

Figure 14.11   Snooping agent TCP in wireless environment.

Table 14.1   Comparison of TCP enhancements for mobility.

Approch	Mechanism	Advantages	Disadvantages
I-TCP	Splits TCP connection into two connections	Isolation of wireless link, simple	Loss of TCP semantics, higher latency at handoff security problem
Snooping TCP	Snoops data and ACKs, local retrans-mission	Transparent for end-to-end connection, MAC integration possible	Insufficient isolation of wireless link, security problem
M-TCP	Splits TCP connection, chokes sender via window size	Maintains end-to-end semantics, handles longterm and frequent disconnections	Bad isolation of wire-less link, overhead due to bandwidth management, security problem
Fast retransmit/fast recovery	Avoids slow-start after roaming	Simple and efficient	Mixed layers, not transparent
Transmission/ time-out freezing	Freezes TCP state at disconnection, resumes after reconnecting	Independent of content, works for longer interruptions	Changes in TCP required, MAC independent
Selective retransmission	Retransmits only lost data	Very efficient	Slightly more complex receiver software, more buffer space required

The current TCP for 2.5G/3G wireless networks describes a profile to optimize TCP for wireless wide-area networks (WWANs) such as GSM/GPRS, UMTS, or cdma2000. The following characteristics have been considered in deploying applications over 2.5/3G wireless links:

- Data rates
- Latency
- Jitter
- Packet loss

Based on these characteristics, the following configuration parameters for TCP in a wireless environment have been suggested:

- *Large window size:* TCP should use large enough window sizes based on the bandwidth delay experienced in wireless systems. A large initial window size (more than the typical 1 segment) of 2 to 4 segments may increase performance particularly for short transmissions (a few segments in total).

- *Limit transmit:* This is an extension of fast retransmission/fast recovery and is useful when small amounts of data are to be transmitted.

- *Large maximum segment size:* The larger the MSS the faster TCP increases the congestion window. Link-layers fragment packet data units (PDUs) for transmit according to their needs and a large MSS may be used to increase performance. MSS path discovery should be used for larger segment sizes instead of assuming the small default MSS.

- *Selective ACK (SACK):* SACK allows the selective retransmission of packets. It is beneficial compared to the standard cumulative scheme.

- *Explicit congestion notification (ECN):* ECN allows a receiver to inform a sender of congestion in the network by setting the ECN-echo flag on receiving an IP packet that has experienced congestion. This scheme makes it easier to distinguish packet loss due to retransmission errors from packet loss due to congestion.

- *Time-stamp:* With the help of time-stamps, higher delay spikes can be tolerated by TCP without experiencing a spurious time-out. The effect of bandwidth oscillation is also reduced.

- *No header compression:* Header compression mechanism does not perform well in the presence of packet losses. It should not be used.

14.5.2   Implementation of Wireless TCP

Traditional TCP schemes suffer from severe performance degradation in a mixed wired and wireless environment. Modifications to standard TCP to remove its deficiency in wireless communications have been proposed.

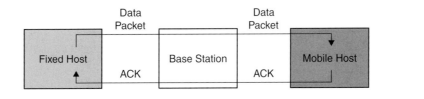

Figure 14.12   End-to-end connection TCP in wireless environment.

The design of wireless TCP should consider the characteristics of a particular type of wireless network and its need; for example, a satellite network has a long propagation delay, and an ad hoc network is infrastructureless.

Wireless TCP algorithms can be designed in either *split mode* or *end-to-end mode*. The split mode divides the TCP connection into a wireless and wired portion, and ACKs are generated for both portions separately (see Figure 14.10). By doing so, the performance of the wired portion is not affected by the relatively unreliable wireless portion. The end-to-end mode treats the route from the sender to the receiver as an end-to-end path, and the sender is acknowledged directly by the receiver (see Figure 14.12). This approach maintains the end-to-end semantics of the original TCP design.

The split mode TCP attempts to shield the wireless portion from the fixed network by separating the flow control at the intermediate router (or a base station), so that the wireless behavior has the least impact on the fixed network. The intermediate router acts as a terminal in both the fixed and wireless portions. Both end hosts communicate with the intermediate router independently, without knowledge of the other end. The intermediate router is provided with functionality to coordinate the transaction between two network portions. In indirect TCP (I-TCP) the mobile support router (MSR) connects the mobile host to the fixed host and establishes two separate TCP connections with the fixed host and mobile host, respectively. The mobile support router communicates with the fixed host on behalf of the mobile host. The congestion window is maintained separately for wireless and fixed connections. When the mobile host switches cells, a new MSR takes over the communication with the fixed host seamlessly. Thus, the fixed host is protected from the unreliable feature of wireless connections.

In split mode, an intermediate router has to reveal the information in the TCP packet and process related data before it reaches its destination. This violates the end-to-end semantics of the original TCP. In the end-to-end approach, only the end hosts participate in flow control. The receiver provides feedback reflecting the network condition, and the sender makes decisions for congestion control.

In the end-to-end approach, the ability to accurately probe for the available bandwidth is the key to better performance. The available bandwidth of a flow is the minimum unused link capacity of the flow's fair share along the path.

The end-to-end approach can have its congestion control mechanism realized in two ways: reactive and proactive. By reactive congestion control, the sender rectifies the congestion window when the network situation becomes marginal or has crossed a threshold. By proactive congestion control, feedback from the network guides the sender to reallocate network resources in order to prevent congestion.

In cellular networks, where the base station interconnects a fixed network and a mobile network, modifications of TCP algorithms focus on cellular characteristics such as handoff and high bit error rate. The end-to-end solution is used to improve TCP performance. It imposes no restrictions on routers and only requires code modifications at the mobile unit or receiver side. This approach addresses the throughput degradation caused by frequent disconnections (and reconnections) due to mobile handoff or temporary blockage of radio signals by obstacles. The assumption is that the mobile has knowledge of radio signal strength, and therefore can predict the impending disconnections. In this approach, the receiver on the mobile proactive sets the window size to zero in the ACK packets in the presence of impending disconnections. The zero window size ACK packet forces the sender into persist mode, where it ceases to send more packets while keeping its sending window unchanged. To prevent the sender from exponentially backing off when it detects the reconnection, the receiver sends several positive ACK packets to the sender acknowledging the last received packet before disconnection so that the transfer can resume quickly at the rate before the disconnection occurs. To implement the scheme proposed in reference [11], cross-layer information is required to be exchanged, and the TCP layer protocol must be exposed to some details of roaming and handoff algorithms implemented by network interface card vendors on the interface devices.

14.6   Mobile IP (MIP) and Session Initiation Protocol (SIP)

Third-generation (3G) mobile networks are designed to provide a variety of IP data services such as voice over IP (VoIP) and instant messaging (IM). Both IPv4 and IPv6 are supported in order to provide future-proof solutions. Mobility is supported through both mobile-specific and IP mechanisms. Mobile IP is a key technology for managing mobility in wireless networks [21–25]. At the same time, the SIP is the key to realizing and provisioning services in IP-based mobile networks. The need for mobility of future real-time service independent of terminal mobility requires SIP to seamlessly interwork with MIP operations.

3G networks introduced support of IP mobility through MIP. In particular, the cdma2000 network specified by 3GPP2 deploys MIP to support terminal mobility between points of attachment to the network. MIP will also be supported in 3GPP networks. The SIP (defined by the IETF) is key to service provisioning in 3G networks beyond plain IP connectivity. 3GPP has defined and standardized a network infrastructure called the IP multimedia service (IMS) based on SIP for supporting a multitude of services to 3G users. Examples are VoIP, instant messaging, and streaming.

There are substantial differences between 3GPP and 3GPP2 packet core networks that have an impact on how SIP services can be provided through an IP multimedia service. In particular, 3GPP2 networks use MIP to support terminal mobility. The adoption of SIP and MIP in 3G networks introduces the need for SIP and the IMS to interwork with MIP.

In this section, we briefly introduce MIP and SIP and discuss the issues related to interworking between SIP and MIP, with a focus on IPv6 and the applicability to 3G networks.

14.6.1   Mobile IP

IP packets do not require mechanisms to set up a dedicated bandwidth or channel. The IP address serves a dual purpose—for routing packets through the Internet and as an end-point identifier for applications in end-hosts. The connections in an IP network use *sockets* to communicate between clients and servers. A socket consists of *source IP address*, *source port*, *destination IP address*, and *destination port*. A TCP connection cannot survive any address change because it relies on the socket to determine a connection. However, when a terminal moves from one network to another, its address changes. A mobile node (MN) is a terminal than can change its location and thus its point of attachment. The partner for communication is called the *correspondent node* that can be either a fixed or an MN.

Mobile IP is designed to support host mobility on the Internet. In order for an MN to move across different connection points while maintaining connectivity with other nodes on the Internet, the MN needs to maintain the same address. Two versions of MIP are defined depending on IP version used in the network: MIPv4 for IPv4 networks and MIPv6 for IPv6 networks.

Mobile IP implies that a user is connected to one or more applications across the Internet, that the user's point of attachment changes dynamically, and that all connections are automatically maintained despite the change. When MN moves its attachment point to another network, it is considered a foreign network for this host. Once the mobile is reattached, it makes its presence known by registering with a network node, typically a router, on the foreign network known as a *foreign agent* (FA). The mobile then communicates with a similar agent on the user's home network, known as a *home agent* (HA), giving the home agent the *care-of address* (CoA) of the mobile node; the care-of address identifies the foreign agent's location. A home agent tracks a mobile host's location. The mobile host is affiliated with a static IP address on the home network and a foreign agent supports mobility on a foreign network by providing routing to a visiting mobile host. Network supporting mobile IP will have to create foreign agents to deliver packets of information to the mobile host. Mobile IP is fundamental for the paradigm to provide the successful model for wireless data, which takes the

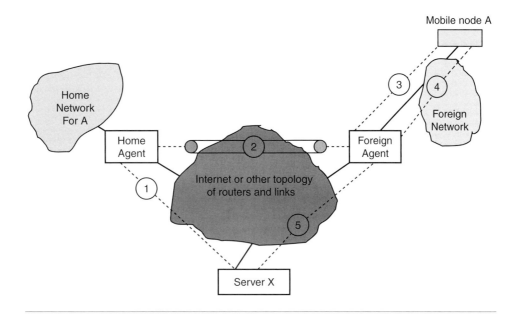

Figure 14.13 Mobile IP scenario.

connection one has into the corporate intranet and makes it wireless. Figure 14.13 shows an MIP scenario that includes the following steps:

1. Server X transmits an IP datagram destined for mobile node A, with A's home address in the IP header. The IP datagram is routed to A's home network.

2. At the home network, the incoming IP datagram is intercepted by the home agent. The home agent encapsulates the entire datagram inside a new IP datagram which has the A's care-of address in the header, and retransmits the datagram. The use of an outer IP datagram with a different destination IP address is known as *tunneling*. This IP datagram is routed to the foreign agent.

3. The foreign agent strips off the outer IP header, encapsulates the original IP datagram in a network-level packet data unit (PDU), and delivers the original datagram to A across the foreign network.

4. When A sends the IP datagram to X, it uses X's IP address. This is a fixed address; that is, X is not a mobile node. Each IP datagram is sent by A to a router on the foreign network to X. Typically, this router is also the foreign agent.

5. The IP datagram from A to X travels directly across the Internet to X, using X's IP address.

In MIPv4, MN registers with an FA that becomes the point of contact for the MN. Subsequently, the MN updates its HA, which is a router on the home network that forwards packets meant for the MN's home address (HoA) to the MN's current point of attachment (i.e., the CoA of the FA). This allows the MN to remain "always on"—always reachable at its HoA.

MIPv6 also supports direct peer-to-peer communication, called route optimization, between the MN and its core networks without having to traverse the HA. In this way, the MN uses the HoA for communication with a core network (CN) and the CoA for routing purposes. Since MIP operates at the network layer, any change of CoA is transparent to the transport protocols and applications. Hence, all applications in the MN and CN can ignore the mobility of the MN and do not have to deal with a change of network attachment.

MIPv4 has been a standard for some years; MIPv6 is currently becoming a standard.

Mobile IP Capabilities

Three basic capabilities of MIP are registration, discovery, and tunneling (see Figure 14.14). These are discussed in more detail in the following paragraphs.

- *Registration:* A mobile node uses an authenticated registration procedure to inform its home agent of its care-of address.

- *Discovery:* A mobile node uses a discovery procedure to identify a prospective home agent and foreign agent.

- *Tunneling:* Tunneling is used to forward IP datagrams from a home address to a care-of address.

The discovery process in MIP is similar to the router advertisement process defined in ICMP. The agent discovery makes use of ICMP router advertisement messages, with one or more extensions specific to MIP. A mobile node is responsible for an ongoing discovery process to determine if it is attached to its network (in which case a datagram may be received without forwarding) or to a foreign network. A transition from the home network to a foreign network can occur at any time without notification to the network layer (i.e., the IP layer). The discovery for a mobile node is a continuous process. Figure 14.15 shows the flow diagram for the agent discovery procedure.

Location management in MIP is achieved via a registration process and an agent advertisement. FAs and HAs periodically advertise their presence using

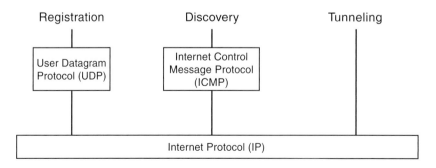

Figure 14.14   Mobile IP capabilities.

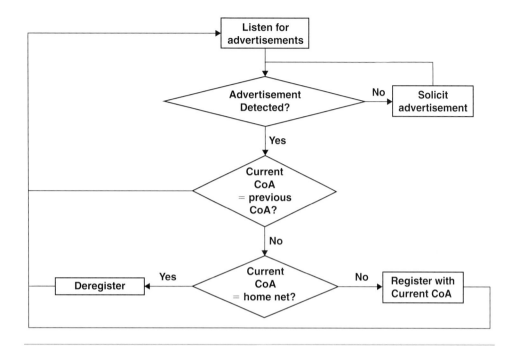

Figure 14.15 Agent discovery procedure.

agent advertisement messages. The same agent may act as both an HA and an FA—mobility extensions to ICMP messages which are used for agent advertisements. The messages contain information about the CoA associated with the FA, whether the agent is busy, whether minimal encapsulation is permitted, whether registration is mandatory, and so on. The agent advertisement packet is a broadcast message on the link. If the mobile node gets an advertisement from its HA, it must deregister its CoA and enable a gratuitous ARP. If a mobile node does not hear any advertisement, it must solicit an agent advertisement using ICMP.

Once an agent is discovered, the MN performs either registration or deregistration with the HA, depending on whether the discovered agent is an HA or an FA. The MN sends a registration request using UDP to HA through the FA (or directly, if it is a co-located CoA). The HA creates a mobility binding between the MN's home address and the current CoA that has a fixed lifetime. The mobile node should register before expiration of the binding. A registration reply indicates whether the registration is successful. A rejection is possible by either the HA or FA for such reasons as insufficient resources, the home agent is unreachable, there are too many simultaneous bindings, failed authentication, and so on.

Each FA maintains a list of visiting mobiles containing the following information:

- Link-layer address of the mobile node
- Mobile node home IP address

- UDP registration request source port
- Home agent IP address
- An identification field
- Registration lifetime
- Remaining lifetime of pending or current registration

IP Tunneling. Tunneling is used to forward IP datagrams from a home address to a care-of-address. Types of IP tunneling are:

- IP-within-IP encapsulation—simplest approach, defined in RFC 2003
- Minimal encapsulation—involves fewer fields, defined in RFC 2004
- *Generic routing encapsulation* (GRE)—procedure developed prior to mobile IP, defined in RFC 1701

IP-within-IP encapsulation (see Figure 14.16): The entire IP datagram becomes the payload in a new IP datagram. The inner, original IP header is

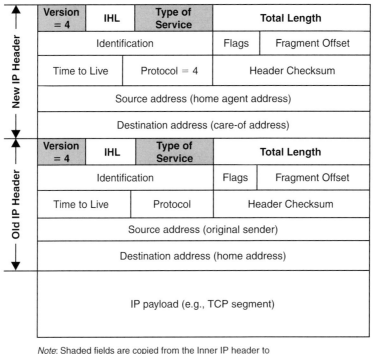

Note: Shaded fields are copied from the Inner IP header to
      the Outer IP header.

Figure 14.16   IP-within-IP encapsulation.

unchanged except to decrement time-to-live (TTL) by 1. The outer header is a full IP header in which:

- Two fields, version number and type of service, are copied from an inner header
- The source address typically is the IP address of the home agent, and the destination address is the CoA for the intended destination

Minimal encapsulation (see Figure 14.17): This results in less overhead and can be used if the mobile node, home agent, and foreign agent all agree to do so. A new header with the following fields is used between the original IP header and the original IP payload:

- Protocol
- Header checksum
- Original destination address
- Original source address

The following fields in the original IP header are modified to form the new outer IP header:

- Total length
- Protocol

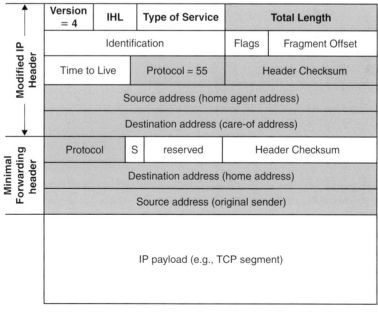

Note: Shaded fields in the inner IP header are copied from the original IP header.
Shaded fields in the outer IP header are modifed from the original IP header.

Figure 14.17   Minimal encapsulation.

- Header checksum
- Source address
- Destination address

The encapsulation (home agent) prepares the encapsulated datagram which is now suitable for tunneling and delivery across the Internet to the care-of address. The fields in the minimal forwarding header are restored to the original IP header and the forwarding header is removed from the datagram. The total length field in the IP header is decremented by the size of the minimal forwarding header and the checksum field is recomputed.

14.6.2 Session Initiation Protocol (SIP)

SIP is used for provisioning services in IP-based mobile networks. SIP specifications define an architecture of user agents and servers (proxy server, redirect server, register) that support communications between SIP peers through user tracking, call routing, and so on. In SIP, each user is uniquely identified by an SIP universal resource indicator, which is used as the identifier to address the called user when the sending session initiation requests. However, an IP address is associated with the user in order to route SIP signaling from the SIP register. A SIP user registers with the SIP register to indicate its presence in the network and its willingness to receive incoming session initiation requests from other users.

A typical session in SIP begins with a user sending an INVITE message to a peer through SIP proxies. When the recipient accepts the request and the initiator is notified, the actual data flow begins, usually taking a path other than the one taken by the SIP signaling messages. An INVITE message typically carries a description of the session parameters. In particular, each media component of the SIP session is described in terms of QoS parameters. The user can modify the parameters regarding an existing session by adding or removing media components or modifying the current QoS using a re-INVITE message. SIP also supports personal mobility by allowing a user to reregister with an SIP register on changing its point of attachment to the network, in particular on changing its IP address. A user could also change point of attachment during an active session provided the user reinvites the session providing the new parameters.

14.7 Internet Reference Model

Although many useful protocols have been developed in the context of OSI, the overall 7-layer model has not flourished. The TCP/IP architecture has come to dominate. There are a number of reasons for this outcome. The most important is that the key TCP/IP protocols were mature and well tested at a time when similar OSI protocols were in the development stage. When businesses began to recognize the need for interoperability across networks, only TCP/IP was available and ready to go. Another reason is that the OSI model is unnecessarily complex, with

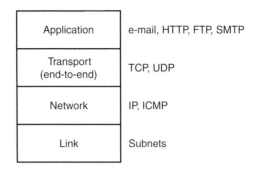

Figure 14.18   Internet reference model.

seven layers to accomplish what TCP/IP does with four layers. The Internet reference model based on TCP/IP is shown in Figure 14.18. The model includes:

- The *application layer* that covers OSI application and presentation layers. Some of the application layer protocols are HTTP, FTP, SMTP, post office protocol (POP), etc.
- The *transport (end-to-end) layer* includes OSI transport and session layers. The end-to-end protocols are TCP and UDP.
- The *network (Internet) layer* corresponds to the upper part of the OSI network layer. The protocol is IP.
- The *link (subnets) layer* includes the lower part of the OSI network layer, link layer, and physical layer.

14.8   Summary

In this chapter, we first presented the workings of the Internet and discussed the addressing scheme used in the Internet. We then discussed IP, TCP, and mobile IP (MIP); presented the IP suite and compared it with the OSI model. We also discussed TCP in a wireless environment.

Problems

14.1   Which two addresses are used in an IP address?

14.2   What are the maximum number of class A, B, and C network IDs?

14.3   Describe four layers of the TCP/IP protocol suite.

14.4   Describe the characteristics of IP in the TCP/IP protocol suite.

14.5   What is the ICMP and what is its role in the TCP/IP protocol suite?

14.6   Describe the IGMP in the TCP/IP protocol suite.

14.7   What are the ARP and RARP in the TCP/IP protocol suite?

14.8   What are the roles of the TCP and UDP in the TCP/IP protocol suite?

14.9   Discuss split-connection and end-to-end implementation in TCP for a wireless environment.

14.10   Discuss IntServ and DiffServ in the Internet.

14.11   Define MIP in the Internet.

14.12   An object of size (O) 1200 kilobytes is transmitted from a server to a client. The maximum segment size (S) is 600 bytes. The round trip time (RTT) is 120 msec. If the transport protocol uses a static window with window size W, what is the maximum possible latency for a transmission rate, R, of 2 Mbps? How many segments are required to achieve this latency?

References

1. Almquist, P. Types of Service in the Internet Protocol Suite. *RFC* 1349, July 1992.

2. Al-Quliti, K., "Mobile IP Overview with an Implementation over CDMA2000 Packet Core Network." Southern Methodist University, Dallas, quliti@yahoo.com.

3. Bakre, A., and Badrinath, B. R. "I-TCP: Indirect TCP for Mobile Hosts." Proc ICDCS'95, May 1995, pp. 136–143.

4. Braden, R. T., ed. Requirements for Internet Hosts—Communication Layers. *RFC* 1122, October 1989.

5. Braden, R. T., ed. Requirements for Internet Hosts—Application and Support. *RFC* 1123, October 1989.

6. Braden, R. T., ed. Extending TCP for Transactions—Concepts. *RFC* 1379, November 1992.

7. Clark, D. D. Window and Acknowledgement Strategy in TCP. *RFC* 813, July 1982.

8. Deering, S. E. Host Extensions for IP Multicasting. *RFC* 1112, August 1989.

9. Deering, S. E., and Hinden, R. Internet Protocol, Version 6 (IPv6) Specification. *RFC* 1883, December 1995.

10. Deering, S. E. ed. ICMP Router Discovery Messages. *RFC* 1256, September 1991.

11. Goff, T., et al. "Freeze-TCP: A true end-to-end enhancement mechanism for mobile environment." Proc. IEEE INFOCOM2000, 2000, pp. 1537–1545.

12. Hiller, T., et al. CDMA 2000 Wireless Data Requirements. *RFC* 3141, June 2001.

13. Jacobson, V. Congestion Avoidance and Control. *Computer Communication Review*, vol. 18, no. 4, August 1988, pp. 314–329.

14. Jacobson, V., Braden, R. T., and Borman, D. A. TCP Extensions for High Performance. *RFC* 1323, May 1992.

15. Kleinrock, L. The Latency/Bandwidth Trade-off in Gigabit Networks. *IEEE Communication Magazine*, vol. 30, no. 4, April 1992, pp. 36–40.

16. LaQuey, T. *The Internet Companion: A Beginner's Guide to Global Networking.* Reading, MA: Addison-Wesley, 1993.

17. Mockapetris, P. V. Domain names: Implementation and Specification. *RFC* 1035, November 1987.

18. Mogul, J. C. IP Network Performance. In *Internet System Handbook*, eds., D. C. Lynch and M. T. Rose. Reading, MA: Addison-Wesley, 1993, pp. 575–675.

19. Nagle, J. Congestion Control in IP/TCP Internetworks. *RFC* 896, January 1984.

20. Partridge, C. *Gigabit Networking*. Reading, MA: Addison-Wesley, 1994.

21. Perkins, C. E. Mobile IP. *IEEE Communication Magazine*, May 1997, pp. 84–99.

22. Perkins, C. E., ed. Mobile IP Joins Forces with AAA. *IEEE Personal Communications*, August 2000.

23. Perkins, C. E. "Mobile Networking Through Mobile IP." http://www.computer.org/ internet/v2n1/perkins.htm.

24. Perkins, C. E. IP Encapsulation within IP. *RFC* 2003, May 1996.

25. Perkins, C. E. Minimal Encapsulation within IP. *RFC* 2004, May 1996.

26. Postel, J. B. User Datagram Protocol. *RFC* 768, August 1980.

27. Postel, J. B. Internet Control Message Protocol. *RFC* 792, September 1981.

28. Postel, J. B., ed. Transmission Control Protocol. *RFC* 793, September 1981.

29. Simpson, W. A. Point-to-Point Protocol (PPP). *RFC* 1548, December 1993.

30. Stevens, W. R. TCP/IP Illustrated Volume 1—Protocols, Reading, MA: Addison-Wesley, 1994.

31. Tanenbaum, A. S. *Computer Network*. Second Edition, Englewood Cliffs, NJ: Prentice-Hall, 1989.

32. Tian, Y., Xu, K., and Ansari, N. TCP in Wireless Environments: Problem and Solution. *IEEE Radio Communications*, March 2005, pp. 527–531.

CHAPTER 15

Wide-Area Wireless Networks (WANs)—GSM Evolution

15.1 Introduction

Third-generation (3G) wireless systems [2,3,9] offer access to services anywhere from a single terminal; the old boundaries between telephony, information, and entertainment services are disappearing. Mobility is built into many of the services currently considered as fixed, especially in such areas as high speed access to the Internet, entertainment, information, and electronic commerce (e-commerce) services. The distinction between the range of services offered via wireline or wireless is becoming less and less clear and, as the evolution toward 3G mobile services speeds up, these distinctions will disappear in the first decade of the new millennium.

Applications for a 3G wireless network range from simple voice-only communications to simultaneous video, data, voice, and other multimedia applications. One of the main benefits of 3G is that it allows a broad range of wireless services to be provided efficiently to many different users.

Packet-based Internet Protocol (IP) technology is at the core of the 3G services. Users have continuous access to on-line information. E-mail messages arrive at hand-held terminals nearly instantaneously and business users are able to stay permanently connected to the company intranet. Wireless users are able to make video conference calls to the office and surf the Internet simultaneously, or play computer games interactively with friends in other locations. Figure 15.1 shows the data rate requirement for various services.

In 1997, the TIA/EIA IS-136 community through the Universal Wireless Consortium (UWC) and the Telecommunications Industry Association (TIA) TR 45.3 adopted a three-part strategy for evolving its IS-136 TDMA-based networks to 3G wireless networks to satisfy International Mobile Telephony-2000 (IMT-2000) requirements. The strategy consists of:

- Enhancing the voice and data capabilities of the existing 30 kHz carrier (IS-136+)

- Adding a 200 kHz carrier for high-speed data (384 kbps) in high mobility applications

- Introducing a 1.6 MHz carrier for very high-speed data (2 Mbps) in low-mobility applications

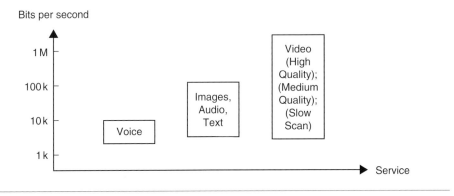

Figure 15.1   User data requirements.

The highlight of UWC strategy was the global convergence of IS-136 time division multiple access (TDMA) with a global system for mobile communications (GSM) through the evolution of the 200 kHz GSM carrier for supporting high-speed data applications (384 kbps) while also improving a 30 kHz carrier for voice and mid-speed data applications.

In this chapter we focus first on GSM evolution to packet data services and present the details of the general packet radio service (GPRS) and the enhanced data for GSM evolution (EDGE) service. We then provide details of 3G systems including wideband code division multiple access (CDMA) (WCDMA) (i.e., universal mobile telecommunications services (UMTS)). We conclude the chapter by outlining the details of high-speed downlink packet access (HSDPA).

15.2   GSM Evolution for Data

From a radio access perspective, adding 3G capabilities to 2G systems mainly means supporting higher data rates. Possible scenarios depend on spectrum availability for the network service provider. Depending on the spectrum situation, two different migration paths can be supported:

- Reframing of existing spectrum bands
- New or modified spectrum bands

Two 3G radio access schemes are identified to support the different spectrum scenarios:

1. Enhanced data rates for GSM evolution (EDGE) with high-level modulation in a 200 kHz TDMA channel is based on plug-in transceiver equipment, thereby allowing the migration of existing bands in small spectrum segments.

2. Universal mobile telecommunications services (UMTS) is a new radio access network based on 5 MHz WCDMA and optimized for efficient support of 3G services. UMTS can be used in both new and existing spectra.

From a network point of view, 3G capabilities implies the addition of packet switched (PS) services, Internet access, and IP connectivity. With this approach, the existing mobile networks reuse the elements of mobility support, user authentication/service handling, and circuit switched (CS) services. With packet switched services, IP connectivity can then be added to provide a mobile multimedia core network by evolving the existing mobile network.

GSM is moving to develop enhanced cutting-edge, customer-focused solutions to meet the challenges of the new millennium and 3G mobile services [29]. When GSM was first introduced, no one could have predicted the dramatic growth of the Internet and the rising demand for multimedia services. These developments have brought about new challenges to the world of GSM. For GSM operators, the emphasis is now rapidly changing from that of instigating and driving the development of technology to fundamentally enabling mobile data transmission to that of improving speed, quality, simplicity, coverage, and reliability in terms of tools and services that will boost mass market take-up.

Users are increasingly looking to gain access to information and services wherever they are and whenever they want. GSM should provide that connectivity. Internet access, web browsing and the whole range of mobile multimedia capability are the major drivers for development of higher data speed technologies.

Current data traffic on most GSM networks is modest, less than 5% of total GSM traffic. But with the new initiatives coming to fruition during the course of the next two to three years, exponential growth in data traffic is forecast. The use of messaging-based applications may reach up to about 90% by the year 2008. GSM data transmission using high-speed circuit switched data (HSCSD) and GPRS may reach a penetration of about 80% by 2008 [1].

GSM operators have two nonexclusive options for evolving their networks to 3G wideband multimedia operation: (1) using GPRS and EDGE in the existing radio spectrum, and in small amounts of the new spectrum; or (2) using WCDMA in the new 2 GHz bands, or in large amounts of the existing spectrum. Both approaches offer a high degree of investment flexibility because roll-out can proceed in line with market demand with the extensive reuse of existing network equipment and radio sites.

In the new 2 GHz bands, 3G capabilities are delivered using a new wideband radio interface that offers much higher user data rates than are available today—384 kbps in the wide area and up to 2 Mbps in the local area. Of equal importance for such services is the high-speed packet switching provided by GPRS and its connection to public and private IP networks.

GSM and digital (D)AMPS (IS-136) operators can use existing radio bands to deliver some of the 3G services, even without the new wideband spectrum by evolving current networks and deploying GPRS and EDGE technologies. In the early years of 3G service deployment, a large proportion of wireless traffic will still be voice-only and low-rate data. So whatever the ultimate capabilities of 3G networks, efficient and profitable ways of delivering more basic wireless services are still needed.

The significance of EDGE for today's GSM operators is that it increases data rates up to 384 kbps and potentially even higher in a good quality radio environment using current GSM spectrum and carrier structures more efficiently. EDGE is both a complement and an alternative to new WCDMA coverage. EDGE also has the effect of unifying the GSM, D-AMPS and WCDMA services through the use of dual-mode terminals.

15.2.1 High Speed Circuit Switched Data

High-speed circuit switched data (HSCSD) [1,4,5] is a feature that enables the co-allocation of multiple full rate traffic channels (TCH/F) of GSM into an HSCSD configuration. The aim of HSCSD is to provide a mixture of services with different air interface user rates by a single physical layer structure. The available capacity of an HSCSD configuration is several times the capacity of a TCH/F, leading to a significant enhancement in air interface data transfer capability.

Ushering faster data rates into the mainstream is the new speed of 14.4 kbps per time slot and HSCSD protocols that approach wireline access rates of up to 57.6 kbps by using multiple 14.4 kbps time slots. The increase from the current baseline of 9.6 kbps to 14.4 kbps is due to a nominal reduction in the error-correction overhead of the GSM radio link protocol (RLP), allowing the use of a higher data rate.

For operators, migration to HSCSD brings data into the mainstream, enabled in many cases by relatively standard software upgrades to base station (BS) and mobile switching center (MSC) equipment. Flexible air interface resource allocation allows the network to dynamically assign resources related to the air interface usage according to the network operator's strategy, and the end-user's request for a change in the air interface resource allocation based on data transfer needs. The provision of the asymmetric air interface connection allows simple mobile equipment to receive data at higher rates than otherwise would be possible with a symmetric connection.

For end-users, HSCSD enables the roll-out of mainstream high-end segment services that enable faster web browsing, file downloads, mobile video-conference and navigation, vertical applications, telematics, and bandwidth-secure mobile local area network (LAN) access. Value-added service providers will also be able to offer guaranteed quality of service and cost-efficient mass-market applications, such as direct IP where users make circuit-switched data calls straight into a GSM network router connected to the Internet. To the end-user, the value-added service provider or the operator is equivalent to an Internet service provider that offers a fast, secure dial-up Internet protocol service at cheaper mobile-to-mobile rates.

HSCSD is provided within the existing mobility management. Roaming is also possible. The throughput for an HSCSD connection remains constant for the duration of the call, except for interruption of transmission during handoff. The handoff is simultaneous for all time slots making up an HSCSD connection. End-users wanting to use HSCSD have to subscribe to general bearer services. Supplementary services applicable to general bearer services can be used simultaneously with HSCSD.

Firmware on most current GSM PC cards needs to be upgraded. The reduced RLP layer also means that a stronger signal strength is necessary. Multiple time slot usage is probably only efficiently available in off-peak times, increasing overall off-peak idle capacity usage. HSCSD is not a very feasible solution for bursty data applications.

15.2.2 General Packet Radio Service

The general packet radio service (GPRS) [6,7] enhances GSM data services significantly by providing end-to-end packet switched data connections. This is particularly efficient in Internet/intranet traffic, where short bursts of intense data communications are actively interspersed with relatively long periods of inactivity. Because there is no real end-to-end connection to be established, setting up a GPRS call is almost instantaneous and users can be continuously on-line. Users have the additional benefits of paying for the actual data transmitted, rather than for connection time. Because GPRS does not require any dedicated end-to-end connection, it only uses network resources and bandwidth when data is actually being transmitted. This means that a given amount of radio bandwidth can be shared efficiently among many users simultaneously.

The next phase in the high-speed road map is the evolution of current short message service (SMS), such as smart messaging and unstructured supplementary service data (USSD), toward the new GPRS, a packet data service using TCP/IP and X.25 to offer speeds up to 115 kbps. GPRS has been standardized to optimally support a wide range of applications ranging from very frequent transmissions of medium to large data volume. Services of GPRS have been developed to reduce connection set-up time and allow an optimum usage of radio resources. GPRS provides a packet data service for GSM where time slots on the air interface can be assigned to GPRS over which packet data from several mobile stations is multiplexed.

A similar evolution strategy, also adopting GPRS, has been developed for DAMPS (IS-136). For operators planning to offer wideband multimedia services, the move to GPRS packet-based data bearer service is significant; it is a relatively small step compared to building a totally new 3G IMT-2000 network. Use of the GPRS network architecture for IS-136+ packet data service enables data subscription roaming with GSM networks around the globe that support GPRS and its evolution. The IS-136+ packet data service standard is known as GPRS-136. GPRS-136 provides the same capabilities as GSM GPRS. The user can access either X.25 or an IP-based data network.

GPRS provides a core network platform for current GSM operators not only to expand the wireless data market in preparation for the introduction of 3G services, but also a platform on which to build IMT-2000 frequencies should they acquire them.

The implementation of GPRS has a limited impact on the GSM core network. It simply requires the addition of new packet data switching and gateway nodes, and an upgrade to existing nodes to provide a routing path for packet data between the wireless terminal and a gateway node. The gateway node provides interworking with external packet data networks for access to the Internet, intranet, and databases.

A GPRS architecture for GSM is shown in Figure 15.2 and network element interfaces in Figure 15.3. GPRS supports all widely used data communications

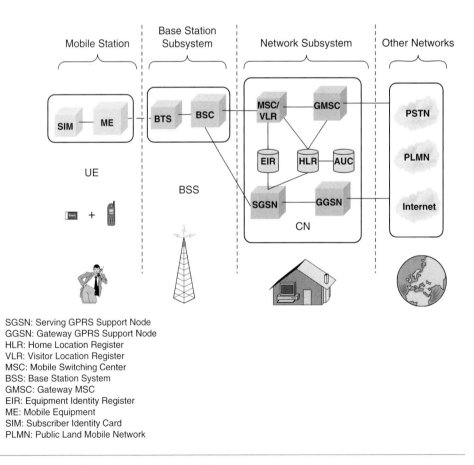

SGSN: Serving GPRS Support Node
GGSN: Gateway GPRS Support Node
HLR: Home Location Register
VLR: Visitor Location Register
MSC: Mobile Switching Center
BSS: Base Station System
GMSC: Gateway MSC
EIR: Equipment Identity Register
ME: Mobile Equipment
SIM: Subscriber Identity Card
PLMN: Public Land Mobile Network

Figure 15.2   A GPRS architecture in GSM.

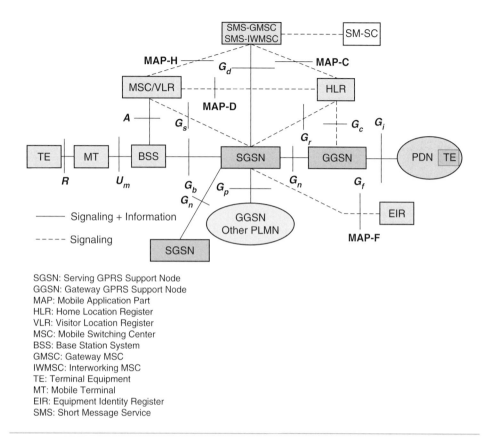

Figure 15.3   GPRS interfaces for different network elements.

protocols, including IP, so it is possible to connect with any data source from anywhere in the world using a GPRS mobile terminal. GPRS supports applications ranging from low-speed short messages to high-speed corporate LAN communications. However, one of the key benefits of GPRS—that it is connected through the existing GSM air interface modulation scheme—is also a limitation, restricting its potential for delivering higher data rates than 115 kbps. To build even higher rate data capabilities into GSM, a new modulation scheme is needed.

   GPRS can be implemented in the existing GSM systems. Changes are required in an existing GSM network to introduce GPRS. The base station subsystem (BSS) consists of a base station controller (BSC) and packet control unit (PCU). The PCU supports all GPRS protocols for communication over the air interface. Its function is to set up, supervise, and disconnect packet switched calls. The packet control unit supports cell change, radio resource configuration, and channel assignment. The base station transceiver (BTS) is a relay station without protocol functions. It performs modulation and demodulation.

The GPRS standard introduces two new nodes, the *serving GPRS support node (SGSN)* and the *gateway GPRS support node (GGSN)*. The home location register (HLR) is enhanced with GPRS subscriber data and routing information. Two types of services are provided by GPRS:

- Point-to-point (PTP)
- Point-to-multipoint (PTM)

Independent packet routing and transfer within the public land mobile network (PLMN) is supported by a new logical network node called the *GPRS support node (GSN)*. The GGSN acts as a logical interface to external packet data networks. Within the GPRS networks, *protocol data units (PDUs)* are encapsulated at the originating GSN and decapsulated at the destination GSN. In between the GSNs, IP is used as the backbone to transfer PDUs. This whole process is referred to as tunnelling in GPRS. The GGSN also maintains routing information used to tunnel the PDUs to the SGSN that is currently serving the mobile station (MS). All GPRS user related data required by the SGSN to perform the routing and data transfer functionality is stored within the HLR. In GPRS, a user may have multiple data sessions in operation at one time. These sessions are called *packet data protocol (PDP) contexts*. The number of PDP contexts that are open for a user is only limited by the user's subscription and any operational constraints of the network.

The main goal of the GPRS-136 architecture is to integrate IS-136 and GSM GPRS as much as possible with minimum changes to both technologies. In order to provide subscription roaming between GPRS-136 and GSM GPRS networks, a separate functional GSM GPRS HLR is incorporated into the architecture in addition to the IS-41 HLR.

The European Telecommunication Standards Institute (ETSI) has specified GPRS as an overlay to the existing GSM network to provide packet data services. In order to operate a GPRS over a GSM network, new functionality has been introduced into existing GSM network elements (NEs) and new NEs are integrated into the existing service provider's GSM network.

The BSS of GSM is upgraded to support GPRS over the air interface. The BSS works with the *GPRS backbone system (GBS)* to provide GPRS service in a similar manner to its interaction with the switching subsystem for the circuit-switched services. The GBS manages the GPRS sessions set up between the mobile terminal and the network by providing functions such as admission control, mobility management (MM), and service management (SM). Subscriber and equipment information is shared between GPRS and the switched functions of GSM by the use of a common HLR and coordination of data between the visitor location register (VLR) and the GPRS support nodes of the GBS. The GBS is composed of two new NEs — the SGSN and the GGSN.

The SGSN serves the mobile and performs security and access control functions. The SGSN is connected to the BSS via frame-relay. The SGSN provides packet routing, mobility management, authentication, and ciphering to and from all GPRS subscribers located in the SGSN service area. A GPRS subscriber may be served by any SGSN in the network, depending on location. The traffic is routed from the SGSN to the BSC and to the mobile terminal via a BTS. At GPRS *attach*, the SGSN establishes a mobility management context containing information about mobility and security for the mobile. At PDP context activation, the SGSN establishes a PDP context which is used for routing purposes with the GGSN that the GPRS subscriber uses. The SGSN may send in some cases location information to the MSC/VLR and receive paging requests.

The GGSN provides the gateway to the external IP network, handling security and accounting functions as well as the dynamic allocation of IP addresses. The GGSN contains routing information for the attached GPRS users. The routing information is used to tunnel PDUs to the mobile's current point of attachment, SGSN. The GGSN may be connected with the HLR via optional interface G_c. The GGSN is the first point of public data network (PDN) interconnection with a GSM PLMN supporting GPRS. From the external IP network's point of view, the GGSN is a host that owns all IP addresses of all subscribers served by the GPRS network.

The PTM-SC handles PTM traffic between the GPRS backbone and the HLR. The nodes are connected by an IP backbone network. The SGSN and GGSN functions may be combined in the same physical node or separated—even residing in different mobile networks.

A special interface (G_s) is provided between MSC/VLR and SGSN to coordinate signaling for mobile terminals that can handle both circuit-switched and packet-switched data.

The HLR contains GPRS subscription data and routing information, and can be accessible from the SGSN. For the roaming mobiles, the HLR may reside in a different PLMN than the current SGSN. The HLR also maps each subscriber to one or more GGSNs.

The objective of the GPRS design is to maximize the use of existing GSM infrastructure while minimizing the changes required within GSM. The GSN contains most of the necessary capabilities to support packet transmission over GSM. The critical part in the GPRS network is the mobile to GSN (MS-SGSN) link which includes the MS-BTS, BTS-BSC, BSC-SGSN, and the SGSN-GGSN link. In particular, the U_m interface including the radio channel is the bottleneck of the GPRS network due to the spectrum and channel speed/quality limitations. Since multiple traffic types of varying priorities are supported by the GPRS network, the quality of service criteria as well as resource management is required for performance evaluation.

The BSC will require new capabilities for controlling the packet channels, new hardware in the form of a packet control unit, and new software for GPRS mobility management and paging. The BSC also has a new traffic and signaling interface from the SGSN.

The BTS has new protocols supporting packet data for the air interface, together with new slot and channel resource allocation functions. The utilization of resources is optimized through dynamic sharing between the two traffic types handled by the BSC.

MS-SGSN Link

The logical link control (LLC) layer (see Figure 15.4) is responsible for providing a link between the MS and the SGSN. It governs the transport of GPRS signaling and traffic information from the MS to the SGSN. GPRS supports three service access points (SAPs) entities: the layer 3 management, subnet dependent convergence, and short message service (SMS). On the MS-BSS link, the radio link control (RLC), the medium access control (MAC), and GSM RF protocols are supported.

The main drawback in implementing GPRS on an existing GSM infrastructure is that the GSM network is optimized for voice transmission (i.e., the GSM channel quality is designed for voice which can tolerate errors at a predefined level). It is therefore expected that GPRS could have varied transmission performance in a different network or coverage area. To overcome this problem, GPRS supports multiple coding rates at the physical layer.

A GPRS could share radio resources with GSM circuit switched (CS) service. This is governed by a dynamic resource sharing based on the capacity of demand criteria. A GPRS channel is allocated only if an active GPRS terminal exists in the

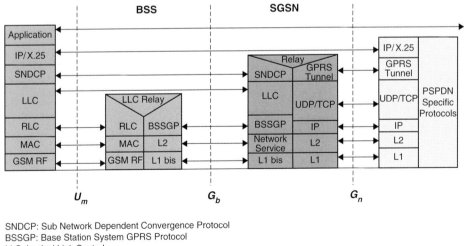

SNDCP: Sub Network Dependent Convergence Protocol
BSSGP: Base Station System GPRS Protocol
LLC: Logical Link Control
RLC: Radio Link Control
MAC: Medium Access Control

Figure 15.4   Protocol stack in GPRS.

network. Once resources are allocated to GPRS, at least one channel will serve as the *master* channel to carry all necessary signaling and control information for the operation of the GPRS. All other channels will serve as *slave* and are only used to carry user and signaling information. If no master channel exists, all the GPRS users will use the GSM *common control channel* (*CCCH*) and inform the network to allocate GPRS resources.

A physical channel dedicated to GPRS is called a *packet data channel* (*PDCH*). It is mapped into one of the physical channels allocated to GPRS (see Figure 15.5). A PDCH can either be used as a *packet common control channel* (*PCCCH*), a *packet broadcast control channel* (*PBCCH*), or a *packet traffic channel* (*PTCH*).

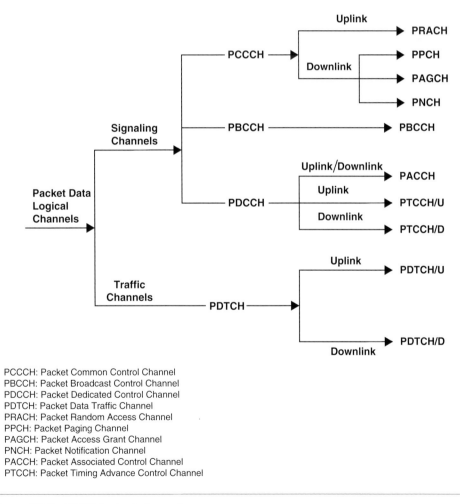

PCCCH: Packet Common Control Channel
PBCCH: Packet Broadcast Control Channel
PDCCH: Packet Dedicated Control Channel
PDTCH: Packet Data Traffic Channel
PRACH: Packet Random Access Channel
PPCH: Packet Paging Channel
PAGCH: Packet Access Grant Channel
PNCH: Packet Notification Channel
PACCH: Packet Associated Control Channel
PTCCH: Packet Timing Advance Control Channel

Figure 15.5 GPRS logical channels.

The PCCCH consists of:

- Packet random access channel (PRACH) — uplink
- Packet access grant channel (PAGCH) — downlink
- Packet notification channel (PNCH) — downlink

On the other hand, the PTCH can either be:

- Packet data traffic channel (PDTCH)
- Packet associated control channel (PACCH)

The arrangement of GPRS logical channels for given traffic characteristics also requires the combination of PCCCHs and PTCHs. Fundamental questions such as how many PDTCHs can be supported by a single PCCCH is needed in dimensioning GPRS.

RLC/MAC Layer

The multiframe structure of the PDCH in which GPRS RLC messages are transmitted is composed of 52 TDMA frames organized into RLC blocks of four bursts resulting in 12 blocks per multiframe plus four idle frames located in the 13th, 26th, 39th, and 52nd positions (see Figure 15.6).

B_0 consists of frames 1, 2, 3 and 4, B_1 consists of frames 5, 6, 7, and 8 and so on. It is important that the mapping of logical channels onto the radio blocks is done by means of an ordered set of blocks (B_0, B_6, B_9, B_1, B_7, B_4, B_{10}, B_2, B_8, B_5, B_{11}). The advantage of ordering the blocks is mainly to spread the locations of the control channels in each time slot reducing the average waiting time for the users to transmit signaling packets. It also provides an interleaving of the GPRS multiframe.

GPRS uses a reservation protocol at the MAC layer. Users that have packets ready to send request a channel via the PRACHs. The random access burst consists of only one TDMA frame with duration enough to transmit an 11-bit signaling message. Only the PDCHs carrying PCCCHs contain PRACHs. The blocks used as PRACHs are indicated by an uplink state flag (USF = free) by the downlink pair channel.

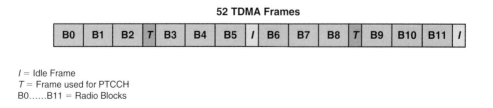

52 TDMA Frames

| B0 | B1 | B2 | T | B3 | B4 | B5 | I | B6 | B7 | B8 | T | B9 | B10 | B11 | I |

I = Idle Frame
T = Frame used for PTCCH
B0......B11 = Radio Blocks

Figure 15.6   Idle frame location in GPRS multiframe.

Alternatively, the first K blocks following the ordered set of blocks can be assigned to PRACH permanently. The access burst is transmitted in one of the four bursts assigned as PRACH. Any packet channel request is returned by a packet immediate assignment on the PRACHs whose locations are broadcast by PBCCH. Optionally, a packet resource request for additional channels is initiated and returned by a packet resource assignment. The persistence of random access is maintained by the traffic load and user class with a back-off algorithm for unsuccessful attempts. In the channel assignment, one or more PTCHs (time slot) will be allocated to a particular user. A user reserves a specific number of blocks on the assigned PTCH as indicated by the USF. It is possible to accommodate more than one user per PTCH. User signaling is also transmitted on the same PTCH using the PAGCH whose usage depends on the necessity of the user.

The performance of the MAC layer depends on the logical arrangement of the GPRS channels (i.e., allocation of random access channels, access grant channels, broadcast channels, etc.) for given traffic statistics. This is determined by the amount of resources allocated for control and signaling compared to data traffic. A degree of flexibility of logical channels is also achieved as the traffic varies. The arrangement of logical channels is determined through the PBCCH.

LLC Layer

The LLC layer is responsible for providing a reliable link between the mobile and the SGSN. It is based on the LAPD (link access protocol D) protocol. It is designed to support variable length transmission in a PTP or PTM topology. It includes the layer function such as sequence control, flow control, error detection, ciphering, and recovery as well as the provision of one or more logical link connections between two layer 3 entities. A logical link is identified by a DLCI (data link control identity) which consists of a service access point identity (SAPI) and terminal equipment identity (TEI) mapped on the LLC frame format. Depending on the status of the logical link, it supports an unacknowledged or an acknowledged information transfer. The former does not support error recovery mechanisms. The acknowledged information transfer supports error and flow control. This operation only applies to point-to-point operations. The LLC frame consists of an address field (1 or 5 octets), control field (2 or 6 octets), a length indicator field (2 octets maximum), information fields (1500 octets maximum), and a frame check sequence of 3 octets. Four types of control field formats are allowed including the supervisory functions (S format), the control functions (U), and acknowledged and unacknowledged information transfer (I and UI).

In the performance evaluation, the objective is to determine delay during the exchange of commands and responses involved in various operations supported by the LLC in relation to the transfer of an LLC PDU. The LLC commands and responses are exchanged between two layer 3 entities in conjunction with a service primitive invoked by the mobile or the SGSN.

Data Packet Routing in the GPRS Network

The following discusses data packet routing for the mobile originated and mobile terminated data call scenarios. In mobile originated data routing, the mobile gets an IP packet from an application and requests a channel reservation. The mobile transmits data in the reserved time slots. The packet switched public data network (PSPDN) PDU is encapsulated into a sub-network dependent convergence protocol (SNDCP) unit that is sent via LLC protocol over the air interface to the SGSN currently serving the mobile (see Figure 15.7).

For mobile terminated data routing (see Figure 15.7), we have two cases: routing to the home GPRS network, and routing to a visited GPRS network. In the first case, a user sends a data packet to a mobile. The packet goes through the local area network (LAN) via a router out on the GPRS context for the mobile. If the mobile is in a GPRS idle state, the packet is rejected. If the mobile is in standby or active mode, the GGSN routes the packet in an encapsulated format to SGSN.

In the second case, the home GPRS network sends the data packet over the inter-operator backbone network to the visiting GPRS network. The visiting GPRS network routes the packet to the appropriate SGSN.

The PTP and PTM applications of GPRS are listed below:

- **Point-to-point**
  - Messaging (e.g., e-mail)
  - Remote access to corporate networks

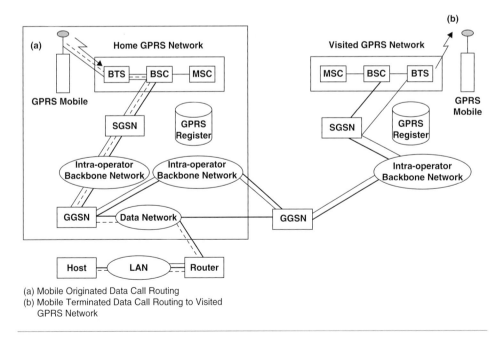

(a) Mobile Originated Data Call Routing
(b) Mobile Terminated Data Call Routing to Visited
    GPRS Network

Figure 15.7   Data call routing in GPRS network.

- Access to the Internet
- Credit card validation (point-of-sales)
- Utility meter readings
- Road toll applications
- Automatic train control

- **Point-to-multipoint**
  - PTM-multicast (send to all)
  - News
  - Traffic information
  - Weather forecasts
  - Financial updates
  - PTM-group call (send to some)
  - Taxi fleet management
  - Conferencing

GPRS provides a service for bursty and bulky data transfer, radio resources on demand, shared use of physical radio resources, existing GSM functionality, mobile applications for a mass application market, volume dependent charging, and integrated services, operation and management.

15.2.3 Enhanced Data Rates for GSM Enhancement

The enhanced data rates for GSM enhancements (EDGE) [8] provides an evolutionary path from existing 2G systems (GSM, IS-136, PDC) to deliver some 3G services in existing spectrum bands. The advantages of EDGE include fast availability, reuse of existing GSM, IS-136 and PDC infrastructure, as well as support for the gradual introduction of 3G capabilities.

EDGE is primarily a radio interface improvement, but it can also be viewed as a system concept that allows GSM, IS-136, and PDC networks to offer a set of new services. EDGE has been designed to improve S/I by using link quality control. Link quality control adapts the protection of the data to the channel quality so that for all channel qualities an optimal bit rate is achieved.

EDGE can be seen as a generic air interface for efficiently providing high bit rates, facilitating an evolution of existing 2G systems toward 3G systems. The EDGE air interface is designed to facilitate higher bit rates than those currently achievable in existing 2G systems. The modulation scheme based on 8-PSK is used to increase the gross bit rate. GMSK modulation as defined in GSM is also part of the EDGE system. The symbol rate is 271 ksps for both GMSK and 8-PSK, leading to gross bit rates per time slot of 22.8 kbps and 69.2 kbps, respectively. The 8-PSK pulse shape is linearized by GMSK to allow 8-PSK to fit into the GSM

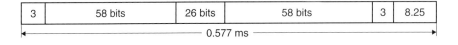

3	58 bits	26 bits	58 bits	3	8.25

‹————————————————— 0.577 ms ——————————————————›

Figure 15.8   Burst format for EDGE with 8-PSK.

spectrum mask. The 8-PSK burst format is similar to GSM (see Figure 15.8). EDGE reuses the GSM carrier bandwidth and time slot structure.

EDGE (also known as the 2.5G system) has been designed to enhance user bandwidth through GPRS. This is achieved through the use of higher-level modulation schemes. Although EDGE reuses the GSM carrier bandwidth and time slot structure, the technique is by no means restricted to GSM systems; it can be used as a generic air interface for efficient provision of higher bit rates in other TDMA systems. In the Universal Wireless Communications Consortium (UWCC) the 136 high speed (136 HS) radio transmission technology (RTT) radio interface was proposed as a means to satisfy the requirements for an IMT-2000 RTT. EDGE was adopted by UWCC in 1998 as the outdoor component of 136 HS to provide 384 kbps data service.

The standardization effort for EDGE has two phases. In the first phase of EDGE the emphasis has been placed on enhanced GPRS (EGPRS) and enhanced CSD (ECSD). The second phase is defined with improvements for multimedia and real-time services.

In order to achieve a higher gross rate, new modulation scheme, quaternary offset quadrature amplitude modulation (QOQAM) has been proposed for EDGE, since it can provide higher data rates and good spectral efficiency. An offset modulation scheme is proposed because it gives smaller amplitude variation than 16-QAM, which can be beneficial when using nonlinear amplifiers. EDGE co-exists with GSM in an existing frequency plan and provides link adaptation (modulation and coding are adapted for channel conditions).

Radio Protocol Design

The radio protocol strategy in EDGE is to reuse the protocols of GSM/GPRS whenever possible, thus minimizing the need for new protocol implementation. EDGE enhances both the GSM circuit-switched (HSCSD) and packet-switched (GPRS) mode operation. EDGE includes one packet-switched and one circuit-switched mode, EGPRS and ECSD, respectively.

Enhanced GPRS (EGPRS). The EDGE radio link control (RLC) protocol is somewhat different from the corresponding GPRS protocol. The main changes are related to improvements in link quality control scheme.

A *link adaptation* scheme regularly estimates the link quality and subsequently selects the most appropriate modulation and coding scheme for the transmission to maximize the user bit rate. The link adaptation scheme offers mechanisms for choosing the best modulation and coding scheme for the radio link. In GPRS only the coding schemes can be changed between two consecutive link layer control (LLC) frames. In the EGPRS even the modulation can be changed. Different coding and modulation schemes enable adjustment for the robustness of the transmission according to the environment.

Another way to handle link quality variations is *incremental redundancy*. In this scheme, information is first sent with very little coding, yielding a high bit rate if decoding is immediately successful. If decoding is not successful, additional coded bits (redundancy) are sent until decoding succeeds. The more coding that has to be sent, the lower the resulting bit rate and the higher the delay.

EGPRS supports combined link adaptation and incremental redundancy schemes. In this case, the initial code rate of the incremental redundancy scheme is based on measurements of the link quality. Benefits of this approach are the robustness and high throughput of the incremental redundancy operation in combination with lower delays and lower memory requirements enabled by the adaptive initial code rate.

In EGPRS the different initial code rates are obtained by puncturing a different number of bits from a common convolutional code $R = 1/3$. The resulting coding schemes are given in Table 15.1. Incremental redundancy operation is enabled by puncturing a different set of bits each time a block is retransmitted, whereby the code rate is gradually decreased toward 1/3 for every new transmission of the block. The selection of the initial modulation and code rate is based on regular measurements of link quality.

Table 15.1 Channel coding scheme in EDGE (PS transmission).

Coding scheme	Gross bit rate (kbps)	Code rate	Modulation	Radio interface rate per time-slot (kbps)	Radio interface rate on 8 time-slots (kbps)
CS-1	22.8	0.49	GMSK	11.2	
CS-2	22.8	0.63	GMSK	14.5	
CS-3	22.8	0.73	GMSK	16.7	
CS-4	22.8	1.0	GMSK	22.8	
PCS-1	69.2	0.329	8-PSK	22.8	182.4
PCS-2	69.2	0.496	8-PSK	34.3	274.4
PCS-3	69.2	0.596	8-PSK	41.25	330.0
PCS-4	69.2	0.746	8-PSK	51.60	412.8
PCS-5	69.2	0.829	8-PSK	57.35	458.8
PCS-6	69.2	1.000	8-PSK	69.20	553.6

Actual performance of modulation and coding scheme together with channel characteristics form the basis for link adaptation. Channel characteristics are needed to estimate the effects of a switch to another modulation and coding combination and include an estimated S/I ratio, but also time dispersion and fading characteristics (that affect the efficiency of interleaving).

EGPRS offers eight additional coding schemes. The EGPRS user has eight modulation and coding schemes available compared to four for GPRS. Besides changes in the physical layer, modifications in the protocol structure are also needed. The lower layers of the user data plane designed for GPRS are the physical, RLC MAC, and LLC layers. With EDGE functionality, the LLC layer will not require any modifications; however, the RLC/MAC layer has to be modified to accommodate features for efficient multiplexing and link adaptation procedures to support the basically new physical layer in the EDGE.

In the case of GSM, EDGE with the existing GSM radio bands offers wireless multimedia, IP-based applications at the rate of 384 kbps with a bit-rate of 48 kbps per time slot and, under good radio conditions, up to 69.2 kbps per time slot.

Enhanced CSD (ECSD). In this case, the objective is to keep the existing GSM CS data protocols as intact as possible. In order to provide higher data rates, multislot solutions as in ECSD are provided in EDGE. This has no impact on link or system performance.

A data frame is interleaved over 22 frames as in GSM, and three new 8-PSK channel coding schemes are defined along with the four already existing for GSM. The radio interface rate varies from 3.6 to 38.8 kbps per time slot (see Table 15.2).

Fast introduction of EGPRS/ECSD services is possible by reusing the existing transponder rate adapter unit (TRAU) formats and 16 kbps channel structure several times on the A_{bis} interface. Since data above 14.4 kbps cannot be rate adapted to fit into one 14.4 kbps TRAU frame, TRAU frames on several 16 kbps

Table 15.2 Channel coding scheme in EDGE (CS transmission).

Channel name	Code rate	Modulation	Radio interface rate/time-slot (kbps)
TCH/F2.4	0.16	GMSK	3.6
TCH/F4.8	0.26	GMSK	6.0
TCH/F9.6	0.53	GMSK	12.0
TCH/F14.4	0.64	GMSK	14.5
ECSD TCS-1 (NT + T)	0.42	8-PSK	29.0
ECSD TCS-2 (T)	0.46	8-PSK	32.0
ECSD TCS-3 (NT)	0.56	8-PSK	38.8

channels are used to meet the increased capacity requirement. In this case the BTS is required to handle a higher number of 16 kbps A_{bis} channels than time slots used on the radio interface. The benefit of using the current TRAU formats is that the introduction of new channel coding does not have any impact on the A_{bis} transmission, but it makes it possible to hide the new coding from the TRAU. On the other hand, some additional complexity is introduced in the BTS due to modified data frame handling.

Instead of reusing the current A_{bis} transmission formats for EDGE, new TRAU formats and rate adaptation optimized for increased capacity is specified. The physical layer can be dimensioned statically for the maximum user rate specified for particular EDGE service or more dynamic reservation of A_{bis} transmission resources can be applied. The A_{bis} resources can even be released and reserved dynamically during the call, if the link adaptation is applied.

The channel coding schemes defined for EDGE in PS transmission are listed in Table 15.1 and in CS transmission in Table 15.2.

Services Offered by EDGE

PS Services. The GPRS architecture provides IP connectivity from the mobile station to an external fixed IP network. For each service, a QoS profile is defined. The QoS parameters include priority, reliability, delay, and maximum and mean bit rate. A specified combination of these parameters defines a service, and different services can be selected to suit the needs of different applications.

CS Services. The current GSM standard supports both transparent (T) and nontransparent (NT) services. Eight transparent services are defined, offering constant bit rates in the range of 9.6 to 64 kbps.

A nontransparent service uses RLP to ensure virtually error-free data delivery. For this case, there are eight services offering maximum user bit rates from 4.8 to 57.6 kbps. The actual user bit rate may vary according to channel quality and the resulting rate of transmission.

The introduction of EDGE implies no change of service definitions. The bit rates are the same, but the way services are realized in terms of channel coding is different. For example, a 57.6 kbps nontransparent service can be realized with coding scheme ECSD TCS-1 (telephone control channel-1) and two time slots, while the same service requires four time slots with standard GSM using coding scheme TCH/F14.4. Thus, EDGE CS transmission makes the high bit rate services available with fewer time slots, which is advantageous from a terminal implementation perspective. Additionally, since each user needs fewer time slots, more users can be accepted which increases the capacity of the system.

Asymmetric Services Due to Terminal Implementation. ETSI has standardized two mobile classes: one that requires only GMSK transmission in the uplink and 8-PSK in the downlink and one that requires 8-PSK in both links. For the first class, the uplink bit rate is limited to that of GSM/GPRS, while the EDGE bit rate is still provided in the downlink. Since most services are expected to require higher bit rates in the downlink than in the uplink, this is a way of providing attractive services with a low complexity mobile station. Similarly, the number of time slots available in the uplink and downlink need not be the same. However, transparent services will be symmetrical.

EDGE Implementation

EDGE makes use of the existing GSM infrastructure in a highly efficient manner. Radio network planning will not be greatly affected since it will be possible to reuse many existing BTS sites. GPRS packet switching nodes will be unaffected, because they function independently of the user bit rates, and any modifications to the switching nodes will be limited to software upgrades. There is also a smooth evolution path defined for terminals to ensure that EDGE-capable terminals will be small and competitively priced.

EDGE-capable channels will be equally suitable for standard GSM services, and no special EDGE, GPRS, and GSM services will be needed. From an operator viewpoint this allows a seamless introduction of new EDGE services — perhaps starting with the deployment of EDGE in the service hot spots and gradually expanding coverage as demand dictates. The roll-out of EDGE-capable BSS hardware can become part of the ordinary expansion and capacity enhancement of the network. The wideband data capabilities offered by EDGE allows a step-by-step evolution to IMT-2000, probably through a staged deployment of the new 3G air interface on the existing core GSM network. Keeping GSM as the core network for the provision of 3G wireless services has additional commercial benefits. It protects the investment of existing operators; it helps to ensure the widest possible customer base from the outset; and it fosters supplier competition through the continuous evolution of systems.

GSM operators who win licenses in new 2 GHz bands will be able to introduce IMT-2000 wideband coverage in areas where early demand is likely to be greatest. Dual-mode EDGE/IMT-2000 mobile terminals will allow full roaming and handoff from one system to the other, with mapping of services between the two systems. EDGE will contribute to the commercial success of the 3G system in the vital early phases by ensuring that IMT-2000 subscribers will be able to enjoy roaming and interworking globally.

Building on an existing GSM infrastructure will be relatively fast and inexpensive, compared to establishing a total 3G system. The intermediate move to GPRS and later to EDGE will make the transition to 3G easier.

While GPRS and EDGE require new functionality in the GSM network, with new types of connections to external packet data networks they are essentially extensions of GSM. Moving to a GSM/IMT-2000 core network is likewise a further extension of this network.

Table 15.3 Comparison of GSM data services.

Service type	Data unit	Max. sustained user data rate	Technology	Resources used
Short message service (SMS)	Single 140 octet packet	9 bps	Simplex circuit	SDCCH or SACCH
Circuit-switched data	30 octet frames	9600 bps	Duplex circuits	TCH
HSCSD	192 octet frames	115 kbps	Duplex circuits	1-8 TCH
GPRS	1600 octet frames	171 kbps	Virtual circuit/ packet switching	PDCH (1-8 TCH)
EDGE		384 kbps	Virtual circuit/ packet switching	1-8 TCH

Note: SDCCH: Stand-alone Dedicated Control Channel; SACCH: Slow Associated Control Channel; TCH: Traffic Channel; PDCH: Packet Data Channel (all refer to GSM logical channels).

EDGE provides GSM operators—whether or not they get a new 3G license—with a commercially attractive solution to develop the market for wideband multimedia services. Familiar interfaces such as the Internet, volume-based charging, and a progressive increase in available user data rates will remove some of the barriers to large-scale take-up of wireless data services. The way forward to 3G services will be a staged evolution from today's GSM data services through GPRS and EDGE.

Increased user data rates over the radio interface requires redesign of physical transmission methods, frame formats, and signaling protocols in different network interfaces. The extent of the modification needed depends on the user data rate requirement, i.e., whether the support of higher data is required or merely a more efficient use of the radio time slot to support current data services is needed.

Several alternatives to cover the increased radio interface data rates on the A_{bis} interface for EGPRS and ECSD can be envisioned. The existing physical structure can be reused as much as possible or a new transmission method optimized for EDGE can be specified. Table 15.3 provides a comparison of GSM data services with GPRS and EDGE.

15.3 Third-Generation (3G) Wireless Systems

The International Telecommunication Union (ITU) began studies on the globalization of personal communications in 1986 and identified the long-term spectrum requirements for the future *third-generation* (3G) mobile wireless telecommunications systems. In 1992, the ITU identified 230 MHz of spectrum in the 2 GHz band to implement the IMT-2000 system on a worldwide basis for satellite and terrestrial components. The aim of IMT-2000 is to provide universal coverage enabling terminals to have seamless roaming across multiple networks. The ITU accepted the overall standardization responsibility of IMT-2000 to define radio interfaces that are applicable in different radio environments including indoor, outdoor, terrestrial, and satellite [23–28, 30–32].

Figure 15.9 provides an overview of the IMT family. *IMT-DS* is the direct spread (DS) technology and includes WCDMA systems. This technology is intended for UMTS terrestrial radio access (UTRA)-FDD and is used in Europe and Japan. *IMT-TC* family members are the UTRA-TDD system that uses time-division (TD) CDMA, and the Chinese TD-synchronous CDMA (TD-SCDMA). Both standards are combined and the third-generation partnership project (3GPP) is responsible for the development of the technology. *IMT-MC* includes multiple carrier (MC) cdma2000 technology, an evolution of the cdmaOne family. 3GPP2 is responsible for standardization. *IMT-SC* is the enhancement of the US TDMA systems. UWC-136 is a single carrier (SC) technology. This technology applies EDGE to enhance the 2G IS-136 standard. It is now integrated into the 3GPP efforts. *IMT-FT* is a frequency time (FT) technology. An enhanced version of the cordless telephone standard digital European cordless technology (DECT) has

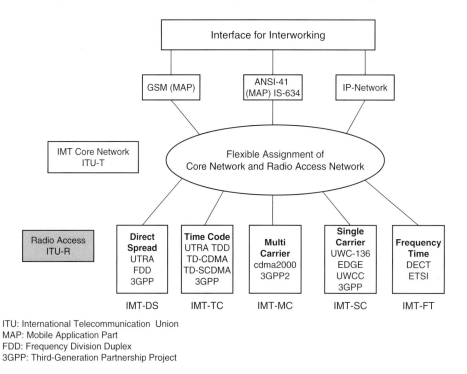

ITU: International Telecommunication Union
MAP: Mobile Application Part
FDD: Frequency Division Duplex
3GPP: Third-Generation Partnership Project

Figure 15.9   IMT family.

been selected for low mobility applications. The ETSI has the responsibility for standardization of DECT.

3G mobile telecommunications systems are intended to provide worldwide access and global roaming for a wide range of services. Standards bodies in Europe, Japan, and North America are trying to achieve harmony on key and interrelated issues including radio interfaces, system evolution and backward compatibility, user's migration and global roaming, and phased introduction of mobile services and capabilities to support terminal mobility. Universal Mobile Telecommunication System (UMTS) studies were carried out by ETSI in parallel with IMT-2000 to harmonize its efforts with ITU. In Japan and North America, standardization efforts for 3G were carried out by the Association of Radio Industries Business (ARIB) and the TIA committee TR45, respectively. Two partnership projects, 3GPP and 3GPP2, are involved in harmonizing 3G efforts in Europe, Japan, and North America.

In Europe, 3G systems are intended to support a substantially wider and enhanced range of services compared to the 2G (GSM) system. These enhancements include multimedia services, access to the Internet, high rate data, and so on. The enhanced services impose additional requirements on the fixed network functions to support mobility. These requirements are achieved through an evolution path to capitalize on the investments for the 2G system in Europe, Japan, and North America.

In North America, the 3G wireless telecommunication system, cdma2000 was proposed to ITU to meet most of the IMT requirements in the indoor office, indoor to outdoor pedestrian, and vehicular environment. In addition, the cdma2000 satisfies the requirements for 3G evolution of 2G TIA/EIA 95 family of standards (cdmaOne).

In Japan, evolution of the GSM platform is planned for the IMT (3G) core network due to its flexibility and widespread use around the world. Smooth migration from GSM to IMT-2000 is possible. The service area of the 3G system overlays with the existing 2G (PDC) system. The 3G system connects and interworks with 2G systems through an interworking function (IWF). An IMT-2000-PDC dual mode terminal as well as the IMT-2000 single mode terminal are deployed.

UMTS as discussed today and introduced in many countries is based on the initial release of UMTS standards referred to as release 99 or R99. This (release) is aimed at a cost-effective migration from GSM to UMTS. After R99 the release of 2000 or R00 followed. 3GPP decided to split R00 into two standards and call them release 4 (Rel-4) and release 5 (Rel-5). The version of all standards finalized for R99 is now referred to as Rel-3 by 3GPP. Rel-4 introduces QoS in the fixed network plus several execution environments (e.g., MExE, mobile execution environment) and new service architectures. Rel-4 was suspended in March 2001. Rel-5 specifies a new core network. The GSM/GPRS-based core network will be replaced by an almost all-IP core network. The content of Rel-5 was suspended in March 2002. This standard integrates IP-based multimedia services (IMS) controlled by the IETF's session initiation protocol (SIP). A high-speed downlink packet access (HSDPA) service with 8 to 10 Mbps was included. Also, a wideband 16 kHz adaptive multirate (AMR) codec was added for better quality. End-to-end QoS and several data compression techniques were added. 3GPP is currently working on Rel-6. This standard includes multiple input multiple output (MIMO) antennas, enhanced multimedia service (MMS), security enhancement, WLAN/WWAN interworking, broadcast/multicast services, enhanced IMS, IP emergency call, and many more management features.

A primary assumption for UMTS is that it is based on an evolved GSM core network. This provides backward compatibility with GSM in terms of network protocols and interfaces (MAP, ISUP (ISDN user part), etc.) The core network supports both GSM and UMTS/IMT-2000 services, including handoff and roaming between the two (see Figure 15.10). The proposed W-CDMA based UMTS terrestrial radio access network (UTRAN) is connected to the GSM-UMTS core network using a new multi-vendor interface (I_u). The transport protocol within the new radio network and the core network will be IP.

There is a clear separation between the services provided by UTRAN and the actual channels used to carry these services. All radio network functions (such as resource control) are handled within the radio access network and clearly separated from the service and subscription functions in the UMTS core network (UCN). The GSM-UMTS network, shown in Figure 15.11, consists of three main entities:

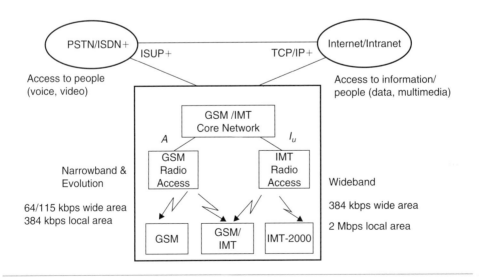

Figure 15.10   Evolution to UMTS/IMT-2000 in a GSM environment.

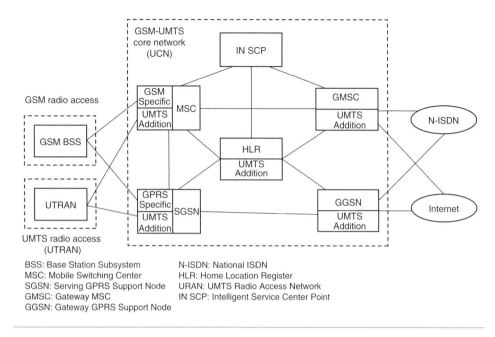

Figure 15.11 General GSM-UMTS network architecture.

- GSM-UMTS core network (UCN)
- UMTS terrestrial radio access network (UTRAN)
- GSM base station subsystem (BSS)

Like the GSM-GPRS core network, the GSM-UMTS core network has two different parts: a circuit-switched MSC and a packet-switched GRPS support node (GSN). The core network access point for GSM circuit-switched connections is the GSM MSC, and for packet-switched connections it is the SGSN.

GSM-defined services (up to and including GSM Phase 2+) are supported in the usual GSM manner. The GSM-UMTS core network implements supplementary services according to GSM principles (HLR-MSC/VLR). New services, beyond Phase 2+ are created using new service capabilities. These service capabilities may be seen as building blocks for application development and include:

- Bearers defined by QoS
- Mobile station execution environment (MExE)
- Telephony value-added services (TeleVAS)
- Subscriber identity module (SIM) toolkit
- Location services
- Open interfaces to mobile network functions

- Down-loadable application software
- Intelligent Network/Customized Applications for Mobile Enhanced Logic (IN/CAMEL) and service nodes.

In addition to new services provided by the GSM-UMTS network itself, many new services and applications will be realized using a client/server approach, with servers residing on service local area networks (LANs) outside the GSM-UMTS core network. For such services, the core network simply acts as a transparent bearer. This approach is in line with current standardization activities, and is important from a service continuity point of view. The core network will ultimately be used for the transfer of data between the end-points, the client and the server.

IN techniques are one way to provide seamless interworking across GSM-UMTS networks. CAMEL already provides the basis for GSM/IN interworking. The IN infrastructure may be shared by fixed and mobile networks, and can support fixed/mobile service integration, as needed by International Mobile Telecommunications (IMT). The inherent support for third-party service providers in IN means such providers could offer all or part of the integrated services. This role of IN is already apparent in services such as virtual private networks (VPNs), regional subscription and one number, which are available as network-independent and customer-driven services.

Service nodes and IN can play a complementary role. IN is suitable for subscription control and group services where high service penetration in a very wide area with frequent service invocation are more important than sophistication. Service nodes are better for providing differentiated user interfaces; e.g., personal call and messaging services that use advanced in-band processing and span several access networks.

To make the most of the new radio access network's capabilities, and to cater to the large increase in data traffic volume, asynchronous transmission mode (ATM) is used as the transport protocol within UTRAN and toward the GSM-UMTS core network. The combination of an ATM cell-based transport network, WCDMA's use of variable-rate speech coding with improved channel coding, and an increased volume of packet data traffic over the air interface will mean a saving of about 50% in transmission costs, compared with equivalent current solutions. ATM, with the newly standardized AAL2 adaptation layer, provides an efficient transport protocol, optimized for delay-sensitive speech services and packet data services. Statistical multiplexing in ATM provides maximum utilization of existing and new transmission infrastructure throughout the entire network.

In the complex multiservice, multivendor, multiprovider environment of 3G wireless services network management is a critical issue. The growth of packet data traffic requires new ways of charging for services and new billing systems to support them. There will continue to be a growing demand for better customer care and cost reductions in managing mobile networks, driven by the need to:

- Provide sophisticated personal communications services
- Expand the customer base beyond the business user base
- Separate the service provider and network operator roles
- Provide "one-stop" billing for a range of services

New operations and management functions are needed to support new services and network functionality. Standardization of interfaces is critical especially for alignment with current management interfaces in the GSM-UMTS core network. Management information needs to be part of standard traffic interfaces.

With the right service strategy and network planning, GSM operators will be able to capitalize on the wideband multimedia market through a staged evolution of their core networks, with the addition of new radio access technology as it becomes available.

15.4 UMTS Network Reference Architecture

A UMTS system can be divided into a set of domains and the reference points that interconnect them. Figures 15.12 and 15.13 show these domains and reference points.

A simplified mapping of functional entities to the domain model is shown in Figure 15.14. Note that this is a reference model and does not represent any physical

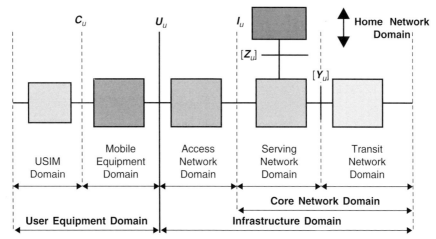

C_U: Reference Point between USIM and UE
I_U: Reference Point between Access and Serving Network Domain
U_U: Reference Point between User Equipment and Infrastructure Domains, UMTS Radio Interface
$[Y_U]$: Reference Point between Serving and Transit Network Domain
$[Z_U]$: Reference Point between Serving and Home Network

Figure 15.12 UMTS domains and reference points.

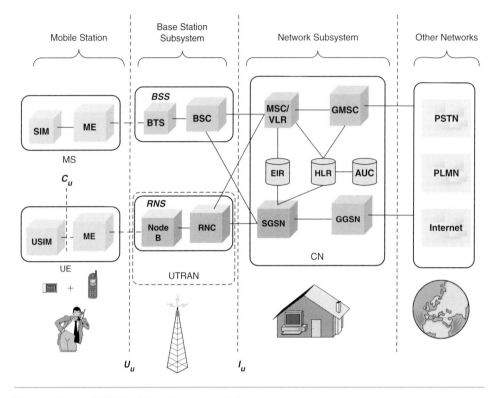

Figure 15.13   UMTS—3G reference architecture.

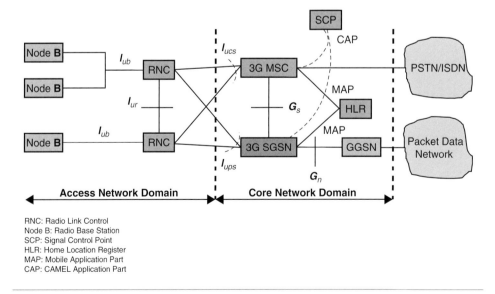

RNC: Radio Link Control
Node B: Radio Base Station
SCP: Signal Control Point
HLR: Home Location Register
MAP: Mobile Application Part
CAP: CAMEL Application Part

Figure 15.14   Simplified UMTS network reference model.

architecture. The I_u is split functionally into two logical interfaces, I_{ups} connecting the packet switched domain to the access network and the I_{ucs} connecting the circuit switched domain to the access network. The standards do not dictate that these are physically separate, but the user plane for each is different and the control plane may be different. The I_{ur} logically connects radio network controllers (RNCs) but could be physically realized by a direct connection between RNCs or via the core network.

15.5   Channel Structure in UMTS Terrestrial Radio Access Network

The UMTS terrestrial radio access network (UTRAN) [10–22] has an *access stratum* and *nonaccess stratum*. The access stratum includes air interface and provides functions related to OSI layer 1, layer 2, and the lower part of layer 3. The non-access stratum deals with communication between user equipment (UE) and core network (CN) and includes OSI layer 3 (upper part) to layer 7.

The radio interface, U_u, is the interface between UE and UTRAN. It consists of three protocol layers: physical layer, data link layer, and network layer (see Figure 15.15). The radio interface provides physical channels to carry data over the radio path and logical channels to carry a particular type of data. There are two types of logical channels: signaling and control, and traffic channel.

The *physical layer* in UTRAN performs the following functions:

- Forward error correction, bit-interleaving, and rate matching
- Signal measurements
- Micro-diversity distribution/combining and soft handoff execution
- Multiplexing/mapping of services on dedicated physical codes

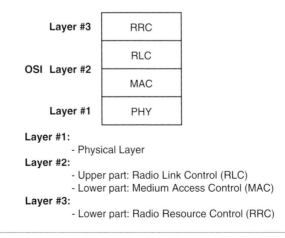

Figure 15.15   OSI layer model and air interface protocols.

- Modulation, spreading, demodulation, despreading of physical channels
- Frequency and time (chip, bit, slot, frame) synchronization
- Fast closed-loop power control
- Power weighting and combining of physical channels
- Radio frequency (RF) processing.

The *medium access control sublayer* is responsible for efficiently transferring data for both real-time (CS) and non-real-time (PS) services to the physical layer. MAC offers services to the radio link control (RLC) sublayer and higher layers. The MAC layer provides data transfer services on logical channels. MAC is responsible for

- Selection of appropriate transport format (basically bit rate) within a predefined set, per information unit delivered to the physical layer
- Service multiplexing on *random access channel (RACH)*, *forward access channel (FACH)*, and *dedicated channel (DCH)*
- Priority handling between data flow of a user as well as between data flows from several users
- Access control on RACH and FACH
- Contention resolution on RACH

Radio link control (RLC) sets up a logical link over the radio interface and is responsible for fulfilling QoS requirements. RLC responsibilities include:

- Segmentation and assembly of the packet data unit
- Transfer of user data
- Error correction through retransmission
- Sequence integrity
- Duplication information detection
- Flow control of data

The *radio resource control (RRC)* layer broadcasts system information, handles radio resources (i.e., code allocation, handover, admission control, and measurement/control report), and controls the requested QoS. The RRC layer offers the following services to the core network:

- General control (GC) service used as an information broadcast service
- Notification (Nt) service used for paging and notification of a selected UE
- Dedicated control (DC) service used to establish/release a connections and transfer messages

The channels in UTRAN are physical, transport, and logical (see Figure 15.16).

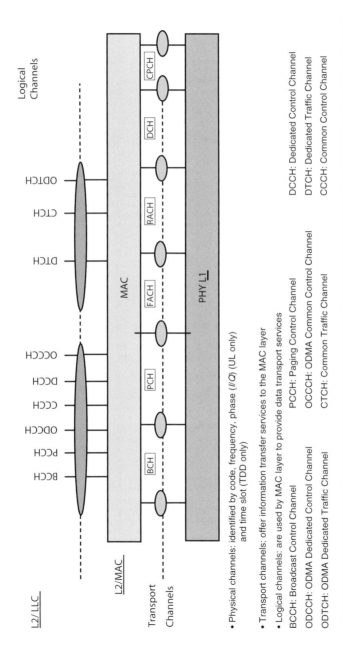

- Physical channels: identified by code, frequency, phase (I/Q) (UL only) and time slot (TDD only)

- Transport channels: offer information transfer services to the MAC layer

- Logical channels: are used by MAC layer to provide data transport services

BCCH: Broadcast Control Channel	PCCH: Paging Control Channel	DCCH: Dedicated Control Channel
ODCCH: ODMA Dedicated Control Channel	OCCCH: ODMA Common Control Channel	DTCH: Dedicated Traffic Channel
ODTCH: ODMA Dedicated Traffic Channel	CTCH: Common Traffic Channel	CCCH: Common Control Channel

Figure 15.16   UTRAN channels.

The functions of *logical control channels* and *logical traffic channels* in UTRAN are listed in Tables 15.4 and 15.5.

In UTRAN *transport channels* can be either common (i.e., shared between users) or dedicated channels. They offer information transfer services to the MAC sublayer. Dedicated transport channels are *dedicated channel (DCH)* (*up link*, *UL* and *down link*, *DL*), *fast uplink signaling channel (FAUSCH)*, and ODMA

Table 15.4 Logical control channel in UTRAN.

Channel	Function
Broadcast control channel (BCCH)	DL channel for broadcasting system and control information
Paging control channel (PCCH)	DL channel to transfer page information, used when: (1) network does not know the location of cell of the mobile, and (2) mobile is in cell connected state (using sleep mode)
Common control channel (CCCH)	Bidirectional channel to transfer control information between network and mobile, it is used: (1) by mobile without RRC connection with the network, and (2) by mobile using common transport channel to access a new cell after cell resection
Dedicated control channel (DCCH)	Point-to-point bidirectional channel to transmit dedicated information between a mobile and network. The channel is established through RRC connection setup procedure
ODMA common control channel (OCCCH)	Bidirectional channel to transmit control information between mobiles
ODMA dedicated control channel (ODCCH)	Point-to-multipoint bidirectional channel to transmit dedicated control information between mobiles. This channel is established through RRC connection setup procedure

Table 15.5 Logical traffic channels in UTRAN.

Channel	Function
Dedicated traffic channel (DTCH)	Point-to-point, dedicated to one mobile to transfer user information. A DTCH can exist in both UL and DL.
ODMA traffic channel (ODTCH)	Point-to-point channel dedicated to one mobile to transfer user information between mobiles. An ODTCH can exist in relay link. A point-to-multipoint unidirectional channel to transfer dedicated user information for all or a group of specified mobiles.

(*opportunity driven multiple access*) *dedicated channel* (ODCH). The common DL transport channels are listed in Table 15.6.

The mapping between logical and transport channels is given below (see Figure 15.17):

- BCCH is connected to BCH
- PCCH is connected to PCH
- CCCH is connected to RACH and FACH

Table 15.6 UTRAN common transport channels.

Channel	Function
Broadcast channel (BCH)	DL channel used to broadcast system- and cell-specific information, transmitted over the entire cell with low fixed bit rate
Forward access channel (FACH)	DL channel transmitted over the entire or only a part of cell using beam-forming antennas, uses slow power control
Paging channel (PCH)	DL channel transmitted over the entire cell, transmission of PCH is associated with the transmission of a physical layer signal, the paging indicator, to support efficient sleep mode procedure
Random access channel (RACH)	UL channel received over the entire cell, characterized by a limited size data field, a collision risk, and by use of open loop power control
Common packet channel (CPCH)	UL channel, contention-based random access channel used for transmission of bursty data traffic, associated with a DCH on DL, which provides power control for the UL CPCH
Downlink shared channel (DSCH)	DL channel shared by several mobiles, associated with a DCH

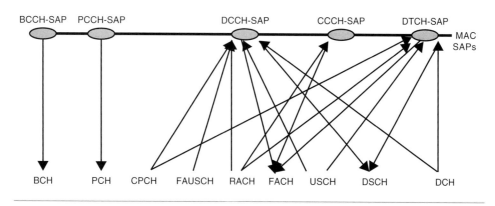

Figure 15.17   Mapping between logical and transport channels in UTRAN.

- DTCH can be connected to either RACH and FACH and DSCH, to DCH and DSCH, to a DCH, to a CPCH (FDD mode only)
- CTCH can be connected to DSCH, FACH, or BCH
- DCCH can be connected to either RACH and FACH, to RACH and DSCH, to DCH and DSCH, to a DCH, a CPCH (TDD mode only), to FAUSCH, CPCH (FDD mode only).

In UTRAN, the basic physical resource is a *physical channel* identified by code and frequency. Physical channels consist of radio frames and time slots (see Figure 15.18). The length of a radio frame is 10 ms and one frame consists of 15 time slots. For DL channels two codes are used, one to identify the cell and the other to identify a particular channel within that cell. For UL a long code is used to identify the channel. The UL channel uses different data streams transmitted on the I and Q branch. A physical channel corresponds to a specific carrier frequency, code(s), and on UL a relative phase (0, $\pi/2$).

The UL dedicated physical channel (DPCH) is a user dedicated, point-to-point channel between UE and node B. These channels carry dedicated channels at various rates up to 2 Mbps. The UL-dedicated physical data channels are I/Q (i.e., DPDCH on I-branch and DPCCH on Q-branch) and code multiplexed. There are two types of DPCH: (1) *dedicated physical data channel (DPDCH)* to carry user data and signaling information generated at layer 2 (there may be none, one, or several DPDCHs); and (2) *dedicated physical control channel (DPCCH)* to carry control information generated at layer 1 (pilot bits, transmit power control (TPC) commands, feedback information (FBI) commands, and optional transport format combination indicator (TFCI)). For each layer 1 connection, there is only one UL DPCCH. DPCCH rate and power remain constant. The UL dedicated physical channel carries 10×2^k ($k = 0, 1, \ldots, 6$) bits per slot and may have a spreading factor (SF) from 256 and 4.

The UL common physical channels are *physical random access channel (PRACH)* used to carry the RACH and fast uplink signaling channel (FAUSCH) and *physical common packet channel (PCPCH)* to carry CPCH.

The DL dedicated physical data channel is time multiplexed. The DL dedicated physical channel carries 10×2^k ($k = 0, 1, \ldots, 7$) bits per slot and may have spreading factor (SF = $512/2^k$) from 512 to 4 (see Figure 15.18).

The DL common physical channels include the following channels:

- *Primary common control physical channel (PCCPCH)* carries BCH, rate 30 kbps, SF = 256; continuous transmission; no power control
- *Secondary common control physical channel (SCCPCH)* carries FACH and PCH, transmitted when data is available; SF range is from 256 to 4
- *Synchronization channel (SCH)* is used for cell search. It consists of two sub-channels: primary SCH transmits a modulated code of 256 chips once every slot, and secondary SCH transmits repeatedly 15 codes of 256 chips.

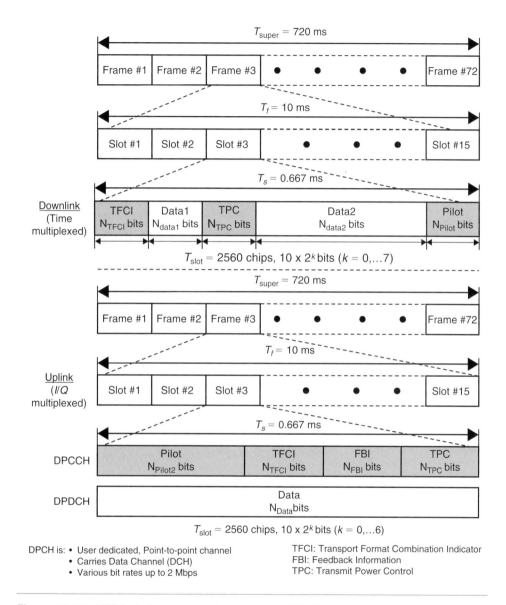

Figure 15.18   UTRA dedicated physical channels (DPCHs) framing.

- *Physical downlink shared channel* (PDSCH) carries DSCH; shared by users based on code multiplexing; associated with DPCH.
- *Acquisition indicator channel* (AICH) carries acquisition indicators.
- *Page indicator channel* (PICH) carries a page for UE; fixed rate, SF = 256

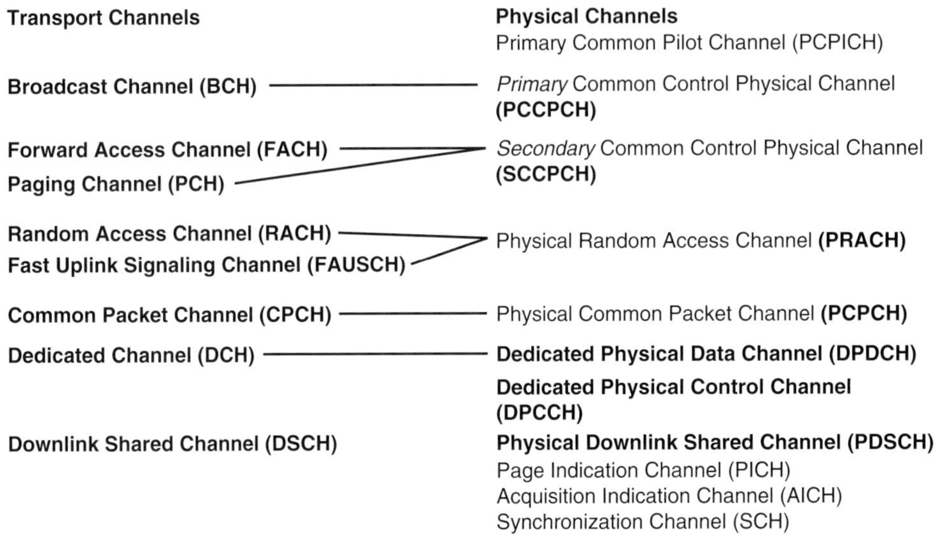

Transport Channels

Physical Channels
Primary Common Pilot Channel (PCPICH)

Broadcast Channel (BCH) ———————— *Primary* Common Control Physical Channel
(PCCPCH)

Forward Access Channel (FACH) ————— *Secondary* Common Control Physical Channel
Paging Channel (PCH) ———— **(SCCPCH)**

Random Access Channel (RACH) ———— Physical Random Access Channel **(PRACH)**
Fast Uplink Signaling Channel (FAUSCH) ———

Common Packet Channel (CPCH) ———————— Physical Common Packet Channel **(PCPCH)**

Dedicated Channel (DCH) ———————— **Dedicated Physical Data Channel (DPDCH)**

**Dedicated Physical Control Channel
(DPCCH)**

Downlink Shared Channel (DSCH)   **Physical Downlink Shared Channel (PDSCH)**
Page Indication Channel (PICH)
Acquisition Indication Channel (AICH)
Synchronization Channel (SCH)

Figure 15.19   Mapping between transport and physical channels in UTRA.

- *Primary common pilot channel* (*PCPICH*) is used as phase reference for SCH, PCCPCH, AICH, PICH, and default phase reference for all other DL physical channels, one per cell and broadcast over entire cell.
- *Secondary common pilot channel* (*SCPICH*) is a continuous channel with the same spreading and scrambling codes, transmitted on different antennas in case of DL transmit diversity; SF = 256. It may be transmitted over only part of the cell.

The mapping between transport and physical channels is shown in Figure 15.19.

15.6   Spreading and Scrambling in UMTS

In a WCDMA system, isolation between users in the downlink is accomplished through the combination of user-specific channelization codes and cell-specific scrambling codes. Channelization codes are generated recursively to form a binary tree structure (see Figure 15.20). Spreading is used in combination with scrambling. Scrambling is used on top of spreading to separate mobile terminals or cells from each other. Scrambling does not change the chip rate nor the bandwidth. Transmissions from a single source are separated by channelization codes based on an orthogonal variable spreading factor (OVSF) technique (see Appendix D) to allow spreading to be changed while maintaining orthogonality between codes.

When a code is intended to be used, no other code generated from the intended code can be used; no code between the intended code and the root code can be used. These restrictions apply to individual sources; they do not apply to different cells in downlink or different mobile stations in the uplink. Table 15.7 lists the characteristics of synchronization, channelization, and scrambling codes used in a WCDMA system.

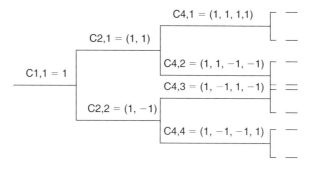

Figure 15.20   Channelization code tree.

Table 15.7 Synchronization, channelization, and scrambling codes of a WCDMA system.

	Synchronization codes	Channelization codes	Scrambling codes, uplink	Scrambling codes, downlink
Type	Gold codes, primary synchronization code (PSC) and secondary synchronization code (SSC)	OVSF, sometimes called Walsh codes	Complex-valued gold code segments (long) or complex-valued S(2) codes (short)	Complex-valued gold code segments. Pseudo noise (PN) codes
Length	256 chips	4–512 chips	38,400 chips/256 chips	38,400 chips
Duration	66.67 μs	1.04 μs–133.34 μs	10 ms/66.67 μs	10 ms
Number of codes	1 primary code/16 secondary codes	Spreading factor: UL: 4–256 DL: 4–512	16,777,216	512 primary/ 15 secondary for each primary code
Spreading	No, does not change bandwidth	Yes, increases bandwidth	No, does not change bandwidth	No, does not change bandwidth
Usage	To enable terminal to locate and synchronize to the cells' main control channels	UL: To separate physical data and control data from same terminal DL: To separate connection to different terminals in a cell	Separation of terminals	Separation of cells

15.7   UMTS Terrestrial Radio Access Network Overview

The UTRAN consists of a set of radio network subsystems (RNSs) (see Figure 15.21). The RNS has two main logical elements: Node B and an RNC. The RNS is responsible for the radio resources and transmission/reception in a set of cells. A cell (sector) is one coverage area served by a broadcast channel.

An RNC is responsible for the use and allocation of all the radio resources of the RNS to which it belongs. The RNC also handles the user voice and packet data traffic, performing the actions on the user data streams that are necessary to access the radio bearers. The responsibilities of an RNC are:

- Intra UTRAN handover
- Macro diversity combining/splitting of I_{ub} data streams
- Frame synchronization
- Radio resource management
- Outer loop power control
- I_u interface user plane setup
- Serving RNS (SRNS) relocation
- Radio resource allocation (allocation of codes, etc.)
- Frame selection/distribution function necessary for soft handover (functions of UMTS radio interface physical layer)
- UMTS radio link control (RLC) sublayers function execution
- Termination of MAC, RLC, and RRC protocols for transport channels, i.e., DCH, DSCH, RACH, FACH
- I_{ub}'s user plane protocols termination

A Node B is responsible for radio transmission and reception in one or more cells to/from the user equipment (UE). The logical architecture for Node B is shown in Figure 15.22.

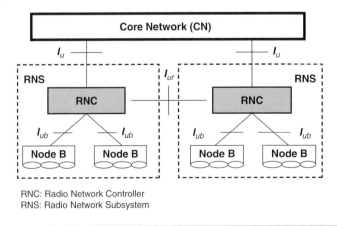

RNC: Radio Network Controller
RNS: Radio Network Subsystem

Figure 15.21   UTRAN logical architecture.

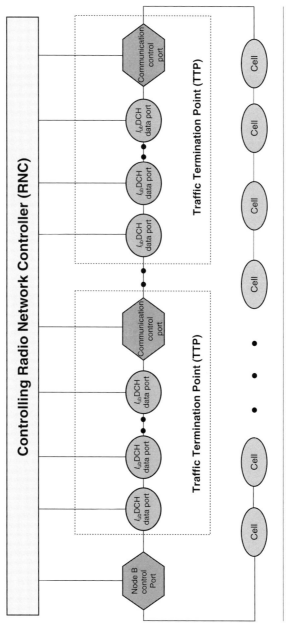

Figure 15.22   Node B logical architecture.

The following are the responsibilities of the Node B:

- Termination of I_{ub} interface from RNC
- Termination of MAC protocol for transport channels RACH, FACH
- Termination of MAC, RLC, and RRC protocols for transport channels: BCH, PCH
- Radio environment survey (BER estimate, receiving signal strength, etc.)
- Inner loop power control
- Open loop power control
- Radio channel coding/decoding
- Macro diversity combining/splitting of data streams from its cells (sectors)
- Termination of U_u interface from UE
- Error detection on transport channels and indication to higher layers
- FEC encoding/decoding and interleaving/deinterleaving of transport channels
- Multiplexing of transport channels and demultiplexing of coded composite transport channels
- Power weighting and combining of physical channels
- Modulation and spreading/demodulation and despreading of physical channels
- Frequency and time (chip, bit, slot, frame) synchronization
- RF processing

15.7.1 UTRAN Logical Interfaces

In UTRAN protocol structure is designed so that layers and planes are logically independent of each other and, if required, parts of protocol structure can be changed in the future without affecting other parts.

The protocol structure contains two main layers, the radio network layer (RNL) and the transport network layer (TNL). In the RNL, all UTRAN-related functions are visible, whereas the TNL deals with transport technology selected to be used for UTRAN but without any UTRAN-specific changes. A general protocol model for UTRAN interfaces is shown in Figure 15.23.

The control plane is used for all UMTS-specific control signaling. It includes the application protocol (i.e., radio access network application part (RANAP) in I_u, radio network subsystem application part (RNSAP) in I_{ur} and node B application part (NBAP) in I_{ub}). The application protocol is used for setting up bearers to the UE. In the three-plane structure the bearer parameters in the application protocol are not directly related to the user plane technology, but rather they are general bearer parameters.

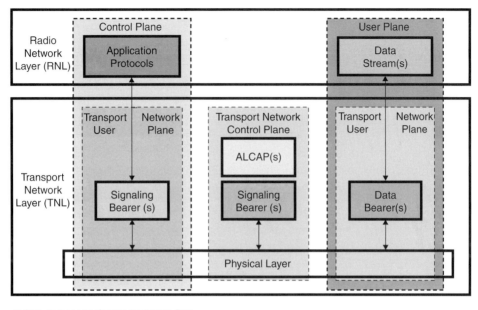

ALCAP: Access Link Control Application Part

Figure 15.23 General protocol model for UTRAN interfaces.

User information is carried by the user plane. The user plane includes data stream(s), and data bearer(s) for data stream(s). Each data stream is characterized by one or more frame protocols specified for that interface.

The transport network control plane carries all control signaling within the transport layer. It does not include radio network layer information. It contains *access link control application part (ALCAP)* required to set up the transport bearers (data bearers) for the user plane. It also includes the signaling bearer needed for the ALCAP. The transport plane lies between the control plane and the user plane. The addition of the transport plane in UTRAN allows the application protocol in the radio network control plane to be totally independent of the technology selected for the data bearer in the user plane.

With the transport network control plane, the transport bearers for data bearers in the user plane are set up in the following way. There is a signaling transaction by application protocol in the control plane that initiates set-up of the data bearer by the ALCAP protocol specific for the user plane technology. The independence of the control plane and user plane assumes that an ALCAP signaling occurs. The ALCAP may not be used for all types of data bearers. If there is no ALCAP signaling transaction, the transport network control plane is not required. This situation occurs when preconfigured data bearers are used. Also, the ALCAP protocols in the transport network control plane are not used to set up the signaling bearer for the application protocol or the ALCAP during real-time operation.

I_u Interface

The UMTS I_u interface is the open logical interface that interconnects one UTRAN to the UMTS core network (UCN). On the UTRAN side the I_u interface is terminated at the RNC, and at the UCN side it is terminated at U-MSC. The I_u interface consists of three different protocol planes—the *radio network control plane* (RNCP), the *transport network control plane* (TNCP), and the *user plane* (UP).

The RNCP performs the following functions:

- It carries information for the general control of UTRAN radio network operations.
- It carries information for control of UTRAN in the context of each specific call.
- It carries user call control (CC) and mobility management (MM) signaling messages.

The control plane serves two service domains in the core network, the packet-switched (PS) domain and circuit-switched (CS) domain. The CS domain supports circuit-switched services. Some examples of CS services are voice and fax. The CS domain can also provide intelligent services such as voice mail and free phone. The CS domain connects to PSTN/ISDN networks. The CS domain is expected to evolve from the existing 2G GSM PLMN.

The PS domain deals with PS services. Some examples of PS services are Internet access and multimedia services. Since Internet connectivity is provided, all services currently available on the Internet such as search engines and e-mail are available to mobile users. The PS domain connects to IP networks. The PS domain is expected to evolve from the GPRS PLMN.

The I_u circuit-switched and packet-switched protocol architecture are shown in Figures 15.24 and 15.25.

The control plane protocol stack consists of RANAP on the top of signaling system 7 (SS7) protocols. The protocol layers are the *signaling connection control part (SCCP)*, the *message transfer part (MTP3-B)*, and *signaling asynchronous transfer mode (ATM) adaptation layer for network-to-network interface (SAAL-NNI)*. The SAAL-NNI is divided into *service-specific coordination function (SSCF)*, the *service-specific connection-oriented protocol (SSCOP)*,

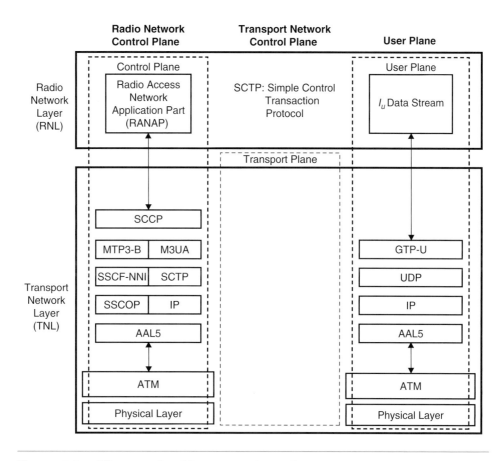

Figure 15.24 PS protocol architecture on I_u interface.

and *ATM adaptation layer 5 (AAL5)* layers. The SSCF and SSCOP layers are specifically designed for signaling transport in ATM networks, and take care of signaling connection management functions. AAL5 is used for segmenting the data to ATM cells.

As an alternative, an IP-based signaling bearer is specified for the I_u PS control plane. The IP-based signaling bearer consists of *SS7-MTP3—user adaptation layer (M3UA)*, *simple control transmission protocol (SCTP)*, IP, and AAL5. The SCTP layer is specifically designed for signaling transport on the Internet.

The *transport network control plane (TNCP)* carries information for the control of transport network used within UCN.

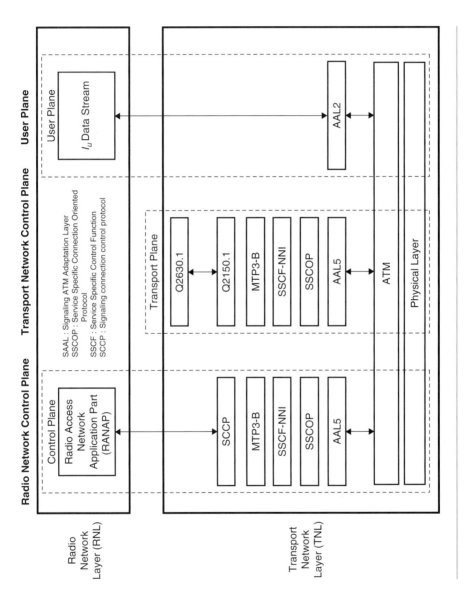

Figure 15.25   CS protocol architecture on I_u interface.

The *user plane* (*UP*) carries user voice and packet data information. AAL2 is used for the following services: narrowband speech (e.g., EFR, AMR); unrestricted digital information service (up to 64 kbps, i.e., ISDN B channel); any low to average bit rate CS service (e.g., modem service to/from PSTN/ISDN). AAL5 is used for the following services: non-real-time PS data service (i.e., best effort packet access) and real-time PS data.

I_{ur} Interface

The connection between two RNCs (serving RNC (SRNC) and drift RNC (DRNC)) is the I_{ur} interface. It is used in soft handoff scenarios when different macro diversity streams of one communication are supported by Node Bs that belong to different RNCs. Communication between one RNC and one Node B of two different RNCs are realized through the I_{ur} interface. Three different protocol planes are defined for it:

- Radio network control plane (RNCP)
- Transport network control plane (TNCP)
- User plane (UP)

The I_{ur} interface is used to carry:

- Information for the control of radio resources in the context of specific service request of one mobile on RNCP
- Information for the control of the transport network used within UTRAN on TNCP
- User voice and packet data information on UP

The protocols used on this interface are:

- Radio access network application part (RANAP)
- DCH frame protocol (DCHFP)
- RACH frame protocol (RACHFP)
- FACH frame protocol (FACHFP)
- Access link control application part (ALCAP)
- Q.aal2
- Signaling connection control part (SCCP)
- Message transfer part 3-B (MTP3-B)
- Signaling ATM adaptation layer for network-to-network interface (SAAL-NNI) (SAAL-NNI is further divided into service specific coordination function for network to network interface (SSCF-NNI), service specific connection oriented protocol (SSCOP), and ATM adaptation layer 5 (AAL5))

The bearer is AAL2. The protocol structure of the I_{ur} interface is shown in Figure 15.26.

Initially, this interface was designed to support the inter-RNC soft handoff, but more features were added during the development of the standard. The I_{ur} provides the following four functions:

1. Basic inter-RNC mobility support

- Support of SRNC relocation
- Support of inter-RNC cell and UTRAN registration area update
- Support of inter-RNC packet paging
- Reporting of protocol errors

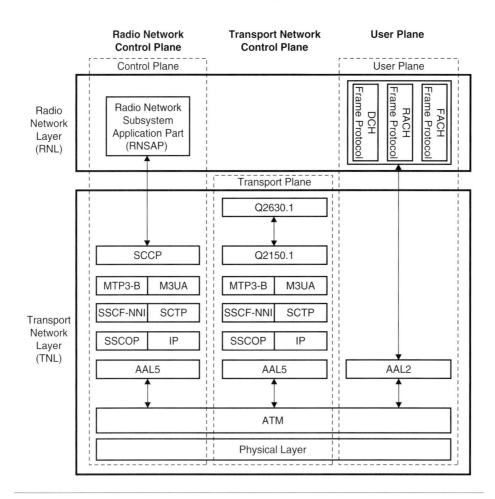

Figure 15.26 Protocol structure of I_{ur} interface.

2. Dedicated channel traffic support

- Establishment, modification, and release of a dedicated channel in the DRNC due to hard and soft handoff in the dedicated channel state
- Setup and release of dedicated transport connections across the I_{ur} interface
- Transfer of DCH transport blocks between SRNC and DRNC
- Management of radio links in the DRNS via dedicated measurement report procedures and power setting procedures

3. Common channel traffic support

- Setup and release of the transport connection across the I_{ur} for common channel data streams
- Splitting of the MAC layer between the SRNC (MAC-d) and DRNC (MAC-c and MAC-sh); the scheduling for downlink data transmission is performed in the DRNC
- Flow control between the MAC-d and MAC-c/MAC-sh

4. Global resource management support

- Transfer of cell measurements between two RNCs
- Transfer of Node B timing between two RNCs

I_{ub} Interface

The connection between the RNC and Node B is the I_{ub} interface. There is one I_{ub} interface for each Node B. The I_{ub} interface is used for all of the communications between Node B and the RNC of the same RNS. Three different protocol planes are defined for it.

- Radio network control plane (RNCP)
- Transport network control plane (TNCP)
- User plane (UP)

  The I_{ub} interface is used to carry:

- Information for the general control of Node B for radio network operation on RNCP
- Information for the control of radio resources in the context of specific service request of one mobile on RNCP
- Information for the control of a transport network used within UTRAN on TCNP
- User CC and MM signaling message on RNCP
- User voice and packet data information on UP

The protocols used on this interface include:

- Node B application part protocol (NBAP)
- DCH frame protocol (DCHFP)
- RACH frame protocol (RACHFP)
- FACH frame protocol (FACHFP)
- Access link control application part (ALCAP)
- Q.aal2
- SSCP or TCP and IP
- MTP3-B
- SAAL-UNI (SSCF-UNI, SSCOP, and AAL5)

When using multiple low-speed links in the I_{ub} interface, Node B supports inverse multiplexing for ATM (IMA).

The bearer is AAL2. The protocol structure for the interface I_{ub} is shown in Figure 15.27.

U_u Interface

The UMTS U_u interface is the radio interface between a Node B and one of its UE. The U_u is the interface through which UE accesses the fixed part of the system.

15.7.2   Distribution of UTRAN Functions

Located in the RNC

- Radio resource control (L3 Function)
- Radio link control (RLC)
- Macro diversity combining
- Active cell set modification
- Assign transport format combination set (centralized data base function)
- Multiplexing/demultiplexing of higher layer PDUs into/from transport block delivered to/from the physical layer on shared dedicated transport channels (used for soft handover)
- L1 function: macro diversity distribution/combining (centralized multipoint termination)
- Selection of the appropriate transport format for each transport channel depending upon the instantaneous source rate—collocate with RRC
- Priority handling between data flows of one user

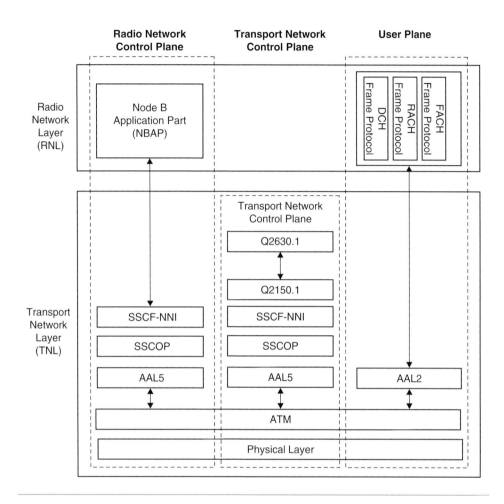

Figure 15.27   Protocol structure of I_{ub} interface.

Located in Node B

- Scheduling of broadcast, paging, and notification messages; location in Node B—to reduce data repetition over I_{ub} and reduce RNC CPU load and memory space

- Collision resolution on RACH (in Node B—to reduce nonconstructive traffic over I_{ub} interface and reduce round trip delay)

- Multiplexing/demultiplexing of higher layer PDUs to/from transport blocks delivered to/from the physical layer on common transport channels

15.8   UMTS Core Network Architecture

Figure 15.28 shows the UMTS core network (UCN) in relation to all other entities within the UMTS network and all of the interfaces to the associated networks.

The UCN consists of a CS entity for providing voice and CS data services and a PS entity for providing packet-based services. The logical architecture offers a clear separation between the CS domain and PS domain. The CS domain contains the functional entities: mobile switching center (MSC) and gateway MSC (GMSC) (see Figure 15.28). The PS domain comprises the functional entities: serving GPRS support node (SGSN), gateway GPRS support node (GGSN), domain name server (DNS), dynamic host configuration protocol (DHCP) server, packet charging gateway, and firewalls. The core network can be split into the following different functional areas:

- Functional entities needed to support PS services (e.g. 3G-SGSN, 3G-GGSN)
- Functional entities needed to support CS services (e.g. 3G-MSC/VLR)
- Functional entities common to both types of services (e.g. 3G-HLR)

Other areas that can be considered part of the core network include:

- Network management systems (billing and provisioning, service management, element management, etc.)
- IN system (service control point (SCP), service signaling point (SSP), etc.)
- ATM/SDH/IP switch/transport infrastructure

Figure 15.29 shows all the entities that connect to the core network—UTRAN, PSTN, the Internet and the logical connections between terminal equipment (MS, UE), and the PSTN/Internet.

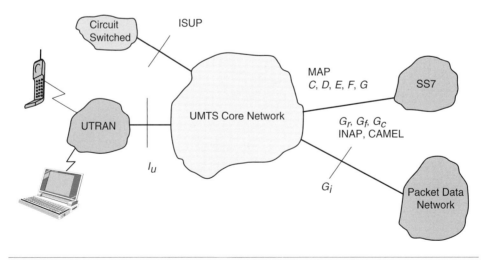

Figure 15.28   UMTS core network architecture.

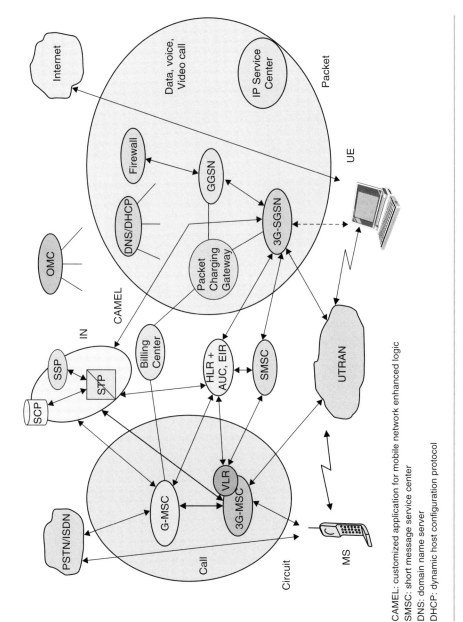

CAMEL: customized application for mobile network enhanced logic
SMSC: short message service center
DNS: domain name server
DHCP: dynamic host configuration protocol

Figure 15.29 Logical architecture of the UMTS core network.

15.8.1 3G-MSC

The 3G-MSC is the main CN element to provide CS services. The 3G-MSC also provides the necessary control and corresponding signaling interfaces including SS7, MAP, ISUP (ISDN user part), etc. The 3G MSC provides the interconnection to external networks like PSTN and ISDN. The following functionality is provided by the 3G-MSC:

- Mobility management: Handles attach, authentication, updates to the HLR, SRNS relocation, and intersystems handover.
- Call management: Handles call set-up messages from/to the UE.
- Supplementary services: Handles call-related supplementary services such as call waiting, etc.
- CS data services: The IWF provides rate adaptation and message translation for circuit mode data services, such as fax.
- Vocoding
- SS7, MAP and RANAP interfaces: The 3G-MSC is able to complete originating or terminating calls in the network in interaction with other entities of a mobile network, e.g., HLR, AUC (Authentication center). It also controls/communicates with RNC using RANAP which may use the services of SS7.
- ATM/AAL2 Connection to UTRAN for transportation of user plane traffic across the I_u interface. Higher rate CS data rates may be supported using a different adaptation layer.
- Short message services (SMS): This functionality allows the user to send and receive SMS data to and from the SMS-GMSC/SMS-IWMSC (Inter working MSC).
- VLR functionality: The VLR is a database that may be located within the 3G-MSC and can serve as intermediate storage for subscriber data in order to support subscriber mobility.
- IN and CAMEL.
- OAM (operation, administration, and maintenance) agent functionality.

15.8.2 3G-SGSN

The 3G-SGSN is the main CN element for PS services. The 3G-SGSN provides the necessary control functionality both toward the UE and the 3G-GGSN. It also provides the appropriate signaling and data interfaces including connection to an IP-based network toward the 3G-GGSN, SS7 toward the HLR/EIR/AUC and TCP/IP or SS7 toward the UTRAN.

The 3G-SGSN provides the following functions:

- Session management: Handles session set-up messages from/to the UE and the GGSN and operates Admission Control and QoS mechanisms.

- I_u and G_n MAP interface: The 3G-SGSN is able to complete originating or terminating sessions in the network by interaction with other entities of a mobile network, e.g., GGSN, HLR, AUC. It also controls/communicates with UTRAN using RANAP.
- ATM/AAL5 physical connection to the UTRAN for transportation of user data plane traffic across the I_u interface using GPRS tunneling protocol (GTP).
- Connection across the G_n interface toward the GGSN for transportation of user plane traffic using GTP. Note that no physical transport layer is defined for this interface.
- SMS: This functionality allows the user to send and receive SMS data to and from the SMS-GMSC /SMS-IWMSC.
- Mobility management: Handles attach, authentication, updates to the HLR and SRNS relocation, and intersystem handover.
- Subscriber database functionality: This database (similar to the VLR) is located within the 3G-SGSN and serves as intermediate storage for subscriber data to support subscriber mobility.
- Charging: The SGSN collects charging information related to radio network usage by the user.
- OAM agent functionality.

15.8.3 3G-GGSN
The GGSN provides interworking with the external PS network. It is connected with SGSN via an IP-based network. The GGSN may optionally support an SS7 interface with the HLR to handle mobile terminated packet sessions.

The 3G-GGSN provides the following functions:

- Maintain information locations at SGSN level (macro-mobility)
- Gateway between UMTS packet network and external data networks (e.g. IP, X.25)
- Gateway-specific access methods to intranet (e.g. PPP termination)
- Initiate mobile terminate Route Mobile Terminated packets
- User data screening/security can include subscription based, user controlled, or network controlled screening.
- User level address allocation: The GGSN may have to allocate (depending on subscription) a dynamic address to the UE upon PDP context activation. This functionality may be carried out by use of the DHCP function.
- Charging: The GGSN collects charging information related to external data network usage by the user.
- OAM functionality

15.8.4   SMS-GMSC/SMS-IWMSC

The overall requirement for these two nodes is to handle the SMS from point to point. The functionality required can be split into two parts. The SMS-GMSC is an MSC capable of receiving a terminated short message from a service center, interrogating an HLR for routing information and SMS information, and delivering the short message to the SGSN of the recipient UE. The SMS-GMSC provides the following functions:

- Reception of short message packet data unit (PDU)
- Interrogation of HLR for routing information
- Forwarding of the short message PDU to the MSC or SGSN using the routing information

The SMS-IWMSC is an MSC capable of receiving an originating short message from within the PLMN and submitting it to the recipient service center. The SMS-IWMSC provides the following functions:

- Reception of the short message PDU from either the 3G-SGSN or 3G-MSC
- Establishing a link with the addressed service center
- Transferring the short message PDU to the service center

Note: The service center is a function that is responsible for relaying, storing, and forwarding a short message. The service center is not part of UCN, although the MSC and the service center may be integrated.

15.8.5   Firewall

This entity is used to protect the service providers' backbone data networks from attack from external packet data networks. The security of the backbone data network can be ensured by applying packet filtering mechanisms based on access control lists or any other methods deemed suitable.

15.8.6   DNS/DHCP

The DNS server is used, as in any IP network, to translate host names into IP addresses, i.e., logical names are handled instead of raw IP addresses. Also, the DNS server is used to translate the access point name (APN) into the GGSN IP address. It may optionally be used to allow the UE to use logical names instead of physical IP addresses.

A dynamic host configuration protocol server is used to manage the allocation of IP configuration information by automatically assigning IP addresses to systems configured to use DHCP.

15.9 Adaptive Multi-Rate Codec for UMTS

The adaptive multi-rate (AMR) codec is the new codec that will be used in UMTS and GSM. The AMR codec has eight source rates; 4.75, 5.15, 5.90, 6.70 (PDC-EFR), 7.40 (IS-641), 7.95 (VSELP), 10.2 and 12.2 kbps (GSM-EFR). The AMR codec rates are controlled by a radio access network and do not depend on speech activity as in the cdma2000. During high cell loading, such as during busy hours, the AMR codec uses lower bit rates to offer higher capacity while providing slightly lower speech quality. Also, if a mobile is running out of the cell coverage area and using maximum transmission power, a lower AMR bit rate is used to extend the cell coverage area. With the AMR speech codec, it is possible to achieve a trade-off between capacity, coverage, and speech quality as per the service provider's requirements. The AMR speech codec is capable of switching the bit rate at every 20-ms frame upon command.

The AMR codec operates on speech frames of 20 ms corresponding to 160 samples at the sampling rate of 8000 samples per second. The AMR uses algebraic code excited linear prediction (ACELP) coding. At every 160 speech samples, the speech signal is analyzed to determine the parameters of the CELP model (i.e., LP filter coefficients, adaptive and fixed codebooks' indices and gains). The speech parameter bits delivered by the speech encoder are rearranged according to their subjective importance before sending them to the network. The rearranged bits are further classified into class Ia, Ib, and II bits based on their sensitivity to errors. Class Ia is the most sensitive; the strongest channel coding is applied to class Ia bits. The AMR codec uses the following functions to provide discontinuous transmission (DTx) activity:

- Voice activity detector (VAD) on the transmission (Tx) side
- Background acoustic noise evaluation on the Tx side to transmit characteristic parameters to the receive (Rx) side
- Comfort noise information transmission to the Rx side through regular use of the silence descriptor (STD) frame
- Comfort noise generation on the Rx side during periods when no normal speech frames are received

DTx prolongs terminal battery life and reduces the average required bit rate, providing a lower interference and an increase in system capacity. The AMR also deals with error concealment. The purpose of frame substitution is to conceal the effect of the lost speech frames. The purpose of muting the output in the case of several lost frames is to indicate the breakdown of the channel to the user and to avoid generating possibly annoying sounds as a result of the frame substitution procedure. The AMR speech codec can tolerate about 1% frame error rate (FER) (about a bit error rate of 10^{-4}) of class Ia bits without any deterioration of speech quality. For class Ib and class II bits, a higher FER is allowed.

15.10   UMTS Bearer Service

Network services are end-to-end, i.e., from terminal equipment (TE) to another TE. An end-to-end service has a certain QoS provided to the user by the network. To realize a certain network QoS, a bearer service with well-defined characteristics and functionality must be set up from the source to the destination of the service. UMTS bearer service layered architecture is shown in Figure 15.30. Each bearer service on layer N offers its service by using the services provided by (N − 1) layers below.

The traffic passes through different services of the network on its way from one piece of TE to another TE (see Figure 15.31). The end-to-end service used by the TE is realized by using a combination of TE/MT local bearer service, a UMTS bearer service and an external bearer service. We focus only on the bearer service that provides the UMTS QoS.

The UMTS bearer service has two parts: the *radio access bearer* (*RAB*) service and the *core network bearer* (*CNB*) service. Both these services are aimed at optimizing the UMTS bearer service over the respective wireless network

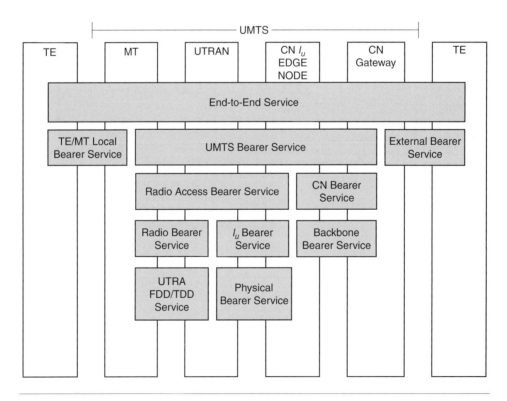

Figure 15.30   UMTS bearer service layered architecture.

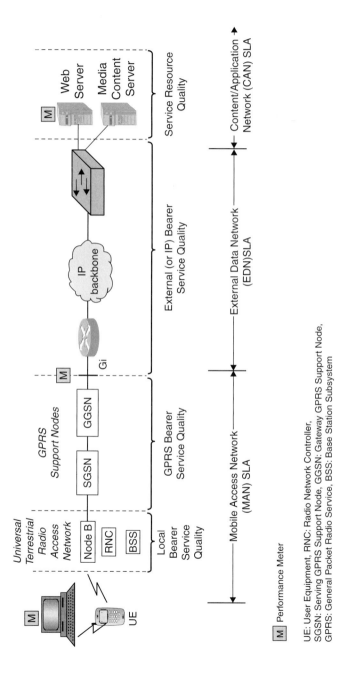

Figure 15.31 End-to-end UMTS bearer service.

M Performance Meter

UE: User Equipment, RNC: Radio Network Controller,
SGSN: Serving GPRS Support Node, GGSN: Gateway GPRS Support Node,
GPRS: General Packet Radio Service, BSS: Base Station Subsystem

topology by taking into consideration aspects such as mobility and mobility subscriber profiles.

The RAB service provides a secured transport of signaling and user data between the MT and CN I_u edge node with QoS adequate to the negotiated UMTS bearer service with default QoS for signaling. The RAB service is based on the characteristics of the radio interface and is maintained for a moving mobile terminal (MT). This service is realized by a radio bearer service and an I_u bearer service.

To support unequal error protection, UTRAN and MT have the ability to segment/reassemble the user flows into different subflows requested by the RAB service. The segmentation/reassemble is provided by the service data unit (SDU) payload format signaled at the RAB establishment. The radio bearer service handles the part of the user flow belonging to one subflow, according to the reliability requirements for that subflow. The I_u bearer service for packet traffic provides a different bearer service for a variety of QoS.

The core network bearer service of the CN connects the CN I_u edge node with the CN gateway to the external network. The role of this service is to effectively control and use the backbone network to provide the contracted UMTS bearer service. The UMTS packet CN supports different backbone bearer services for different QoS.

The CN bearer service uses a generic network service. The backbone network service covers the layer 1/layer 2 functionality and is selected according to the service provider's choice to satisfy the QoS requirements of the CN bearer service.

15.11 QoS Management

UMTS allows a user/application to negotiate bearer parameters that are most appropriate to carry the information. During an active session, it is possible to change bearer properties using a bearer renegotiation procedure. Bearer negotiation is initiated by an application, whereas renegotiation may be initiated either by the application or by the network. An application-initiated renegotiation is similar to a negotiation procedure that occurs in the bearer establishment phase. The application requests a bearer depending on its needs, the network checks the available resources and the user's service class subscription to respond. The user either accepts or rejects the offer. The bearer class, bearer parameters, and parameter values are directly related to an application as well as to the networks that lie between the sender and the receiver.

15.11.1 Functions for UMTS Bearer Service in the Control Plane

Control plane QoS management functions support the establishment and modification of a UMTS bearer service by signaling/negotiation with the UMTS external services and by the establishment or modification of all UMTS internal services with the required characteristics. The control plane QoS management functions include:

- The service manager coordinates the functions of the control plane for establishing, modifying, and maintaining the service. It provides all user-plane QoS management functions with the relevant attributes.
- The translation function performs a conversion between UMTS bearer service attributes and QoS parameters of the external networks service control protocol.
- Admission/capability control maintains information about all available resources of a network entity and about all resources allocated to UMTS bearer services. It determines for each UMTS bearer service request whether the required resources can be provided by this entity. It reserves the resources if allocated to the UMTS bearer service.
- Subscription control checks the administrative rights of the UMTS bearer service user to use the requested service with the specified QoS attributes.

15.11.2 Functions for UMTS Bearer Service in the User Plane

The user plane QoS management function maintains the signaling and user data traffic within certain limits, defined by specific QoS attributes. UMTS bearer service with different QoS values are supported by the QoS management functions. These functions ensure the provision of the QoS negotiated for a UMTS bearer service and include:

- The mapping function provides each data unit with the specific marking required to receive the intended QoS at the transfer by a bearer service.
- The classification function assigns data units to the established service of an MT according to the related QoS attributes if the MT has multiple UMTS bearer services established. The appropriate UMTS bearer service is derived from the data unit header or from traffic characteristics of data.
- The resource manager distributes the available resources between all services sharing the same resources. The resource manager distributes the resources according to the required QoS.
- The traffic conditioner provides conformance between the negotiated QoS for a service and the data unit traffic. Policing or traffic shaping performs traffic conditioning. The policing function compares the data unit traffic with the related QoS attributes. The traffic shape forms the data unit traffic according to the QoS of the service.

15.12   Quality of Service in UMTS

The QoS in the UMTS network must take into account the restrictions and limitations of the air interface. In CDMA systems, all users use the same bandwidth at the same time and, therefore, users interfere with one another. Because of the propagation mechanism, the signal received by the base station from a user close to the base station will be stronger than the signal received from another user located at the cell boundary. Hence, the close users will dominate the distant users. This situation becomes much worse if the close user happens to be a high-speed data user. To achieve considerable capacity and QoS, all signals should arrive at the base station with the same mean power, irrespective of distance. A solution to this problem is precise power control, which attempts to achieve constant mean power for each user. The UTRA/WCDMA air interface uses power control on both uplink and downlink at 1500 Hz.

15.12.1   QoS Classes

The QoS mechanisms for the wireless network must be robust and capable of providing reasonable QoS resolution. UMTS defines four different QoS classes. These are *conversational class*, *streaming class*, *interactive class*, and *background class*. The main distinguishing factor between these classes is how delay sensitive the traffic is. The conversational class is meant for traffic, which is very delay sensitive, whereas the background class is the most delay insensitive traffic class.

The conversational and streaming classes are mostly used to carry real-time traffic flows. The main difference between them is based on how delay sensitive the traffic is. The conversational real-time services, such as video telephony and speech are the most delay sensitive applications. They require data streams to be carried in conversational class.

The interactive class and background class are mainly used for traditional applications such as www, e-mail, Telnet, FTP, and news. Due to less stringent delay requirements compared to the conversational and streaming classes, both classes provide a better error rate by means of channel coding and retransmission. The main difference between the interactive and background class is that the interactive class is mainly used for interactive applications (e.g., e-mail or interactive web browsing), and the background class is meant for background traffic (e.g., background of e-mails or to background the downloading). Separating interactive and background applications ensure response for the interactive applications and background applications. Traffic in the interactive class has a higher priority in scheduling than the background class traffic. The background applications use transmission resources only when interactive applications do not need them.

15.12.2   QoS Attributes

The defined UMTS bearer attributes ranges and radio access bearer attributes ranges and their relationship for each bearer class is summarized in Tables 15.8 and 15.9.

Table 15.8 UMTS bearer attributes for each bearer class.

Traffic class	Conversational class	Streaming class	Interactive class	Background class
Maximum bit rate (kbps)	<2000	<2000	<2000 overhead	<2000 overhead
Delivery order	Yes/No	Yes/No	Yes/No	Yes/No
Maximum SDU size (octets)	<1500	<1500	<1500	<1500
SDU format information	(1)	(1)	(1)	(1)
SDU error ratio	10^{-2}, 10^{-3}, 10^{-4} 10^{-5}	10^{-2}, 10^{-3}, 10^{-4} 10^{-5}	10^{-3}, 10^{-4} 10^{-5}	10^{-3}, 10^{-4} 10^{-5}
Residual bit error ratio	$5*10^{-2}$, 10^{-2}, 10^{-3}, 10^{-4}	10^{-2}, 10^{-3}, 10^{-4}, 10^{-5}	$4*10^{-3}$, 10^{-5}, $6*10^{-8}$	$4*10^{-3}$, 10^{-5}, $6*10^{-8}$
Delivery of erroneous SDUs	Yes/No	Yes/No	Yes/No	Yes/No
Transfer delay (ms)	100, maximum	500, maximum		
Guaranteed bit rate (kbps)	<2000	<2000		
Traffic handling priority			TBD	

Table 15.9 Radio access bearer attributes for each bearer class.

Traffic class	Conversational class	Streaming class	Interactive class	Background class
Maximum bit rate (kbps)	<2000	<2000	<2000 overhead	<2000 overhead
Delivery order	Yes/No	Yes/No	Yes/No	Yes/No
Maximum SDU size (octets)	<1500	<1500	<1500	<1500
SDU format information	(1)	(1)	(1)	(1)
SDU error ratio	10^{-2}, 10^{-3}, 10^{-4} 10^{-5}	10^{-2}, 10^{-3}, 10^{-4} 10^{-5}	10^{-3}, 10^{-4} 10^{-5}	10^{-3}, 10^{-4} 10^{-5}
Residual bit error ratio	$5*10^{-2}$, 10^{-2}, 10^{-3}, 10^{-4}	$5*10^{-2}$, 10^{-2}, 10^{-3}, 10^{-4}, 10^{-5}	$4*10^{-3}$, 10^{-5}, $6*10^{-8}$	$4*10^{-3}$, 10^{-5}, $6*10^{-8}$
Delivery of erroneous SDUs	Yes/No	Yes/No	Yes/No	Yes/No

(Continued)

Table 15.9 Radio access bearer attributes for each bearer class.   (*continued*)

Traffic class	Conversational class	Streaming class	Interactive class	Background class
Transfer delay (ms)	80, maximum	500, maximum		
Guaranteed bit rate (kbps)	<2000	<2000		
Traffic handling priority			TBD	

Note: (1) Definition of possible values of exact SDU sizes for which UTRAN can support transparent RLC protocol mode.

15.13   High-Speed Downlink Packet Access (HSDPA)

In third-generation partnership project (3GPP) standards, Release 4 specifications provide efficient IP support enabling provision of services through an all-IP core network (see Figures 15.32 and 15.33). Release 5 specifications focus on HSDPA to provide data rates up to approximately 8–10 Mbps to support packet-based multimedia services. Multi input and multi output (MIMO) systems are the work item in Release 6 specifications, which will support even higher data transmission rates of up to 20 Mbps. HSDPA is evolved from and backward compatible with Release 99 WCDMA systems.

HSDPA is based on the same set of technologies as high data rate (HDR) to improve spectral efficiency for data services—such as shared downlink packet data channel and high peak data rates—using high-order modulation and adaptive modulation and coding, hybrid ARQ (HARQ) retransmission schemes, fast scheduling and shorter frame sizes.

HSDPA marks a similar boost for WCDMA that EDGE does for GSM. It provides a two-fold increase in air interface capacity and a five-fold increase in data speeds in the downlink direction. HSDPA also shortens the round-trip time between the network and terminals and reduces variance in downlink transmission delay. The improvements in performance are achieved by:

- Bringing some key functions, such as scheduling of data packet transmission and processing of retransmissions (in case of transmission errors) into the base station—that is, closer to the air interface.
- Using a short frame length to further accelerate packet scheduling for transmission.
- Employing incremental redundancy for minimizing the air-interface load caused by retransmissions.

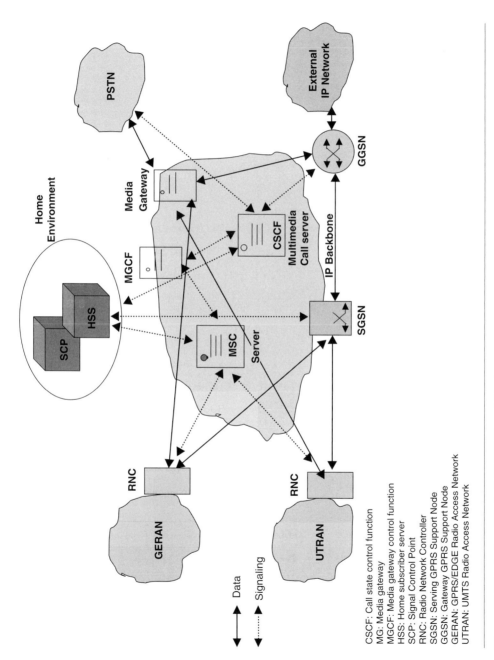

CSCF: Call state control function
MG: Media gateway
MGCF: Media gateway control function
HSS: Home subscriber server
SCP: Signal Control Point
RNC: Radio Network Controller
SGSN: Serving GPRS Support Node
GGSN: Gateway GPRS Support Node
GERAN: GPRS/EDGE Radio Access Network
UTRAN: UMTS Radio Access Network

Figure 15.32   A simplified all-IP UMTS architecture.

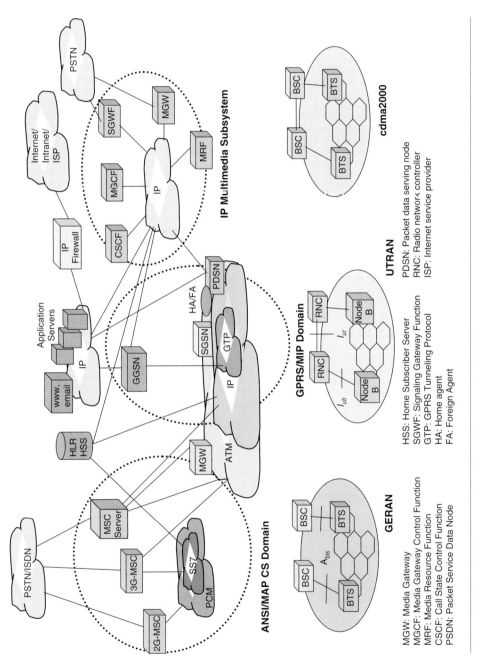

Figure 15.33   All-IP core network architecture for UMTS.

- Adopting a new transport channel type, known as *high-speed downlink shared channel* (HS-DSCH) to facilitate air interface channel sharing between several users.
- Adapting the modulation and coding scheme according to the quality of the radio link

The primary objective behind HSDPA is to provide a cost-effective, high-bandwidth, low-delay, packet-oriented service within UMTS. Backward compatibility is critical, so the HSDPA architecture adheres to an evolutionary philosophy. From an architectural perspective, HSDPA is a straightforward enhancement of the UMTS Release '99 (R99) architecture, with the addition of a repetition/scheduling entity within the Node B that resides below the R99 media-access control (MAC) layer. From a cellular-network perspective, all R99 techniques can be supported in a network supporting HSDPA, since HSDPA mobile terminals (UEs) are designed to coexist with R99 UEs. HSDPA is particularly suited to extremely asymmetrical data services, which require significantly higher data rates for the transmission from the network to the UE, than they do for the transmission from the UE to the network.

HSDPA introduces enablers for the high-speed transmission at the physical layer like the use of a shorter transmission time interval (TTI) (2 ms), and the use of adaptive modulation and coding. HS-DPCCH is used to carry the acknowledgment signals to Node B for each block. It is also used to indicate channel quality (CQI) used for adaptive modulation and coding. HS-DSCH uses 2 ms TTI to reduce trip time, to increase the granularity in the scheduling process, and to track the time varying radio channel better.

The basic operational principles behind HSDPA are relatively simple. The RNC routes data packets destined for a particular UE to the appropriate Node B. Node B takes the data packets and schedules their transmission to the mobile terminal over the air interface by matching the user's priority and estimated channel operating environment with an appropriately chosen coding and modulation scheme (that is, 16-QAM vs. QPSK).

The UE is responsible for acknowledging receipt of the data packet and providing Node B with information regarding channel condition, power control, and so on. Once it sends the data packet to the UE, Node B waits for an acknowledgment. If it does not receive one within a prescribed time, it assumes that the data packet was lost and retransmits it.

HSDPA continuously strives, with some modest constraints, to give the maximal bandwidth to the user with the best channel conditions. The data rates achievable with HSDPA are more than adequate for supporting multimedia-streaming services (refer to Table 15.10).

Although conceptually simple, HSDPA's implementation within the context of a Node B does raise some architectural issues for the designer. In a typical

Table 15.10 HSDPA data rates.

		Chip rate = 3.84 Mcps, frame size = 3 slots		
Modulation	Coding rate	Throughput with 5 codes	Throughput with 10 codes	Throughput with 15 codes
16-QAM	1/2	2.4 Mbps	4.8 Mbps	7.2 Mbps
16-QAM	3/4	3.6 Mbps	7.2 Mbps	10.8 Mbps
16-QAM	4/4	4.8 Mbps	9.6 Mbps	14.4 Mbps
QPSK	1/4	600 kbps	1.2 Mbps	1.8 Mbps
QPSK	1/2	1.2 Mbps	2.4 Mbps	3.6 Mbps
QPSK	3/4	1.8 Mbps	3.6 Mbps	5.4 Mbps

network deployment, the Node B radio cabinet sits in proximity to the radio tower and the power cabinet. For indoor deployments the radio cabinet may be a simple rack, while in outdoor deployments it may be an environmental-control unit. The guts of the radio cabinet are an antenna interface section (filters, power amplifiers, and the like), core processing chassis (RF transceivers, combiner, high-performance channel cards, network interface and system controller card, timing card, back-plane, and so on), plus mechatronics (power supply, fans, cables, etc.) and other miscellaneous elements. The core processing chassis is the cornerstone of Node B and bears most of the cost. It contains the RF transceiver, combiner, network interface and system controller, timing card, channel card and back-plane. Of the core processing chassis elements, only the channel card needs to be modified to support HSDPA.

The typical UMTS channel card comprises a general-purpose processor that handles the miscellaneous control tasks, a pool of digital signal processor (DSP) resources to handle symbol-rate processing and chip-rate assist functions, and a pool of specialized ASIC (application specific intergrated circuit) devices to handle intensive chip-rate operations such as spreading, scrambling, modulation, rake receiving, and preamble detection.

To support HSDPA, two changes must be made to the channel card. First, the downlink chip-rate ASIC must be modified to support the new 16-QAM modulation schemes and new downlink slot formats associated with HSDPA. In addition, the downlink symbol-rate processing section must be modified to support HSDPA extensions.

The next change requires a new processing section, called the MAC-h$_s$, which must be added to the channel card to support the scheduling, buffering, transmission, and retransmission of data blocks that are received from the RNC. This is the most intrusive augmentation to the channel card because it requires the introduction of a programmable processing entity together with a retransmission buffer.

Since the channel card already contains both a general-purpose processor and a DSP, one can make convincing arguments that the MAC-h_s could be effectively realized using either of the two types of devices. Nonetheless, many designers are finding that, because of the close ties between the MAC-h_s function and the lower-layer symbol and chip-rate functions, the DSP is the more practical choice. Simulations have shown that a retransmission buffer of approximately 2.5 Mbits in size is adequate to handle the buffering requirement of a standard cell with 75 or so users.

The new channels introduced in HSDPA are *high-speed downlink shared channel (HS-DSCH)*, *high-speed shared control channel (HS-SCCH)*, and *high-speed dedicated physical control channel (HS-DPCCH)*. The HS-DSCH is the primary radio bearer. Its resources can be shared among all users in a particular sector. The primary channel multiplexing occurs in a time domain, where each TTI consists of three time slots (each 2 ms). TTI is also referred to as a sub-frame. Within each 2 ms TTI, a constant spreading factor (SF) of 16 is used for code multiplexing, with a maximum of 15 parallel codes allocated to HS-DSCH. Codes may all be assigned to one user, or may be split across several users. The number of codes allocated to each user depends on cell loading, QoS requirements, and UE code capabilities (5, 10, or 15 codes).

The HS-SCCH (a fixed rate 960 kbps, SF = 128) is used to carry downlink signaling between Node B and UE before the beginning of each scheduled TTI. It includes UE identity, HARQ-related information and the parameters of the HS-DSCH transport format selected by the link-adaptation mechanism. Multiple HS-SCCHs can be configured in each sector to support parallel HS-DSCH transmissions. A UE can be allocated a set of up to four HS-SCCHs, which need to be monitored continuously.

The HS-DPCCH (SF = 256) carries ACK/NACK signaling to indicate whether the corresponding downlink transmission was successfully decoded, as well as a channel quality indicator (CQI) to be used for the purpose of link adaptation. The CQI is based on a common pilot channel (CPICH) and is used to estimate the transport block size, modulation type, and number of channelization codes that can be supported at a given reliability level in downlink transmission. The feedback cycle of CQI can be set as a network parameter in predefined steps of 2 ms.

UE capabilities include the maximum number of HS-DSCHs supported simultaneously (5, 10, or 15), minimum TTI time (minimum time between the beginning of two consecutive transmissions to the UE), the maximum number of HS-DSCH transport block (TB) bits received within an HS-DSCH TTI, the maximum number of soft channel bits over all HARQ and supported modulations (QPSK only or both QPSK and 16-QAM). Table 15.11 gives UE categories.

Table 15.11 HSDPA UE categories.

Category	Codes	Inter-TTI	TB size (bits)	Total soft bits	Modulation	Data rate (Mbps)
1	5	3	7300	19,200	QPSK/QAM	1.2
2	5	3	7300	28,800	QPSK/QAM	1.2
3	5	2	7300	28,800	QPSK/QAM	1.8
4	5	2	7300	38,400	QPSK/QAM	1.8
5	5	1	7300	57,600	QPSK/QAM	3.6
6	5	1	7300	67,200	QPSK/QAM	3.6
7	10	1	14,600	115,200	QPSK/QAM	7.2
8	10	1	14,600	134,400	QPSK/QAM	7.2
9	15	1	20,432	172,800	QPSK/QAM	10.2
10	15	1	28,776	172,800	QPSK/QAM	14.4
11	5	2	3650	14,400	QPSK	0.9
12	5	1	3650		QPSK	1.8

15.14   Freedom of Mobile multimedia Access (FOMA)

The NTT DoCoMo's FOMA (Freedom Of Mobile multimedia Access) service in Japan, offers a revolutionary third-generation (3G) mobile communications platform. The advanced new service provides voice transmission quality on a par with fixed-line communications, with minimal interference and noise, and supports diverse multimedia content. Based on the WCDMA system, FOMA is providing the dramatic evolution of i-mode, other web-connection services, and innovative data-rich applications. FOMA supports full-motion video image transmission, music and game distribution, and other high-speed, large-capacity data communications. The service also offers roaming in various countries around the world.

With packet data transfer speeds of 64 to 384 kbps, FOMA service supports a variety of applications including Internet access, e-mail, file transfer, remote login, and Internet phone applications. Also available are ISDN-type services that function on a streamed basis, effectively using n $\times$ 64 kbps channels and supporting applications such as telephony, video-conferencing, and Group 4 fax. High-quality voice services at up to 12.2 kbps are also supported.

NTT DoCoMo's 3G network utilizes ATM technology that manages packet switching and circuit switching on the same network node, enabling use of various traffic types for true multimedia connectivity. Also supported is the use of asymmetric communications with differing characteristics for the uplink and downlink, possible because ATM can handle these resources separately. Moreover, ATM makes possible flexible connection services and both point-to-point and point-to-multipoint connections.

Through 3G technologies, the IN provides quicker response to customer needs. Development costs are also more easily controlled and lowered, and global roaming (virtual home environment) becomes a practical possibility.

FOMA's ability to smoothly and flexibly handle a wide range of high-quality content, including sound, still images, and video images, owes much to two key technologies employed in WCDMA:

- Spread spectrum technology transmits radio signals over a wider frequency band than that used by conventional mobile phone systems.
- Multirate technology selects the most suitable communication speed and transmission channel for data based on its type and size.

In addition to introducing sophisticated new terminal functions and further expanding its service area, NTT DoCoMo's future plans include the reduction of terminal weight to less than 100 g and the extension of terminal battery life to more than 300 hours. Dual terminals compatible with both GSM and FOMA networks will be launched, with full-scale global roaming service. This dual network service will allow subscribers to use their mobile phones all over the world, and use of a variety of video-based services in areas supporting WCDMA. To accelerate the expansion of the FOMA user base in Japan, with a priority on indoor coverage in public areas and population centers, NTT DoCoMo aims to extend coverage to 99% of the populated areas of Japan.

Within the next several years, the FOMA system will allow anyone with a cellular phone to retrieve vital business and personal information from anywhere around the world. Company personnel will be able to conduct video conferencing with branch offices and business partners, and send/receive customer and sales data while on the move.

Various models also allow the transmission and display of still and moving images, while e-mail with attached images and movie files, video distribution, and videophone capabilities are also made possible through advanced multimedia technologies such as MPEG4. All models offer the ability to download software through a highly advanced and secure implementation of Java technology, all offer access to the popular i-mode service, and all are compatible with PCs and PDAs.

15.15 Summary

In this chapter we discussed the wide-area wireless systems used in Europe and Japan. We presented the evolution path for GSM. We discussed the architecture of the UMTS including its domains and interfaces. We outlined the channel structure in the UTRAN. The channel structure in the UTRAN is a three-tier system which includes logical, transport, and physical channels. A detail description of the control, transport, and user plane in the I_u, I_{ur}, I_{ub} interfaces was provided. The role of each network entity in UMTS was presented. We discussed the UMTS bearer services and UMTS QoS. We concluded the chapter by discussing the high-speed downlink packet access (HSDPA) service.

Problems

15.1 Discuss two evolution paths for the GSM to offer 3G services.

15.2 What is the high-speed circuit switched data (HSCSD) in the GSM?

15.3 What is the role of the general packet radio service (GPRS) in the GSM?

15.4 Define roles of two new network entities in the GPRS.

15.5 Name the physical channels of the GPRS and discuss their functions.

15.6 What are some of the point-to-point (PTP) and point-to-multipoint (PTM) applications of the GPRS?

15.7 How are higher data rates achieved in the enhanced data rates for GSM enhancement (EDGE)? Discuss.

15.8 What are the modulation and coding schemes that are used for the packet mode in the EDGE?

15.9 Discuss the roles of 3G systems.

15.10 What is the UMTS? List important features of the UMTS air interface.

15.11 What are three channel types that are used in the UMTS? Discuss the role of each channel type.

15.12 How is isolation between users in the downlink accomplished in a WCDMA system?

15.13 What are the three main entities of the UMTS network? Discuss their functions.

15.14 Discuss the responsibilities of the RNC in the UMTS network.

15.15 What are the responsibilities of Node B in the UMTS network?

15.16 Discuss the role of the access link control application part (ALCAP) in the UMTS.

15.17 Discuss I_u, I_{ur}, and I_{ub} interfaces in the UMTS.

15.18 The core network of the UMTS is divided into three different functional areas. Name these areas and discuss their roles.

15.19 What is adaptive multi-rate (AMR) codec?

15.20 Discuss the UMTS bearer service layered architecture.

15.21 What are the QoS classes in the UMTS?

15.22 Discuss QoS attributes used in the UMTS.

15.23 Discuss briefly HSDPA for WCDMA.

References

1. Austin, M., Buckley, A., Coursey, C. et al. Service and System Enhancements for TDMA Digital Cellular Systems. *IEEE Personal Communications*, vol. 6, no. 3, June 1999, pp. 20–33.

2. CDMA for Next Generation Mobile Communications Systems. *IEEE Communications Magazine*, vol. 36, no. 9, September 1998.

3. Dahlman, E., Gumudson, B., Nilsson, M., and Skold, J. UMTS/IMT-2000 Based Wideband CDMA. *IEEE Communications Magazine*, vol. 36, no. 9, September 1998.

4. Digital Cellular Telecommunication System (Phase 2+), High Speed Circuit Switched Data (HSCSD)—Stage 1, Draft ETSI document, GSM 02.34 version 0.1.0, February 1995.

5. Digital Cellular Telecommunication System (Phase 2+), High Speed Circuit Switched Data (HSCSD), Service Description, Stage 2, GSM 03.34.

6. ETSI, TS 03 64 V5.10 (1997-11), Digital Cellular Telecommunication System (Phase 2+); General Packet Radio Service (GPRS). Overall Description of the GPRS Radio Interface; Stage 2, GSM 03.64 version 5.1.0.

7. ETSI Technical Specification GSM 02.60 GPRS Service Description—Stage 1, version 5.2.1, July 1998.

8. ETSI Tdoc SMG2 95/97, EDGE Feasibility Study, Work Item 184; Improved Data Rates through Optimized Modulation version 0.3, December 1997.

9. Garg, V. K., Halpern, S., and Smolik, K. F. Third Generation (3G) Mobile Communication Systems. 1999 IEEE International Conference on Personal Wireless Communications, February 1999, Jaipur, India.

10. 3G TS 25.321 MAC Protocol Specification.

11. 3G TS 25.322 RLC Protocol Specification.

12. 3G TS 25.331 RRC Protocol Specification.

13. 3GPP Technical Specification 25.401, UTRAN Overall Description.

14. 3GPP Technical Specification 25.410, UTRAN I_u Interface: General Aspects and Principles.

15. 3GPP Technical Specification 25.420, UTRAN I_{ur} Interface: General Aspects and Principles.

16. 3GPP Technical Specification 25.430, UTRAN I_{ub} Interface: General Aspects and Principles.

17. 3GPP Technical Specification 25.211, Physical Channels and Mapping of Transport Channels onto Physical Channels (FDD).

18. 3GPP Technical Specification 25.212, Multiplexing and Channel Coding (FDD).

19. 3GPP Technical Specification 25.213, Spreading and Modulation (FDD).

20. 3GPP Technical Specification 25.214, Physical Layer Procedures (FDD).

21. 3GPP Technical Specification 23.107, QoS Concept and Architecture.

22. Holma, H., and Toskala, A. *WCDMA for UMTS*, New York: John Wiley and Sons, LTD, 2000.

23. Ihrfors, H. *3G Wireless:* What Does it Mean for GSM Core Networks? *Mobile Communication International*, September 1998, pp. 35–38.

24. Knisley, D., Quinn, L., and Ramesh, N. cdma2000: A Third Generation Radio Transmission Technology. *Bell Labs Technical Journal*, vol. 3, no. 3, J1. September 1998.

25. Ojanpera, T., and Prasad, R. An Overview of Third-Generation Wireless Personal Communications. European Perspective. *IEEE Personal Communications*, vol. 5, no. 6, December 1998, pp. 59–65.

26. Prasad, N. R. GSM Evolution towards Third Generation UMTS/IMT2000. Third ICPWC99, February 17–19, 1999, Jaipur, India.

27. Rao, Y. S., and Kripalani, A. cdma2000 Mobile Radio Access for IMT-2000. 1999 IEEE International Conference on Personal Wireless Communications, February 1999, Jaipur, India.

28. Shanker, B., McClelland, S. Mobilising the Third-Generation [Cellular Radio]. *Journal of Telecommunications (International Edition)*, August 1997.

29. Sollenberger, N. R., Seshadri, N., and Cox, R. The Evolution of IS-136 TDMA for Third-Generation Wireless Services. *IEEE Personal Communications*, vol. 6, no. 3, June 1999, pp. 8–18.

30. Universal Mobile Telecommunications System (UMTS). Requirements for the UMTS Terrestrial Radio Access System (UTRA). ETSI Technical Report, UMTS 21.01 version 3.0.1, November 1997.

31. Universal Mobile Telecommunications System (UMTS). Selection Procedure for the Choice of Radio Transmission Technologies of the UMTS. ETSI Technical Report, UMTS 30.03 version 3.1.0, November 1997.

32. Universal Mobile Telecommunications System (UMTS). Requirements for the UMTS Terrestrial Radio Access System (UTRA) Concept Evaluation. ETSI Technical Report, UMTS 30.06 version 3.0.0, December 1997.

CHAPTER 16

Wide-Area Wireless Networks — cdmaOne Evolution

16.1 Introduction

While the primary objective of 2G networks was to provide mobile circuit-switched voice and low rate data services, a key goal of the evolution to 3G networks was the introduction of connectivity to packet data networks via cellular systems while at the same time increasing voice capacity.

Early definitions of 3G systems sought to boost the gross bit rates over radio, and to introduce support for quality of service (QoS) classes to improve packet switched bearer services. As demand for both greater capacity and more packet data services has grown, the industry has sought to improve 3G data throughput while at the same time enhancing voice performance over the same RF carrier. The cdma2000 1X EV-DV technology was developed to meet requirements and is first included in IS-2000 Revision C.

The cdma2000 radio transmission technology (RTT) is a wideband, spread spectrum radio interface that uses CDMA (IS-95) technology to satisfy the needs of 3G wireless communication systems. The RTT meets all requirements specified in the ITU circular letter and the corresponding documents of the IMT-2000. The service requirements are satisfied for indoor office, indoor-to-outdoor/pedestrian, and vehicular environments. The cdma2000 system is backward compatible with the current cdmaOne (IS-95) family of standards. cdmaOne is the family of standards of CDMA technology consisting of IS-95, IS-95A (J-STD-008) and IS-95B. The cdma2000 is a resultant system that evolves from 2G IS-95, satisfying the requirement of the IMT-2000 system with input from RTT 1X/3X submitted by the TR 45.5 subcommittee of TR 45 of the TIA to the ITU. IS-856 supports high-speed packet data services up to 2 Mbps. This is also referred to as high data rate (HDR) or 3G1X EV-DO (evolution for data-only systems). The development of the 3G1X EV-DV (evolution for data and voice) system has been completed in 3GPP2. Figure 16.1 shows the evolution of IS-95. Figure 16.2 gives the cdma2000 evolution path [1–9].

The cdma2000 system provides a wide range of implementation options to support data rates (both circuit switched and packet switched) starting from a TIA/EIA-95B compatible rate of 9.6 kbps up to greater than 2 Mbps. The cdma2000

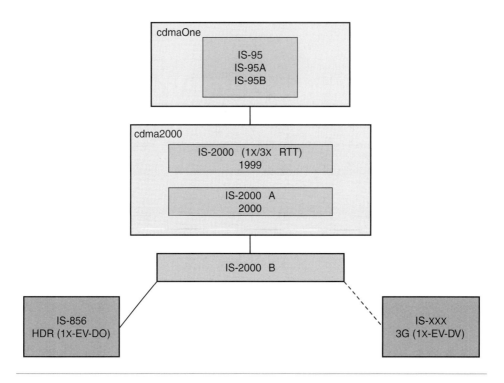

Figure 16.1   Evolution of IS-95.

system provides maximum flexibility to carriers in making engineering trade-offs between:

- Channel sizes of 1, 3, 6, 9, and 12 channels (each of 1.25 MHz)
- Support for advanced antenna technologies
- Cell sizes (e.g., the cdma2000 system's increased performance can be realized in terms of increased range to permit carriers to reduce the total number of cell sites)
- Higher data rates that can be supported in all channel sizes
- Support for advanced services possible or practical in other systems (e.g., high-speed circuit data, B-ISDN, or H.224/223 teleservices)

The cdma2000 system can be operated economically in a wide range of environments including outdoor megacells (cell > 35 km radius), outdoor macrocells (cell 1–35 km radius), indoor/outdoor microcells (up to 1 km radius), and indoor/outdoor picocells (<50 m radius).

The cdma2000 system can be deployed in indoor/outdoor environments, wireless local loops (WLL), vehicular environments, and mixed vehicular and indoor/outdoor environments.

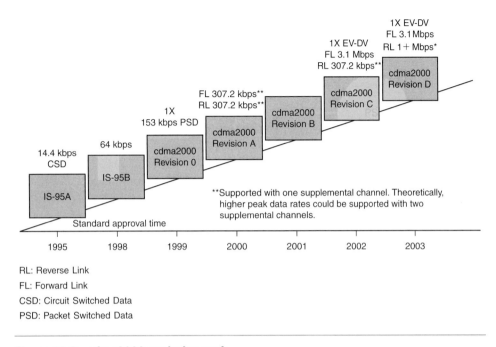

Figure 16.2 cdma2000 evolution path.

The cdma2000 system mobility is variable, ranging from fixed wireless to high speeds of up to 300 mph. cdma2000 provides a layered structure to support the integration of the bottom two layers of the radio transmission technology (RTT) into systems that implement any network standards (e.g., ITU-T defined signaling services). It also provides backward compatibility to TIA/EIA-95B signaling and call control models. An extended cdma2000 upper layer signaling structure is capable of supporting a wide range of advanced services (e.g., multimedia) in an optimized and efficient manner.

cdma2000 supports the 3G wireless intelligent networking (WIN) services and services defined by the ITU or other international standards organizations and provides a graceful evolution from existing 2G TIA/EIA-95B technology. It includes the support for overlay configurations, support for backward compatibility to TIA/EIA-95B signaling and network, support for a graceful and gradual upgrade from the 2G system to the 3G system, and sharing of common channels with an underlay TIA/EIA-95B system during transition periods.

cdma2000 provides an evolutionary path by reusing existing TIA/EIA-95B standards including:

- TIA/EIA-95-B: Mobile station and radio interface specifications
- IS-707: Data services (packet, async, and fax)
- IS-127: Enhanced variable rate codec (EVRC) 8 kbps speech coder

- IS-733: 13 kbps speech coder
- IS 637: Short message service (SMS)
- IS 638: Over-the-air-activation and parameter administration (supporting the configuration and service activation of mobile stations over the radio interface)
- IS 97 and 98 (minimum performance specifications)
- The basic TIA/EIA-95B channel structure
- Extensions to TIA/EIA-95B fundamental/supplemental channel structure, multiplex layer, and signaling to support higher rate operation, common broadcast channels (pilot, paging, and sync)
- IS-634: No significant changes are expected for cdma2000; the layered structure of cdma2000 integrates smoothly with the component structure of IS-634
- TIA/EIA-41D: No significant changes needed for the cdma2000; the layered structure of cdma2000 offers the potential for easy integration with enhanced network services (WIN).

In Chapter 15, we discussed the evolution of GSM to UMTS/WCDMA (wideband code division multiple access). In this chapter, we focus on the evolution of cdmaOne (see Chapter 11) to cdma2000. We provide typical end-to-end call flows for cdma2000 and the WCDMA system. We conclude the chapter by summarizing differences between cdma2000 and WCDMA.

16.2   cdma2000 Layering Structure

16.2.1   Upper Layer

Figure 16.3 shows the layer structure of cdma2000. The upper layers open system interconnection (OSI layers 3–7) contain three basic services [10–13]:

- **Voice services.** Voice telephony services, including public switched telephone network (PSTN) access, mobile-to-mobile voice services, and Internet telephony.
- **End user data-bearing services.** Services that deliver any form of data on behalf of the mobile end user, including packet data (e.g., IP service), circuit-switched data services (e.g., B-ISDN emulation services), and SMS. Packet data services conform to industry standard connection-oriented and connectionless packet data including IP-based protocols (e.g., transmission control protocol (TCP) and user datagram protocol (UDP) and OSI connectionless interworking protocol (CLIP)). Circuit-switched data services that emulate international standards-defined, connection-oriented services such as asynchronous (async) dial-up access, fax, V.120 rate-adapted ISDN, and B-ISDN services.
- **Signaling.** Services that control all aspects of the operation of the mobile.

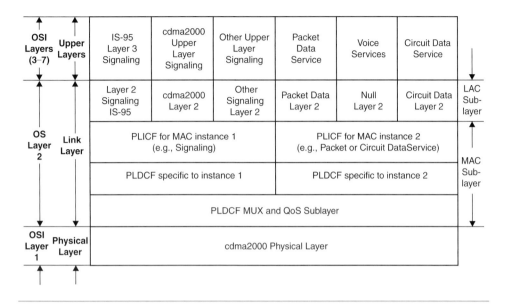

Figure 16.3 cdma2000 layering structure.

16.2.2 Lower Layers

The link layer provides varying levels of reliability and QoS characteristics according to the needs of the specific upper layer service. It gives protocol support and control mechanisms for data transport services and performs all functions necessary to map the data transport needs of the upper layers into specific capabilities and characteristics of the physical layer. The link layer is divided into two sublayers: link access control (LAC) and medium access control (MAC) (refer to Figure 16.4).

The LAC sublayer manages point-to-point communication channels between peer upper layer entities and provides framework to support a wide range of different end-to-end reliable link layer protocols.

The cdma2000 system includes a flexible and efficient MAC sublayer that supports multiple instances of an advanced-state machine, one for each active packet or circuit data instance. Together with a QoS control entity, the MAC sublayer realizes the complex multimedia, multiservice capabilities of 3G wireless systems with QoS management capabilities for each active service. The MAC sublayer provides three important functions.

- **MAC control state.** Procedures for controlling the access of data service (packet and circuit) to the physical layer (including contention control between multiple services from a single user as well as between competing users).
- **Best effort delivery.** Reasonably reliable transmission over the radio link with *radio link protocol* (RLP) providing a best-effort level of reliability.

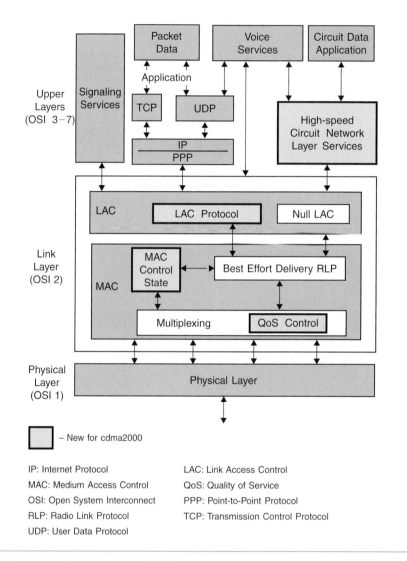

Figure 16.4   cdmaOne and cdma2000 layering structure.

- **Multiplexing and QoS control.** Enforcement of negotiated QoS levels by mediating conflicting requests from competing services and appropriately prioritizing access requests.

The MAC sublayer provides differing QoS to the LAC sublayer (e.g., different modes of operation). It may be constrained by backward compatibility (e.g., for IS-95B signaling layer 2) and it may have to be compatible with other link layer protocols (e.g., for compatibility with non-IS-95 air interfaces or for compatibility with future ITU-defined protocol stacks). The MAC sublayer is subdivided into a *physical layer independent convergence function (PLICF)* and a *physical layer dependent convergence function (PLDCF)* that is further subdivided into instance-specific PLDCF and PLDCF MUX (multiplexing) and a QoS sublayer.

The PLICF provides service to the LAC sublayer and includes all MAC operational procedures and functions that are not unique to the physical layer. Each instance of PLICF maintains service status for the corresponding service. PLICF uses services provided by PLDCF to implement actual communications activities in support of MAC sublayer service. Services used by PLICF are defined as a set of logical channels that carry different types of control or data information. The PLICF data service consists of the following states/substates (see Figure 16.5).

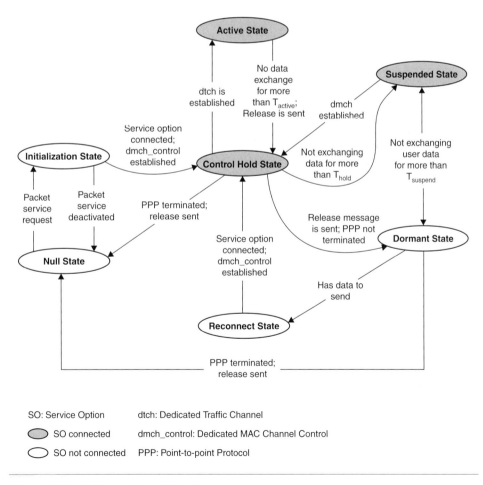

Figure 16.5 Data services state diagram.

- Null state
- Initialization state
- Control hold state
  - Normal substate
  - Slotted substate
- Active state
- Suspended state
  - Virtual traffic substate
  - Slotted substate
- Dormant state
  - Dormant/idle substate
  - Dormant/burst substate

The null state is considered to be a default state prior to the activation of the packet data service. After the packet service is invoked, a transition to the initialization state occurs during which an attempt is made to connect the packet service.

Traffic, power control, and control channels are assigned in the active state. In the control hold state, a dedicated control channel is maintained between the user and the base station on which any MAC control command (for example, the command to begin a high-speed data burst) can be transmitted with virtually no latency. Power control is also maintained so that a high-speed burst operation can begin with no delay due to the stabilization of power control.

In the suspended hold state, there are no dedicated channels maintained to or from the user. However, the state information for RLP is maintained, and the base station and user maintain a virtual active set that allows either one of them to know which base station can best be used (accessed by the user or paged by the base station) in the event that packet data traffic occurs for the user. This state also supports a slotted substate that permits the user's mobile device to preserve power in a highly efficient manner.

A short data burst mode is added to the cdma2000 dormant state to support the delivery of short messages without incurring the overhead of a transition from a dormant state to the active state. Transitions between MAC states can be indicated by MAC control signaling or by the expiration of timers. By properly selecting the values for the timers, the cdma2000 MAC can be adapted to a wide variety of data services and operating environments.

The states are categorized as either connected or not connected, depending on the status of the data service option. The data service option is connected in the control hold state, active state, and suspended hold state. The data service option is not connected in the null state, initialization state, dormant state, or reconnect state. Figure 16.5 shows the state diagram for the PLICF data service option.

The PLDCF performs mapping of logical channels from the PLICF to logical channels supported by the specific physical layer. The PLDCF performs multiplexing, demultiplexing, and consolidation of control information with bearer data from the control and traffic channels from multiple PLICF instances in the same mobile. The PLDCF implements QoS capabilities, including resolution of priorities between competing PLICF instances, and maps QoS requests from PLICF instances into the appropriate physical layer service requests to deliver the desired QoS. The major functions of this sublayer are:

- Perform any required mapping of the simpler logical channels from the PLICF into the logical channels supported by the physical layer.

- Perform any (optional) automatic repeat request (ARQ) protocol functions that are tightly integrated with the physical layer.

- Perform some of the physical layer specific low-level functions of IS-95B RLP.

For cdma2000, four PLDCF specific protocols are defined.

1. **Radio link protocol (RLP).** This protocol provides a highly efficient streaming service that makes a best effort to deliver data between peer PLICF entities. RLP provides both transparent and nontransparent modes of operation. In the nontransparent mode, RLP uses ARQ protocol to retransmit data segments that were not delivered properly by the physical layer. In the nontransparent mode, RLP can introduce some delay. In the transparent mode, RLP does not retransmit missing data segments. However, RLP maintains byte synchronization between the sender and receiver and notifies the receiver of the missing parts of the data stream. Transparent RLP does not introduce any transmission delay, and is useful for implementing voice services over RLP.

2. **Radio burst protocol (RBP).** This protocol provides a mechanism for delivering relatively short data segments with best-effort delivery over a shared access common traffic channel (CTCH). This capability is useful for delivering small amount of data without incurring the overhead of establishing a dedicated traffic channel (DTCH).

3. **Signaling radio link protocol (SRLP).** This protocol provides a best-effort streaming service for signaling information analogous to RLP, but is optimized for the dedicated signaling channel (DSCH).

4. **Signaling radio burst protocol (SRBP).** This protocol provides a mechanism to deliver signaling messages with best-effort delivery analogous to RBP, but is optimized for signaling information and a common signaling channel (CSCH).

The PLDCF includes a radio link access control (RLAC) function that abstracts the RLP and RBP from the PLICF and coordinates the transmission of data (traffic or signaling) between the RLP and RBP according to the current operational state of MAC (e.g., restrict the use of the RBP to cases in which the PLICF is in the packet data dormant state).

The PLDCF MUX and QoS sublayer coordinates multiplexing and demultiplexing of code channels from multiple PLICF instances. It implements and enforces QoS differences between instances and maps the data streams and control information on multiple logical channels from different PLICF instances into requests for logical channels, resources, and control information from the physical layer.

16.3 Forward Link Physical Channels of cdma2000

Forward dedicated channels carry information between the base station and a specific mobile; common channels carry information from the base station to a set of mobiles in a point-to-multipoint manner (see Figures 16.6–16.8). Table 16.1 lists these channels.

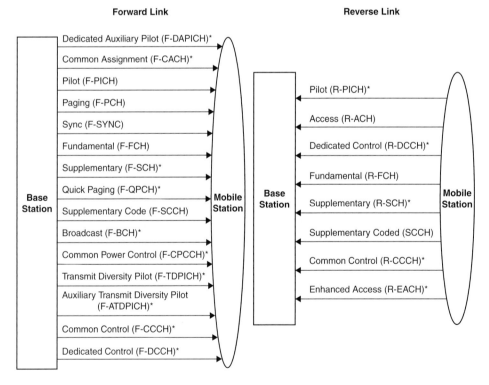

* New to cdma2000

Figure 16.6 cdma2000 physical channels.

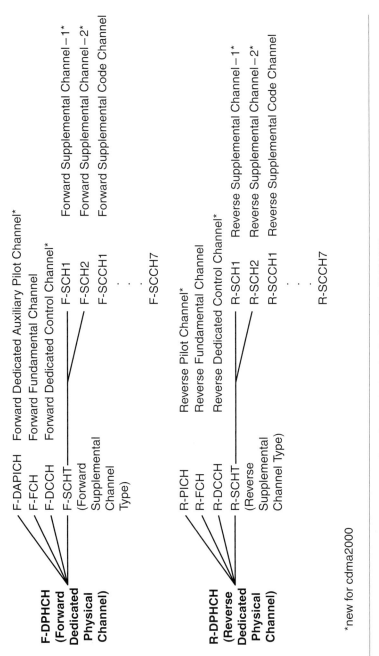

F-DPHCH (Forward Dedicated Physical Channel)

- F-DAPICH — Forward Dedicated Auxiliary Pilot Channel*
- F-FCH — Forward Fundamental Channel
- F-DCCH — Forward Dedicated Control Channel*
- F-SCHT (Forward Supplemental Channel Type)
  - F-SCH1 — Forward Supplemental Channel – 1*
  - F-SCH2 — Forward Supplemental Channel – 2*
  - F-SCCH1 — Forward Supplemental Code Channel
  - . . .
  - F-SCCH7

R-DPHCH (Reverse Dedicated Physical Channel)

- R-PICH — Reverse Pilot Channel*
- R-FCH — Reverse Fundamental Channel
- R-DCCH — Reverse Dedicated Control Channel*
- R-SCHT (Reverse Supplemental Channel Type)
  - R-SCH1 — Reverse Supplemental Channel – 1*
  - R-SCH2 — Reverse Supplemental Channel – 2*
  - R-SCCH1 — Reverse Supplemental Code Channel
  - . . .
  - R-SCCH7

*new for cdma2000

Figure 16.7   cdma2000 overview of dedicated physical channels (DPHCH).

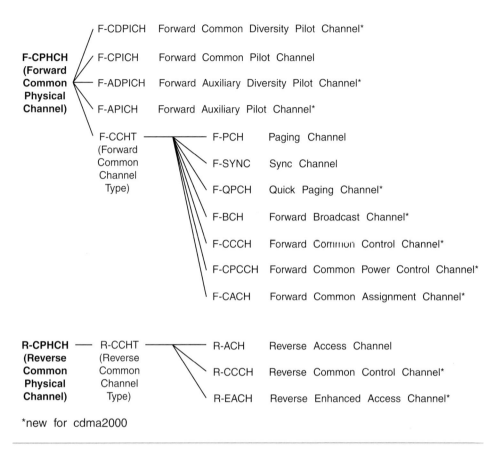

*new for cdma2000

Figure 16.8   cdma2000 overview of common physical channels (CPHCH).

Table 16.1   Forward link channels of cdma2000.

	Physical channel	Channel name
Forward Common Physical Channels (control and over-head channels)	Forward Pilot Channel	F-PICH
	Forward Paging Channel	F-PCH
	Forward Synch Channel	F-SYNCH
	Forward Common Control Channel	F-CCCH
	Forward Quick Paging Channel	F-QPCH
	Forward Transmit Diversity Pilot Channel	F-TDPICH
	Forward Auxiliary Transmit Diversity Pilot Channel	F-ATDPICH
	Forward Common Power Control Channel	F-CPCCH

(Continued)

	Physical channel	Channel name
	Forward Common Assignment Channel	F-CACH
	Forward Broadcast Channel	F-BCH
Forward Dedicated Physical Channels	Forward Dedicated Auxiliary Pilot Channel	F-DAPICH
	Forward Dedicated Control Channel	F-DCCH
	Forward Traffic -Fundamental -Supplementary -Supplementary Coded	F-FCH F-SCH F-SCCH

16.4   Forward Link Features

The forward link supports chip rates of $N \times 1.2288$ Mcps (where $N = 1, 3, 6, 9, 12$). For $N = 1$, the spreading is similar to IS-95 B; however, QPSK modulation and fast closed loop power control are used. For chip rates with $N > 1$ the multicarrier (see Figure 16.9) option is used. The multicarrier approach demultiplexes modulation symbols onto N separate 1.25 MHz carriers ($N = 3, 6, 9, 12$). Each carrier is spread with 1.2288 Mcps rate. There are nine radio configurations for the forward traffic channels. Table 16.2 provide the details.

16.4.1   Transmit Diversity

Transmit diversity can reduce the required E_b/N_0 (or required transmit power per channel) and thus enhance the system capacity. Transmit diversity can be implemented in the following ways:

- **Multi-carrier transmit diversity.** Antenna diversity can be implemented in a multicarrier forward link with no impact on the subscriber terminal, where a subset of carriers is transmitted on each antenna. The main characteristics of a multicarrier approach are
  - Coded information symbols are demultiplexed among multiple 1.25 MHz carriers.
  - Frequency diversity is equivalent to spreading the signal over the entire bandwidth.
  - Both time and frequency diversity are captured by convolutional coder/ symbol repetition and interleaver.
  - The rake receiver captures signal energy from all bands.
  - Each forward link channel may be allocated an identical Walsh code on all carriers.
  - Fast power control is used.

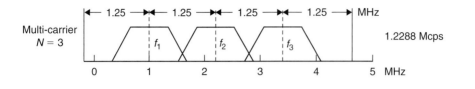

Figure 16.9   Multi-carrier approach in cdma2000.

Table 16.2 Forward traffic channel spreading for different radio configurations.

Radio configuration	Spreading rate	Rate, modulation, coding, and diversity
RC1	IS-95B compatible	Based on RS1 (9.6 kbps), coding rate (r) = 1/2, BPSK prespreading symbols.
RC2	IS-95B compatible	Based on RS2 (14.4 kbps), coding rate (r) = 1/2, BPSK prespreading symbols.
RC3	3G1X, SR-1	1.5–153.6 kbps with r = 1/4, QPSK prespreading symbols, orthogonal transmit diversity (OTD) allowed.
RC4	3G1X, SR-1	1.5–307.2 kbps with r = 1/4, QPSK prespreading symbols, OTD allowed.
RC5	3G1X, SR-1	1.8–230.4 kbps with r = 1/2, QPSK prespreading symbols, OTD allowed.
RC6	3G3X, SR-3	1.5–307.2 kbps with r = 1/6, QPSK prespreading symbols, multicarriers (MC) modes, OTD allowed.
RC7	3G3X, SR-3	1.5–614.4 kbps with r = 1/6, QPSK prespreading symbols, MC modes, OTD allowed.
RC8	3G3X, SR-3	1.8–460.8 kbps with r = 1/4, (20 ms frame) or r = 1/3 (5 ms frame), QPSK prespreading symbols, MC modes, OTD allowed.
RC9	3G3X , SR-3	1.8–1036.8 kbps with r = 1/2, (20 ms frame) or r = 1/3 (5 ms frame), QPSK prespreading symbols, MC modes, OTD allowed.

In a 3X 1.25 MHz multicarrier transmitter, the serial coded information symbols are divided into three parallel data streams, and each data stream is spread with a Walsh code and a long pseudo-random noise (PN) sequence at a rate of 1.2288 Mcps. At the output of the transmitter, there are three carriers: A, B, and C (see Figure 16.10).

After processing the serial coded information symbols with parallel carriers, the multicarrier will be transmitted by a multi-antenna, called multi-carrier transmit diversity (MCTD). In the MCTD, the total carriers are divided into subsets, then each subset of the carriers is transmitted on each antenna, where frequency filtering provides near-perfect orthogonality between antennas. This provides improved frequency diversity and hence increases forward link capacity.

- **Direct-spread transmit diversity.** Orthogonal transmit diversity (OTD) may be used to provide transmit diversity for direct spread ($N = 1$). Coded bits are split into two data streams and are transmitted via separate antennas. A different orthogonal code is used per antenna for spreading. This maintains the orthogonality between the two output streams, and hence self-interference is eliminated in flat fading. Note that by splitting the coded bits into two separate streams, the effective number of spreading codes per user is the same as the case without OTD. An auxiliary pilot is introduced for the additional antenna.

16.4.2 Orthogonal Modulation

To reduce or eliminate intracell interference, each forward link physical channel is modulated by a Walsh code. To increase the number of usable Walsh codes,

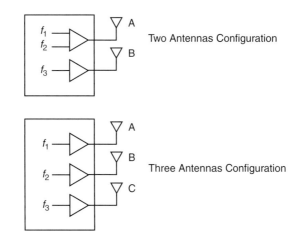

Figure 16.10   3X 1.25 MHz multi-carrier transmitter.

quadrature phase shift keying (QPSK) modulation is used before spreading. Every two information bits are mapped to a QPSK symbol. As a result, the available number of Walsh codes is increased by a factor of two, relative to binary phase shift keying (BPSK) (prespreading) symbols. Walsh code length varies to achieve different information bit rates. The forward link may be interference limited or Walsh code limited depending on the specific deployment and operating environment. When a Walsh code limit occurs, additional codes may be generated by multiplying Walsh codes by the masking functions. The codes generated in this way are called *quasi-orthogonal functions*. The quasi-orthogonal functions are not totally orthogonal.

16.4.3   Power Control

In IS-95, forward link power control was not considered to be as important and demanding as the one required by the reverse link power control. This was supported by the argument that the base station (BS) transmits all the channels coherently in the same RF carrier, and they all fade together as the composite signal arrives at the mobile. This argument is valid as long as the thermal and background noise is negligible. For this reason the initial IS-95A forward power control requirements were not as stringent as the ones for the reverse link. However, a particular mobile may be near a significant source of interference, or may suffer a large path loss such that the arriving composite signal is of the order of the background noise. For this reason, a faster power control is also required on the forward link.

In IS-95, the forward link power control is slow and is performed at a rate not faster than 50 Hz. Its implementation depends on the rate set of the traffic channel. For RS1 the mobile collects the frame error rate (FER) and reports statistics to the BS, which compares the FER to a target value. The BS increases or decreases the forward link power depending on whether the FER is higher or lower than the target FER. Typically, the BS receives an FER measurement message once every one hundred frames, and takes action every two seconds on average, thus providing a forward power control that runs at a rate of about 0.5 Hz for RS1.

For RS2 the mobile sends an erasure indicator bit (EIB) to the BS in every frame, indicating the quality of the frame received previously by the mobile. The BS uses the EIB to estimate the forward link FER and increases or decreases the forward link traffic channel power depending on the relationship of the FER estimated to the FER target value. The RS2 power control runs at a rate of 50 Hz, which represents a performance improvement over its predecessor algorithm used in the RS1.

Although there are substantial performance improvements in the forward link power control when updating the RS1 algorithm with the RS2 algorithm, the performance is still limited by the inability to cope with fast Rayleigh fading and fast changing RF conditions. The main reason for this is the slow rate at which the forward link power control operates. Therefore, to improve the performance (i.e., to lower the BS average transmit power per voice channel) and to increase the forward link call capacity, a faster power control loop is required to mitigate fast Rayleigh fading.

The 3G forward link power control functions are similar to the reverse link power control in IS-95. The algorithm consists of two loops running at an effective rate of 800 Hz. This enhancement is expected to increase the user capacity of the forward link.

The forward link power control in 3G is fundamentally different from IS-95. Its main objective is to increase the voice call capacity in the forward link by a series of new enhancements, including:

- High speed forward power control
- Closed loop with fast time response
- Variable power step size controlled by the BS

The forward link power control operates at a high rate to track and compensate accurately the fast Rayleigh fading on the forward link. The accurate tracking minimizes the average power transmitted by the BS to the mobile and, as a consequence, increases the forward link call capacity.

The rate of the forward link power control is increased by replacing the slow FER-based algorithm of IS-95 with a closed loop based on E_b/N_t measurements similar to the reverse link in IS-95. While the IS-95 procedure of FER statistics is a slow process with a response time of many frames, the E_b/N_t measurements are fast and easy to perform at a subframe interval. This allows the 3G forward link power control to operate at high speed.

The increase in forward link capacity is expected mostly for mobiles moving at low speed and in simplex mode (i.e., not in soft handoff) where fast Rayleigh fading can be substantial and can be mitigated effectively. For mobiles travelling at high speed, the fast power control cannot track the fast Rayleigh fading accurately, and, therefore, a large increase in capacity may not be expected. In addition, the capacity increase for mobiles in soft handoff is less due to the already existing path diversity of soft handoff, which reduces Rayleigh fading.

The variable power step size has not been standardized in IS-2000, and therefore the BS has the flexibility to adapt the step size depending on the speed of the mobile and soft handoff status of the call. This adaptability permits minimization of the peak to average power ratio, and decreases the overall interference, which increases the forward link capacity.

The new fast forward power control (FFPC) algorithm on the forward link and power control for the F-FCH and F-SCH is used in cdma2000. The standards specify a fast closed loop power control at 800 Hz. Two schemes of power control for the F-FCH and F-SCH have been proposed.

- **Single channel power control.** This is based on the performance of the higher rate channel between the F-FCH and F-SCH. The gain setting for the lower rate channel is specific deployment and operating environment. When a Walsh code limit occurs, additional codes may be generated by multiplying Walsh codes by the masking function.

- **Independent power control.** In this case, gains for the F-FCH and F-SCH are determined separately. The mobile runs two separate outer loop algorithms (with different E_b/N_t targets) and sends two forward error bits to the BS.

16.4.4 Walsh Code Administration

IS-95 A/B uses fixed length 64-chip Walsh codes. The new rate sets in cdma2000 require variable length Walsh codes for traffic channels. The Walsh codes used are from 128 chips to 2 chips in length. The F-FCH Walsh code is fixed (128 chips for RS3 and RS5, and 64 chips for RS4 and RS6), whereas the length of the Walsh codes for F-SCH decreases as the information rate increases to maintain a constant bandwidth of the modulated signal. In addition to different Walsh code lengths, the coordination of the allocation of Walsh codes across 2G and 3G systems is necessary for overlay systems.

The algorithm must ensure that Walsh codes assigned for different rate supplemental channels are always orthogonal to each other as well as to the fundamental traffic channels, paging channels, synch channel, and pilot channel. For example, if an all 0s 4-chip Walsh code (0 0 0 0) is assigned, then there are two 8-chip Walsh codes that are not to be assigned at the same time (0 0 0 0 0 0 0 0, 0 0 0 0 1 1 1 1); the remaining six 8-chip codes can be used since they are all orthogonal to it. By induction, four 16-chip, eight 32-chip, sixteen 64-chip, and thirty-two 128-chip codes must also be set aside to maintain orthogonality.

The 2G and 3G Walsh code assignments must be coordinated to ensure that assigning the longer-length codes does not block out all of the shorter codes.

16.4.5 Modulation and Spreading

The SR1 system can be deployed in a new spectrum or as a backward-compatible upgrade anywhere an IS-95B forward link is deployed in the same RF channel. The new cdma2000 channels can coexist in an orthogonal manner with the code channels of the existing IS-95B system. The SR1 spreading is shown in Figure 16.11. First, the user data is scrambled by the user-long PN code followed by I and Q mapping, channel gain, power control puncturing, and Walsh spreading. The power control bits may or may not be punctured onto the forward link depending on the specific logical-to-physical channel mapping. Next, as shown in Figure 16.11, the signal is a complex PN spread, followed by baseband filtering and frequency modulation.

The multicarrier system can be deployed in a new spectrum or as a backward-compatible upgrade anywhere an IS-95B forward link is deployed in the same N RF channels. The new cdma2000 channels can coexist in an orthogonal manner with the code channels of the existing IS-95B system.

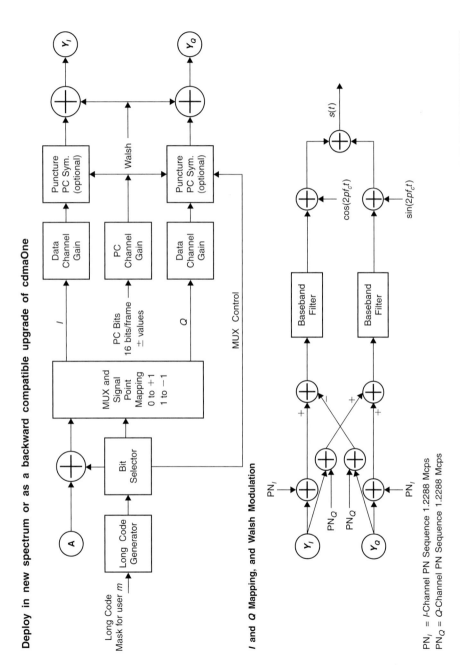

Figure 16.11   SR1 PN spreading, baseband filtering, and frequency modulation.

The overall structure of the multicarrier CDMA channel is shown in Figure 16.12. After scrambling with the long PN code corresponding to user m, the user data is demultiplexed into N carriers, where $N = 3, 6, 9,$ or 12. On each carrier, the demultiplexed bits are mapped onto I and Q followed by Walsh spreading. When applicable, power control bits, for reverse closed loop power control, may be punctured onto the forward link channel at a rate of 800 Hz. The signal on each carrier is orthogonally spread by the appropriate Walsh code function in such a manner as to maintain a fixed chip rate of 1.2288 Mcps per carrier, where the Walsh code may differ on each carrier. The signal on each carrier is then a complex PN spread (see Figure 16.11) followed by baseband filtering, and frequency modulation.

Figure 16.13 provides a comparison between the forward physical channels used in IS-95 A/B and cdma2000.

The key characteristics of the forward link are summarized as follows:

- Channels are orthogonal and use Walsh codes. Different-length Walsh codes are used to achieve the same chip rate for different information bit rates.
- QPSK modulation is used before spreading to increase the number of usable Walsh codes.
- Forward error correction (FEC) is used:
  - Convolutional codes ($k = 9$) are used for voice and data
  - Turbo codes ($k = 4$) are used for high data rates on SCHs
- Supports nonorthogonal forward link channelization
  - These are used when running out of orthogonal space (insufficient number of Walsh codes)
  - Quasi-orthogonal functions are generated by masking existing Walsh functions
- Synchronous forward link
- Forward link transmit diversity
- Fast-forward-power-control (closed loop) 800 times per second
- Supplemental channel active set, subset of fundamental channel active set. The maximum data rate supported for RS3 and RS5 for a supplemental channel is 153.6 kbps (raw data rate). RS4 and RS6 will be supported only for voice calls with the fundamental channel rates up to 14.4 kbps (raw data rate).
- Frame lengths
  - 20-ms frames are used for signaling and user information
  - 5-ms frames are used for control information

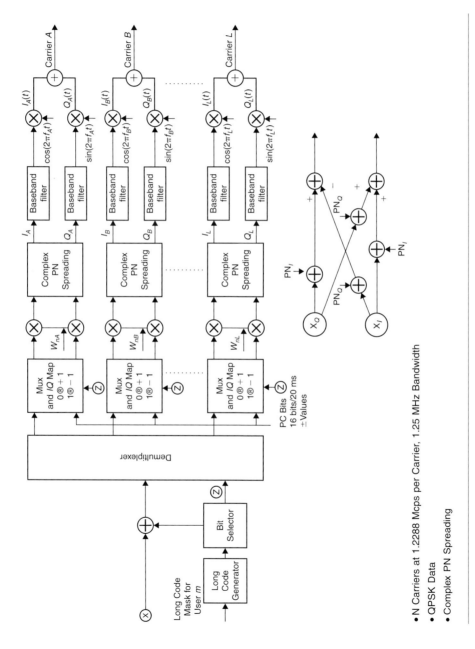

- N Carriers at 1.2288 Mcps per Carrier, 1.25 MHz Bandwidth
- QPSK Data
- Complex PN Spreading

Figure 16.12 Multicarrier cdma2000 forward link structure.

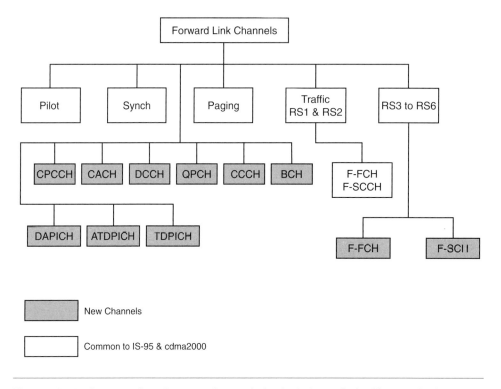

Figure 16.13   A comparison between forward physical channels for IS-95 and cdma2000.

16.5   Reverse Link Physical Channels of cdma2000

Reverse link physical channels (see Figure 16.6) include dedicated channels to carry information from a single mobile to the base station and common channels to carry information from multiple mobiles to the base station. Table 16.3 lists the reverse physical channels.

There are six radio configurations for the reverse traffic channels. A mobile station supports the operation in RC1, RC3, or RC5, and it may support the operation in RC2, RC4, or RC6. A mobile station supporting the operation in RC2 supports RC1, while a mobile station supporting the operation in RC4 supports RC3, and a mobile station supporting the operation in RC6 supports RC5. A mobile station does not use RC1 or RC2 simultaneously with RC3 or RC4 on the reverse traffic channels (see Table 16.4).

Figure 16.14 provides a comparison between the reverse physical channels used in IS-95 A/B and cdma2000. Reverse link physical layer characteristics are as follows:

- **Continuous waveform.** A continuous pilot and continuous data-channel waveform are used for all data rates. This continuous waveform minimizes

Table 16.3 Reverse physical channels of cdma2000.

	Physical channels	Channel name
Reverse Common Physical Channel	Reverse Access Channel	R-ACH
	Reverse Enhanced Access Channel	R-EACH
	Reverse Common Control Channel (9.6 kbps only)	R-CCCH
Reverse Dedicated Physical Channel	Reverse Pilot Channel	R-PICH
	Reverse Dedicated Control Channel	R-DCCH
	Reverse Traffic Channel -Fundamental -Supplemental -Supplemental Code	R-FCH R-SCH R-SCCH

Table 16.4 Reverse traffic channel spreading for various radio configurations.

Radio configuration	Spreading rate	Rate, coding, modulation, and vocoder
RC1	IS-95B Compatible	Based on RS1 (9.6 kbps), coding rate (r) = 1/3, 64-ary orthogonal, 8 kbps vocoder
RC2	IS-95B Compatible	Based on RS2 (14.4 kbps), coding rate (r) = 1/2, 64-ary orthogonal, 8 kbps vocoder
RC3	3G1X, SR1	1.2–153.6 kbps, r = 1/4, 307.2 kbps with r = 1/2 BPSK modulation with a pilot, 8 kbps vocoder
RC4	3G1X, SR1	1.8–230.4 kbps with r = 1/4 BPSK modulation with pilot, 8 kbps vocoder
RC5	3G3X, SR3	1.2–153.6 kbps, r = 1/4, 307.2 kbps with r = 1/3 BPSK modulation with a pilot, 8 kbps vocoder
RC6	3G3X, SR3	1.8–460.8 kbps with r = 1/4, 1.0368 Mbps with r = 1/2 BPSK modulation with a pilot, 8 kbps vocoder

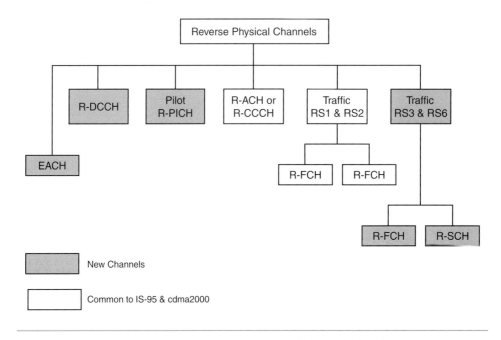

Figure 16.14   A comparison between reverse physical channels for IS-95 and cdma2000.

interference to biomedical devices such as hearing aids and pacemakers and permits a range increase at lower transmission rates. The continuous waveform also enables the interleaving to be performed over the entire frame rather than just the portions that are not gated off. This enables the interleaving to achieve the full benefit of the frame time diversity. The base station uses the pilot for multipath searches, tracking, coherent demodulation as well as to measure the quality of the link for power-control purposes. Separate orthogonal channels for the pilot and each of the data channels are used. Thus, the relative levels of the pilot and physical data channels can easily be adjusted without changing the frame structure or power levels of some symbols of a frame.

- **Orthogonal spreading with different length Walsh sequences.** The mobile uses orthogonal spreading when transmitting on the R-PICH, EACH, R-CCCH, or R-TCH with RC3 through RC6. Table 16.5 specifies the Walsh functions that are applied to the reverse link channels.

W_m^N represents a Walsh function of length N that is serially constructed from the m-row of an $N \times N$ Hadamard matrix with the 0th row being Walsh function 0, the first row being Walsh function 1, etc. Within Walsh function m, Walsh

Table 16.5 Walsh function for reverse link channels.

Channel type	Walsh function
R-PICH	W_0^{32}
EACH	W_2^8
R-CCCH	W_2^8
R-DCCH	W_8^{16}
R-FCH	W_4^{16}
R-SCH1	W_1^2 or W_2^4
R-SCH2	W_2^4 or W_6^8

chips are transmitted serially from the mth row left to right. The Walsh function spreading sequence repeats with a period of $N/1.2288\,\mu$s for SR1 and with a period of $N/3.6864\,\mu$s for SR3.

- **Rate matching.** Several approaches are needed to match the data rates to the Walsh spreader input rates. These include adjusting the code rate using puncturing, symbol repetition, and sequence repetition. The design approach is to first try to use a low rate code, but not to reduce the rate below $r = 1/4$ since gains of smaller rates would be small and the decoder implementation complexity would increase significantly.

- **Low spectral sidelobes.** The cdma2000 system achieves low spectral side-lobes with nonideal mobile power amplifiers by splitting the physical channels into the I and Q channels and using a complex-multiply-type PN spreading approach.

- **Independent data channels.** Two types of physical data channels (R-FCH and R-SCH) are used on the reverse link that can be adapted to a particular type of service. The use of R-FCH and R-SCH enables the system to be optimized for multiple simultaneous services. These channels are separately coded and interleaved and may have different transmit power level and FER set points.

16.5.1 Reverse Link Power Control

The primary objective of power control in the reverse link is to resolve the near-far problem, where mobiles that are near the BS have a better signal path than mobiles that are far away. Due to this effect, near mobiles can, in principle, raise the RF interference levels that screen out mobiles located far from the BS. This problem is resolved by controlling the reverse transmit power of each mobile, and requiring that the signal-to-noise ratio (SNR) of each mobile is the same at the BS. The reverse power control must be dynamic in order to compensate for the time-dependent variations of the RF environment. In IS-95, the reverse power control has been designed to solve the following reverse link problems.

1. Choosing the initial mobile transmit power
2. Power compensation due to slow varying and lognormal shadowing effects where there is a correlation between the forward and reverse link fades
3. Measure and maintain a target SNR at the BS for each mobile
4. Compensation of power fluctuation due to fast Rayleigh fading

The first and second items are resolved using open loop power control. The third and fourth items are resolved using a closed loop, which consists of an inner and an outer loop nested together.

In IS-95 the open loop power control is purely a mobile-controlled operation and does not involve the BS. The mobile estimates the forward path loss, and based on this measurement determines the transmit power. Since forward and reverse links operate at different carrier frequencies, the open loop power control is inadequate and too slow to compensate for fast and frequency-dependent Rayleigh fading.

The closed loop power control is used to compensate fast Rayleigh fading. The process involves both the BS and the mobile station (MS). Once the MS is on the traffic channel, the open and closed loops work together so that the slower open loop is able to include the faster closed loop correction.

The IS-2000 reverse power control scheme is a generalization of the IS-95 version. The system supports the same open and closed loops, but it integrates the functionality of the power control of the fundamental and supplemental channels in a simple scheme—easy to expand. The key factor in the simplification is the introduction of the R-PICH, which is used as a reference for measurement and scaling in the open and inner loops, and then translates them to corrections that apply to F-FCH. The scaling is performed per channel and per data rate, so that rate equalization can be performed easily. This simplifies the design of the outer loop of the R-FCH, since all rates can be treated equally.

In the reverse link open loop power control, the mobile estimates the transmitted power of the reverse link channels based on the measurement of the received aggregate power. As in IS-95, the open loop functionality is located at the mobile, however the BS must provide the value of the parameters the mobile needs to operate the open loop. This is done via a signaling message in the overhead channels and in the forward traffic channel. Open loop power control compensates for the path loss from the mobile to the base station and handles very slow fading.

IS-2000.2 provides formulas for the mean output power at the mobile of all the reverse channels. The formulas for the R-ACH and R-TCHs with RC1 and RC2 are identical to the ones in IS-95. Thus, the cell support of the open loop power control for R-ACH and R-FCH with RC1 and RC2 is the same as the one used in 2G systems.

The 3G reverse closed loop power control is based on the 2G model with two main components: the inner and outer loop. Although the main functionality of the 2G and 3G closed loop power control appears to be similar, the 3G closed loop has some differences.

In 3G, the mobile transmits the R-FCH in continuous fashion for all subrates, while in 2G the reverse traffic channel is gated by the mobile depending on the subrate. In 2G, the mobile reduces the average power for subrate frames by gating off certain power control groups (PCGs) (8 for 1/2 rate, 12 for 1/4 rate, and 14 for 1/8 rate). In this way the power level of a full rate frame for a nongated PCG is the same as a half, quarter, or one-eighth rate frame. The inner loop at the BS measures the reverse E_b/N_t of the traffic channel for every PCG. For the gated PCGs the BS sends, most probably, power-up commands to the MS. Since the MS knows which are the gated PCGs, it ignores these commands. For nongated PCGs, the BS measures the E_b/N_t of the traffic channel, which is independent of the rate of the frame.

In 3G, the reverse link transmission is continuous, and no PCG is gated. The mobile obtains a reduction in the average power of the subrate frames by repeating the PCGs more times, and reducing the power per PCG, such that the energy received at the cell is rate independent. For this reason the inner loop cannot use the E_b/N_t per PCG measurement of the R-FCH because it is rate dependent. Since the rate of the frame is known after decoding, the 3G inner loop cannot use the R-FCH measurement, and must use the R-PICH strength measurement.

Continuous transmission provides several benefits to 3G power control. The 3G inner loop operates at a genuine 800 Hz, whereas the 2G inner loop rate is lower depending on the subrate. For example, the 2G inner loop runs at 100 Hz for a 1/8 rate frame. In addition, the power variance of the R-FCH in 3G is reduced. These effects improve the coherent detection, and increase the reverse link call capacity.

Since the inner loop power control uses R-PICH signal measurements, and the output of the R-FCH outer loop is the full rate E_b/N_t set point for the R-FCH, the output of the R-FCH must be converted to reverse pilot strength units that can be used by the inner loop.

The input to the reverse outer loop algorithm is the reverse frame errors detected by the BS in every frame. The output is the R-FCH E_b/N_t set point at full rate, which is converted by the BS to the R-PICH SNR set point via offsets. The SNR set point value is mapped to an R-PICH energy threshold using a set of numerical tables. The SNR is compared to the R-PICH energy measurement in determining if the reverse power control bit is to be sent to the mobile on forward power control subchannel.

The key characteristics of the reverse link are summarized as follows:

- Channels are primarily code multiplexed
- Separate channels are used for different QoS and physical layer characteristics
- Transmission is continuous to avoid EMI (Electro-magnetic interference)
- Channels are orthogonalized by Walsh functions and I/Q split so that performance is equivalent to BPSK
- Hybrid combination of QPSK and BPSK
- Coherent reverse link with continuous pilot
- Forward power control information is time-multiplexed with the pilot
- By restricting alternate phase changes of the complex scrambling, power peaking is reduced and sidelobes are narrowed
- Independent fundamental and supplemental channels with different transmit power and FER target
- Forward error correction:
  - Convolutional codes ($k = 9$) are used for voice and data
  - Parallel turbo codes ($k = 4$) are used for high data rates on supplemental channels
- Fast reverse power control: 800 times per second
- Frame lengths
  - 20-ms frames are used for signaling and user information
  - 5-ms frames are used for control information

16.6   Evolution of cdmaOne (IS-95) to cdma2000

With a high data rate, cdmaOne operators look for a host of new network capabilities enabling them to offer new value-added services that can exploit present and future generations of technology. With the Internet and corporate intranet becoming more essential to daily business activities, the rush is on to create the wireless office that can easily tie mobile workers to the enterprise. Further, there is great potential for push technologies that deliver news and other information directly to a wireless device. This could create entirely new revenue streams for operators.

Although cdmaOne networks were not the first to offer data access, these networks are uniquely designed to accommodate data. To start with, these networks handle data and voice transmissions in much the same way. cdmaOne's inherent variable rate transmission capability allows data rate determination to accommodate the amount of information being sent, so system resources are used only as needed. Because cdmaOne systems employ a packetized backbone for voice, packet data capabilities are already inherent in the equipment. cdmaOne packet data transmission technology uses a TCP/IP-compliant cellular digital packet data (CDPD) protocol stack to enable seamless connectivity with enterprise networks and to expedite third-party application development.

Adding data to the cdmaOne network allows an operator to continue using its existing radios, back-haul facilities, infrastructure, and handsets while merely implementing a software upgrade with an interworking function. The upgrade to IS-95B allows for code or channel aggregation to provide data rates of 64–115 kbps, as well as offering improvements in soft handoffs and interfrequency hard handoffs. Equipment manufacturers have already announced IS-707 packet data, circuit-switched data, and digital fax capabilities on its cdmaOne infrastructure equipment.

Mobile IP, the proposed Internet standard for mobility, is an enhancement to basic packet data services. Mobile IP lets users maintain a continuous data connection and retain a single IP address while traveling between base station controllers (BSCs) or roaming on other CDMA networks.

One of the key objectives of ITU IMT-2000 is the creation of standards to encourage a worldwide frequency band to promote a high degree of design commonality and to support high-speed data services. IMT-2000 will utilize small pocket terminals, an expanded range of operation environments, and the deployment of an open architecture that allows the graceful introduction of newly created technology. Furthermore, 3G systems promise to deliver wireless voice services with wireline quality levels, along with the speed and capacity needed to support multimedia and high-speed data applications. Location-based services, on-board navigation, emergency assistance, and other advanced services will also be supported.

The evolution of 3G will open the door of the WLL to the PSTN and public data network access, while providing more convenient control of applications and network resources. It will also provide global roaming, service portability, zone-based ID and billing, and global directory access. 3G technology is even expected to support seamless satellite interworking.

One of the technical requirements for cdma2000 includes cdmaOne backward compatibility for voice services, vocoders, and signaling structure, as well as for privacy, authentication, and encryption capabilities.

Phase one of the cdma2000 effort, also known as 3G1X, employs 1.25 MHz of bandwidth and delivers a peak data rate of 144 kbps for stationary or mobile applications. Phase two of cdma2000, called 3G3X, will use a 5 MHz bandwidth and is expected to deliver a peak data rate of 144 kbps for mobile and vehicular applications, and up to 2 Mbps for fixed applications. Industry insiders predict that the 3G3X phase will eventually yield up to 1 Mbps for each traffic or Walsh channel. By aggregating or bundling two channels, users can achieve the 2 Mbps peak data rate targeted for IMT-2000.

The primary difference between phase one and phase two of cdma2000 is bandwidth and resulting throughput speed, or peak data rate capability. Phase two will introduce advanced multimedia capabilities and lay the foundation for popular 3G voice services and vocoder, such as voice-over IP. Since the 3G1X and 3G3X standards essentially share the same baseband radio elements, operators can take a major step toward full 3G capabilities by implementing 3G1X. cdma2000 phase two will include detailed descriptions of signal protocols, data management, and expected upscale requirements for moving from 5-MHz to 10- and 15-MHz radios in future interactions.

By migrating from the current IS-95 CDMA air interface technology to 3G1X of the cdma2000 standard, operators can achieve a two-fold increase in radio capacity and ability to handle up to 144 kbps of packet data. Phase one capabilities of cdma2000 include a new physical layer for 1× and 3× 1.25 MHz channel sizes; support for multicarrier forward link 3X options; and definitions for the 1X and 3X numerology. Operators also enjoy voice service enhancements that will produce two times the voice capacity.

In the area of extended battery life, phase one employs a quick paging channel and gated transmission of 1/8 rate to produce gains of two times the battery lives currently available. Hard-handoff enhancements between 2G and 3G systems and power control enhancements are also the key factors in the improvements of voice service.

Data services are also improved with the advent of cdma2000 phase one. Phase one features a MAC framework and packet data RLP definition to support packet data rates of at least 144 kbps.

Implementation of cdma2000 phase two will bring a host of new capabilities and service enhancements. Phase two will support all channel sizes (6×, 9× and 12×) and associated numerology and a framework for advanced cdma2000 3G voice services and vocoders—including voice-over IP. With phase two, true multimedia services will be available. To bring additional revenue opportunities for operators, multimedia services will be made possible through enhanced packet data MAC, full support for packet data services up to 2 Mbps, RLP support for all data rates up to 2 Mbps, and the advanced multimedia call model.

In the area of signaling and services, phase two cdma2000 will bring a native 3G cdma2000 signaling structure to the LAC and upper layer signaling structure. This structure will provide support for enhanced privacy, authentication, and encryption functionality. An operator's existing architecture and network equipment can greatly affect the ease of this migration. A network built on an open, advanced architecture with a clear upward migration pathway can attain 3G1X capabilities with a simple modular upward movement of the H-band operation of the radio. Networks with a less flexible architecture may be required to take the more costly steps of replacing the entire BTS. To achieve the expected 144 kbps peak data rate performance, operators can make software upgrades to networks and base stations to support 3G1X data protocol.

Packet data service node (PDSN) will be required to support data connectivity to the Internet/intranet. Many equipment vendors already offer solutions that incorporate PDSN elements, thus opening a smooth upward pathway to 3G technologies.

The recent agreement between Qualcomm and Ericsson proposes three optional CDMA modes and eventual development of a global standard that is compatible with both ANSI IS-41 and GSM mobile application part (MAP). This approach envisions the use of multimode handsets and various market-driven solutions as the surest pathway to a unified CDMA 3G standard in the next generation of wireless communications. As subscribers demand greater wireless power and convenience, the migration to 3G technology will benefit operators by supporting higher capabilities, lowering network costs, and increasing overall profitability. Figure 16.15 shows the timeline of cdmaOne evolution.

cdmaOne operators will be able to upgrade to 3G without acquiring an additional spectrum, a key component to minimum time to market without additional, significant investment. The design of cdma2000 allows for deployment of the 3G enhancements while maintaining existing 2G support for cdmaOne in the spectrum an operator has today.

Both cdma2000 phase one and cdma2000 phase two can be intermingled with cdmaOne to maximize the effective use of the spectrum according to the needs of an individual operator's customer base. For example, an operator that has a strong demand for high-speed data service may choose to deploy a combination

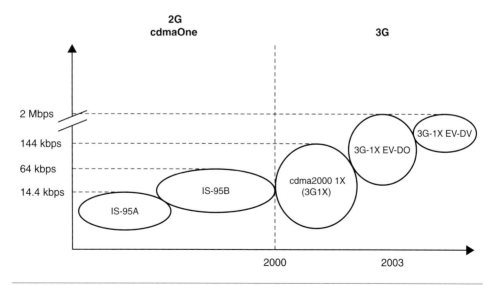

Figure 16.15   cdmaOne evolution timeline.

of cdma2000 phase one and cdmaOne that uses more channels for cdma2000 (see Figure 16.16). In another market, users may not be as quick to adopt high-speed data services and more channels will remain dedicated to cdmaOne services. As cdma2000 phase two capabilities become available, an operator has even more choices of ways in which to use the spectrum to support the new services (see Figures 16.17 and 16.18).

The network standards for the cdma2000 network consist of the interoperability specification (IOS) and ANSI-41. The IOS defines the interfaces within the radio access network (RAN), between RAN and CS-CN, and between RAN

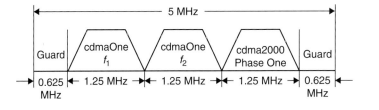

Figure 16.16   Intermixing of cdmaOne and cdma2000 phase one (3G1X).

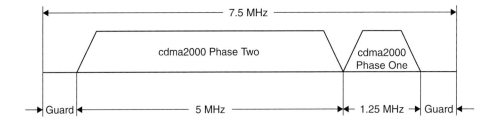

Figure 16.17   Intermixing of cdma2000 phase one (3G1X) and cdma2000 phase two (3G3X).

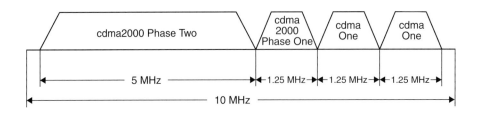

Figure 16.18   Intermixing of cdmaOne, cdma2000 phase one (3G1X) and cdma2000 phase two (3G3X).

and PS-CN. ANSI-41 defines the interfaces between different circuit-switched core network components. The components are mobile switching center (MSC), visitor location register (VLR), home location register (HLR), and authentication center (AC). In addition to these standards, cdma2000 networks leverage many Internet standards such as remote authentication dial-in user service (RADIUS), Mobile IP, etc. The Internet standards are used only for packet data services. The following interfaces are defined (see Figure 16.19):

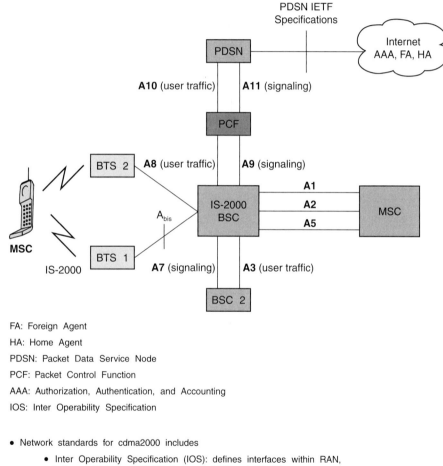

FA: Foreign Agent

HA: Home Agent

PDSN: Packet Data Service Node

PCF: Packet Control Function

AAA: Authorization, Authentication, and Accounting

IOS: Inter Operability Specification

- Network standards for cdma2000 includes
  - Inter Operability Specification (IOS): defines interfaces within RAN, between RAN and CS-CN, between RAN and PS-CN
  - ANSI-41
- IOS interfaces between BSC and MSC are A1, A2, and A5
- IOS interfaces between BSCs are A3 and A7
- IOS interfaces between RAN and PDSN are A10 and A11

Figure 16.19   cdma2000 interfaces.

- A1—signaling connection between the MSC and BSC
- A2—64 kbps DS0 to carry full-rate speech or circuit-oriented data
- A3—carries user traffic (voice/data) using ATM AAL2 and signaling using ATM AAL5 between BS and a service distribution unit (SDU); includes separate signaling and traffic subchannels
- A5—user traffic connection that carries a circuit data call between the SDU and the interworking function (IWF). This interface is only necessary when SDU and IWF are in separate pieces of equipment
- A7—signaling connection using ATM AAL5 between two BSCs to support efficient inter-BSC packet mode soft handoff
- A8—user traffic connection between source BSC and packet control function (PCF) for packet data services
- A9—signaling connection between source BSC and PCF for packet data services
- A10—user traffic path between a PCF and a PDSN
- A11—signaling connection between a PCF and PDSN for packet data services

Figure 16.20 shows the possible architecture for a cdma2000 network. The BSC and BTS provide the RAN function of the IMT functional model. The MSC/VLR and HLR are the main entities of the CS CN. The CS CN is based on ANSI-41 protocol. Other functional entities of the CS CN are AC, voice mail (VM), and message center (MC). The PS CN of the cdma2000 network offers connectivity to the Internet. The PDSN is the primary element of the PS CN. It provides an interface to the RAN and routes packets from the Internet to the RAN and vice versa. The PS CN uses mobile IP protocol to provide mobility for packet data services. The mobile IP entities are the home agent (HA) and foreign agent (FA). The AAA server provides authentication, authorization, and accounting functionality in PS CN. Figure 16.21 shows the Internet access to the end user.

16.6.1   cdma2000 1X EV-DO

cdma2000 1X EV-DO (also called high data rate (HDR)) is the high-speed, high-capacity wireless data-only technology that provides up to 2.4 Mbps in a 1.25 MHz channel. In the HDR, the system is optimized for packet data services, with a decentralized architecture based on the IP protocol/platform. The HDR is compatible with CDMA IS-95 networks. In conjunction with an existing voice network, or stand-alone data network, the HDR offers a high-speed, cost-effective data solution. Its aim is to maximize throughput subjected to an acceptable delay. The HDR optimizes forward (down) link for packet data. Unequal delays and data rates depend on users' locations. The reverse (up) link structure is similar to that of cdma2000 1X.

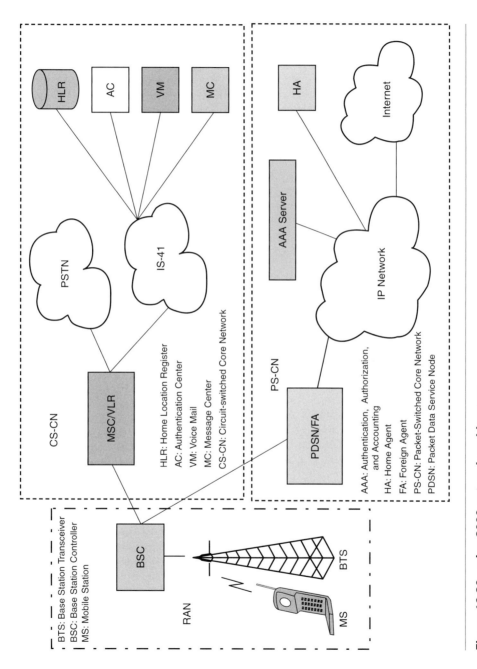

Figure 16.20 cdma2000 network architecture.

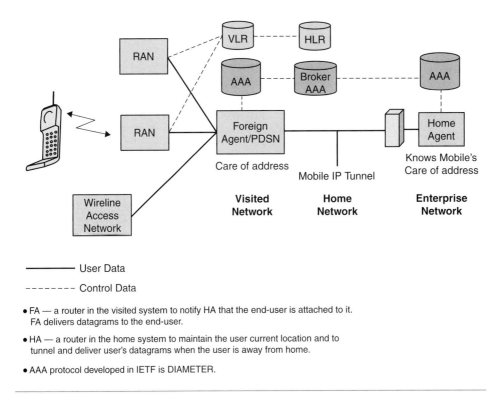

—————— User Data

- - - - - - - Control Data

- FA — a router in the visited system to notify HA that the end-user is attached to it.
  FA delivers datagrams to the end-user.

- HA — a router in the home system to maintain the user current location and to
  tunnel and deliver user's datagrams when the user is away from home.

- AAA protocol developed in IETF is DIAMETER.

Figure 16.21   Internet access to end-user in the cdma2000 network.

The HDR does not rely on any element in the CDMA voice network to provide service, mobility, or roaming. The operator does not require MSC or network elements such as HLR and VLR. As a result, HDR can be deployed by any voice operator irrespective of the voice technology it is currently using. All that is needed to launch HDR service is 1.25 MHz of paired spectrum. An HDR network has three key elements: radio nodes (RNs), a radio network controller (RNC), and a packet data serving node (PDSN). Each radio node typically supports three sectors and serves one cell site. A dedicated transceiver in each sector terminates the HDR airlink between the user equipment (UE) and RN. Higher layers of the HDR protocol are processed at the RNC. The RNC also manages handoffs and passes user data between the RNs and the PDSN. The PSDN is a wireless edge router that connects the radio network to the Internet.

In addition to the RNC and PDSN, an HDR data center has an aggregation router, an element manager system (EMS), and several Internet service provider (ISP) servers. The aggregate router terminates IP traffic from the RNs and passes it to the RNC. The EMS manages the radio access network with commonly used ISP servers. It includes standard IP servers for the domain name system (DNS), dynamic host configuration protocol (DHCP) and authentication, authorization, and accounting (AAA).

The basic unit of transmission on the downlink is the airlink frame. Airlink frames destined for different users in the same sector are time division multiplexed (TDM). The radio nodes transmit each airlink frame at the link rate, which can vary between 38.4 kbps and 2.4 Mbps. The link rate depends on the SNR at the subscriber's location. SNR may vary significantly within a cell. Since the link between the network and the subscribers is shared, depending on system load, the actual data rate experienced by a subscriber can be less than link rate. The number of subscribers receiving data at the same time in the same sector determines system load.

On the uplink, subscribers can transmit at data rates ranging from 9.6 to 153.6 kbps. Unlike the downlink, in which a scheduler time-division multiplexes airlink frames over the channel, the uplink uses CDMA, which allows multiple users to transmit at the same time. The uplink sector throughput represents the total link capacity of the uplink. Computed as sum of the uplink data rates of all simultaneously transmitting subscribers, it is about 250 kbps per sector or about 1/5 of the downlink sector throughput. HDR radio nodes equitably share this throughput among all active subscribers by using a rate control mechanism.

The HDR downlink allows variable delays within the bounds of maximum/minimum delay equal to 8 achieved by scheduling at higher layers. The rate allocation is based on SNR measurements reported by mobiles. Turbo codes are used for increased coding gains. There is no power control and no soft or softer handoff on downlink.

Downlink packets are time multiplexed and transmitted at full available power. Data rates and slot lengths vary according to user channel conditions. Short pilot bursts and periodic transmission of control information are used for users without data (see Figure 16.22). Access point (AP, e.g., BS) uses pilot bursts to estimate accurately and rapidly channel conditions. The AP also determines E_b/N_t to predict effective received SNR. SNR is mapped to the maximum data rate value. Uplink data rate request channel (DRC) (see Figure 16.23) is sent to AP with a 4-bit value that maps SNR into one of the data rate modes (see Table 16.6).

Each point-to-point protocol (PPP link) is carried by a separate radio link protocol (RLP) stream. This gives flexibility to allow for finer control of the QoS. The model for the transport of signaling streams is based on PPP. Signaling is partitioned into link control protocol (LCP) and stream control protocol (SCP) (see Figure 16.24). LCP is used at the start of the session and to control the radio link during the session. For example, LCP is used at the start of the session to negotiate the link layer authentication type that will be used for the duration of the session. SCP is used to carry stream-specific signaling messages.

Figure 16.25 shows the HDR and 3G1X mobile IP (MIP) architecture. Note that the changes are the addition of MIP, software upgrade only, and that HDR

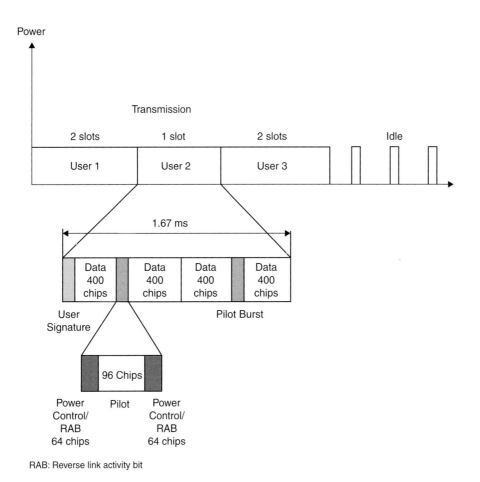

Figure 16.22   Access point transmission.

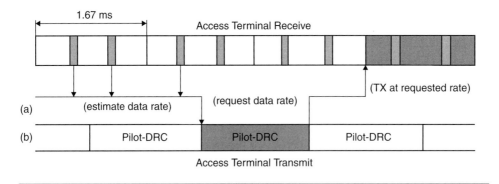

Figure 16.23   (a) Channel estimation and (b) DRC timing diagram.

Table 16.6 Data rates of 3G1X EV-DO in downlink direction.

Data rate (kbps)	Encoder packet duration (ms)	FEC rate	Modulation	PN chips per bit
36.8	26.67	1/5	QPSK	32
76.8	13.33	1/5	QPSK	16
153.6	6.66	1/5	QPSK	8
307.2	3.33	1/5	QPSK	4
614.4	1.67	1/5	QPSK	2
307.2	6.66	1/3	QPSK	4
614.4	3.33	1/3	QPSK	2
1228.8	1.67	1/3	QPSK	1
921.6	3.33	1/3	8PSK	1.33
1843.2	1.67	1/3	QPSK	0.67
1228.8	3.33	1/3	16-QAM	1
2457.6	1.67	1/3	16-QAM	0.5

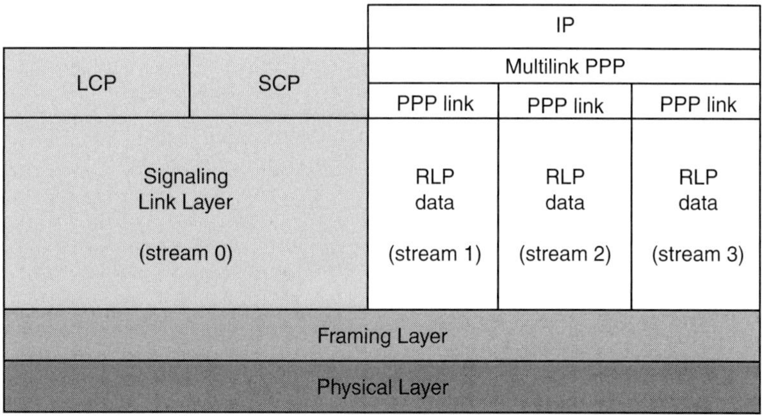

- IP: Internet Protocol
- PPP: Point-to-point Protocol
- LCP: Link Control Protocol
- SCP: Stream Control Protocol
- RLP: Radio Link Protocol

Figure 16.24   Protocol stack in HDR.

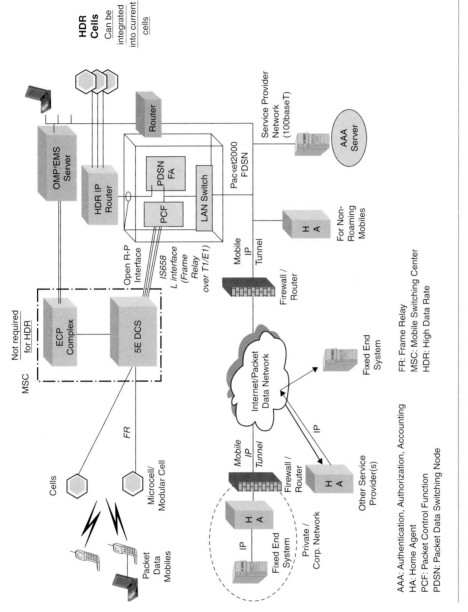

AAA: Authentication, Authorization, Accounting
HA: Home Agent
PCF: Packet Control Function
PDSN: Packet Data Switching Node

FR: Frame Relay
MSC: Mobile Switching Center
HDR: High Data Rate

Figure16.25   HDR and cdma2000 1X mobile IP architecture.

access to the Packet 2000 PDSN is via separate backhaul. The HDR radios may coexist with currently installed cells. The L2TP tunneling is replaced with MIP tunnels—one of the differences from simple IP.

Table 16.7 shows a comparison of high-speed downlink packet access (HSDPA) and the cdma2000 1X EV-DO.

16.6.2 cdma2000 1X EV-DV

The cdma2000 1X EV-DV is an enhancement to the cdma2000 1X data-carrying capability that is designed to deliver rates that are significantly higher than the cdma2000 revision 0 system. Revision C of the standard specifies higher data rates on the forward link (FL), whereas revision D addresses increased rates on the reverse link (RL).

The cdma2000 1X EV-DV system is designed to deliver greater spectrum usage efficiencies, backward compatibility for legacy handsets, and all previous versions of IS-95 and cdma2000 (including existing channels and signaling structure) and support for the broadest possible range of applications. 1X EV-DV enhancements occur at the physical layer, MAC, RLP, and layer 3.

cdma2000 1X EV-DV is designed to deliver real-time circuit-switched voice services and high rate packet data services in the same RF carrier. The cdma2000 1X EV-DV delivers a peak data rate of 3.09 Mbps and a typical sector throughput of 1 Mbps in a 1.25 MHz frequency channel. 1X EV-DV delivers a system-wide packet call user throughput that ranges from 420 kbps to 1.7 Mbps, depending on traffic and channel models, and up to 451.2 kbps peak reverse link data rate.

Table 16.7 Comparison of HSDPA and cdma2000 1X EV-DO.

Features	HSDPA	cdma2000 1× EV-DO
Downlink frame size	2 ms transmission time interval (TTI) (3 slots)	1.25, 2.5, 5, 10 ms variable frame size (1.25 ms slot size)
Channel feedback	Channel quality reported at 2 ms rate or 500 Hz	SNR feedback at 800 Hz (every 1.25 ms)
Data user multiplexing	TDM/CDM	TDM/CDM (variable frame)
Adaptive modulation and coding	QPSK and 16-QAM mandatory	QPSK, 8-PSK and 16-QAM
Hybrid ARQ	Incremental redundancy	Async. incremental redundancy
Spreading factor (SF)	SF = 16 using OVSF channelization codes	Walsh code length 32
Control channel approach	Dedicated channel pointing to shared channel	Common control channel

Because it supports concurrent voice and data services, 3G1X EV-DV does not require new spectrum or dedicated spectrum for data-only use, a feature that presents CDMA operators with a more attractive cost structure than other evolutionary options.

The cdma2000 1X EV-DV evolutionary model provides operators with exceptional flexibility in meeting daily fluctuations in capacity demands. When deployed for voice, data, or in a combined voice/data environment, 1X EV-DV delivers needed voice resources during peak-usage commuting hours, while at the same time providing carriers with the resources that will be needed to meet growing demands for data capacity during data busy hours.

The cdma2000 1X EV-DV is not an overlay system, but is in fact a true upgrade from cdma2000 1X, and provides backward compatibility for IS-95 and cdma2000 revision 0 mobile devices. cdma2000 1X EV-DV also reuses existing cell sites, and thus maintains current coverage in cdma2000 networks. Figures 16.26 and 16.27 provide comparisons of peak data rate and aggregate throughput of cdma2000 1X and cdma2000 1X EV-DV.

cdma2000 1X EV-DV New Channels

The cdma2000 1X EV-DV specifications incorporate three new control channels and one new traffic channel. In the forward link, there are two new channels added:

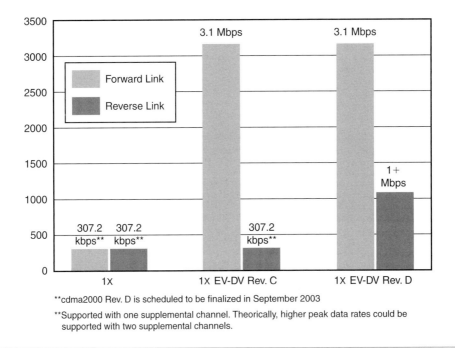

Figure 16.26 Peak data rate comparison.

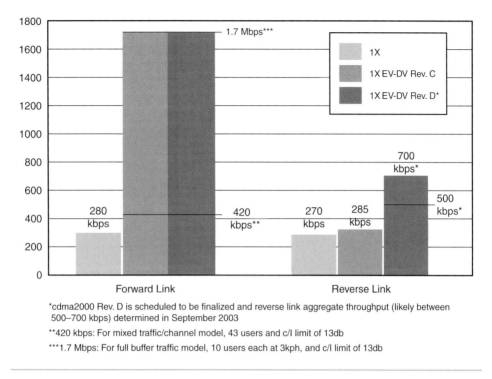

Figure 16.27   Aggregate throughput comparison.

(1) a new control channel, forward packet data control channel (F-PDCCH), and (2) a new traffic channel, forward packet data channel (F-PDCH). Two new control channels are provided in the reverse link: (1) the reverse channel quality (R-CQICH), and (2) reverse ACK channel (R-ACKCH). Operators can benefit from the fact that cdma2000 1X EV-DV channels are integrated with current cdma2000 1 × channels to cover areas that need a higher data rate.

cdma2000 1X EV-DV Features
The following are the new features of 1X EV-DV.

- **Forward link capacity.** 1X EV-DV leverages a number of improvements including adaptive modulation and channel coding schemes, the use of both time division multiplexing (TDM) and code division multiplexing (CDM), a forward packet data channel, and hybrid automatic repeat request (HARQ) at the physical layer to deliver forward link data rates up to 3.09 Mbps and average sector throughput of 1 Mbps.

- **Backward compatibility.** Because it incorporates a number of relevant cdma2000 standards such as the reuse of 1X reverse link channels, IOS interoperability, cdma2000 MAC and signaling layer procedures, and radio channel handoff, cdma2000 1X EV-DV provides seamless support for voice and legacy services. This backward compatibility means consumers can use an operator's entire network with a single terminal. For operators, this strength extends the useful life of the existing infrastructure while opening a smooth evolutionary path from the current cdma2000 1X infrastructure.

- **Concurrent voice and data.** cdma2000 1X EV-DV supports voice and data services in the forward and reverse links. The 1X EV-DV network can support 1X voice users and 1X EV-DV data users on the same channel at the same time, or 1X EV-DV users can have 1X voice and DV data sessions active at the same time. This capability gives operators improved spectrum flexibility, the ability to share the voice and data spectrum, and the freedom to provide concurrent voice and data.

- **Multiple traffic types.** The specification supports multiplexing of signal and user data over the forward packet data channel and multiple concurrent data sessions. This capability means users can use more than one application—such as voice and data at the same time. For operators this capability translates into the ability to generate revenue from multiple simultaneous applications, without the need to dedicate a separate channel for each application.

- **TDM/CDM.** cdma2000 1X EV-DV is the only standard in the cdma2000 development path that enables both TDM and CDM scheduling, favoring TDM where the technology works best and supporting CDM for wireless application protocol (WAP), voice over IP (VoIP), streaming video, and other data services. Because data is sent in bursts rather than on a continuous basis, TDM/CDM statistical multiplexing enables the system to maximize throughput gain. Operators can leverage this feature to enjoy optimum flexibility in a demanding marketplace.

- **Hybrid ARQ.** By migrating ARQ from the upper layer to the physical layer, cdma2000 1X EV-DV maintains high bandwidth through the very rapid retransmission of frames received in error. Rather than discarding failed transmission attempts, the HARQ technique combines those failed attempts with the current attempt. This supports a higher tolerance to selection errors for faster adaptive modulation and coding (AMC), and creates a more potent code. Two basic forms of combining techniques incorporated with HARQ are the chase combining and incremental redundancy (IR).

- **Adaptive modulation and coding.** To provide real time adaptation in a changing RF environment, 1X EV-DV uses forward link modulation and coding variation to adoptively assign users the best modulation and coding rate under current channel conditions, a capability known as "channel sensitive" or "opportunistic" scheduling to maximize the multiuser diversity gain. By varying the number of bits per RF frame and the coding algorithm, and by providing continuous feedback to continually maximize link throughput, this approach delivers higher data rate performance and supports services that could not be provided using previous-generation CDMA technology.

- **Cell selection:** Cell selection takes advantage of micro diversity by allowing the handset to select the best serving sector.

Subscriber Benefits

By delivering improved overall data rates, and by allowing users to perform simultaneous voice and data tasks on the same device (i.e., holding a voice conversation while at the same time downloading e-mail messages) takes consumers an important step toward the promise of the 3G future.

End users will also benefit from the low latency response of the 1X EV-DV system. So far, with even small amounts of data transferred, end users will see a much faster response by the network. This fast response time makes applications such as web browsers much more usable with 1X EV-DV than with cdma2000 1X networks.

The efficiencies supported by 3G1X EV-DV will allow operators to deliver data-intense services more quickly, more effectively, and at a lower cost. Those capabilities translate directly into better services and lower prices for subscribers.

cdma2000 1X EV-DV in the Marketplace

Real time gaming applications typically generate a relatively small volume of delay-sensitive traffic, which arrives at a fairly regular rate as updates to the game engine state. The popular on-line game requires about 1.6 kbps on each of the forward and reverse links. While the packets are not as continuous as voice, they do arrive in a reasonably regular fashion at a rate of roughly one packet per 40 ms. To maintain playability of the game, the application round trip delay must be less than 160 ms, and for this reason the traffic must obtain immediate radio resources. Since cdma2000 1X EV-DO systems do not meet these requirements effectively, 1X EV-DV is ideally positioned to exploit the on-line gaming market.

On the forward link, with slot sizes of 1.25, 2.5, and 5 ms, and TDM/CDM capability, 1X EV-DV serves more users per cell. On the reverse link, 1X EV-DV frame size is 20 ms for revision C and 10 ms for revision D, which will meet real-time gaming delay requirements.

The cdma2000 1X EV-DV evolutionary pathway also provides significant support for VoIP. IP connectivity is expected to dominate the terrestrial transmission network for cellular systems, with end users expecting seamless connectivity to services delivered exclusively via IP networks. The all-IP feature will enable transcoder-free operation and the use of wideband speech coding, and all-IP voice efficiency is supported by the IP-optimized voice codec.

The cdma2000 1X EV-DV requires a lower end-to-end infrastructure investment compared to cdma2000 1X EV-DO, and enables simultaneous voice and data services in either a voice capacity enhancement or all-IP deployment. Due to its mixed-traffic capabilities (supporting voice, video, gaming, WAP, HTTP, and FTP) 1X EV-DV also delivers the superior QoS performance required to ensure a reliable VoIP revenue stream. When deployed in a voice, data, or dual voice/data environment, 1X EV-DV provides an assured level of voice resources during high-volume commuting hours, as well as greater data resources when those applications take off. 1X EV-DV can be deployed for data-only services to deliver performance at least on a par or better when compared to 1X EV-DO; and when implemented in a voice/data environment, 1X EV-DV enables a class of services that 1X EV-DO cannot support.

16.7   Technical Differences between cdma2000 and WCDMA

We conclude the chapter by listing major technical differences between the two wideband CDMA technologies (see Table 16.8) and including their deployments

Table 16.8 Major technical differences between cdma2000 and WCDMA.

	cdma2000	WCDMA
Core network	ANSI-41 MAP	GSM MAP
Channel bandwidth	1.25 MHz (1X), 3.75 MHz (3X)	5.0 MHz
Channelization codes	4-128 (1X), 4-256 (3X)	4-256
Chip rate	1.2288 Mcps (1X), 3.6864 Mcps (3X)	4.096 Mcps (DOCOMO) 3.84 Mcps (UMTS)
Synchronized base station	Yes	No; but synchronized BS is optional
Frame length	5 ms (signaling), 20, 40, 80 ms physical layer frames	10 ms for physical layer, 10, 20, 40, and 80 ms for transport layer
Multi-carrier spreading option	Yes, but in cdma2000 1X (direct spread)	No (direct spread)
Modulation	QPSK (forward link), BPSK (reverse link)	QPSK (both links)
Modes of operation	FDD	FDD and TDD
Source identification code for sector	One PN code (32,768 chips), 512 unique offsets are generated using PN offsets	512 unique scrambling codes each identifying a sector (38,400 chips)
Source identification code for mobile	One long PN code (2^{42}) chips, unique offsets are generated based on ESN, not assigned by sector	Unique scrambling codes assigned by sector

(Continued)

	cdma2000	WCDMA
Channel coding	Convolutional and turbo code	Convolutional and turbo code
Power control	Both links (800 Hz)	Both links (1500 Hz)
Circuit-switched core network	IS-41, MSC/HLR/ AC	GSM-MAP, MSC/ HLR/AUC
Packet-switched core network	IETF based, PDSN/ AAA/HA/FA	GPRS based, SGSN/ GGSN
Voice coder	EVRC	AMR
Peak data rate	614 kbps	2 Mbps
Multimedia services	Yes	Yes
Overhead	Low (because of shared pilot code channel)	High (because of non-shared pilot code channel)

(see Figure 16.28). We also provide end-to-end call flows in the two systems (see Figures 16.29 and 16.30) [14–16].

Figure 16.31 shows the layout of the cdma2000 1X and cdma2000 1X EV-DO network. The main differences in the GSM/UMTS architecture are in the packet domain where the PDSN is used. It has a similar role to the SGSN and GGSN in UMTS. Mobility management with 3GPP2 is based on mobile IP (RFC 2002) instead of GPRS mobility management in the GPRS/UMTS packet-switched network.

16.8 Summary

The ongoing growth in demand for high-speed data services and multimedia applications over mobile wireles networks has set up new requirements and objectives for the next generation of air interface protocols and network architectures. Although the channelization, signaling, and access protocols of 2G cellular systems were designed to efficiently support symmetric circuit-switched data and voice traffic, most of the new data applications are IP-based with highly asymmetric packet-switched traffic. This asymmetric and bursty nature of multimedia packet data traffic along with the variability of data rates and packet sizes and the complexity of QoS management makes conventional voice-oriented channelization and access protocols of 2G systems inefficient.

3G systems use new physical and logical channelization schemes with enhanced media and link access protocols. Also, to maximize the spectrum efficiency, the physical layer designs utilize advanced coding, link adaptation, and diversity schemes as well as power and interference control mechanisms.

The scope of the 3GPP2 activities was to harmonize different variations of cdma2000 in a single family of standards that is based on the evolution of cdmaOne air interface. This scope was also expanded to include the development of a data-optimized

Service	2G	2.5G	2.5G (Step 2)	3G			
				WCDMA	1x EV-DO	1x EV-DV	3x RTT
Service	GSM, TDMA, CDMA, iDEN	GPRS, cdma2000		WCDMA	1x EV-DO	1x EV-DV	3x RTT
Speed	9.6 kbps and 14.4 kbps	20 kbps, 40 kbps		2.4 mbps			3x RTT
Europe & World	GSM			→ UMTS			
VoiceStream AT & T Wireless	GSM	→ GPRS		WCDMA	1XEV-DO		→ 3x RTT
Cingular Wireless	GSM	→ GPRS	→ EDGE	WCDMA	1x EV-DO	1x EV-DV	3x RTT
	GSM	→ GPRS		WCDMA	1x EV-DO	1x EV-DV	3x RTT
AT & T Wireless	TDMA	→ GPRS	→ EDGE	WCDMA	1x EV-DO	1x EV-DV	3x RTT
Sprint PCS	CDMA	→ CDMDA2000 1x RTT					→ 3x RTT
Verizon Wireless	CDMA	→ CDMA2000 1x RTT					→ 3x RTT
Nextel	iDEN	→ CDMA2000 1x RTT				→ 1x EV-DO	→ 3x RTT

(1) Started out with low speed Global System Mobile (GSM), TDMA services, CDMA, and Nextel's iDEN (integrated Digital Enhanced Network) that was enhanced TDMA service.

(2) Moved next to General Packet Radio Service (GPRS)—a packet service that is suitable for non-real-time (email, fax, asymmetric web browsing) internet applications. It is a circuit sharing and time slot sharing service that is bursty but better than an earlier service called HSCSD—High Speed Circuit Switched Data service that allowed users to concatenate time slots in a TDMA service thus giving users more throughput and higher data transfer speeds in the early systems.

(3) cdma2000, following a different path than the GSM standard approach was taken by some of the vendors. 1× means one times the single CDMA bandwidth being used (1.25 MHz radio channel), RTT means Radio Transrrission Technology.

Figure 16.28   Deployment of 3G systems.

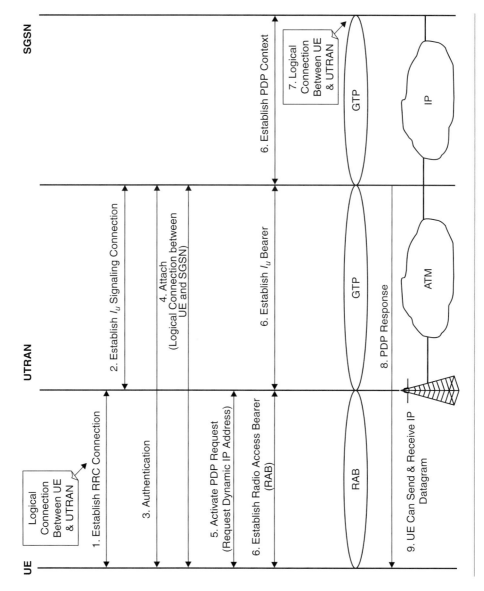

Figure 16.29   End-to-end call flow in WCDMA.

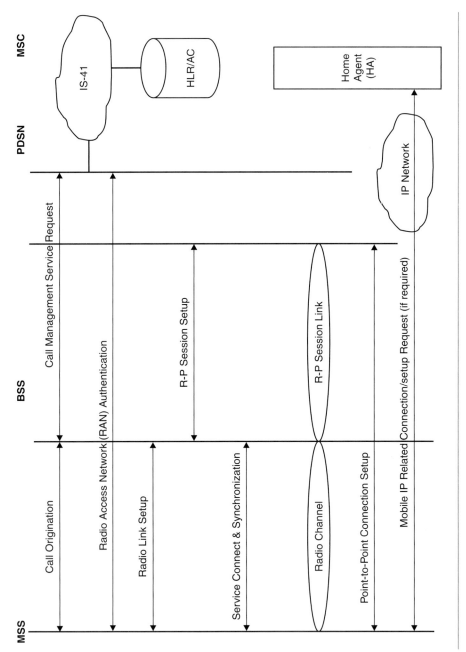

Figure 16.30   End-to-end call flow in cdma2000.

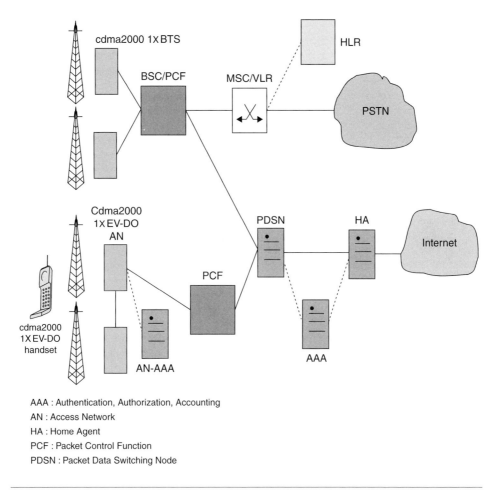

Figure 16.31   cdma2000 1X and cdma2000 1X EV-DO network.

air interface called the high-rate data (HDR) system. In the development of cdma2000 systems the core network specifications are based on an evolved ANSI-41 and IP network; however, the specifications also include the necessary capabilities for operation with an evolved GSM-MAP-based core network.

The preliminary release of cdma2000 proposed as an RTT to ITU was IS2000 Release 0, which is also referred to as the 1X RTT system. The next cdma2000 release was Release A that was published in the year 2000. It includes both the narrowband (1X) and wideband (3X) multicarrier (MC) modes occupying 1.25 and 3.75 MHz, respectively. Shortly after release A, release B of cdma2000 was issued with a few signaling protocol improvements. In 2002, cdma2000 Release

C, called the 1X EV-DV system was published. cdma2000 Release D was published by 3GPP2 in 2004.

In this chapter, we first presented the cdma2000 layering structure including logical and physical channels. We focused on the cdma2000 physical layer and provided features of forward and reverse links. Next we outlined the evolution plan for cdmaOne to cdma2000 and presented details of the 1X EV-DO and 1X EV-DV systems. We concluded the chapter by providing the major technical differences between cdma2000 and UMTS/WCDMA and discussing the differences between cdma2000 1X EV-DO and high-speed downlink packet access (HSDPA).

Problems

16.1   What is cdma2000?

16.2   Discuss the layering structure used in cdma2000.

16.3   What are the PLDCF-specific protocols in the cdma2000? Discuss.

16.4   Discuss forward and reverse link channels in the cdma2000.

16.5   Briefly discuss the forward and reverse link features in cdma2000.

16.6   How is orthogonal modulation used in cdma2000?

16.7   Why is power control used in cdma2000 and WCDMA?

16.8   Discuss briefly cdma2000 1X EV-DO—a derivative of cdma2000.

16.9   What are the major differences between 1X EV-DO and HSDPA?

16.10   Discuss briefly the cdma2000 1X EV-DV system.

References

1. CDMA for Next Generation Mobile Communications Systems. *IEEE Communications Magazine*, vol. 36, no. 9, September 1998.

2. Dahlman, E., Gumudson, B., Nilsson, M., and Skold, J. UMTS/IMT-2000 Based Wideband CDMA. *IEEE Communications Magazine*, vol. 36, no. 9, September 1998.

3. Garg, V. K., Halpern, S., and Smolik, K. F. Third Generation (3G) Mobile Communication Systems. 1999 IEEE International Conference on Personal Wireless Communications, February 1999, Jaipur, India.

4. Holma, Harri, and Toskala, Antti. *WCDMA for UMTS*. New York: John Wiley and Sons, LTD, 2000.

5. Knisley, D., Quinn, L., and Ramesh, N. cdma2000: A Third Generation Radio Transmission Technology. *Bell Labs Technical Journal*, vol. 3, no. 3, J1. September 1998.

6. Ojanpera, T., and Prasad, R. "An Overview of Third-Generation Wireless Personal Communications: European Perspective." *IEEE Personal Communications*, vol. 5, no. 6, December 1998, pp. 59–65.

7. Rao, Y. S., and Kripalani, A. cdma2000 Mobile Radio Access for IMT-2000. 1999 IEEE International Conference on Personal Wireless Communications, February 1999, Jaipur, India.

8. Shanker, B., McClelland, S. Mobilising the Third-Generation [Cellular Radio], *Journal of Telecommunications (International Edition)*, August 1997.

9. Sollenberger, N. R., Seshadri, N., and Cox, R. "The Evolution of IS-136 TDMA for Third-Generation Wireless Services." *IEEE Personal Communications*, vol. 6, no. 3, June 1999, pp. 8–18.

10. TR 45.5 *The cdma2000 ITU-RTT Candidate Submission*. TR 45-ISD/98.06.02.03, May 15, 1998.

11. TIA/EIA IS-2000-1. "Introduction to cdma2000 Spread Spectrum Systems." November 1999.

12. TIA/EIA IS-2000-2. "Physical Layer Standard for cdma2000 Spread Spectrum Systems." November 1999.

13. TIA/EIA IS-2000-3. "Medium Access Control (MAC) Standard for cdma2000 Spread Spectrum Systems." November 1999.

14. Universal Mobile Telecommunications System (UMTS). Requirements for the UMTS Terrestrial Radio Access System (UTRA). ETSI Technical Report, UMTS 21.01 version 3.0.1, November 1997.

15. Universal Mobile Telecommunications System (UMTS). Selection Procedure for the Choice of Radio Transmission Technologies of the UMTS. ETSI Technical Report, UMTS 30.03 version 3.1.0, November 1997.

16. Universal Mobile Telecommunications System (UMTS). Requirements for the UMTS Terrestrial Radio Access System (UTRA) Concept Evaluation. ETSI Technical Report, UMTS 30.06 version 3.0.0, December 1997.

CHAPTER 17

Planning and Design of Wide-Area Wireless Networks

17.1 Introduction

The wireless network planning process can be divided into three phases: the initial planning (dimensioning) phase, the detailed radio network planning, and the operation and optimization phase. Each phase requires additional support functions such as propagation measurements, key performance indicator (KPI) definitions, and so on. Both uplink as well as downlink have to be analyzed. In 2G systems with mostly voice traffic, the links tend to be in balance; however, in 3G systems with both voice and data traffic, one of the links can be loaded more than the other, so that either link could be the limiting factor in determining the cell capacity or coverage. The radio waves propagation calculation is basically the same for all radio access technologies, with the exception that different propagation models (see Chapter 3) could be used. Another common feature is the interference analysis. In the time division multiple access/frequency division multiple access (TDMA/FDMA) system, the interference analysis is required for frequency allocation, whereas in code division multiple access (CDMA) it is needed for cell loading and sensitivity analysis.

The dimensioning phase is used to estimate the approximate number of cell sites required, base station configurations, and number of network elements in order to determine the projected costs and associated investments.

The detailed planning phase takes into account the real site locations, real propagation conditions, and real user distributions based on the operator's traffic forecasts. After the detailed planning has been performed, network coverage, capacity, and other key performance indicators representing the network can be analyzed.

The network operation and optimization phase deals with individual cell and system testing. It is used to verify key performance indicators in the detailed planning phase to optimize the network.

In this chapter, we focus on the dimensioning phase and demonstrate steps to determine the number of cell sites, the number of radio network controllers (RNCs) and other network elements.

17.2 Planning and Design of a Wireless Network

The planning of a wireless network is a multidiscipline task in which competing requirements must be balanced. Network plans change throughout the life of a wireless network. During the initial phase, coverage is the primary issue and traffic demand is secondary. Too small a network initially will require expansion later; in the first few years, human resources need to be devoted to the roll-out of the network, not to re-engineering the existing network.

To establish the size of a cellular network, we need to know the:

- Network topology: base-station transceiver-base station controller-mobile switching center (BTS-BSC-MSC)
- Link capacity: BTS-BSC, BSC-MSC
- BSC sizing
- MSC sizing

Cell and frequency planning in a cellular network requires initial traffic estimation, a traffic growth plan, the selection of an initial design, an initial cell plan, a network expansion plan, a selection of reuse pattern, the development of an initial theoretical plan, computer simulation, localized performance measurements, and final network design. We need the following data to develop a traffic plan for a cellular network:

- The number of users
- Users' behavior, i.e., heavy or light users
- Busy-hour traffic as a percentage of the total traffic
- Users' distribution over the service area
- Division of the service area into zones of different traffic density

A cellular network growth plan involves time to roll out the network, requirements of additional cells, requirements of microwave link sites, and the cost to rework network versus initial deployment costs.

Basic traffic engineering of a cellular network is based on coverage. In this case, the maximum cell radius based on the required in-building coverage criterion is used. Often a maximum cell radius of less than 10 km is used to determine the required number of cells in the coverage area. For some in-building coverage in large urban areas, cell radius may be 1 km or even less to achieve good in-building coverage.

To design for capacity, market forecasts of average subscriber call minutes per month are used to determine the number of busy hours per day. This is then used to calculate the average busy-hour traffic per subscriber. The frequency reuse plan is established from the available spectrum. The traffic capacity of a cell is determined based on a grade of service (GoS) criterion. Market forecasts for total number of subscribers for a given geographic zone are used to establish the number of base stations (BSs) in the zone. The geographic area for each zone is quantified, and the number of base stations and cell sites for each zone is calculated. During the first few years the network design is coverage driven, and in later years is traffic- demand driven. To limit expanding cell sites in the first few years when human resources are focused on network roll-out, the initial deployment is based on a four- to five-year design period.

Example 17.1

Using the following data for a GSM1800 network, calculate (1) average busy-hour traffic per subscriber, (2) traffic capacity per cell, (3) required number of base stations per zone, and (4) the hexagonal cell radius for the zone.

- Subscriber usage per month = 150 minutes
- Days per month = 24
- Busy hours per day = 6
- Allocated spectrum = 4.8 MHz
- Frequency reuse plan = 4/12
- RF channel width = 200 kHz (full rate)
- Present number of subscribers in the zone = 50,000
- Subscriber growth = 5% per year
- Area of the zone = 500 km^2
- Initial installation based on a four-year design
- Capacity of a base station transceiver (BTS) = 30 Erlangs

Solution

Erlangs per subscriber $= \dfrac{150}{24 \times 6 \times 60} = 0.0174$

Total number of RF carriers $= \dfrac{4.8 \times 1000}{200} = 24$

RF carriers per cell with a reuse factor of 4 $= \dfrac{24}{4 \times 3} = 2$ (3 cells per BTS)

Traffic channels per cell (assuming two control channels per cell) $= 2 \times 8 - 2 = 14$

Traffic capacity of a GSM cell at 2% GoS (using Erlang-B table in Appendix A) = 8.2 Erlangs

Traffic per BTS $= 8.2 \times 3 = 24.6 < 30$ Erlangs

Maximum number of subscribers per BTS $= \dfrac{24.6}{0.0174} \approx 1414$

Number of subscribers at initial installation $= 50,000(1 + 0.05)^4 = 60,775$

$$\text{Number of BTSs in a zone} = \frac{60,775}{1414} \approx 43$$

$$\text{Average hexagonal cell radius} = \sqrt{\frac{500}{43 \times 2.6}} = 2.1\,\text{km}$$

Example 17.2

Using the following data for a GSM network, estimate the voice and data traffic per subscriber. If there are 40 BTS sites, calculate voice and data traffic per cell.

- Subscriber usage per month: 150 minutes
- Days per month: 24
- Busy hours per day: 6
- Allocated spectrum: 4.8 MHz
- Frequency reuse plan: 4/12
- RF channel width: 200 kHz (full rate)
- Present number of subscribers in a zone: 50,000
- Subscriber growth per year: 5%
- Network roll-over period: 4 years
- Number of packet calls per session (NPCS): 5 (see Figure 17.1)
- Number of packets within a packet call (NPP): 25
- Reading time between packet calls (T_r): 120 s
- Packet size (NBP): 480 bytes
- Time interval between two packets inside a packet call (T_{int}): 0.01 s
- Total packet service holding time during one hour (T_{tot}): 3000 s
- Busy hour packet sessions per subscriber: 0.15

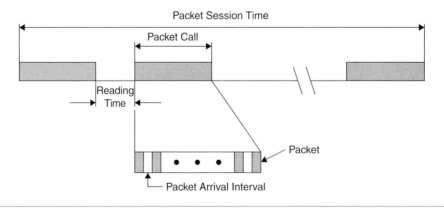

Figure 17.1 Packet session.

- Penetration of data subscribers: 25%
- Data rate of each subscriber: 48 kbps
- Packet transmission time: 10 s

Solution

No. of bits transmitted during packet session time $= \dfrac{\text{NPCS} \times \text{NPP} \times \text{NBP} \times 8}{1000}$

$$= \frac{5 \times 25 \times 480 \times 8}{1000} = 480 \text{ kilobits}$$

Packet transmission time $= \dfrac{\text{NPCS} \times \text{NPP} \times \text{NBP} \times 8}{1000 \times \text{Rate}} = \dfrac{480}{48} = 10 \text{ s}$

Refer to Figure 17.1:

Packet session time $=$ Packet transmission time $+ T_r (\text{NPCS} - 1) + T_{\text{int}} (\text{NPP} - 1)$

$$= 10 + 120(5 - 1) + 0.01(25 - 1) = 490.24 \text{ s}$$

Number of packet sessions per subscriber during busy hour $= \dfrac{0.15 \times 3000}{490.24}$

$$= 0.918$$

Bits per subscriber per second $= \dfrac{480 \times 0.918}{3600} = 0.1224 \text{ kbps}$

Voice Erlang per subscriber $= \dfrac{150}{24 \times 6 \times 60} = 0.0174$

Number of subscribers at initial installation $= 50{,}000 \times (1 + 0.05)^4 = 60{,}775$

Data subscribers $= 0.25 \times 60{,}775 = 15{,}194$

Total voice traffic during busy hour $= 60{,}775 \times 0.0174 = 1058$ Erlangs

Total data throughput during busy hour $= 15{,}194 \times 0.1224 = 1860 \text{ kbps}$

With 40 BTS sites per zone

Voice traffic per BTS $= \dfrac{1058}{40} = 26.45$ i.e., 8.82 Erlangs per cell (sector)

Data throughput per BTS $= \dfrac{1860}{40} = 46.5$ i.e., 15.5 kbps per cell (sector)

Example 17.3

Using traffic data per cell for a GSM/GPRS network from Example 17.2, calculate (a) data Erlangs, (b) time slot (TS) utilization, and (c) TS capacity. Use the following data:

- Average call holding time during busy hour: 120 seconds
- No. of transceivers (TX$_s$) per cell: 3
- No. of TS$_s$ per cell for signaling: 3
- Radio link control (RLC) efficiency: 80%
- Total numbers of transmitted radio blocks: 9000
- TS$_s$ allocated for data traffic per cell: 3
- Data throughput per cell (from Example 17.2): 15.5 kbps
- Voice traffic per cell (from Example 17.2): 8.82 Erlangs

Solution

(a) Data Erlangs $= \dfrac{\text{No. of radio blocks} \times \text{Duration of block}}{\text{Duration of busy hour}} = \dfrac{9000 \times 0.02}{120} = 1.5$

(b) TS utilization $= \dfrac{1.5}{3} = 0.5$

(c) Throughput per TS $= \dfrac{15.5}{3} \times 0.8 = 4.13$ kbps

TS capacity $= \dfrac{4.13}{0.5} = 8.26$ kbps

TS available for voice traffic = No. of TX$_s$ × TS$_s$/TX − TS$_s$ for signaling − TS$_s$, for data = 3 × 8 − 3 − 3 = 18

Voice traffic @ 2% GoS = 11.49 Erlangs (Erlang-B Table, Appendix A) > 8.82 Erlangs

17.3 Radio Design for a Cellular Network

Several factors are considered in the design of a cellular network. The extent of radio coverage, the quality of service for different environments, efficient use of spectrum, and the evolution of the network are some of the key factors that need to be carefully evaluated by all prospective service providers. Often these factors are further complicated due to the constraints imposed by the operating environment and regulatory issues. A system designer must carefully balance all trade-offs to ensure that the network is robust and of high service quality.

17.3.1 Radio Link Design

The first important step in the design of a cellular system is to design the radio link. This is required to determine the base station (BS) density in different environments as well as the corresponding radio coverage. For a cellular network to provide good quality of service (QoS) in indoor and outdoor environments, flexibility and resilience should be incorporated into the design.

The transmit power of handsets is the determining factor for a cellular system with balanced up/downlink power. Although the mobile antenna gain does not affect the balancing of the link budget, it is an important factor in the design of the power budget for the handset coverage. From a user point of view, a cellular network should imply that there is little restriction on making or receiving calls within a building or traveling in a vehicle using handsets. A system should be designed to allow the antenna of a handset to be placed in nonoptimal positions. In addition, the antenna may not even be extended when calls are being made or received. In normal system designs, it is assumed that the gain of a mobile antenna is 0 dBi. However, allowing for the handset antenna to be placed in suboptimal positions, a more conservative gain of -3 dBi should be used. In reality, mobile antenna gain could be as low as -6 to -8 dBi because of the placing of an antenna in an arbitrary position or with the antenna retracted into the handset housing, depending on specific handsets and their corresponding housing designs.

17.3.2 Coverage Planning

The most important design objective of a cellular network is to provide near-ubiquitous radio coverage. In the radio coverage planning process, the selection of the propagation model (see Chapter 3) is one of the most important considerations. The accuracy of the prediction by a particular model depends on its ability to account for the details of terrain, vegetation, and buildings. This accuracy is important in determining path loss and, hence, cell sizes and infrastructure requirements of a cellular network. An overestimation of path loss will lead to an inefficient use of network resources, whereas an underestimation will result in poor radio coverage. Propagation models generally tend to oversimplify real-life propagation conditions and may be grossly inaccurate in complex metropolitan urban environments. The empirical propagation models provide only general guidelines; they are too simplistic for accurate network design. Accurate field measurements should be made to provide information on radio coverage in an urban environment. Measured data can be used either directly in the planning process to access the feasibility of the individual cell site or to determine the coefficients of the empirical propagation model to achieve better characterization of a specific environment.

Another critical factor that affects radio coverage is penetration loss for both buildings and vehicles. If radio coverage for the outer section of a building is sufficient, then an assumed penetration loss of 10 to 15 dB should be adequate. However, if calls are expected to be received and originated within the inner core of the building, a penetration loss of about 16 to 20 dB should be used. Similarly, for in-vehicle coverage, penetration loss is equally important. A car could experience a penetration loss of 3 to 6 dB, whereas vans and buses may have larger variations. The penetration loss at the front of a van should be no more than that experienced in a car, but the loss at the back of the van could be as high as 10 to 12 dB, depending on the amount of window space. Thus, for design purposes, a high penetration loss should be assumed to ensure good service quality. For an urban environment, because building penetration loss is the dominant factor, in-vehicle penetration will generally be sufficient. Other losses such as body, cables, and connectors, should also be accounted for in the link budget calculations.

Radio propagation is subjected to shadowing (see Chapter 3). To ensure that the signal level in 90% of the cell area is equal to or above the specified threshold, a shadow margin, which is dependent on the standard deviation of the signal level, must be included in the link budget calculations. For a typical urban environment, a shadow fading margin of 8 to 9 dB should be used based on the assumption that path loss follows an inverse exponent law.

17.4   Receiver Sensitivity and Link Budget

The minimum received signal power to satisfy a certain bit error rate (BER) needs to be calculated. This is referred to as *receiver sensitivity* or S_{min}. For voice transmission, a BER of 10^{-3} is usually used. We should also know under which conditions (additive white Gaussian noise (AWGN) or fading) the BER is evaluated. When the holding time for a radio channel is relatively long compared to the fading duration as in the case of voice transmission, we evaluate the BER under fading conditions. The fade margin is determined from the statistics of a large-scale signal variation based on the lognormal distribution. On the other hand, when the holding time of a radio channel is relatively short as in the case of data transmission, we evaluate the BER under static conditions. In this case, fade margin is determined from the combined statistics of both large-scale and small-scale signal variations based on the lognormal and Rayleigh distribution (see Chapter 3), respectively.

17.4.1   Link Budget for the GSM1800 System

A link budget is used to compute cell coverage by accounting for all the factors that determine the cell coverage to balance the system cost against the required cell capacity. Link budget factors controlled by the RF engineer include transmitter-radiated power, antenna gain, noise figures, and co-channel interference (i.e., reuse factor N). Factors that cannot be controlled directly, but must be considered in the link budget, include propagation path loss and system bandwidth. We show link budget calculations for uplink and downlink of a GSM1800 system.

Example 17.4

Using the following data for GSM1800, develop downlink (see Figure 17.2) and uplink (see Figure 17.3) budgets and determine the cell radius.

- Base station transmit power (P_t): 4 W = 36 dBm
- Mobile station transmit power (P_m): 24 dBm
- Mobile station (hand-held unit) noise figure: 8 dB

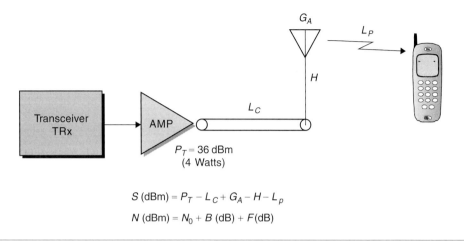

$$S\,(\text{dBm}) = P_T - L_C + G_A - H - L_p$$
$$N\,(\text{dBm}) = N_0 + B\,(\text{dB}) + F\,(\text{dB})$$

Figure 17.2   Downlink link budget for GSM1800.

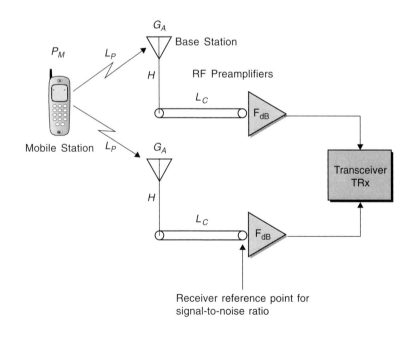

$$S\,(\text{dBm}) = P_M - L_p + G_A - H - L_c + F$$
$$N\,(\text{dBm}) = N_0 + B\,(\text{dB}) + F\,(\text{dB})$$

Figure 17.3   Uplink link budget for GSM1800.

- Base station noise figure: 5 dB
- Base station transmit and receive antenna gain (G_A): 18 dBi
- Mobile antenna gain: 0 dBi
- Required signal-to-noise ratio (SNR): 12 dB
- BS transmit antenna cable, connector, and filter losses (L_c): 5 dB
- BS receiver antenna cable, connector, and filter losses (L_c): 2 dB
- Orientation/body losses at mobile: 3 dB
- Shadow fading: 10.2 dB
- Thermal noise density: −174 dBm/Hz
- Antenna diversity gain at BS: 5 dB

Solution

Mobile station noise floor:

$$N = -174 + 10\log(200{,}000) + \text{Noise figure} = -174 + 53 + 8 = -113\,\text{dBm}$$

$$S_{\min} = N + \left(\frac{S}{N}\right)_{\text{reqd}} = -113 + 12 = -101\,\text{dBm}$$

Mean signal required at mobile station:

$$S_{\text{mean}} = S_{\min} + \text{shadow fading} + (\text{orient})/(\text{body}) = -101 + 10.2 + 3 = -87.8\,\text{dBm}$$

$$L_p = P_t - L_c + G_A - S_{\text{mean}} = 36 - 5 + 18 - (-87.8) = 136.8\,\text{dB}$$

Base station noise floor:

$$N = -174 + 10\log(200{,}000) + 8 = -116\,\text{dBm}$$

$S_{\min}$ *for acceptable quality with diversity:*

$$S_{\min} = N + \left(\frac{S}{N}\right)_{\text{reqd}} - \text{diversity} = -116 + 12 - 5 = -109\,\text{dBm}$$

Mean signal required at base station:

$$S_{\text{mean}} = S_{\min} + \text{shadow fading} + (\text{orient})/(\text{body}) = -109 + 10.2 + 3 = -95.8\,\text{dBm}$$

$$L_p = P_m - S_{\text{mean}} + G_A - L_c = 24 - (-95.8) + 18 - 2 = 135.8\,\text{dB}$$

Next we choose the link with the least acceptable path loss to calculate the cell radius. In this case the coverage will be limited by uplink. Hand-held units inside a building experience an additional path loss that is dependent on building type, building material, and placement of the user within the building. Changing antenna height, gains, or use of antenna diversity will also affect coverage under noise-limited conditions.

We use uplink path loss and Hata model to calculate cell radius as:

$$L_p = 133.2 + 33.8 \log R; \text{ where } R \text{ is the cell radius in kilometers.}$$

$$\therefore R = 10^{(135.8-133.2)/33.8} = 1.2 \, \text{km}$$

17.4.2 Pole Capacity of a CDMA Cell

To determine the pole capacity of a CDMA cell, we consider a user with the received signal strength of S at the base station and account for interference from $(M - 1)$ users in its own cell, I_{own} and M users in other cells, I_{other} along with thermal power N_{th}. We assume the same signal strength at the base station from other users.

$$I_{\text{own}} = (M - 1)Sv_f$$

where:

v_f = channel activity
$N_{\text{th}} = N_0 B_w$

Total interference including thermal noise power:

$$I_t = I_{\text{own}} + I_{\text{other}} + N_{\text{th}} = [(1 + \beta)M - 1]v_f S + N_0 B_w$$
$$= \left[\frac{M}{\eta} - 1\right]Sv_f + N_0 B_w \tag{17.1}$$

where:

$$\eta = \frac{1}{1 + \beta} = \text{Total same-cell power/(Total same-cell power + Other cells}$$
power)

$$\text{Cell loading} = \rho = \frac{M}{M_{\text{max}}} \approx \frac{I_{\text{own}} + I_{\text{other}}}{I_{\text{own}} + I_{\text{other}} + N_{\text{th}}} \tag{17.2}$$

$$\frac{E_b}{I_t} = G_p \cdot \frac{S}{N_0 B_w + (M - 1)v_f(S/\eta_c)(1 + \beta)} \tag{17.3}$$

Solving Equation 17.3 for M, we get

$$M = 1 + G_p \left[\frac{\eta_c}{(E_b/I_t)v_f\,(1 + \beta)} \right] - \frac{N_0 B_w}{S v_f (1 + \beta)} \qquad (17.4)$$

and solving Equation 17.3 for S, we get

$$S = \frac{(E_b/I_t)N_0}{\dfrac{1}{R} - \dfrac{(M - 1)v_f(1 + \beta)(E_b/I_t)}{B_w \eta_c}} \qquad (17.5)$$

$$\therefore M_{\max} \approx \frac{G_p}{d} \cdot \frac{1}{(1 + \beta)} \cdot \frac{\eta_c}{vf}; \qquad (17.6)$$

where:
    η_c = power control efficiency (often 80% to 85%)
    β = interference factor due to other cells
    $d = (E_b/I_t)_{\text{reqd}}$
    G_p = processing gain
    N_0 = noise density
    R = information rate

Equation 17.6 is known as the *pole capacity* of a cell. We rewrite Equation 17.5 as

$$\frac{S\eta_c}{N_0 B_w} = \frac{1}{M_{\max}\, v_f\,(1 + \beta)(1 - \rho)} \qquad (17.7)$$

Note when ρ tends to 1, $S\eta_c$ tends to infinity because interference tends to infinity.

17.4.3   Uplink Radio Link Budget for a CDMA System

We consider an ith mobile in a given cell. The required $(E_b/I_t)_i = d_i$ for the ith mobile is given as:

$$\frac{R_c}{R_i} \cdot \frac{S_i}{I_{\text{own}} - S_i + I_{\text{other}} + N_{\text{th}}} = \frac{R_c}{R_i} \cdot \left(\frac{S_i}{I_{\text{own}} - S_i + \beta I_{\text{own}} + N_{\text{th}}} \right) \geq d_i,\ i = 1, 2, \ldots M \ (17.8)$$

where:
    R_c = chip rate
    R_i = information rate for ith mobile
    I_{own} = interference for ith mobile from its own cell

$I_{\text{other}} = \beta I_{\text{own}} = $ interference from other cells

$N_{\text{th}} = $ thermal noise power for ith mobile $= N_f k T_\phi R_c$

$d_i = (E_b/N_t)_{\text{reqd}}$ for ith mobile

$N_f = $ receiver noise figure for ith mobile

$k = $ Boltzmann constant

$T_\phi = $ Absolute temperature

Solving the inequalities in Equation 17.8 as equalities means solving for the minimum required power (sensitivity) S_i as

$$S_i = \left(\frac{1}{1 + \frac{R_c}{R_i d_i}}\right) \cdot (1 + \beta) I_{\text{own}} + \frac{1}{1 + \frac{R_c}{R_i d_i}} N_{\text{th}} \quad i = 1, 2, \ldots M \qquad (17.9)$$

If Equation 17.9 is summed over the mobile stations, M, connected to a particular base station, then

$$\sum_{i=1}^{M} S_i = \left[\sum_{i=1}^{M} \left(\frac{1}{1 + \frac{R_c}{R_i d_i}}\right) \cdot (1 + \beta)\right] \cdot \sum_{i=1}^{M} S_i + \sum_{i=1}^{M} \frac{1}{1 + \frac{R_c}{R_i d_i}} \cdot N_{\text{th}} \qquad (17.10)$$

$$\therefore \sum_{i=1}^{M} S_i (1 + \beta) = \frac{N_{\text{th}}\left[\sum_{i=1}^{M} \frac{1}{1 + \frac{R_c}{R_i d_i}} \cdot (1 + \beta)\right]}{1 - \left[\sum_{i=1}^{M} \frac{(1 + \beta)}{1 + \frac{(1 + \beta)}{1 + \frac{R_c}{R_i d_i}}}\right]} \qquad (17.11)$$

we define the uplink loading as:

$$\eta_{UL} = \left[\sum_{i=1}^{M} \frac{1}{1 + \frac{R_c}{R_i d_i}}\right] \cdot (1 + \beta) \qquad (17.12)$$

where:

$\eta_{UL} = $ uplink loading

Next, we modify Equation 17.12 to include the effect of sectorization (sectorization gain, λ, number of sector, N_s) and channel activity factor v_i then

$$\eta_{UL} = \left[\sum_{i=1}^{M} \frac{1}{1 + \frac{R_c}{R_i d_i} \cdot \frac{1}{v_i}}\right] \cdot \left(1 + \beta \cdot \frac{N_s}{\lambda}\right) \qquad (17.13)$$

Example 17.5

Calculate the uplink cell load-factor and number of voice users for a WCDMA system using the following data. What is the pole capacity of the cell?

- Information rate (R_i): 12.2 kbps
- Chip rate (R_c): 3.84 Mcps
- Required E_b/N_t: 4 dB
- Required interference margin: 3 dB
- Interference factor due to other cells: 0.5
- Channel activity factor: 0.65

Solution

From Equation 17.13:

$$\text{Load-factor} = (1 + \beta) \cdot \sum_{i=1}^{M} \frac{1}{1 + \dfrac{R_c}{R_i} \cdot \dfrac{1}{(E_b/N_t)_{\text{reqd}}} \cdot \dfrac{1}{v_i}}$$

$$\text{Load-factor per voice user} = (1 + 0.5) \cdot \frac{1}{1 + \dfrac{3840}{12.2} \cdot \dfrac{1}{2.512} \cdot \dfrac{1}{0.65}} = 000774$$

$$\text{Required interference margin} = 3\,\text{dB} = 2 = \frac{1}{1 - \rho}$$

$$\text{Cell loading} = \rho = 0.5$$

$$\text{Number of voice users} = \frac{0.5}{0.00774} \approx 64 \text{ per cell}$$

Using Equation 17.6 and assuming $\eta_c = 1$ (for this case), we get the pole capacity of the cell as

$$M_{\max} = \frac{3.84 \times 10^6}{12.2 \times 10^3} \cdot \frac{1}{0.65} \cdot \frac{1}{(1 + 0.5)} \cdot \frac{1}{2.52} = 128$$

Example 17.6

Calculate the uplink throughput for data service only for a WCDMA cell using the following information:

- Required E_b/N_t: 1 dB
- Required interference margin: 3 dB (cell loading = 0.5)

- Interference factor due to other cells: 0.5
- Channel activity factor: 1.0

Solution

$$\text{Load factor} = M(1 + \beta) \cdot \frac{1}{1 + \dfrac{R_c}{R_i} \cdot \dfrac{1}{(E_b/N_t)_{\text{reqd}}} \cdot \dfrac{1}{v_i}} \approx M(1 + \beta) \cdot \frac{1}{\dfrac{R_c}{R_i} \cdot \dfrac{1}{(E_b/N_t)_{\text{reqd}}} \cdot \dfrac{1}{v_i}}$$

$$\text{Throughput} = R_i \times M = \frac{\text{cell loading} \times R_c}{(E_b/N_t)_{\text{reqd}} \cdot (1 + \beta)} = \frac{0.5 \times 3840}{1.259 \times 1.5} = 1016 \text{ kbps}$$

17.4.4 Downlink Radio Link Budget for a CDMA System

Downlink dimensioning uses the same method as uplink [7]. For a selected range the total base station transmit power must be estimated. In this estimation the soft handoff connections must be included. If the power exceeds that estimated, either the cell range should be limited or the number of users in the cell should be reduced.

$$\eta_{DL} = \sum_{i=1}^{I} \left[\frac{R_i d_i v_i}{R_c} \left\{ \xi_i + \sum_{n=1, n \neq m}^{N} \frac{L_{pmi}}{L_{pni}} \right\} \right] \tag{17.14}$$

where:

L_{pmi} = link loss from serving BS m to ith mobile station (MS)
L_{pni} = link loss from the other BS n to ith MS
d_i = transmit (E_b/N_t) requirement for ith MS including the soft handoff combining gain and average power rise caused by fast power control
N = number of base stations
I = number of connections in a sector
ξ_i = orthogonality factor ($0 \leq \xi_i \leq 1$); $\xi_i = 1$, fully orthogonal

$$i_{DL} = \sum_{n=1, n \neq m}^{N} \frac{L_{pmi}}{L_{pni}} \text{ defines the average other-to-own cell interference in}$$

downlink

The total base station transmit power estimation must take into account multiple communication links with average distance $(\overline{L}_{pmi})$ from the serving BS. In the radio link budget calculations in the uplink, the limiting factor is MS transmit power; in the downlink the limiting factor is total base station transmit power. While balancing the uplink and downlink service areas both links must be considered.

The interference degradation margin to be taken into account in the link budget due to certain loading ρ (in either uplink or downlink) is

$$L = 10\log(1 - \rho) \tag{17.15}$$

Some margin is required in MS transmission power to maintain adequate closed-loop fast power control in unfavorable propagation conditions such as near the cell edge. This applies especially to pedestrian users, where the E_b/N_t to be maintained is more sensitive to closed-loop power control.

Example 17.7

Calculate the downlink cell load-factor and number of voice users per cell for a WCDMA system using the following data.

- Information rate (R_i): 12.2 kbps
- Chip rate (R_c): 3.84 Mcps
- Required E_b/N_t: 4 dB
- Average interference factor due to other cells: 0.5
- Orthogonality factor (ξ): 0.6
- Interference margin: 3 dB

Solution

$$\eta_{DL} = (\xi + i_{DL}) \cdot \sum_{i=1}^{I} \frac{1}{\dfrac{R_c}{R_i} \cdot \dfrac{1}{(E_b/N_t)_{reqd}} \cdot \dfrac{1}{v_i}}$$

$$\text{Load-factor per voice user} = (0.6 + 0.5) \cdot \frac{1}{\dfrac{3840}{12.2} \cdot \dfrac{1}{2.512} \cdot \dfrac{1}{0.65}} = 0.0057$$

$$\text{Interference margin} = 3\,\text{dB} = 2 = \frac{1}{1 - \rho}; \rho = 0.5$$

$$\text{Number of voice users} = \frac{0.5}{0.0057} \approx 87 \text{ per cell}$$

Example 17.8

Determine the minimum signal power required for the acceptable quality of voice at the base station receiver of an IS-95 CDMA system. What is the maximum allowable path loss? Use the following data and refer to Figure 17.4.

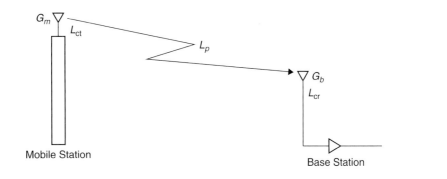

Figure 17.4   Uplink transmission from MS to BS.

- Noise density (N_0) = -174 dBm/Hz
- Channel bandwidth (B_c) = 1.25 MHz
- Chip rate (R_c) = 1.2288 Mcps
- Receiver noise figure (N_f) = 6 dB
- Effective radiated power of the mobile = 0.5 Watt (27 dBm)
- Transmitter cable and connector loss (L_{ct}) = 0.5 dB
- Body loss (L_{body}) = 1.5 dB
- Receiver cable and connector loss (L_{cr}) = 2 dB
- Interference margin (m_{inf}) = 0 dB
- Fast fading margin (m_{fading}) = 2 dB
- Penetration losses (L_{pent}) = 8 dB
- Transmitter antenna gain (G_m) = 0 dBi
- Receiver antenna gain (G_b) = 12 dBi
- Fade margin (f_m) = 8 dB
- Required E_b/N_t = 7 dB

Solution

Base station noise floor:

$$N_{th} = N_0 + N_f = -174 + 6 = -168 \text{ dBm/Hz}$$

Required S/N_t:

$$(S/N_t)_{reqd} = (E_b/N_t)_{reqd} + 10\log(R_c/B_c) = 7 + (10)\log\left(\frac{1.2288}{1.25}\right) = 6.93 \text{ dB}$$

Minimum signal power required:

$$S_{min} = (S/N_t)_{reqd} + 10\log R_c + N_{th} = 6.93 + 10\log(1.2288 \times 10^6) - 168$$
$$= -100.17\,\text{dBm}$$

We refer to Figure 17.4:

Maximum allowable path loss:

$$L_{pmax} = (P_m - S_{min}) + (G_b + G_m) - (L_{body} + L_{ct} + L_{cr} + f_m + L_{pent}) - m_{int} - m_{fading}$$
$$= 27 - (-100.17) + (12 + 0) - (1.5 + 0.5 + 2.0 + 8 + 8) - 2 = 117.17\,\text{dB}$$

Example 17.9

Develop a radio link budget for uplink and downlink of a WCDMA system using the following data (including Table 17.1):

- Chip rate: 3.84 Mcps
- Data rate: 16 kbps
- UL loading: 50%
- DL loading: 90%
- Required E_b/N_t uplink: 4 dB
- Required E_b/N_t downlink: 6 dB
- Mobile antenna gain: 0 dBi
- Base station antenna gain: 18 dBi

Solution

The Okumura-Hata model for an urban macro-cell with a base station antenna height of 25 m, a mobile antenna height of 1.5 m, and a carrier frequency of 1950 MHz gives

$$L_p = 138.5 + 35.7\log R$$

where:

L_p = allowable path losses (dB)
R = hexagonal cell radius (km)

Table 17.1 Radio link budget (RLB) for WCDMA uplink (UL) and downlink (DL).

	UL	DL	Unit		Remark
Tx power	125	1372.97	mw	a	
	20.97	31.38	dBm	b	
Tx antenna gain	0.0	18.0	dBi	c	
Cable/body losses	2.0	2.0	dB	d	
Tx ERIP	18.97	47.38	dBm	e	$e = b + c - d$
Thermal density	−174	−174	dBm/Hz	f	
Receiver noise figure	5	8	dB	g	
Receiver noise density	−169	−166	dBm/Hz	h	$h = f + g$
Receiver noise power	−103.16	−100.16	dBm	i	$i = h + 10\log(3.84 \times 10^6)$
Interference margin UL loading (ρ) 50%; DL loading 90%	−3.01	−10.0	dB	j	$L = 10\log(1-\rho)$
Required E_b/N_t	−16.79	−7.8	dB	k	$k = 10\log[(E_b/N_t)/G_p] - L$
					$UL = 10\log[10^{0.4}/(3840/16)] - L$ $= -19.8 - (-3.01) = -16.79$
					$DL = 10\log[10^{0.6}/(3840/16)] - L$ $= -17.8 - (-10) = -7.8$
Required signal power	−119.95	−107.96	dB	l	$l = i + k$
Rx antenna gain	18	0	dBi	m	
Cable/body losses	2.0	2.0	dB	n	
HO gain (including any micro-diversity combining gain at cell edge)	0.0	2.0	dB	o	
Slow fading margin	−7.27	−7.27	dB	p	
Slow fading + HO gain	−7.27	−5.27	dB	q	$q = p + o$
Indoor losses	8.0	8.0	dB	r	
Allowable path loss	139.65	140.07	dB	s	$s = e - l + m - n + q - r$

What is the cell radius based on the calculated allowable path loss?

$$139.65 = 138.5 + 35.7 \log R$$

$$\therefore R = 1.077 \, \text{km}$$

Area covered by hexagonal cell = $2.6 \times R^2 = 3.02 \sim 3.0 \, \text{km}^2$
Number of BTSs required to cover an area of $2400 \, \text{km}^2 = 800$

Example 17.10

During a busy hour an average user downloads 10 megabits with 384 kbps, 2 megabits with 144 kbps, and 1 megabit with 64 kbps. Data has to be retransmitted 1.2 times the average because of network conditions. Calculate the number of users that can be supported on the downlink of the WCDMA network. Use the following data:

- Spreading rate (R_c): 3.84 Mcps
- Noise rise: 3 dB
- Orthogonality factor (α): 0.8
- Interference from other cells (β): 0.55
- $(E_b/N_0)_{\text{reqd}}$: 4 dB
- Sector efficiency (λ): 0.85
- Power control efficiency (η_c): 0.80

Solution

Service rate	*Average data rate*
• $1.2 \times 10 \times 10^3/3600$	3.33 kbps
• $1.2 \times 2 \times 10^3/3600$	0.67 kbps
• $1.2 \times 1 \times 10^3/3600$	0.33 kbps
Total:	4.33 kbps

$$\text{Noise rise} = 3 \, \text{dB} = 2 = \frac{1}{1 - \rho}$$

$$\rho = 0.5$$

$$\text{DL capacity} = \frac{R_c}{(E_b/N_0)_{\text{reqd}}} \cdot \frac{\eta_c}{(\alpha + \beta)} \cdot \frac{\lambda}{v_f} = \frac{3.84 \times 10^3}{2.5} \cdot \frac{0.80}{1.35} \cdot \frac{0.85}{1} = 774 \, \text{kbps}$$

Allowable DL capacity = $\rho \times \text{DL capacity} = 0.5 \times 774 = 387 \, \text{kbps}$

$$\text{Number of users} = \frac{387}{4.33} \approx 89$$

17.5 cdma2000 1X EV-DO

In cdma2000 1X EV-DO the data traffic is asymmetric. A much higher forward link (or downlink) rate is needed from the base station than that generated by the user terminal in the reverse (or uplink) direction. We introduced cdma2000 1X EV-DO in Chapter 16; in this section we present concepts behind 1X EV-DO and discuss a procedure for determining the reverse link (or uplink) capacity.

cdma2000 1X EV-DO provides data services in wide-area mobile and fixed networks [1]. It is based on the high data rate (HDR) concept introduced in [2,8]. The 1X EV-DO system achieves high spectral efficiency due to several techniques that make it uniquely suited for high data rate transmission. These techniques include the long-range channel estimation and rate prediction, incremental redundancy, hybrid automatic request control (HARQ), turbo-coding, and scheduling that exploits multiuser diversity in a fast fading environment.

17.5.1 1X EV-DO Concept

In the 1X EV-DO users are separated into N classes according to their signal-to-noise ratio (SNR) levels, and corresponding instantaneous rate levels supportable [8]. Thus, user class i can receive slots at rate R_i bits per second, where, $i = 1, 2, 3,..., N$ and the relative frequency of user packets P_i of class i. Suppose slots are assigned one at a time successively to each user, then the average rate (or throughput) of the downlink will be:

$$R_{av} = \sum_{i=1}^{N} R_i P_i \qquad (17.16)$$

Equation 17.16 implies that lower-data-rate users will have proportionately higher latency. However, if we want all users to have essentially the same latency irrespective of the R_i, then we should allocate slots inversely proportional to rate, i.e.,

$$S_i = \frac{k}{R_i} \qquad (17.17)$$

where:
S_i = number of slots allocated to user class i
k = a constant

In this case, the average throughput will be:

$$\overline{R}_{av} = \frac{\sum_{i=1}^{N} P_i R_i S_i}{\sum_{i=1}^{N} P_i S_i} = \frac{k\sum_{i=1}^{N} P_i}{k\sum_{i=1}^{N} P_i/R_i} = \frac{1}{\sum_{i=1}^{N} P_i/R_i} \qquad (17.18)$$

In this case the latency of user classes will be the same. For the general case of N classes and latency ratio $L_{max}/L_{min} = \rho_L$, the maximum achievable throughput C is given as:

$$C = \frac{\displaystyle\sum_{i=1}^{K} P_i + \sum_{i=K+1}^{N} P_i \rho_L}{\displaystyle\sum_{i=1}^{K} \frac{P_i}{R_i} + \sum_{i=K+1}^{N} \left(\frac{P_i}{R_i}\right)\rho_L} \tag{17.19}$$

where:

$R_1 < R_2 < R_3 \cdots < R_N$; K is such that $R_i \leq C$ for all $i \leq K$, while $R_i > C$ for all $i > K$.

In this case each user's latency is either L_{max} (for those $R_i < C$) or L_{min} (for those $R_i > C$).

In the equal access situation we assign a number of slots to each user based on the requested data rate. The average throughput will be:

$$R_{av} = \frac{\displaystyle\sum_{i=1}^{N} S_i P_i R_i}{\displaystyle\sum_{i=1}^{N} S_i P_i} \tag{17.20}$$

where:

S_i = number of slots assigned to the R_i user

Example 17.11

Consider a data system in which $P_1 = 1/2$, $P_2 = 1/3$, and $P_3 = 1/6$. The data rates are $R_1 = 16\,\text{kbps}$, $R_2 = 64\,\text{kbps}$, and $R_3 = 1024\,\text{kbps}$, respectively. The assigned slots are $S_1 = 16$, $S_2 = 8$ and $S_3 = 2$. What is the average throughput? Compare it with the equal latency and $\rho_L = 4$ cases.

Solution

Using Equation 17.20:

$$R_{av} = \frac{16 \times 16 \times \frac{1}{2} + 8 \times 64 \times \frac{1}{3} + 2 \times 1024 \times \frac{1}{6}}{16 \times \frac{1}{2} + 8 \times \frac{1}{3} + 2 \times \frac{1}{6}} = 58.2\,\text{kbps}$$

For equal latency, using Equation 17.18:

$$R_{av} = \frac{1}{\frac{1}{2} \times \frac{1}{16} + \frac{1}{3} \times \frac{1}{64} + \frac{1}{6} \times \frac{1}{1024}} = 27.3\,\text{kbps}$$

For $\rho_L = 4$, using Equation 17.19:

$$C = \frac{\frac{1}{2} + \frac{1}{3} + 4 \times \frac{1}{6}}{\frac{1}{2 \times 16} + \frac{1}{3 \times 64} + \frac{1}{6 \times 1024}} = 40.96\,\text{kbps}$$

It can be observed that equal access provides the highest average throughput.

17.5.2   Details of cdma2000 1X EV-DO

Figure 17.5 shows an overview of downlink channels of the cdma2000 1X EV-DO system. The pilot, traffic, medium access, and control channels are time division multiplex (TDM) channels. The medium access channel consists of two code division multiplex (CDM) channels, and reverse activity and reverse power control channels. The control channel is used for system acquisition, system parameters broadcast, and service negotiations during call setup. The pilot channel is used to provide coherent reception, soft handoff, channel estimation, and long-range prediction and rate selection. The reverse activity channel is used for uplink overload and rate control to indicate the uplink interference loading of the sector. The reverse power control channel is used for fast power control of the existing uplink connections. The traffic channel is used to transmit data to the multiple users in a TDM fashion. The time slots have a duration of 1.67 ms (see Figure 16.23) and could be assigned to any user as determined by the scheduling algorithm. The data on the traffic channel can be transmitted at 38.4, 76.8, 153.6, 307.2, 614.4, 921.6, 1228.8, 1843.2, and 2457.6 kbps. The higher data rates are obtained using a combination of higher order modulation (QPSK, 8-PSK, 16-QAM), forward

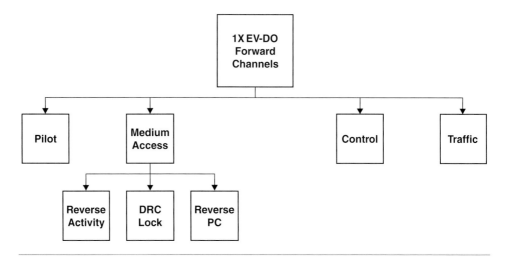

Figure 17.5   cdma2000 1X EV-DO downlink (forward link) channels.

error correction coding ($r = \frac{1}{5}$ and $\frac{1}{3}$), and spreading. Transmission of one encoder packet can occupy from 1 to 16 time slots. The coding rate, modulation, and packet duration is one of 12 rate configurations given in Table 16.6.

Besides the traffic channel, the reverse link (or uplink) of 1X EV-DO includes a number of channels used for signaling (see Figure 17.6). These are the pilot, reverse rate indicator (RRI), ACK, and data rate control (DRC) channels. The RRI channel indicates the data rate of the reverse traffic channel. The ACK channel indicates success or failure on the forward link traffic channel. The DRC indicates the rate at which the mobile station can receive the forward traffic channel and the sector from which the mobile station wishes to receive the forward traffic channel. All active mobiles transmit these channels continuously, with the exception of ACK. Therefore, the impact of these channels on the uplink capacity must be included in the analysis. The uplink traffic channel may transmit at one of the following rates: 9.6, 19.2, 38.4, 76.8, or 153.6 kbps, or may be discontinued (DTx) where there is no data to send.

The uplink power control maintains a constant minimal ratio of the pilot chip energy (E_c) to total noise and interference power spectral density (N_t) at the base station receiver for each user's uplink signal. In the uplink throughput analysis, the required E_c/N_t for a minimum of two paths is assumed since diversity-receive antennas used at the cell site guarantee the presence of least two paths.

For the traffic channel rate of 9.6 to 76.8 kbps, we use the required E_c/N_t per antenna of -23 dB and for the traffic channel rate of 153.6 kbps, we use E_c/N_t of -22 dB, since the 153.6 kbps traffic channel uses a weaker turbo-coding than other traffic channels ($\frac{1}{2}$ rate versus $\frac{1}{4}$ rate). An implementation margin of 10% is added to these values for the uplink throughput evaluation.

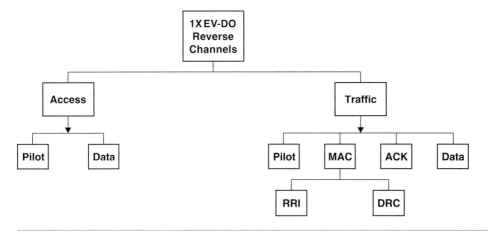

Figure 17.6 cdma2000 1X EV-DO uplink (reverse link) channels.

The total received power, S, at the base station from a mobile is the sum of the pilot channel power (p_p), the DRC channel power (p_d), the traffic channel power (p_t), and the ACK channel power (p_a). To simplify the analysis, we neglect the contribution of the ACK channel, because only one mobile station at most transmits the ACK channel at a time. The ratio of the total received power, S, from the mobile to the received pilot power can be expressed as [3]:

$$S \approx p_p + p_d + p_t \tag{17.21}$$

$$\frac{S}{p_p} = 1 + 10^{G_D/10} + v_f 10^{G_T/10} \tag{17.22}$$

where:

   G_D = DRC channel gain in dB with respect to pilot
   G_T = traffic channel gain in dB with respect to pilot
   v_f = traffic channel activity factor

The required ratio of the pilot chip energy to total noise plus interference power spectral density E_c/N_t will be:

$$\frac{E_c}{N_t} = \frac{(p_p/B_w)}{N_f N_0 + \dfrac{(1+\beta)(M-1)S}{B_w}} = d \tag{17.23}$$

where:

   β = interference factor due to other cells (~ 0.85)
   M = number of users in a sector
   B_w = system bandwidth
   N_f = noise figure of base station receiver
   d = allowable E_c/N_t

We rewrite Equation 17.23 and get:

$$M = \frac{p_p}{d(1+\beta)S} + 1 - \frac{N_f N_0 B_w}{(1+\beta)S} \tag{17.24}$$

$$\therefore M_{\max} = 1 + \frac{p_p}{d(1+\beta)S} \tag{17.25}$$

where:

   $M_{\max}$ = the finite limit on capacity (pole capacity).

Substituting Equation 17.22 into Equation 17.25 we get:

$$M_{\max} = \left(\frac{1}{1 + 10^{G_D/10} + v_f 10^{G_T/10}} \right) \cdot \left[\frac{1}{d(1+\beta)} \right] \tag{17.26}$$

Table 17.2 DL traffic channel gains relative to pilot [3].

Traffic channel rate (kbps)	Traffic channel gain (dB) relative to pilot
9.6	3.75
19.2	6.75
38.4	9.75
76.8	13.25
153.6	18.5

The sector loading can be expressed as a fraction of the pole capacity M_{max}. This is typically 70% of the pole capacity.

The dependency of the traffic channel gain G_T on the data rate is given in Table 17.2 and is based on the recommendation in [1]. The DRC channel gain is a function of the number of slots in which the same value of DRC is repeated. A DRC length of 2 slots provides an acceptable trade-off between the up- and downlink performance [5]. The recommended value of the DRC channel gain, G_D, is -1.5 dB.

Example 17.12

Find the allowable throughput of the reverse link in cdma2000 1X EV-DO if the average rate on the reverse traffic channel is 9.6 kbps. Use the following data:

- Allowable E_c/N_t: -23 dB
- DRC gain with respect to pilot: -1.5 dB
- Traffic channel gain with respect to pilot: 3.75 (see Table 17.2)
- Interference factor due to other cells (β): 0.85

Solution

$$M_{max} = \frac{1}{(1 + 10^{-0.15} + 10^{0.375})} \cdot \frac{1}{10^{-2.3} \times 1.85} = 26.4$$

$$M_{allowable} = 0.7 \times 26.4 \approx 18$$

$$\text{Reverse link throughput} = 173 \text{ kbps}$$

17.6 High-Speed Downlink Packet Access

We discussed high-speed downlink packet access (HSDPA) in Chapter 15. In this section we present the link budget and throughput calculations of the high-speed downlink shared channel (HS-DSCH) [6,10].

HSDPA will require investments to the R99/R4 UMTS (WCDMA) network and will affect radio network hardware and software, core network software, and transmission network hardware. It can be deployed using small upgrades, not required for all base stations and radio network controllers. HSDPA does not require a completely new network structure and protects the current investments made to the network.

In HSDPA, all users' traffic is carried on an HS-DSCH. The HSDPA requires a different approach for dimensioning network resources. The most critical dimensioning input is the average throughput expected from the HS-DSCH. The amount of throughput that an HS-DSCH can carry depends on many variables, the most important being its power allocation. Because HSDPA is an add-on feature to existing BTSs of WCDMA, a fixed amount of power is allocated for use by the HS-DSCH. The designer's target during the dimensioning phase is to specify the exact amount of that power. HSDPA power should be sufficient to meet expected throughput demand. On the other hand, enough power should be reserved for all dedicated channel (DCH) traffic in WCDMA. For a mixed R99 and HSDPA HS-DSCH power should be around 4 to 7 W, whereas for a dedicated HSDPA carrier the power should be 10 to 12 W, assuming that the maximum transmission power of the BTS is 20 W.

HSDPA can enable a wider coverage than R99 due to adaptive modulation, coding, and fast sheduling in the base station. In HSDPA, throughput is not a single concept. It consists of average cell throughput for a single user, minimum throughput at the cell edge for a single user, and average user throughput. The average cell throughput and minimum throughput at the cell edge for a single user are not dependent on the number of simultaneous HSDPA users in the cell. These are commonly used to set the dimensioning target for HSDPA coverage. HSDPA throughput depends directly on radio channel conditions. It changes rapidly all the time due to fast fading of the radio channel (BS uses a link adaptation procedure). The achieved throughput is different in every transmission time interval (TTI) of 2 ms. Average throughput can be estimated if the average signal-to-interference-plus-noise ratio (SINR) is known (see Figure 17.7) [9].

Since HSDPA is to be overlaid on the WCDMA network, the objective of the HSDPA link budget is to estimate the maximum data rate achievable in the downlink at the cell edge, assuming that coverage is uplink limited. Table 17.3 gives a sample HSDPA link budget. The HSDPA peak data rate calculations are based on the following parameters:

- Path loss as calculated by the HSDPA link budget
- Percent of BTS power allocated to HSDPA data users
- Scheduler margin—For HSDPA, the adaptive modulation and coding (AMC) combination allows different operating points to be selected based on initial transmission power. It is therefore possible to use either a more conservative approach that uses more power but results in fewer

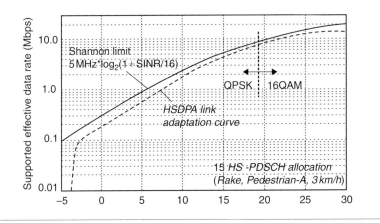

Figure 17.7   Supported effective data rate versus instantaneous per TTI SINR.

Table 17.3 HSDPA link budget.

	Item	Downlink	Unit	Remark
Base Station				
Tx power HSDPA	a	37.4	dBm	5.5 W
CPICH Tx power	b	33	dBm	2 W
Total BTS Tx power	c	43	dBm	20 W
Transmit antenna gain	d	18	dBi	
Cable + connector loss	e	4	dB	
DSCH EIRP	f	51.4	dBm	$f = a + d - e$
CPICH EIRP	g	47	dBm	$g = b + d - e$
Mobile Station				
Noise figure	h	8	dB	
Thermal noise	i	-108	dB	
Downlink loading	j	0.87		
Interference margin	k	8.9	dB	$k = 10 \log[1/(1-j)]$
Noise + interference floor	l	-91.1	dB	$l = h + i + k$
Orthogonality factor (α)	m	0.2		
Geometry factor (G)	n	-4.4	dB	G = ratio of own interference to other interference + noise
DSCH spreading factor	o	12	dB	$10 \log 16$
SINR	p	1.4	dB	See (Equation 17.27)

(Continued)

	Item	Downlink	Unit	Remark
Receiver sensitivity	q	−101.7	dB	q = l + p − o
Receive antenna gain	r	0.0	dB	
Body loss	s	2	dB	
DL fast fading margin	t	0	dB	
Soft handover gain	u	0	dB	
Building penetration loss	v	16.0	dB	
Lognormal fading margin	w	10.0	dB	
Maximum path loss	x	125.1	dB	x = f − q + r − s − t + u − v − w

retransmissions, or a more aggressive strategy that uses less power but results in more retransmissions.

- Orthogonality factor—This accounts for nonorthogonal interference received by the serving cell because of a multipath.

The link budget provides a user channel quality estimate that is then mapped to a maximum achievable physical layer data rate. The resulting HSDPA data rates can vary from 900 kbps to 14.4 Mbps.

17.6.1 HSDPA SINR Calculation

The use of E_b/N_0 in link budget requires the knowledge of bit rate or number of codes. Since bit rate and number of codes change every TTI (2 ms) in the HSDPA, we use average SINR instead of E_b/N_0 in link budget calculations. The average SINR does not depend on bit rate or number of codes of HSDPA and can be calculated when HSDPA power, BS total power, orthogonality, and G factor are known.

$$\text{SINR} = SF_{\text{DSCH}} \cdot \frac{(P_{\text{tx}})_{\text{DSCH}}}{(P_{\text{tx}})_{\text{Tot}} \times (1 - \alpha + 1/G)} \quad (17.27)$$

where:

SF_{DSCH} = spreading factor of DSCH; fixed at 16
$(P_{\text{tx}})_{\text{DSCH}}$ = transmit power of the DSCH
$(P_{\text{tx}})_{\text{Tot}}$ = Total base station power, including HS-PDSCH and HS-SCCH
α = downlink orthogonality factor
G = geometry factor

Example 17.13

Estimate the average SINR of HSDPA when the maximum transmit power of DSCH is 5.5 W and total base station power is 18 W. Use α and G as 0.2 and 0.363, respectively.

Solution

From Equation 17.27:

$$\text{SINR} = 16 \times \frac{5.5}{18\left(1 - 0.2 + \dfrac{1}{0.363}\right)} = 1.3753 \ (1.4\,\text{dB})$$

17.7 I_{ub} Interface Dimensioning

In the UMTS, the I_{ub} interface (see Chapter 15) is dimensioned to carry expected user traffic between a BTS (node B) and an RNC including overheads from signaling, O&M, ATM, etc.

$$I_{ub} \ \text{bandwidth} = \text{expected traffic} \times (1 + \text{burstiness}) \times (1 + \text{signaling overhead}$$
$$+ \ \text{O\&M overhead}) \times (1 + \text{ATM overhead})$$
$$= (1 + 0.25) \times (1 + 0.1 + 0.1) \times (1 + 0.2) \sim 1.8 \ \text{expected traffic}$$

Example 17.14

If each BTS supports 350 users and generates a busy hour traffic of 1.785 kbps including soft handoff and if an RNC supports 180 BTSs, what is the required bandwidth of the I_{ub} interface?

Solution

$$I_{ub} \ \text{bandwidth} = \frac{1.8 \times 1.785 \times 350}{1000} = 1.125 \ \text{Mbps (use one T1 providing}$$
1.5 Mbps)

Total I_{ub} bandwidth $= 180 \times 1.125 = 202.42$ Mbps

17.8 Radio Network Controller Dimensioning

In the initial planning phase, it is assumed that sites are distributed uniformly across an RNC area and carry roughly the same amount of traffic. The purpose of RNC dimensioning is to establish the number of RNCs needed to support the estimated traffic. Several limitations on RNC capacity exist and any of the following Equations 17.28 to 17.30 must be taken into account, of which the most demanding one should be selected.

$$\text{numRNCs} = \frac{\text{numCells}}{\text{cellsRNC} \times \text{fillrate1}} \tag{17.28}$$

where:

numCells = number of cells in the coverage area
cellsRNC = maximum number of cells that can be connected to one RNC
fillrate1 = margin used as a backoff from the maximum capacity

$$\text{numRNCs} = \frac{\text{numBTSs}}{\text{btsRNC} \times \text{fillrate2}} \tag{17.29}$$

where:

numBTSs = number of BTSs in the area to be dimensioned

btsRNC = maximum number of BTSs that can be connected to one RNC

fillrate2 = margin used as a backoff from the maximum capacity

The number of RNCs to support the I_{ub} throughput is calculated as

$$\text{numRNCs} = \frac{\text{VoiceTP} + \text{CSdataTP} + \text{PSdataTP}}{\text{tpRNC} \times \text{fillrate3}} \times \text{numSubs} \tag{17.30}$$

where:

tpRNC = maximum I_{ub} capacity

fillrate3 = margin used as backoff from the maximum capacity

numSubs = expected number of simultaneously active subscribers

VoiceErl, CSdataErl = expected amount of Erlangs per subscriber

avePSdata/PSoverhead = Layer 2 data rate + overhead introduced by frame protocol (FP) [including retransmission overhead (10%) and layer 2 + FP overhead (5%)]

$\text{SHO}_{\text{Voice}} + \text{SHO}_{\text{CSdata}} + \text{SHO}_{\text{PSdata}}$ = overhead due to SHO, typically 30–40%

$\text{VoiceTP} = \text{VoiceErl} \times \text{bitrate}_{\text{Voice}} \times (1 + \text{SHO}_{\text{Voice}})$

$\text{CSdataTP} = \text{CSdataErl} \times \text{bitrate}_{\text{CSdata}} \times (1 + \text{SHO}_{\text{CSdata}})$ and

$\text{PSdataTP} = \text{PSdata/PSoverhead} \times (1 + \text{SHO}_{\text{PSdata}})$

Example 17.15

In a certain area of a WCDMA system, there are 800 BTSs. Each BTS has three sectors with two frequencies used per sector. Find the required RNCs using the following data:

- Maximum capacity of cellRNC: 1152 cells
- One RNC can support btsRNC: 384 BTSs
- Voice service: VoiceErl = 25 mErl/subscriber; $\text{bitrate}_{\text{Voice}}$=16 kbps
- CS data service 1: CSErl = 10 mErl/subscriber; $\text{bitrate}_{\text{CSdata}}$ = 32 kbps
- CS data service 2: CSErl = 5 mErl/subscriber; $\text{bitrate}_{\text{CSdata}}$ = 64 kbps
- PS data service: avePSdata = 0.2 kbps per subscriber, PSoverhead = 15%
- SHO factor = 40%
- Total subscribers = 350,000
- Maximum I_{ub} capacity of tpRNC = 196 Mbps
- fillrate1, fillrate2, and fillrate3 = 90%

Solution

$$RNC_{required} = \frac{800 \times 3 \times 2}{1152 \times 0.9} = 4.63$$

$$RNC_{required} = \frac{800}{384 \times 0.9} = 2.32$$

$$RNC_{required} = \frac{\left(0.025 \times 16 + 0.01 \times 32 + 0.005 \times 64 + \frac{0.2}{0.85}\right) \cdot (1 + 0.4) \cdot 350,000}{196,000 \times 0.9}$$

$$= 4.63 \approx 5$$

Use 5 RNCs.

17.9   Summary

In this chapter we discussed the network planning aspects of the GSM, cdma2000, WCDMA, cdma2000 1X EV-DO, and the HSDPA network. We demonstrated procedures for calculating cell coverage based on link budget calculations. We focused mainly on the dimensioning phase that is used to estimate the approximate number of cell sites required, base station configurations, and number of other network elements to determine the projected costs and associated investments. We concluded the chapter by providing additional details for cdma2000 1X EV-DO and UMTS HSDPA systems. Due to the significant reverse link overhead incurred to improve the efficiency of the forward link in cdma2000 1X EV-DO, in practical scenarios the reverse link may become the limiting factor. In such scenarios, the conventional methods for load estimation, e.g., based on total received power plus the thermal noise floor used to drive a slow rate control, are not suitable. A more reliable fast method (based on DRC) for fast rate control should be used.

Problems

17.1   Why is average SINR useful in HSDPA DL link budget calculations?

17.2   What is the pole capacity of a cell?

17.3   A GSM network serving 100K subscribers in an urban area is to be replaced by a WCDMA network. In the GSM, each subscriber generates 0.01 Erlang voice traffic during busy hour. Determine the number 3 sector cell sites for WCDMA at 2% GoS using the following data:

- Allowable E_b/N_t for voice: 4 dB
- Information rate: 14.4 kbps
- Interference due to other cells: 0.67
- Power control efficiency: 0.80

- Channel activity factor: 0.5
- 3-sector antenna gain: 2.56
- Total available spectrum for the WCDMA network: 10 MHz
- Spreading rate in WCDMA: 3.84 Mcps

17.4 A 3-sector WCDMA BTS supports 80 subscribers per cell (sector). During a busy hour on an average each subscriber transmits 2.5 megabytes data at 384 kbps and makes two voice calls with an average call holding time of 120 seconds. Determine the uplink loading of the BTS using the following data: chip rate = 3.84 Mcps, required (E_b/N_t) for voice = 4 dB, required (E_b/N_t) for data = 2 dB, channel activity for voice = 0.5 and for data = 1.0, information rate for voice = 12.2 kbps, interference factor from other cells = 0.67, and 3-sector antenna gain = 2.55.

17.5 Calculate downlink throughput for data service only for a WCDMA cell using the following information: required (E_b/N_t) = 1 dB, required interference margin = 3 dB, average interference factor from other cells = 0.5, orthogonality factor = 0.6, and channel activity factor = 1. How many users can be supported if the average data traffic per user during the busy hour is 48 kbps?

17.6 A WCDMA system is required to serve 80% voice and 20% data users simultaneously. The cell loading is to be maintained at 70%. Using the following data, calculate the number of voice and data users in the cell:

- Spreading rate: 3.84 Mcps
- Information rate for voice: 14.4 kbps
- Information rate for data: 64 kbps
- Channel activity for voice: 0.6
- Interference factor from other cells: 0.67
- Power control accuracy: 0.80
- E_b/N_t required for voice service: 3 dB
- E_b/N_t required for data service: 2 dB
- 3-sector antenna gain: 2.55
- Total spectrum available: 5 MHz

17.7 What is the maximum path loss for HSDPA service using the following data?

- HSDPA transmission power, $(P_{tx})_{DSCH}$ = 5 W
- Base station antenna gain = 16.5 dBi
- Base station cable and connector loss = 3 dB
- Mobile receiver noise figure = 7 dB

- Planned DL load = 75%
- Minimum SINR required = 2.5 dB
- Service processing gain = 12 dB
- Rx antenna gain = 2 dBi
- Body loss = 2 dB
- Lognormal fading margin = 10 dB
- Building penetration loss = 16 dB

17.8 Your mobile phone company is given a contract to build a WCDMA network to cover an urban area of 1000 km^2. The following are the system data:

- Maximum transmit power of base station: 20 W
- Maximum transmit power of mobile station: 200 mW
- Base station antenna gain: 18 dBi
- Mobile antenna gain: 0 dBi
- Base station cable + connector loss: 2 dB
- Base station noise figure: 6 dB
- Mobile station noise figure: 8 dB
- Fast fading margin: 4 dB
- Soft handover gain: 2 dB
- Lognormal fading: 8.5 dB
- Building penetration loss: 0 dB
- Required E_b/N_0: 4 dB
- Thermal noise density: −174 dBm/Hz
- Information rate per subscriber: 14.4 kbps
- Average call holding time during busy hour: 100 sec per subscriber
- Average busy-hour traffic per subscriber: 0.03 Erlangs
- GoS: 2% blocking
- Roll-out time for network: 5 years
- Present population: 200 K
- Market penetration: 50%
- Projected population growth rate: 5% per year
- One-km intercept path loss in urban area: 125 dB/km
- Slope of path-loss curve (γ): 4.5

  Assume the following data:
- Channel activity factor = 0.5
- Power control accuracy: 80%

- Interference factor due to other cells: 0.67
- Sector efficiency for a 3-sector antenna: 0.85
- 10% overlaying of cells to allow cell breathing
- 1.0 handover per call with 70% intra- and 30% inter-RNC handover

  The following are the equipment limits:

- MSC capacity 200 K BHCA; maximum connections 6 RNCs
- RNC capacity 2400 Erlangs; maximum connections 20 BTSs
- HLR: maximum capacity 400 K customers
- VLR: maximum capacity 300 K customers

Note: VLR can be shared by MSCs.

Determine the cell radius and number of required BTSs, RNCs, MSCs, VLRs, and HLRs for the network.

References

1. 3G Partnership Project 2 (3GPP2). "cdma2000 High Rate Packet Data Air Interface Specification." TIA/EIA/IS-856, version 2.0, C.S0024, October. 27, 2000.

2. Bender, P., et al. CDMA/HDR: A Bandwidth-Efficient High-Speed WirelessData Service for Nomadic Users. *IEEE Communication Magazine.*, 38(1)7, 2000, pp. 70–78.

3. Bi, Q., et al. Performance of 1X EV-DO Third-Generation Wireless High-Speed Data Systems. *Bell Labs Technical Journal* 7(3), 2003, pp. 97–107.

4. Bi, Q, and Vitebsky, S. Performance Analysis of 3G-1X EV-DO High Data Rate System. IEEE Wireless Communication and Networking Conference, Orlando, FL, 2002, vol.1, pp. 379–395.

5. Cui, D., et al. "Reverse DRC Channel Performance Analysis for 1X EV-DO: Third-Generation High-speed Wireless Data Systems." IEEE 56th Vehicular Technology Conference, Vancouver, BC, 2002, pp. 137–140.

6. Griparis, T., and Dinan, E. HSDPA Network Dimensioning Challenges and Key Performance Parameters. *Bethtel Telecommunication Technical. Journal*, vol. 4, no. 2, June 2006.

7. Holma, and Toskala. *WCDMA for UMTS*. New York: John Wiley & Sons, 2004.

8. Jalali, A., Padovani, R., and Pankaj, R. Data Throughput of CDMA-HDR a High Efficiency-High Rate Personal Communication Wireless System. IEEE Vehicular Technology Conference (VTC2000 Spring), May 2000, Tokyo, pp. 1854–1858.

9. Kahtava, J. "WCDMA Evolution with HSDPA." CIC2004, Korea.

10. Qualcomm, CDMA Technologies. "HSDPA for Improved Downlink Data Transfer." October 2004.

CHAPTER 18

Wireless Application Protocol

18.1 Introduction

Wireless application protocol (WAP)[1, 2] is a set of protocols that allows wireless devices such as cell phones, PDAs, and two-way radios to access the Internet. It is designed to work with small screens and with limited interactive controls. WAP incorporates wireless markup language (WML)[15], which is used to specify the format and presentation of text on the screen. WAP is a standard developed by the WAP forum (now the open mobile alliance (OMA)), and defines a communications protocol as well as an application environment. It is a standardized technology for cross-platform, distributed computing. WAP is very similar to the combination of hyper text markup language (HTML) and hyper text transport protocol (HTTP) except that it adds one very important feature: optimization for low-bandwidth, low-memory, and low-display capability environment. These types of environments include PDAs, wireless phones, pagers, and virtually any other communications device.

The imode is a proprietary mobile Internet service provider (ISP) and portal service from NTT DoCoMo, Japan. In this chapter we focus on WAP and discuss WAP programming model, WAP protocol stack, and WAP architecture. The goal of imode and its protocol stack is also presented and compared with the WAP.

18.2 WAP and the World Wide Web (WWW)

From a certain viewpoint, the WAP approach to content distribution and the web approach are virtually identical in concept. Both concentrate on distributing content to remote devices using inexpensive, standardized client software. Both rely on back-end servers to handle user authentication, database queries, and intensive processing. Both use markup languages derived from standard generalized markup language (SGML) for delivering content to the client. In fact, as WAP continues to grow in support and popularity, it is highly likely that WAP application developers will make use of their existing web infrastructure (in the form of application servers) for data storage and retrieval. WAP allows a further extension of the concept as existing "server" layers can be reused and extended to reach out to the vast array of wireless devices in business and personal use today. Extensible markup language (XML), as opposed to HTML, contains no screen formatting instructions; instead, it concentrates on returning structured data that the client can use as it sees fits.

18.3   Introduction to Wireless Application Protocol

WAP has become the defacto global industry standard for providing data to wireless hand-held mobile devices[3–9]. WAP takes a client server approach and incorporates a relatively simple microbrowser into the mobile phone, requiring only limited resources on mobile phones. WAP puts the intelligence in the WAP Gateways while adding just a microbrowser to the mobile phones themselves. Microbrowser-based services and applications reside temporarily on servers, not permanently in phones. The WAP is aimed at turning mass-market phones into a *network-based smart phone*. The philosophy behind WAP's approach is to use as few resources as possible on the hand-held device and compensate for the constraints of the device by enriching the functionality of the network.

WAP specifies a thin-client microbrowser using a new standard called *wireless markup language* (*WML*) that is optimized for wireless hand-held mobile devices. WML is a stripped down version of HTML.

WAP specifies a proxy server that acts as a gateway between the wireless network and the wireline Internet, providing protocol translation and optimizing data transfer for the wireless handset. WAP also specifies a computer-telephony integration *application programming interface* (*API*), called *wireless telephony application interface* (*WTAI*), between data and voice. This enables applications to take full advantage of the fact that this wireless mobile device is most often a phone and a mobile user's constant companion. On-board memory on a WAP phone can be used for off-line content, enhanced address books, bookmarks, and text input methods.

The importance of WAP can be found in the fact that it provides an evolutionary path for application developers and network operators to offer their services on different network types, bearers, and terminal capabilities. The design of the WAP standard separates the application elements from the bearer being used. This helps in the migration of some applications from short message service (SMS) or circuit-switched (CS) data to general packet radio service (GPRS), for example. WAP 1.0 was optimized for early WAP-phones.

The wireless application protocol cascading style sheet (WAP CSS) is the mobile version of a cascading style sheet. It is a subset of CSS2 (the cascading style sheet language of the world wide web) plus some WAP specific extensions. CSS2 features and properties that are not useful for mobile Internet applications are not included in WAP CSS. WAP CSS is the companion of XHTML Mobile Profile (XHTML MP). Both of them are defined in the WAP 2.0 specification, which was created by the WAP forum. XHTML MP is a subset of XHTML, which is the combination of HTML and XML. There are many WAP 2.0-enabled cell phones on the market currently.

Before creating WAP 2.0, developers used WML to build WAP sites and HTML/XHTML/CSS to build web sites. Now with WAP 2.0 they can make use of the same technologies to create both web sites and WAP sites. Documents written in XHTML MP/ WAP CSS are viewable on ordinary PC web browsers, since XHTML MP and WAP CSS are just the subsets of XHTML and CSS.

The following are the goals of wireless application protocol:

- Independent of wireless network standards
- Interoperability: Terminals from different manufacturers must be able to communicate with services in the mobile network
- Adaptation to bounds of wireless networks: Low bandwidth, high latency, less connection stability
- Adaptation to bounds of wireless devices: Small display, limited input facilities, limited memory and CPU, limited battery power
- Efficient: Provide quality of service (QoS) suitable to the behavior and characteristics of the mobile world
- Reliable: Provide a consistent and predictable platform for deploying services
- Secure: Enable services to be extended over potentially unprotected mobile networks while preserving the integrity of data
- Applications scale across transport options
- Applications scale across device types
- Extensible over time to new networks and transport

WAP is envisaged as a comprehensive and scalable protocol designed for use with:

- Any mobile device from those with a one-line display to a smart phone
- Any existing or planned wireless service such as the SMS, circuit-switched data, unstructured supplementary services data (USSD) and GPRS
- Any mobile network standard such as code division multiple access (CDMA), global system of mobile communications (GSM), or universal mobile telephone system (UMTS). WAP has been designed to work with all cellular standards and is supported by major worldwide wireless leaders such as AT&T wireless and NTT DoCoMo.
- Multiple input terminals such as keypads, keyboards, touch-screens, etc.

18.4 The WAP Programming Model

Before presenting the WAP programming model, we briefly discuss the world wide web (WWW) model that is the basis for the WAP model.

18.4.1 The WWW Model

The Internet WWW architecture provides a flexible and powerful programming model. Applications and content are presented in standard data formats, and are browsed by applications known as web browsers. The web browser is a network application, i.e., it sends requests for named data objects to a network server and the network server responds with encoded data using the standard formats.

The WWW standards specify several mechanisms necessary to build a general-purpose application environment which includes:

- **Standard naming model.** All servers and content on the WWW are named with an Internet-standard *Uniform Resource Locator (URL)*
- **Content typing.** All content on the WWW is given a specific type, thereby allowing web browsers to correctly process the content based on its type
- **Standard content formats.** All web browsers support a set of standard content formats. These include (HTML), JavaScript scripting language (ECMAScript, JavaScript), and a large number of other formats
- **Standard protocols.** Standard networking protocols allow any web browser to communicate with any web server. The most commonly used protocol on the WWW is the HTTP. This infrastructure allows users to easily reach a large number of third-party applications and content services. It also allows application developers to easily create applications and content services for a large community of clients.

The WWW protocols define three classes of servers:

1. *Origin server*—The server on which a given resource (content) resides or is to be created
2. *Proxy*—An intermediary program that acts as both a server and a client for the purpose of making requests on behalf of other clients. The proxy typically resides between clients and servers that have no means of direct communication, e.g., across a firewall. Requests are either serviced by a proxy program or passed on with possible translation to other servers. A proxy must implement both the client and server requirements of WWW specifications.
3. *Gateway*—A server which acts as an intermediary for some other server. Unlike a proxy, a gateway receives requests as if it were the origin server for the requested resource. The requesting client may not be aware that it is communicating with a gateway.

18.4.2 The WAP Model

The WAP programming model (see Figure 18.1) is similar to the WWW programming model. This provides several benefits to the application developer community, including a familiar programming model, a proven architecture,

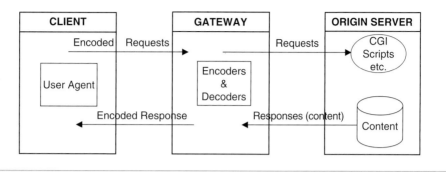

Figure 18.1   WAP programming model.

and the ability to leverage existing tools (e.g., web servers, XML tools, etc.). Optimization and extensions have been made in order to match the characteristics of the wireless environment. Wherever possible, existing standards have been adopted or have been used as the starting point for the WAP technology.

WAP content and applications are specified in a set of well-known content formats based on WWW content formats. Content is transported using a set of standard communication protocols based on WWW communication protocols. A microbrowser in the wireless terminal coordinates the user interface and is analogous to a standard web browser. WAP defines a set of standard components that enable communication between mobile terminals and network servers, including:

- **Standard naming model.** WWW-standard URLs are used to identify WAP content on origin servers. WWW-standard URLs are used to identify local resources in a device, e.g., call control functions

- **Content typing.** All WAP content is given a specific type consistent with WWW typing. This allows WAP user agents to correctly process the content based on its type.

- **Standard content formats.** WAP content formats are based on WWW technology and include display markup, calendar information, electronic business card objects, images, and scripting language.

- **Standard protocols.** WAP communication protocols enable the communication of browser requests from the mobile terminal to the network web server. The WAP content types and protocols have been optimized for mass market, hand-held wireless devices. WAP utilizes proxy technology to connect between the wireless domain and the WWW.

The WAP proxy typically comprises the following functionality:

- **Protocol gateway:** The protocol gateway translates requests from the WAP protocol stack to the WWW protocol stack (HTTP and TCP/IP).

• **Content encoders and decoders:** The content encoders translate WAP content into compact encoded formats to reduce the size of data over the network. This infrastructure ensures that mobile terminal users can browse a wide variety of WAP content and applications, and that the application author is able to build content services and applications that run on a large base of mobile devices. The WAP proxy allows content and applications to be hosted on standard WWW servers and to be developed using proven WWW technologies such as cell global identity (CGI) scripting. While the nominal use of WAP includes a web server, WAP proxy, and WAP client, WAP architecture can easily support other configurations. It is possible to create an origin server that includes WAP proxy functionality. Such a server might be used to facilitate end-to-end security solutions, or applications that require better access control or a guarantee of responsiveness.

18.5   WAP Architecture

WAP architecture [10–14] provides a scalable and extensible environment for the application development of mobile communication devices. This is achieved using a layered design of the entire protocol stack (see Figure 18.2). Each of the layers of architecture is accessible by layers above, as well as by other services and applications. WAP layered architecture enables other services and applications to utilize the features of the WAP stack through a set of well-defined interfaces. External applications may access the session, transaction, security, and transport layers directly.

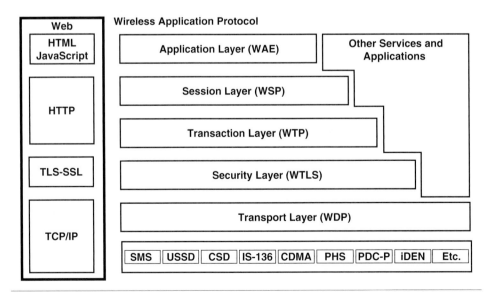

Figure 18.2   WAP architecture.

The layered design and functions of the WAP protocol stack resemble the layers of the open system interconnection model. In Figure 18.3, WAP and Internet protocol stacks are compared. The following sections provide a description of the various elements of the WAP protocol.

18.5.1 Wireless Application Environment

The uppermost layer in the WAP stack, the wireless application environment (WAE), is a general-purpose application environment based on a combination of WWW and mobile telephony technologies. The primary objective of the WAE is to establish an interoperable environment that allows operators and service providers to build applications and services that can reach a wide variety of

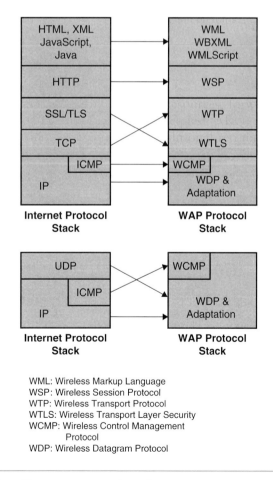

Figure 18.3   WAP and Internet protocol stacks.

different wireless platforms in an efficient and useful manner. Various components of the WAE are:

- *Addressing model:* WAP uses the same addressing model as the one used on the Internet, i.e., URL. A URL uniquely identifies a resource on a server that can be retrieved using well-known protocols. WAP also uses Uniform Resource Identifiers (URI). A URI is used for addressing resources that are not necessarily accessed using known protocols. An example of using a URI is local access to a wireless device's telephony functions.

- *Wireless markup language:* WML is WAP's analogy to HTML and is based on XML. It is WAP's answer to problems such as creating services that fit on small hand-held devices, low bandwidth wireless bearers, etc. WML uses a deck/card metaphor to specify a service. A card is typically a unit of interaction with the user, i.e., either presentation of information or request for information from the user. A collection of cards is called a deck, which usually constitutes a service. This approach ensures that a suitable amount of information is displayed to the user simultaneously since inter-page navigation can be avoided to the maximum possible extent. Key features of WML include variables, text formatting features, support for images, support for soft-buttons, navigation control, control of browser history, support for event handling (e.g., telephony services) and different types of user interactions (e.g., selection lists and input fields). One of the key advantages of WML is that it can be binary encoded by the WAP Gateway/Proxy in order to save bandwidth in the wireless domain.

- *WMLScript:* WMLScript is used for enhancing services written in WML. WMLScript can be used for validation of user input. Since WML does not provide any mechanisms for achieving this, a round trip to the server would be needed in order to determine whether user input is valid if scripting was not available. Access to local functions in a wireless device is another area in which WMLScript is used (for example, access to telephony related functions). WMLScript libraries contain functions that extend the basic WMLScript functionality. This provides a means for future expansion of functions without having to change the core of WMLScript. Just as with WML, WMLScript can be binary encoded by the WAP Gateway/Proxy in order to minimize the amount of data sent over the air.

18.5.2   Wireless Telephony Application

The wireless telephony application (WTA) environment provides a means to create telephony services using WAP. WTA utilizes a user-agent, which is based on the WML user-agent, but extends its functionality that meets special requirements for telephony services. This functionality includes:

- *Wireless telephony application interface (WTAI)*: An interface toward a set of telephony-related functions in a mobile phone that can be invoked from WML and/or WMLScript. These functions include call management, handling of text messages, and phone book control. WTAI enables access to functions that are not suitable for allowing common access to them, e.g., setting up calls and manipulating the phone book without user acknowledgment.
- *Repository:* Since it is not feasible to retrieve content from a server every now and then, repository makes it possible to store WTA services in the device in order to enable access to them without accessing the network.
- *Event Handling:* Typical events in a mobile network are incoming calls, call disconnect, and call answered. The event handling within WTA enables WTA services stored in the repository to be started in response to such events.
- *WTA service indication:* It is a content type that allows a user to be notified about events of different kinds (e.g., new voice mails) and be given the possibility to start the appropriate service to handle the event. In the most basic form, the WTA service indication makes it possible to send a URL and a message to a wireless device.

The WTA framework relies on a dedicated WTA user-agent to carry out these functions. Only trusted content providers should be able to make content available to the WTA user-agent. Thus, it must be possible to distinguish between servers that are allowed to supply the user-agent with services containing these functions and those who are not. To accomplish this, the WTA user-agent retrieves its services from the WTA domain, which, in contrast to the Internet, is controlled by the network operator. WTA services and other services are separated from each other using WTA access control based on port numbers.

18.5.3 Wireless Session Protocol

Wireless session protocol (WSP) provides a means for the organized exchange of content between cooperating client/server applications. Its functions are to:

- Establish a reliable session from the client to the server and release the session in an orderly manner.
- Agree on a common level of protocol functionality using capability negotiation.
- Exchange content between client and server using compact encoding.
- Suspend and resume the session.
- Provide HTTP 1.1 functionality.
- Exchange client and server session headers.

- Interrupt transactions in process.
- Push content from server to client in an unsynchronized manner.
- Negotiate support for multiple, simultaneous asynchronous transactions.

The core of the WSP design is a binary form of HTTP. Consequently, all methods defined by HTTP 1.1 are supported. In addition, capability negotiation can be used to agree on a set of extended request methods, so that full compatibility to HTTP applications can be retained. HTTP content headers are used to define content type, character set encoding, language, etc., in an extensible manner. However, compact binary encoding is defined for the well-known headers to reduce protocol overhead.

The life cycle of a WSP is not tied to the underlying transport protocol. A session can be suspended while the session is idle to free up network resources or save battery power. A lightweight session re-establishment protocol allows the session to be resumed without the overload of full-blown session establishment. A session may be resumed over a different bearer network.

WSP allows extended capabilities to be negotiated between peers (as an example this allows for both high-performance, feature-full implementation as well as simple, basic, and small implementations). WSP provides an optimal mechanism for attaching header information to the acknowledgment of a transaction. It also optionally supports asynchronous requests so that a client can submit multiple requests to the server simultaneously. This improves utilization of air time and latency as the result of each request can be sent to the client when it becomes available.

18.5.4 Wireless Transaction Protocol

Wireless transaction protocol (WTP) does not have security mechanisms. WTP has been defined as a light-weight transaction-oriented protocol that is suitable for implementation in "thin" clients and operates efficiently over wireless datagram networks. Reliability is improved through the use of unique transaction identifiers, acknowledgments, duplicate removal, and retransmissions. There is an optional user-to-user reliability function in which WTP user can confirm every received message. The last acknowledgment of the transaction, which may contain out-of-band information related to the transaction, is also optional. WTP has no explicit connection set-up or tear-down phases. This improves efficiency over connection-oriented services. The protocol provides mechanisms to minimize the number of transactions being replayed as a result of duplicate packets.

Wireless transaction protocol is designed for services oriented toward transactions, such as browsing. The basic unit of interchange is an entire message and not a stream of bytes. Concatenation may be used, where applicable, to convey multiple packet data units (PDUs) in one service data unit (SDU) of the datagram transport. WTP allows asynchronous transactions. There are three classes of transaction service:

- Class 0: Unreliable "send" with no result message. No retransmission if the sent message is lost.
- Class 1: Reliable "send" with no result message. The recipient acknowledges the sent message; otherwise the message is resent.
- Class 2: Reliable "send" with exactly one reliable result message. A data request is sent and a result is received which is finally acknowledged by the initiating part.

Note: For reliable "send," both success and failure is reported

18.5.5 Wireless Transport Layer Security

The purpose of wireless transport layer security (WTLS) is to provide transport layer security between a WAP client and the WAP Gateway/Proxy. WTLS is a security protocol based on the industry standard transport layer security (TLS) protocol with new features such as datagram support, optimized handshake, and dynamic key refreshing. The WTLS layer is modular and depends on the required security level of the given application, or characteristics of the underlying network, whether it is used or not. WTLS is optional and can be used with both the connectionless and the connection mode WAP stack configuration. In addition, WTLS provides an interface for managing secure connections. The primary goal of WTLS is to provide the following features between two communicating applications:

- *Data integrity:* WTLS contains facilities to ensure that data sent between the terminal and an application server is unchanged and not corrupted.
- *Privacy:* WTLS contains facilities to ensure that data transmitted between the terminal and an application server is private and cannot be understood by any intermediate parties that may have intercepted the data stream.
- *Authentication:* WTLS contains facilities to establish the authenticity of the terminal and application server.
- *Denial-of-service protection:* WTLS contains facilities for detecting and rejecting data that is replayed or not successfully verified. WTLS makes many typical denial-of-service attacks harder to accomplish and protects the upper protocol layers.

WTLS protocol is optimized for low-bandwidth bearer networks with relatively long latency. These features make it possible to certify that the sent data have not been manipulated by a third party, that privacy is guaranteed, that an author of a message can be identified, and that both parties cannot falsely deny having sent their messages.

18.5.6 Wireless Datagram Protocol

Wireless datagram protocol (WDP) offers a consistent service to the upper layer protocols of WAP and communicates transparently over one of the available bearer services. The services offered by WDP include application addressing by port numbers, optional segmentation and reassembly, and optional error detection.

WDP supports several simultaneous communication instances from a higher layer over a single underlying WDP bearer service. The port number identifies the higher layer entity above the WDP. Reusing the elements of the underlying bearers and supporting multiple bearers, WDP can be optimized for efficient operation within the limited resources of a mobile device.

The WDP adaptation layer is the layer of the WDP that maps the WDP functions directly onto a specific bearer. The adaptation layer is different for each bearer and deals with the specific capabilities and characteristics of that particular bearer service. At the gateway, the adaptation layer terminates and passes the WDP packets on to a WAP proxy/server via a tunneling protocol.

If WAP is used over a bearer UDP, the WDP layer is not needed. On other bearers, such as GSM SMS, the datagram functionality is provided by WDP. This means that WAP uses a datagram service, which hides the characteristics of different bearers and provides port number functionality.

Processing errors can occur when the WDP datagrams are sent from one WDP provider to another. For example, a wireless data gateway is unable to send a datagram to the WAP gateway, or the receiver does not have enough buffer space to receive large messages. The wireless control message protocol (WCMP) provides an efficient error handling mechanism for WDP.

18.5.7   Optimal WAP Bearers

The WAP is designed to operate over a variety of different services, including SMS, circuit switched data (CSD), and packet switched data (PSD). The bearers offer differing levels of quality of service with respect to throughput, error rate, and delays. The WAP is designed to compensate for or tolerate these varying levels of service:

- *Short message service (SMS):* Given its limited length of 160 characters per short message, the overhead of the WAP that would be required to be transmitted in an SMS message would mean that even for the simplest of transactions several SMS messages may have to be sent.

- *Circuit switched data (CSD):* Most of the trial-based services use CSD as the underlying bearer. CSD lacks immediacy—a dial-up connection taking about 10 seconds is required to connect the WAP client to the WAP gateway; and this is the best case scenario when there is a complete end-to-end digital call.

- *Unstructured supplementary services data (USSD):* USSD is a means of transmitting information or instructions over a GSM network. In USSD a session is established and the radio connection stays open until the user, application, or time-out releases it. USSD text messages can be up to 182 characters in length. USSD can be an ideal bearer of WAP on GSM networks. USSD is preferable due to the following advantages:

- Turnaround response times for interactive applications are shorter for USSD.

- Users need not access any particular phone menu to access services with USSD but they can enter the command directly from the initial mobile phone screen.

- *General packet radio service (GPRS):* GPRS is a new bearer because it is immediate, relatively fast, and supports virtual connectivity, allowing relevant information to be sent from the network as and when it is generated. There are two efficient means of delivering proactively sending (pushing) content to a mobile phone: by SMS (which is, of course, one of the WAP bearers) or by the user maintaining more or less a permanent GPRS session with the content server. WAP incorporates two different connection modes—WSP connection mode or WSP connection protocol. This is similar to the two GPRS point-to-point services—connection oriented and connectionless. For the interactive menu-based information exchanges that WAP anticipates, GPRS and WAP can be ideal bearers for each other.

18.6   Traditional WAP Networking Environment

WAP allows the presentation and delivery of information and services to wireless devices such as mobile telephone or hand-held computer. The major players in WAP space are the wireless service provider (WSP) and the enterprise (see Figure 18.4). The WSP is the wireless equivalent of an Internet service provider (ISP). The role of the WSP is to provide access to back-end resources for wireless users. The WSP provides additional service because wireless users must transition from a wireless to a wired environment (unlike an Internet environment where the user is already "on" on the Internet). The WSP's space contains a modem bank, remote access service (RAS) server, router, and potentially a WAP gateway. The environment is similar to the wired environment, where all connection type services are provided by the WSP. The WSP handles the processing associated with incoming WAP communications, including the translation of wireless communication from the WAP device through the transmission towers to a modem bank and RAS and on to the WAP gateway. The modem bank receives incoming phone calls from the user's mobile device, the RAS server translates the incoming calls from a wireless packet format to a wired packet format, and the router routes these packets to correct destinations. The WAP gateway is used to translate

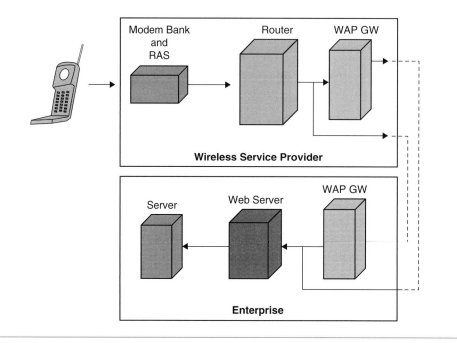

Figure 18.4   Traditional WAP networking environment.

the WAP into traditional Internet protocol (TCP/IP). The WAP gateway is based on proxy technology. Typical WAP gateways provide the following functionality:

- Domain name server (DNS) service, for example, to resolve domain names used in URLs.

- A control point for management of fraud and service utilization.

- Act as a proxy, translating the WAP protocol stack to the Internet protocol stack.

Many gateways also include a transcoding function that will translate an HTML page into a WML page that is suited to a particular device type. The enterprise space contains the backed web and application servers that provide the enterprises's transactions. Generally, the WSP maintains and manages the WAP gateway, but there are circumstances under which this is not desirable. This is due to the presence of an encryption gap caused by the ending of the WTLS session at the gateway. The data is temporarily in clear text on the gateway until it is re-encrypted under the SSL session established with the enterprise's web server. In such cases, the WAP gateway should be maintained at the enterprise. The problem with this solution is the absence of the DNS client at the mobile device, which would require the storage of profiles for every target on the mobile device. This also requires that the enterprise set up a relationship with service provider whereby all incoming packets destined for the enterprise (identified by the IP address) are immediately routed by the WSP directly to the enterprise and are never sent to the WSP's gateway.

18.7 WAP Advantages and Disadvantages

The following are the advantages of WAP:

- Implementation near to the Internet model
- Most modern mobile telephone devices support WAP
- Real-time send/receive data
- Multiplatform functionality (little change is needed to run on any web site since XML is used)
- No hardware obsolescence

The following are some of the advantages of using WAP CSS on mobile Internet sites:

- Because of WAP 2.0 (XHTML MP/WAP CSS), web programming and WAP programming converge. Learning WAP programming does not require much effort if you already know how to program the web. Web developers can continue to use their familiar web authoring tools and PC web browsers for building mobile Internet sites. This is one major advantage of XHTML MP/WAP CSS over WML.
- You can have greater control on the appearance of WAP pages with WAP CSS than with WML. For example, you can specify the colors, fonts, background, borders, margins, and padding of various elements with WAP CSS.
- If you apply a single cascading style sheet to the whole mobile Internet site, a mobile device will download the cascading style sheet only once the first time the mobile Internet site is visited. The cascading style sheet will then be stored in the cache and it can be accessed later without connecting to the server.
- The file sizes of XHTML MP documents can become smaller if the layout and formatting information is moved to an external WAP CSS style sheet. A small file size has the advantage of a shorter download time.
- Using WAP CSS has the advantage that the content and presentation can be separated. This means you can:
  - Match the layout and style of the same content to the characteristics of different wireless devices easily
  - Match the layout and style of the same content for different user agents easily

- Minimize the effort to maintain a WAP site. When new mobile phone models come onto the market, you can write new WAP CSS style sheets to optimize the layout of the WAP site on these new mobile phones. The content files do not need to be modifed.
- Apply a single WAP CSS style sheet to multiple WAP pages. Later if you want to change the look and feel of the whole WAP site, just modify the WAP CSS.
- Reuse the style code in multiple projects
- Remotely divide work—someone can focus on look and feel WAP, whereas others can concentrate on contents.

Some of the disadvantages of the WAP are the following:

- Low speeds, security, and very small user interface
- Not very familiar to the users
- Business model is expensive
- Forms are hard to design
- Third party is included

Some of the disadvantages of using WAP CSS style sheets on mobile Internet sites are:

- Different WAP browsers have varied levels of support for WAP CSS. One property supported on one WAP browser may not be available on another WAP browser.
- An external WAP CSS style sheet can increase the time required for a page to be completely loaded the first time the WAP site is visited because of the following reasons:
  - The external WAP CSS style sheet does not exist in the cache of the mobile phone at the first visit, which means the mobile phone has to download it from the server.
  - An XHTML MP document and its external WAP CSS style sheet have to be downloaded in separate requests.
  - If you make use of a single WAP CSS file to specify all the presentation information about the mobile Internet site, the file size of the WAP CSS file can be quite large.
  - The WAP browser needs to parse the cascading style sheet in addition to the XHTML MP document.

18.8   Applications of WAP

The first and foremost application of WAP is accessing the Internet from mobile devices. This is already in use in many mobile phones. This application is gaining popularity daily, and many web sites already have a WAP version of their site. An application, which is out, is sending sale offers to mobile customers through WAP. The user's phone will be able to receive any sale prices and offers from the web site of a store.

Games can be played from mobile devices over wireless devices. This application has been implemented in certain countries and is under development in many others. This is an application which has been predicted to gain high popularity.

Application to access time sheets and filing expenses claims via mobile handsets are currently being developed. These applications, when implemented, will be a breakthrough in the business world.

Applications to locate WAP customers geographically have been developed. Applications to help users who are lost or stranded by guiding them using their locations are under consideration.

WAP also provides short messaging, e-mail, weather, and traffic alerts based on the geographic location of the customer. These applications are available in some countries but will soon be provided in all countries.

One of the biggest applications of WAP under consideration is banking from mobile devices. These applications will be very popular if they are implemented in a secure manner.

The mobile industry appears to be moving forward, putting aside the issues of network and air interface standards, and instead concentrating on laying the foundations for service development, regarded by many as the key driver to multimedia on the move and third-generation mobile systems. From that point of view, in the near future WAP and Bluetooth will play fundamental roles.

18.9 imode

imode is a proprietary mobile ISP and portal service from NTT DoCoMo, Japan, with about 50 million subscribers. For imode, DoCoMo adopted the Internet model and protocol. imode uses compact HTML (cHTML) as a page description language. The structure of cHTML means that the user can view traditional HTML and imode sites can be inspected with ordinary Internet web browsers. This is in contrast to WAP, where HTML pages must be translated to WML. imode provides Internet service using personal digital cellular-packet (PDC-P) and a subset of HTML 3.0 for content description.

imode is a packet-switched service (always connected, as long as the user's handset is reached by imode signal) which includes images, animated images, and colors. In imode, users are charged per packet of downloaded information. imode allows application/content providers to distribute software to cellular phones and also permits users to download applets (e.g., games). imode uses packet-switched technology for the wireless part of the communication. The wired part of the communication is carried over TCP/IP.

Packet-switched services send and receive information by dividing messages into small blocks called packets and adding headers containing address and control information to each packet. This allows multiple communications to be carried on a communication channel, giving efficient channel usage with low cost. Dopa, DoCoMo's dedicated data communications service, offers connections to location area network (LAN) and Internet service providers. The mobile packet communications system has a network configuration in which a packet communications function is added and integrated into PDC, the digital system for portable and automobile telephones. DoCoMo has developed a data transmission protocol specific to imode. This protocol is used with the PDC-P system. The PDC-P network includes a mobile message packet gateway (M-PGW) to handle conversions between the two protocol formats. Connection between the imode server and the Internet uses TCP/IP. The imode server is a regular web server which can reside at NTT DoCoMo or at the enterprise. DoCoMo has been acting as a 1717 portal and normally maintains the imode server. Figure 18.5 shows the imode protocol stack.

imode relies on Internet security as provided by SSL/TLS and does not have the ability to handle server-side authenticated SSL sessions. imode phones are pre-configured with root collision avoidance keys from public key infrastructure. This will allow for establishment of a server-side authenticated SSL session between the imode device and imode server hosted by the enterprise. imode does not have the

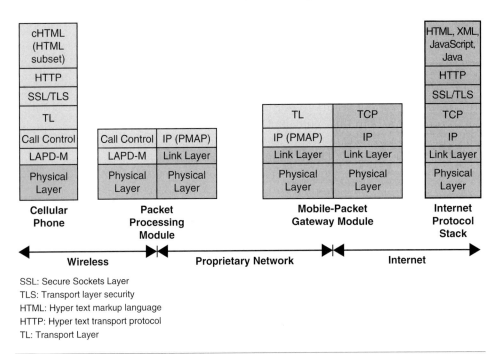

Figure 18.5   imode protocol stack.

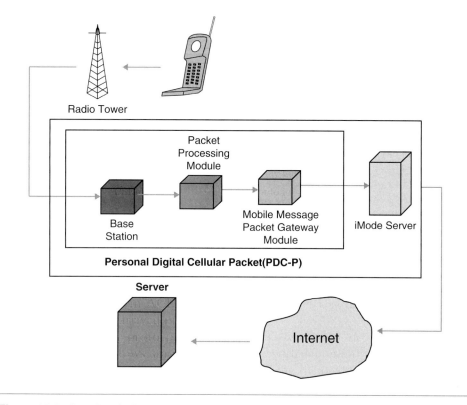

Figure 18.6   imode wireless networking environment.

capability of handling client-side certificates which means that nonrepudiation is not possible with the current implementation of imode. Figure 18.6 shows a typical imode wireless networking environment.

18.10   imode versus WAP

imode is available only in Japan, whereas Europe and other big markets for 3G mobile service providers are completely WAP-based. In the United States, most service providers have chosen WAP.

The most basic difference between imode and WAP is the different graphic capabilities; imode only supports simple graphics, which is far more than WAP allows. The imode packet-switched data network is more suited for transferring data than the WAP circuit-switched network.

Another major difference is the "always-on" capabilities of imode. Since users are not charged for the time they spend on-line, it is more convenient, and also less expensive. Since there is no need to dial up before using the various IP-based services, e-mail becomes as instant as SMS.

imode uses cHTML, a subset of HTML, while WAP uses WML, a subset of XML. cHTML, while certainly easier to develop from a web-designer stand-point, has its limitations. The downside of WML, on the other hand, is similarly obvious—currently a WAP gateway is required to translate between HTML and WML for almost every data transfer. On the other hand, since WML is derived from XML, it is much more extensible. XML allows for more dynamic content and various different applications. In the future, a WML-based service will be of more advantage than an HTML-based one. So while WAP may currently require more complicated technology, in the long run, it may enable the user to do more with his or her device.

18.11 Summary

In this chapter we discussed wireless application protocol (WAP). WAP specifies a thin-client microbrowser using a new standard called wireless markup language (WML) that is optimized for wireless hand-held mobile devices. WML is a stripped-down version of hyper text markup language (HTML). NTT DoCoMo imode provides Internet service using PDC-P and a subset of HTML 3.0 for content description. imode allows application/content providers to distribute software to cellular phones. We concluded the chapter by providing basic differences between WAP and imode.

Problems

18.1 What is WAP? Discuss briefly.

18.2 Discuss WAP architecture.

18.3 What is wireless transport layer security in WAP?

18.4 What is imode? Discuss briefly.

18.5 What is WAP 2.0?

18.6 What are the major differences between WAP and imode?

References

1. Buckingham, S. "Introduction to WAP." http://www.gsmworld.com/technology/yes2wap.htm.

2. Heijden, M. Van Der, and Taylor, M. *Understanding WAP, Wireless Applications, Devices and Services*. Artech House Publishers, 2000.

3. http://www.wapforum.org/.

4. http://www.wap.de/.

5. http://www.w3c.org/.

6. http://www.wap.com/.

7. Kanjilal, J. "WAP: Internet over Wireless Networks." http://www.Infocommworld .com//99sep/cover02.htm.

8. Laurence, Lee Min En. "Introduction to WAP Architecture." http://www.fit.qut .edu.au//DataComms/itn540/gallery/a100/lee/INTROD~1.HTM.

9. Shah, R. "Wireless Application Protocol set to take over." http://www.sunworld .com/sunworldonline/swol-01-2000/swol-01-connectivity.htm.

10. Simpson, W. Ed. "The Point-to-Point Protocol (PPP)." *STD 50, RFC1661, Day-dreamer,* July 1994.

11. Simpson, W. Ed. "PPP in HDLC Framing." *STD 51, RFC1662, Day-dreamer,* July 1994.

12. WAP Forum. "Wireless Application Protocol White Paper." http://www.wapforum .org/what/WAP white pages.pdf.

13. Wireless Application Protocol Forum. "Wireless Application Protocol (WAP)." *Bluetooth Special Interest Group,* version 1.0, 1998.

14. Wireless Application Protocol Forum. "WAP Conformance." Draft version 27, May 1998.

15. Wireless Markup Language (WML). http://www.cellular.co.za/wml.htm.

CHAPTER 19

Wireless Personal Area Network—Bluetooth

19.1 Introduction

Several technical committees of IEEE are responsible for developing standards of the local area network (LAN), wireless local area network (WLAN), and wireless personal area network (WPAN). Table 19.1 summarizes them to familiarize the readers.

In a WPAN, the residence is connected to a public switched telephone network (PSTN) for telephone services, Internet for web access, and cable network for multichannel television services. In this chapter, we first introduce the concept of WPAN and discuss the roles of the IEEE 802.15[6] committee. We then provide details of Bluetooth wireless technology[1,3–5].

Bluetooth enables users to connect a wide range of computing and telecommunication devices without any additional or proprietary cables. Ultrawide band (UWB-IEEE 802.15.3a) is the IEEE standard for a high-data-rate WPAN designed to provide sufficient quality of service for real-time distribution of content such

Table 19.1 IEEE 802 working groups.

	Working group
802.1	Higher Layer LAN Protocol
802.2	Logical Link Control (LLC)
802.3	Ethernet
802.11	WLAN
802.15	WPAN
802.16	Broadband Wireless Access (BWA)
802.17	Resilient Packet Ring
802.18	Radio Regulatory TAG
802.20	Mobile Broadband Wireless Access (MBWA)
Link Security	Executive Committee Study Group
802 Handoff	Executive Committee Study Group

as video and music. It is ideally suited for a home multimedia wireless network. We will discuss UWB in Chapter 22.

The Bluetooth wireless specification includes RF, link layer, and application layer definitions for product developers for data, voice, and content-centric applications. The specification contains the information necessary to ensure that diverse devices supporting Bluetooth wireless technology can communicate with each other worldwide.

19.2   The Wireless Personal Area Network

Within the home, computers and printers are connected to the Internet through voice band modems, XDSL services, or cable modems. The number of home networks in the United States is expected to almost double each year. The industry has two distinct segments—home access and home distribution. Home access technology uses different wireless and wired alternatives to secure a broadband Internet access to the home gateway to be distributed to the user's information appliances. The home distribution or WPAN interconnects all home appliances and connects them to the Internet through a home gateway. It is expected that more than 80% of U.S. households will have a broadband data access by the year 2008+.

The WPAN provides an infrastructure to interconnect a variety of home appliances and enables them to access the Internet through a central home gateway. Home computing equipment used for computing and Internet transaction interface access includes PCs, laptops, printers, scanners, and web cameras. A home computing network allows multiple computers as well as multiple devices to connect with a network protocol.

A wireless personal area network (WPAN) is a short-distance (typically <10 m but as far as 20 m) wireless network specially designed to support portable and mobile computing devices such as PCs, PDAs, printers, storage devices, cell phones, pagers, set-up boxes, and a variety of consumer electronic equipment. Bluetooth (IEEE 802.15.1), UWB (IEEE 802.15.3a), and ZigBee (IEEE 802.15.4) are examples of WPANs that allow devices within close proximity to join together in wireless networks in order to exchange information. Many cell phones already have two radio interfaces—one for the cellular network and the other for PAN connections.

WPANs such as Bluetooth provide enough bandwidth and convenience to make data exchange practical for certain mobile devices requiring data exchanges at rates up to 1 Mbps. At the other end of the scale, UWB will provide the capability of streaming video signals at data rates up to 1 Gbps. However, many control and command applications require much lower data rates and also the lowest possible cost, thus ZigBee. At this time most WPAN applications target cable replacements (e.g., Bluetooth headsets) but these technologies are capable of forming elegant peer-to-peer networks, offering more enhanced application capabilities such as sensor networks and so on (see Chapter 22).

The IEEE 802.15 committee has the responsibility for developing standards for short distance wireless networks used in the networking of portable and mobile computing devices such as PCs, PDAs, cell phones, printers, speakers, microphones, and other consumer electronics. IEEE 802.15.1 and 802.15.4 focus on the devices with the following characteristics:

- Power management: low current consumption
- Range: 0–10 m
- Rate: 19.2–100 kbps
- Size: 0.5 in^3 without antenna
- Low cost relative to target device
- Should allow overlap of multiple networks in the same area
- Network supports a minimum of 16 devices

The IEEE 802.15 committee consists of the following task groups (see Figure 19.1):

- **Task group I:** It is for Bluetooth and defines physical (PHY) and medium access control (MAC) specifications for wireless connectivity with fixed, portable, and moving devices within or entering a personal operating space (POS). A POS is the space around a person or object that typically extends up to 10 m in all directions and envelops the person whether stationary or in motion.

- **Task group II:** It focuses on the coexistence of WPAN and IEEE 802.11 WLANs. The goal of the WPAN group is to achieve a level of interoperability that allows the data transfer between a WPAN device and an IEEE 802.11 device.

- **Task group III:** It works on PHY and MAC layers for high-rate WPANs that operate at a data rate of more than 20 Mbps, and will provide for

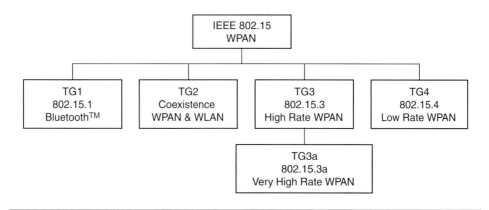

Figure 19.1   IEEE 802.15 task groups.

low-power, low-cost solutions to address the needs of portable consumer digital imaging and multimedia applications. The standards aim at providing compatibility with Bluetooth specifications.

- **Task group IV:** It investigates an ultra-low complexity, ultra-low power consuming, ultra-low-cost PHY and MAC layer for data rates of up to 200 kbps. Potential applications are sensors, interactive toys, smart badges, remote controls, and home automation.

19.3   Bluetooth (IEEE 802.15.1)

In 1994, the Swedish telecommunication company Ericsson decided to honor old, weird Herald I. Bluetooth, king of Denmark between 940 and 985 AD, by naming its new wireless networking standard after him. Table 19.2 outlines the evolution of Bluetooth technology.

Bluetooth provides short-range, low-cost (<$10 per device) connectivity between portable devices. Bluetooth is limited in range (<10 meters) and bandwidth (780 kbps compared to Home RF that goes to 1–2 Mbps and IEEE 802.11b that goes to 11 Mbps with 150 m and greater distance).

Bluetooth radio characteristics [8–10] include low power, short range, and medium transmission speed. The low power consumption makes Bluetooth ideal for small, battery-powered devices like mobile phones and pocket PCs. Bluetooth is poised to capitalize on the emerging market of small mobile devices that is expected to grow. Bluetooth's short range (<10 m) is ideal for the concept of "personal operating space" and integrates the notion of using a device carried or worn on the body or otherwise located within immediate reach. Bluetooth's transmission speed of 780 kbps works well for transferring small to medium-sized files.

Bluetooth enables users to connect a wide range of computing and telecommunication devices easily and simply, without the need to buy additional or proprietary cables. The cable solution is complicated since it may require a cable specific to the device that is being connected. The infrared solution eliminates the cable, but requires line of sight. To solve all these problems the Bluetooth standard has been developed.

The Bluetooth system operates in the 2.4 GHz Industrial Scientific Medicine (ISM) band. In a vast majority of countries around the world the range of this

Table 19.2 Evolution of Bluetooth technology.

1998	Special Interest Group (SIG)[2] was formed including 3Com, Ericsson, IBM, Intel, Lucent, Microsoft, Motorola, Nokia, and Toshiba.
1999	The first open Bluetooth specification 1.0 released.
2000	The first certified Bluetooth products on market.
2001	The latest protocol 1.1 released.

frequency band is 2.4–2.4835 GHz. The ISM band is open to any radio system such as cordless phones, garage door openers, and microwaves, and therefore is susceptible to strong interferences.

A WPAN can be formed with a Bluetooth-enabled pocket PC to dial into an ISP and access the Internet. After downloading a file, you could walk within 10 m of a Bluetooth-enabled printer and send the file from the pocket PC to the printer to have it printed. These examples show how Bluetooth can eliminate the need for a cable to the phone or printer.

A Bluetooth WPAN involves up to eight devices, located within a 10-m radius personal operating space, that unite to exchange information or share services. Because it can be done spontaneously according to immediate need, it is known as *ad hoc networking*. Because a WPAN involves directly networking between different points, without the use of network infrastructure, it is also referred to as a "point-to-point network" [7,11,12].

The Bluetooth market focuses on four categories of users: professional and field workers who need to travel off-site but still require access to corporate communications and information; technology-savvy electronic consumers; industrial and retail workers involved in automated processes where cables can get in the way; and office workers whose worksites are outfitted with Bluetooth devices.

Bluetooth is an ideal solution for connecting the increasing number of devices designed to be held in the hand or worn on the body. These devices are being embraced by industries using mobile automation systems. For example, at a car rental facility, an employee could use a Bluetooth-enabled pocket PC equipped with a bar code scanner to scan a vehicle identification number, enter mileage and fuel data, then instantly transmit a receipt to a Bluetooth-enabled portable printer worn on the hip. Because of the GSM subscriber identity module (SIM), industry experts foresee Bluetooth-enabled GSM phones to be used like credit cards. For example, one could use the Bluetooth-enabled mobile phone at a Bluetooth-enabled vending machine to charge one's account and buy a drink.

One of Bluetooth's greatest advantages is that it can be used absolutely anywhere that at least two Bluetooth devices share a 10-m range. It is possible because Bluetooth is designed for direct point-to-point networking between devices and does not require proximity to infrastructure stations like signal towers or access points.

Bluetooth radio technology provides a universal bridge to existing data networks, a peripheral interface, and a mechanism to form small private ad hoc groupings of connected devices away from fixed network infrastructures. Bluetooth radio uses a fast acknowledgment and frequency hopping scheme to make the link robust. Bluetooth typically hops faster and uses shorter packets. Short packets and fast hopping limit the impact of interference from other radio systems that use the same frequency band. Use of a forward error correction (FEC) scheme limits the

Table 19.3 Bluetooth air interface details.

Feature	Values	Notes
Frequency range:		
• USA, Europe, and most other countries	2.4–2.4835 GHz	79 RF channels
• Spain	2.445–2.475 GHz	23 RF channels
• France	2.4465–2.4835 GHz	23 RF channels
Bandwidth of each RF channel	1 MHz	
Gross data rate	1 Mbps (initial) 2 Mbps (latter)	
Time slot duration	626 μs	Time division multiplexing (TDM) is used to divide the channel into time slots. Transmission occurs in packets that occupy an odd number of slots (up to 5).
One-to-one connection allowable maximum data rate	721 kbps	3 voice channels
Signal modulation	Gaussian frequency shift keying (GFSK)	
Piconet access		FH-TDD-TDMA
Scatternet access		FH-CDMA
Frequency hopping rate	1600 hops/second	
Range	10 meters	

impact of random noise on long-distance links. Table 19.3 lists the parameters of Bluetooth air interface.

Bluetooth is being used in mobile computers, bar code laser scanners, cash registers, vending machines, GPS receivers, slide projectors, printers, digital cameras, digital camcorders, test and measurement equipment, and LAN access points. IEEE now has a formalized standards development in process for Bluetooth known as 802.15.1. The IEEE is also exploring the enhancement of 802.15.1 with a high data rate Bluetooth standard: 802.15.3.

Transmitter equipment is classified into three power classes (see Table 19.4). A power control is required for power class 1 equipment, whereas power controls for power class 2 and 3 equipment is optional. The power control is used for limiting the transmitted power over 0 dBm. Power control capability under 0 dBm is optional and could be used for optimizing the power consumption and overall interference level. The power steps form a monotonic sequence, with a maximum step size of 8 dB and a minimum step size of 2 dB. Class 1 equipment with a maximum transmit power of 20 dBm must be able to control its transmit power down to 4 dBm or less. Equipment with power control capability optimizes the output power in a link.

Table 19.4 Bluetooth transmitter characteristics.

Power class	Maximum output power (Pmax)	Nominal output power	Minimum output power*	Power control
1	100 mW (20 dBm)	N/A	1 mW (0 dBm)	Pmin <4 dBm to Pmax Optional: Pmin** to Pmax
2	2.5 mW (4 dBm)	1 mW (0 dBm)	0.25 mW (−6 dBm)	Optional: Pmin** to Pmax
3	1 mW (0 dBm)	N/A	N/A	Optional: Pmin** to Pmax

* Minimum output power at maximum power setting.

** The lower power limit Pmin < −30 dBm is suggested but it is not mandatory, and may be chosen according to application needs.

The actual sensitivity level is defined as an input level for which a raw bit error rate (BER) of 0.1% is met. The requirement for a Bluetooth receiver is an actual sensitivity level of −70 dBm or better. The receiver must achieve a −70 dBm sensitivity level with any Bluetooth transmitter compliant to the transmitter specification given in Table 19.4.

19.4   Definitions of the Terms Used in Bluetooth

- **Piconet.** A collection of devices connected via Bluetooth technology in an ad hoc fashion. A piconet starts with two connected devices, such as a PC and cellular phone, and may grow to eight connected devices. All Bluetooth devices are peer units and have identical implementations. However, when establishing a piconet, one unit will act as a master for synchronization purposes, and the other(s) as slave(s) for the duration of the piconet connection.

- **Scatternet.** Two or more independent and nonsynchronized piconets that communicate with each other. A slave as well as a master unit in one piconet can establish this connection by becoming a slave in the other piconet.

- **Master unit.** The device in the piconet whose clock and hopping sequence are used to synchronize all other devices in the piconet.

- **Slave units.** All devices in a piconet that are not the master (up to seven active units for each master).

- **MAC address.** A 3-bit medium access control address used to distinguish between units participating in the piconet.

- **Parked units.** Devices in a piconet which are time-synchronized but do not have MAC addresses.

- **Sniff and hold mode.** Devices that are synchronized to a piconet, and which have temporarily entered power-saving mode in which device activity is reduced.

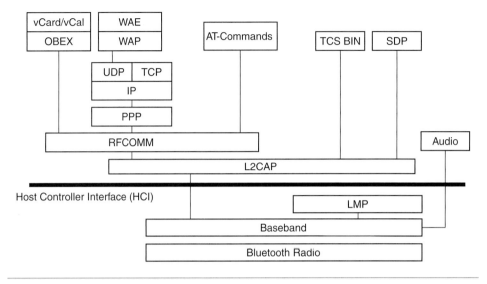

Figure 19.2   Bluetooth protocol stack.

19.5   Bluetooth Protocol Stack

The Bluetooth protocol stack allows devices to locate, connect, and exchange data with each other and to execute interoperable, interactive applications against each other. The Bluetooth protocol stack can be placed into three groups: transport protocol group, middleware protocol group, and application group (see Figure 19.2).

19.5.1   Transport Protocol Group

The protocols in this group are designed to allow Bluetooth devices to locate and connect to each other. These protocols carry audio and data traffic between devices and support both synchronous and asynchronous transmission for telephony-grade voice communication. Audio traffic is treated with high priority in Bluetooth. Audio traffic bypasses all protocol layers and goes directly to the baseband layer which then transmits it in small packets directly over Bluetooth's air interface.

The protocols in this group are also responsible for managing the physical and logical links between the devices so that the layers above and applications can pass data through the connections. The protocols in this group are radio, baseband, link manager, logical link, and host controller interface (HCI).

- **Logical link control and adaptation protocol (L2CAP) layer.** All data traffic is routed through the logical link control and adaptation protocol layer. This layer shields the higher layers from the details of the lower layers. The higher layers need not be aware of the frequency hops occurring at the radio and baseband level. It is also responsible for segmenting larger packets from higher layers into smaller packets, which are easier to handle by the lower layer. The L2CAP layer in two peer devices facilitates the maintenance of the desired grade of service. The L2CAP layer is responsible for admission control based on the requested level of service and for coordinating with the lower layers to maintain this level of service.

- **Link manager layer (LML).** The link manager layers in communicating devices are responsible for negotiating the properties of the Bluetooth air interface between them. These properties may be anything from bandwidth allocation to support services of a particular type to periodic bandwidth reservation for audio traffic. This layer is responsible for supervising device pairing. Device pairing is the creation of a trust relationship between the devices by generating and storing an authentication key for future device authentication. This is an important step in establishing a communication between two devices. If this fails, the communication link may get severed. The link managers are also responsible for power control and may request adjustments in power levels.

- **Baseband and radio layers.** The baseband layer is responsible for the process of searching for other devices and establishing a connection with them. It is also responsible for assigning the master and slave roles. This layer also controls the Bluetooth unit's synchronization and transmission frequency hopping sequence. This layer also manages the links between the devices and is responsible for determining the packet types supported for synchronous and asynchronous traffic.

- **Host controller interface (HCI) layer.** The HCI allows higher layers of the stack, including applications, to access the baseband, link manager, etc., through a single standard interface. Through HCI commands, the module may enter certain modes of operation. Higher layers are informed of certain events through the HCI. The HCI is not a required part of the specification. It has been developed to serve the purpose of interoperability between host devices and Bluetooth modules. Bluetooth product implementation need not be HCI compliant to support a fully compliant Bluetooth air interface.

19.5.2 Middleware Protocol Group

This group comprises the protocols needed for existing applications to operate over Bluetooth links. The protocols in this group can be third party and industry standard protocols and protocols developed specifically by the Special Interest Group (SIG)[2] for Bluetooth wireless communication. The protocols in this group can include TCP, IP, PPP, etc. A serial port emulator protocol, RFCOMM, enables applications that normally would interface with a serial port to operate with Bluetooth links. A packet-based telephony control protocol for advanced telephony operations is also present. This group has a service discovery protocol (SDP), which lets devices discover each other's services.

- **RFCOMM layer.** Serial ports are the most common communication interface in use today. These serial ports invariably involve the use of cable. Bluetooth's prime aim is to eliminate cables and provide support for serial communication without cables. RFCOMM provides a virtual serial port to applications. The advantage provided by this layer is that it is easy for applications designed for cabled serial ports to migrate to Bluetooth. The applications can use RFCOMM much like a serial port to accomplish scenarios like dial-up networking, etc. RFCOMM is an important part of the protocol stack because of the function it performs.

- **Service discovery protocol (SDP) layer.** In Bluetooth wireless communications any two devices can start communicating on the spur of the moment. Once a connection is established there is a need for the devices to find and understand the services the other devices have to offer. This is taken care of in this layer. The SDP is a standard method for Bluetooth devices to discover and learn about the services offered by the other device. Service discovery is important in providing value to the end-user.

- **Infrared data association (IrDA) interoperability protocols.** The SIG has adopted some IrDA protocols to ensure interoperability between applications. IrDA and Bluetooth share some important attributes. The Infrared Object Exchange Protocol is designed to enable units supporting infrared communication to exchange a wide variety of data and comments.

- **Object exchange (OBEX) protocol.** IrOBEX (in short, OBEX) is a session protocol developed by the Infrared Data Association to exchange objects in a simple and spontaneous manner. OBEX provides the same basic functionality as HTTP but in a much lighter fashion. It uses a client-server model and is independent of the transport mechanism and transport application programming interface (API), provided it realizes a reliable transport base. In addition, the OBEX protocol defines a folder-listing object, which is used to browse the contents of folders on a remote device.

- **Networking layers.** Bluetooth wireless communication uses a peer-to-peer network topology rather than an LAN type topology. Dial-up networking uses the attention (AT) command layer. In most cases the network that is being accessed is an IP network. Once a dial-up connection is established to an IP network, then standard protocols like TCP, UDP, and HTTP can be used. A device can also connect to an IP network using a network access point. The Internet PPP is used to connect to the access point.

The specification does not define a profile that uses the TCP/IP directly over Bluetooth links.

- **Telephone control specification (TCS) layer and audio.** This layer is designed to support telephony functions, which include call control and group management. These are associated with setting up voice calls. Once a call is established a Bluetooth audio channel can carry the call's voice content. TCS can also be used to set up data calls. The TCS protocols are compatible with ITU specifications. The SIG is also considered a second protocol called TCS-AT, which is a modem control protocol. AT commands over RFCOMM are used for some applications. Audio traffic is treated separately in Bluetooth. Audio traffic is isochronous, meaning that it has a time element associated with it. Audio traffic is routed directly to the baseband. Special packets called synchronous connection-oriented are used for audio traffic. Bluetooth audio communication takes place at a rate of 64 kbps using one of the two data encoding schemes—8-bit logarithmic pulse code modulation or continuous variable slope delta modulation.

19.5.3 Application Group

This group consists of actual applications that make use of Bluetooth links and refers to the software that exists above the protocol stack. The software uses the protocol stack to provide some function to the user of the Bluetooth devices. The most interesting applications are those that substantiate the Bluetooth profiles. The Bluetooth-SIG does not define any application protocols nor does it specify any API. Bluetooth profiles are developed to establish a base point for use of a protocol stack to accomplish a given usage case.

19.6 Bluetooth Link Types

The Bluetooth baseband technology supports two link types: a synchronous connection oriented (SCO) type (used primarily for voice) and an asynchronous connectionless (ACL) type (used primarily for packet data).

Different master-slave pairs of the same piconet can use different link types and the link type may change arbitrarily during a session. Each link type supports up to sixteen different packet types. Four of these are control packets and are common for both SCO and ACL links. Both link types use a time division duplex (TDD) scheme for full-duplex transmission.

The SCO link is symmetric and typically supports time-bounded voice traffic. SCO packets are transmitted over reserved intervals. Once the connection is established, both master and slave units may send SCO packets without being polled. The SCO link type supports circuit-switched, point-to-point connections and is used often for voice traffic. The data rate for SCO links is 64 kbps.

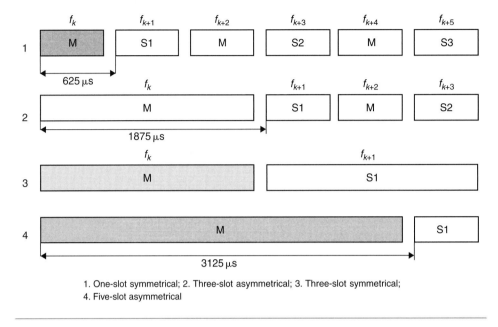

1. One-slot symmetrical; 2. Three-slot asymmetrical; 3. Three-slot symmetrical;
4. Five-slot asymmetrical

Figure 19.3   Bluetooth packets.

The ACL link is packet oriented and supports both symmetric and asymmetric traffic. The master unit controls the link bandwidth and decides how much piconet bandwidth is given to each slave, and the symmetry of the traffic. Slaves must be polled before they can transmit data. The ACL link also supports broadcast messages from the master to all slaves in the piconet. Multislot packets can be used in ACL and they can reach maximum data rates of 721 kbps in one direction and 57.6 kbps in the other direction if no error correction is used.

Data packets are protected by an automatic retransmission query (ARQ) scheme. Thus, when a packet arrives, a check is performed on it. If there is an error detected, the receiving unit indicates this in the return packet. In this way, retransmission is done only for the faulty packets. Retransmission is not feasible for voice so better error protection is used. Figure 19.3 shows the Bluetooth packets and Table 19.5 provides the details.

A symmetric 1-slot DH1 link between the master and slave carries 216 bits per slot at a rate of 800 slots per second in each direction. The associated rate is $216 \times 800 = 172.8$ kbps.

The asymmetric DM5 link uses a 5-slot packet carrying 1792 bits per packet by the master and a 1-slot packet carrying 136 bits per packet by the slave terminal. The number of packets per second in each direction is 1600/6. The data rate of the master is $1792 \times 1600/6 = 477.8$ kbps and the data rate of the slave terminal is $136 \times 1600/6 = 36.3$ kbps.

Table 19.5 Bluetooth packet types.

Type	Link	Name	No. of slots	Description
0000	Common	Null	1	No payload; used to return link information to source about the success of previous transmission, or status of Rx buffer (flow); no ACK
0001	Common	Poll	1	No payload, used by master to poll slave; ACK
0010	Common	FHS	1	Special control packet for revealing device address and the clock of sender; used in page master response, inquiry response, and frequency hop synchronization; $2/3$ FEC encoded
0011	Common	DM1	1	Support control messages and can also carry user data, 16-bit CRC, $2/3$ FEC encoded
0100	ACL	DH1	1	Carries 28 information bytes (header + payload) + 16 bit-CRC; not FEC encoded; used for high-speed data services
0101	SCO	HV1	1	Carries 240 bits for user voice samples, used for 64 kbps voice, no FEC encoding
0110	SCO	HV2	1	Carries 160 bits for user voice samples and 80 bits of parity for $1/3$ FEC encoding
0111	SCO	HV3	1	Carries 80 bits for user voice samples and 160 bits of parity for $2/3$ FEC encoding
1000	SCO	DV	1	Combined data (150 bits) and voice (50 bits) packet data field, $2/3$ FEC encoded
1001	ACL	AUX1	1	Carries 30 information bytes, no CRC or FEC, used for high-speed data services
1010	ACL	DM3	3	Carries 123 information bytes + 16-bit CRC, $2/3$ FEC encoded
1110	ACL	DH3	3	Carries 185 information bytes + 16-bit CRC, not FEC encoded
1110	ACL	DM5	5	Carries 226 information bytes + 16-bit CRC, $2/3$ FEC encoded
1111	ACL	DH5	5	Carries 341 information bytes + 16-bit CRC, not FEC encoded

The asymmetric DH5 link uses a 5-slot packet carrying 2712 bits per packet by the master and a 1-slot packet carrying 216 bits per packet by the slave terminal. The number of packets in each direction is 1600/6 packets per second. The data rate of the master is $2712 \times 1600/6 = 723.2$ kbps and the data rate of the slave terminal is $216 \times 1600/6 = 57.6$ kbps. Table 19.6 lists the ACL packet types and their data rates.

Table 19.6 ACL packet types and data rates.

Link	Symmetric (kbps)	Asymmetric (kbps)	
		Master	Slave
DM1	108.8	108.8	108.8
DH1	172.8	172.8	172.8
DM3	256	384	54.4
DH3	384	576	86.4
DM5	286.7	477.8	36.3
DH5	432.6	723.2	57.6

Example 19.1

What is the hopping rate of Bluetooth, and how many bits are transmitted in one slot? If each frame of the HV3 voice packet in Bluetooth carries 80 bits of sample speech, what is the efficiency of the packet transmission? How often do HV3 packets have to be sent to support 64 kbps voice in each direction?

Solution

- Hopping rate = 1600 hops per second, 240 bits in one slot packet
- $\eta = \dfrac{80}{240} = 0.3333$
- Let x be the number of times a packet is sent

  $x \times 80 = 64,000$
  $\therefore x = 800$

Example 19.2

A symmetric 1-slot DM1 link between a master and a slave carries 136 bits per slot at a rate of 800 slots per second (every other slot) in each direction. Find the associated data rate.

Solution

- Associated data rate $= \dfrac{136 \times 800}{1000} = 108.8\,\text{kbps}$

19.7   Bluetooth Security

Bluetooth security supports authentication and encryption. These features are based on a secret link key that is shared by pair of devices. A pairing procedure is

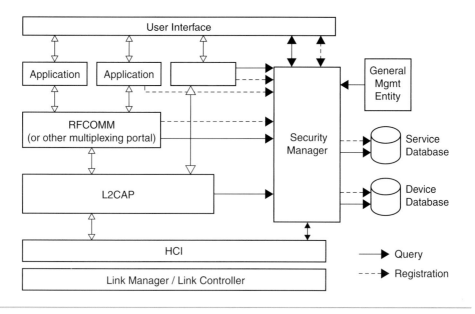

Figure 19.4 Bluetooth security architecture.

used when two devices communicate for the first time to generate this key. There are three security modes to a device:

- **Non-secure.** A device will not initiate any security procedure.
- **Service level enforced security.** A device does not initiate security procedures before channel establishment at the L2CAP level. This mode allows different and flexible access policies for applications, especially running applications with different security requirements in parallel.
- **Link level enforced security.** A device initiates security procedures before the link set-up at the LMP is completed.

Figure 19.4 shows Bluetooth security architecture.

19.7.1 Security Levels

There are two kinds of security levels: authentication and authorization.

- **Authentication** verifies who is at the other end of the link. In Bluetooth this is achieved by the authentication procedure based on the stored link key or by the pairing procedure. To meet different requirements on availability of services without user intervention, authentication is performed after determining what the security level of the requested service is. Thus, authentication cannot be performed when the ACL link is established.

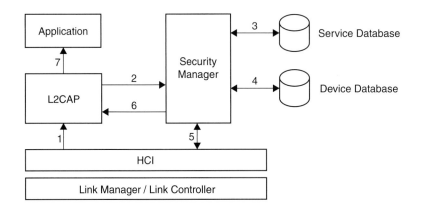

Figure 19.5   Authentication procedures.

The authentication is performed when a connection request to a service is submitted. The following procedure is used (see Figure 19.5):

1. The connect request to L2CAP is sent.
2. L2CAP requests access from the security manager.
3. The security manager enquires the service database.
4. The security manager enquires the device database.
5. If necessary, the security manager enforces the authentication and encryption procedure.
6. The security manager grants access, and L2CAP continues to set up the connection.

Authentication can be performed in both directions: client authenticates server and vice versa.

- **Authorization.** When one device is allowed to access the other, the concept of trust comes into existence. Trusted devices are allowed access to services. While, on the contrary, untrusted devices may require authorization based on user interaction before access to services is granted. There are two kinds of device trust levels:

   1. *Trusted device:* A device with a fixed relationship (paired) that has trusted and unrestricted access to all services.
   2. *Untrusted device:* This device has been previously authenticated, a link key is stored, but the device is not marked as trusted in the device database.
   3. *An unknown device* is also an untrusted device. No security information is available for this device.

For services, the requirement for authorization, authentication, and encryption are set independently (although some restrictions apply). The access requirements define three security levels:

- Services that require authorization and authentication—automatic access is only granted to trusted devices. Other devices need a manual authorization.
- Services that require authentication only—authorization is not necessary.
- Services open to all devices—authentication is not required, no access approval is required before service access is granted.

A default security level is defined to serve the needs of legacy applications. This default policy will be used unless other settings are found in a security database related to a service.

19.7.2   Limitations of Bluetooth Security

Only a device is authenticated, and not its user. There is no mechanism to preset authorization per service. However, a more flexible security policy can be implemented with the present architecture without a need to change the Bluetooth protocol stack. Also, it is not possible to enforce unidirectional traffic.

19.8   Network Connection Establishment in Bluetooth

Before any connection in a piconet is created, all devices are in *STANDBY* mode. In this mode, an unconnected unit periodically listens for messages every 1.28 seconds. Each time a device wakes up, it listens on a set of 32 hop frequencies defined for that unit. The number of hop frequencies varies in different geographic regions.

The connection procedure is initiated by any one of the devices, which then becomes master. A connection is made by a *PAGE* message if the address is already known, or by an *INQUIRY* message followed by a subsequent *PAGE* message if the address is unknown (see Figure 19.6).

In the initial *PAGE* state, the master unit sends a train of 16 identical page messages on 16 different hop frequencies defined for the device to be paged (slave unit). If no response is received, the master transmits a train on the remaining 16 hop frequencies in the wake-up sequence. The maximum delay before the master reaches the slave is twice the wake-up period (2.56 seconds) while the average delay is half the wake-up period (0.64 seconds).

The *INQUIRY* message is typically used for finding Bluetooth devices, including public printers, fax machines, and similar devices with an unknown address. The *INQUIRY* message is similar to the *PAGE* message, but may require one additional train period to collect all responses.

A power saving mode can be used for connected units in a piconet if no data needs to be transmitted. The master unit can put slave units into *HOLD* mode, where only the internal timer is running. Slave units can also demand to be put

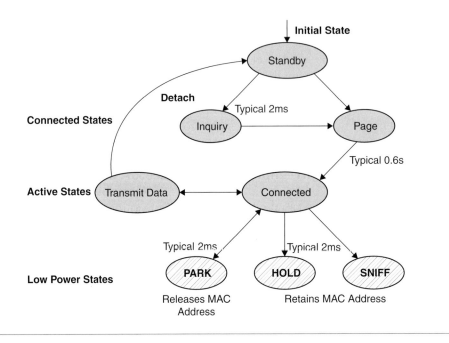

Figure 19.6 Device states in Bluetooth.

into *HOLD* mode. Data transfer restarts instantly when units transition out of *HOLD* mode. The *HOLD* is used when connecting several piconets or managing a low-power device such as a temperature sensor.

Two more low-power modes are also available: the *SNIFF* mode and *PARK* mode. In the *SNIFF* mode, a slave device listens to the piconet at a reduced rate, thus reducing its duty cycle. The *SNIFF* interval is programmable and depends on the application. In the *PARK* mode, a device is still synchronized to the piconet but does not participate in the traffic. Parked devices have given up their MAC address and occasionally listen to the traffic of the master to resynchronize and check on broadcast messages.

If we list the modes in increasing order of power efficiency, then the *SNIFF* mode has the higher duty cycle, followed by the *HOLD* mode with a lower duty cycle, and the *PARK* mode with the lowest duty cycle.

19.9 Error Correction in Bluetooth

Three error correction schemes are defined for the Bluetooth baseband controller:

- $1/3$ rate forward error correction (FEC) code
- $2/3$ rate forward error correction code
- Automatic repeat request (ARQ) scheme for data

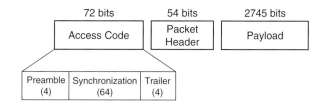

Figure 19.7   Packet format in Bluetooth.

The purpose of the FEC scheme on the data payload is to reduce the number of retransmissions. However, in a reasonably error-free environment, FEC creates unnecessary overhead that reduces the throughput. Therefore, the packet definitions have been kept flexible as to whether or not to use FEC in the payload. The packet header is always protected by a 1/3 rate FEC. It contains link information and should survive bit errors. An unnumbered ARQ scheme is applied in which data transmitted in one slot is directly acknowledged by the recipient in the next slot. For a data transmission to be acknowledged, both the header error check and the cyclic redundancy check must be satisfied, otherwise a negative acknowledgment is returned.

19.10   Network Topology in Bluetooth

Bluetooth devices can create both point-to-point and point-to-multipoint connections. A connection with two or several (maximum 8 devices) devices is a piconet where all devices follow the same frequency-hop scheme. To avoid interference between devices, one of the devices automatically becomes a master of the piconet. In each slot, a packet can be exchanged between the master (M) and one of the slaves (S). Packets have a fixed format (see Figure 19.7).

A Bluetooth packet format is based on one packet per hop and a basic 1-slot packet of $625\,\mu s$ that can be extended to 3-slot ($1875\,\mu s$) and 5-slot ($3125\,\mu s$). A frame format allows the master to poll multiple slaves.

Each packet begins with a 72-bit access code that is derived from the master identity and is unique for the channel. Every packet exchanged on the channel is preceded by this access code. Recipients on the piconet compare incoming signals with the access code. If the two do not match, the received packet is not considered valid on the channel and the rest of its contents are ignored. Besides packet identification, the access code is also used for synchronization and compensating for offset. The access code is robust and resistant to interference.

Two or several piconets can communicate with each other and are then called a scatternet.

19.11   Bluetooth Usage Models

In this section we discuss some of the Bluetooth usage models. Each usage model has one or more profiles.

- **Three-in-one phone.** The three-in-one phone usage model describes how a telephone handset may connect to three different service providers. The telephone may act as a cordless telephone connecting to the PSTN at home, charged at a fixed line charge. The telephone can also connect directly to other telephones acting as a walkie-talkie or handset extension. Finally, the telephone may act as a cellular phone connecting to the cellular infrastructure.

- **File transfer.** The file transfer usage model offers the capability to transfer data objects from one Bluetooth device to another. Files, entire folders, directories, and streaming media formats are supported in this model. The model also offers the possibility of browsing the contents of the folders on a remote device.

- **Synchronization.** The synchronization usage model provides the means for automatic synchronization between, for instance, a desktop PC, a portable PC, a PDA, and a notebook. The synchronization requires business card, calendar, and task information to be transferred and processed by computers, cellular phones, and PDAs utilizing a common protocol and format.

- **Internet bridge.** The Internet bridge usage model describes how a mobile phone or cordless modem provides a PC with dial-up networking capabilities without the need for physical connection to the PC. This networking scenario requires a two-piece protocol stack, one for AT-commands to control the mobile phone and another stack to transfer payload data.

- **Ultimate handset.** The ultimate handset usage model defines how a Bluetooth-equipped wireless handset can be connected to act as a remote unit's audio input and output interface. This unit is probably a mobile phone or a PC for audio input and output.

19.12 Bluetooth Applications

The following are some of the areas where Bluetooth can be used:

- Replacing serial cables with radio links
- Wearable networks/WPANs
- Desktop/room wireless networking
- Hot-spot wireless networking
- Medical: Transfer of measured values from training units to analytical systems, patient monitoring
- Automotive: Remote control of audio/video equipment, hands-free telephony
- Point-of-sale payments: Payments by mobile phone

19.13   WAP and Bluetooth

Bluetooth can be used with WAP like any other wireless network. Bluetooth wireless networks can be used to transport data from a WAP client to a WAP server. The WAP client can make use of Bluetooth's SOP to find the WAP server/gateway. This is very useful when the WAP device is a mobile phone and when it comes into the range of a WAP server, it can use Bluetooth's SDP to discover the gateway. The Bluetooth SDP must be able to provide some details about the WAP server to the WAP client.

The other feature that can be supported is the reverse of the above. The WAP server can periodically check for the availability of WAP-enabled clients in its range. It can use Bluetooth's SDP to do this. If there are any clients, the server can push any data to the client. The client of course is not required to accept the data pushed to it.

19.14   Summary

In this chapter we discussed Bluetooth technology that allows for replacing many proprietary cables that connect one device to another with one universal short-range radio link. Bluetooth radio technology provides a universal bridge to existing data networks, a peripheral interface, and a mechanism to form small, private ad hoc groupings of connected devices away from fixed network infrastructures. The Bluetooth technology has a number of advantages including minimal hardware dimensions, low cost of components, and low power consumption. These advantages make it possible to introduce Bluetooth in many types of devices at a low cost. The 720 kbps data capability provided by Bluetooth can be used for cable replacement and several other applications, such as LAN.

Problems

19.1   What are piconet and scatternet in Bluetooth?

19.2   Discuss transport protocol group in Bluetooth.

19.3   Define link types in Bluetooth.

19.4   How is security achieved in Bluetooth?

19.5   The hopping rate of 1600 hops per second is used in Bluetooth that carries 240 bits in a 1-slot packet. If each frame of the HV2 voice packet carries 160 bits of sample speech, what is the efficiency of packet transmission? How often are HV2 packets sent to support 64 kbps voice in each direction?

19.6   A symmetric 1-slot DH1 link of Bluetooth between a master and a slave carries 216 bits per slot. What is the associated rate?

19.7   The asymmetric DM5 link of Bluetooth uses a 5-slot packet to carry 1792 bits per packet by the master terminal and a 1-slot packet to carry 136 bits per packet by the slave terminal. What are the data rates of the master and slave terminal?

References

1. Bisdikian, C., and Miller, B. *Blue Tooth Revealed*. Upper Saddle River, NJ: Prentice Hall, 2000.

2. Bluetooth Special Interest Group (SIG). "Specifications of the Bluetooth System," vol. 1 v.1.1, "Core" and vol. 2 v. 1.0 B "Profiles," 2000.

3. Bluetooth Specification Release 1.0, section F:4, "Interoperability Requirements for Bluetooth as WAP Bearer."

4. Bray, J., and Sturman, C. F. *Bluetooth: Connect without Cables*. Upper Saddle River, NJ: Prentice Hall, 2001.

5. http://www.bluetooth.com/.

6. IEEE 802.15 Working Group. http://grouper.ieee.org/groups/802/15/.

7. Moran, P. Ed. "Bluetooth LAN Access Profile using PPP." *Bluetooth Special Interest Group*, version 1.0, 1999.

8. Muller, N. J. *Bluetooth Demystified*. New York: McGraw Hill, 2000.

9. Pahlavan, K., and Krishnamurthy, P. *Principle of Wireless Networks*. Upper Saddle River, NJ: Prentice Hall, 2002.

10. Sand Kjell, Tik-111.550 Seminar on Multimedia: "Bluetooth," March, 4, 1999.

11. Simpson, W. Ed. "The Point-to-Point Protocol (PPP)." *STD 50, RFC1661. Day-dreamer,* July 1994.

12. Simpson, W. Ed. "PPP in HDLC Framing." *STD51, RFC1662. Day-dreamer,* July 1994.

CHAPTER 20

Wireless Personal Area Networks: Low Rate and High Rate

20.1 Introduction

Wired sensor networks have been around for decades, with an array of gauges measuring temperature, fluid levels, humidity, and other attributes on pipelines, pumps, generators, and manufacturing lines. Many of these run as separately wired networks, sometimes linked to a computer but often to a control panel that flashes lights or sounds an alarm when a temperature rises too high or a machine vibrates too much. Also wired in are actuators, which let the control panel slow down a pump or start a fan in response to the sensor data.

Now advances in silicon radio chips, coupled with cleverly designed routing algorithms and network software are promising to eliminate those wires and their installation and maintenance costs. Mesh network topologies will let these wireless networks route around nodes that fail or whose radio signal is degraded by interference from heavy equipment. A gateway will create a two-way link with legacy control systems, hosts, wired local area networks (WLANs), or the Internet [12,13].

Wireless sensor networks can use several different wireless technologies, including IEEE 802.11 WLANs, Bluetooth, and radio frequency identification (RFID). But at present most of the applications are of low-power radios having a range of about 30 to 200 feet and data rates of up to around 300 kbps. IEEE 802.15.4 [4] is the approved low-rate standard for a simple, short-range wireless network whose radio components could run several years on a single battery. Figure 20.1 gives a summary of wireless personal access network (WPAN) standards. We discussed IEEE 802.15.1 (WPAN-Bluetooth) in Chapter 19. In this chapter we focus on IEEE 802.15.3 (WPAN-LR) and IEEE 802.15.3a (WPAN-HR) standards [5].

20.2 Wireless Sensor Network

A wireless sensor network contains a large number of tiny *sensor nodes* that are densely deployed either inside the phenomenon to be sensed or very close to it. Sensor nodes consist of sensing, data processing, and communicating components. The position of sensor nodes need not be engineered or predetermined. This allows

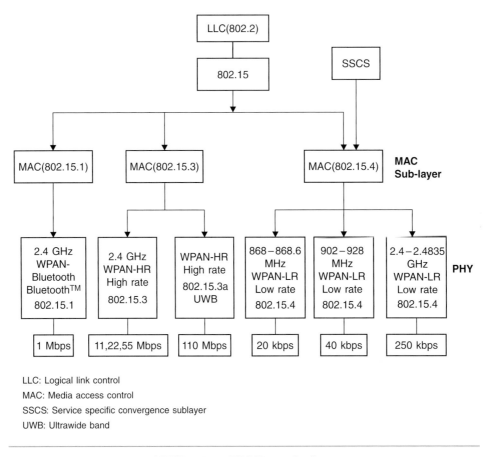

Figure 20.1   A summary of IEEE 802.15 WPAN standards.

random deployment in inaccessible terrain or disaster relief operations. This also means that sensor network protocols and algorithms must possess self-organizing capabilities. Another unique feature of sensor networks is the cooperative effort of sensor nodes. Sensor nodes are fitted with an inboard processor. Instead of sending the raw data to the nodes responsible for the fusion, they use their processing abilities to locally carry out simple computations and transmit only required and partially processed data. In the following sections we introduce wireless sensor networks briefly. Interested readers should refer to references [1 and 2] for more details.

Wireless sensor network applications require wireless ad hoc networking techniques. Although many protocols and algorithms have been proposed for traditional wireless ad hoc networks, they are not well suited for the unique features and application requirements of wireless sensor networks. The differences between wireless sensor networks and traditional wireless ad hoc networks are listed here [22]:

- The number of sensor nodes in a wireless sensor network can be several orders of magnitude higher than the nodes in a wireless ad hoc network.
- In a wireless sensor network, sensor nodes are densely deployed.
- Sensor nodes are prone to failure.
- The topology of a wireless sensor network changes very frequently.
- Sensor nodes mainly use broadcast communication paradigms whereas most traditional ad hoc networks are based on point-to-point communications.
- Sensor nodes are limited in power, computational capabilities, and memory.
- Sensor nodes may not have global identification because of the large amount of overhead and large number of sensors.
- Another factor that distinguishes wireless sensor networks from traditional mobile ad hoc networks (MANETs) is that the end goal is the detection/estimation of some event(s) of interest, and not just communication. To improve detection performance, it is often quite useful to fuse data from multiple sensors [23]. Data fusion requires the transmission of data and control messages. This need may impose constraints on network architecture.
- The large number of sensing nodes may congest the network with information. To solve this problem, some sensors, such as cluster heads, can aggregate the data, perform some computation (e.g., average, summation, highest value, etc.), and then broadcast the summarized new information.

Since large numbers of sensor nodes are densely deployed, neighbor nodes may be very close to each other. Hence, multihop communication in wireless sensor networks is expected to consume less power than traditional single hop communication. Furthermore, the transmission power level can be kept low, which is highly desirable in covert operations. Multihop communication can effectively overcome some of the signal propagation effects experienced in long-distance wireless communication.

One of the most important constraints on sensor nodes is the low power consumption requirement. Sensor nodes carry limited, generally irreplaceable power sources. Therefore, while traditional networks aim to achieve high quality of service (QoS) provisions, wireless sensor network protocols must focus primarily on power conservation. They must have built-in trade-off mechanisms that give the end-user the option of prolonging network lifetime at the cost of lower throughput or higher transmission delay.

20.3   Usage of Wireless Sensor Networks

Wireless sensor networks have been useful in a variety of areas. The primary areas in which these networks are deployed follows.

- **Environmental observations.** Wireless sensor networks can be used to monitor environmental changes. An example is water pollution detection in a lake that is located near a factory that uses chemical substances. Sensor nodes are randomly deployed in unknown and hostile areas and relay the exact origin of a pollutant to a centralized authority which takes appropriate measures to limit the spreading of pollution. Other examples include forest fire detection, air pollution, and rainfall observation in agriculture.

- **Military monitoring.** The military uses sensor networks for battlefield surveillance. Sensors can monitor vehicular traffic, track the position of the enemy, or even safeguard the equipment of side deploying sensors.

- **Building monitoring.** Wireless sensor networks can also be used in large buildings or factories to monitor climate changes. Thermostats and temperature sensor nodes are deployed all over the building's area. In addition, sensors can be used to monitor vibration that can damage the structure of a building.

- **Health care.** Sensors can be used in biomedical applications to improve the quality of provided care. Sensors are implanted in the human body to monitor medical problems such as cancer and help patients maintain their health.

20.4   Wireless Sensor Network Model

A wireless sensor network consists of hundreds or thousands of low-cost nodes, which could either have a fixed location or be randomly deployed to monitor the environment. Due to their small size, they have a number of limitations. Sensors usually communicate with each other using a multi hop approach. The flow of data ends at special nodes called *base stations* (sometimes they are referred to as *sinks*) (see Figure 20.2). A base station links the sensor network to another network (like a gateway) to disseminate the data sensed for further processing. Base stations have enhanced capabilities over simple sensor nodes since they must do complex data processing; this justifies the fact that base stations have workstation/laptop class processors, and of course enough memory, energy, storage, and computational power to perform their tasks well. Usually, the communication between base stations is initiated over high bandwidth links.

One of the biggest problems of sensor networks is power consumption, which is greatly affected by communication between nodes. To solve this, *aggregation points* are introduced in the network. This reduces the total number of messages exchanged between nodes and saves some energy. Usually, aggregation

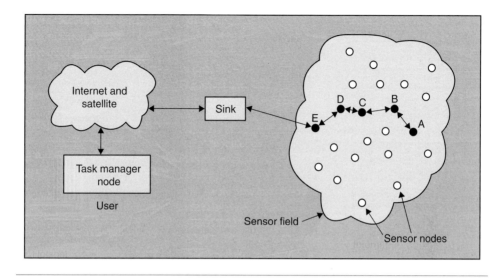

Figure 20.2 Wireless sensor network.

points are regular nodes that receive data from neighboring nodes, perform some kind of processing, and then forward the filtered data to the next hop. Similar to aggregation points is *clustering*. Sensor nodes are organized into clusters, each cluster having a "cluster head" as the leader. The communication within a cluster must travel through the cluster head, which is then forwarded to a neighboring cluster head until it reaches its destination, the base station. Another method to save energy is to set the nodes to go idle (into *sleep mode*) if they are not needed, and wake up when required. The challenge is to find a pattern at which energy consumption is made evenly for all the nodes in the network.

The design of the sensor network as described by Figure 20.2 is influenced by many factors, including fault tolerance, scalability, production costs, operating environment, sensor network topology, hardware constraints, transmission media, and power consumption.

- **Fault tolerance.** Some sensor nodes may fail or be blocked because of lack of power, physical damage, or environmental interference. The failure of sensor nodes should not affect the overall task of the sensor network. This is the reliability or fault tolerance issue. Fault tolerance is the ability to sustain sensor network functionality without any interruption due to sensor node failures. The reliability $R_k(t)$ of a sensor node is usually modeled using Poisson's distribution:

$$R_k(t) = e^{-\lambda_k t} \tag{20.1}$$

where:

λ_k = the failure rate of sensor node k

t　= the time period

- **Scalability.** The number of sensor nodes deployed to study a phenomenon may be on the order of hundreds or thousands. Depending on the application, the number may reach an extreme value of millions. New schemes must be able to work with this number of nodes and must also utilize the high density of the sensor networks. The density can range from a few sensor nodes to a few hundred sensor nodes in the region, which can be less than 10 m in diameter. The density μ can be calculated as

$$\mu(R) = \frac{N\pi R^2}{A} \tag{20.2}$$

where:

N = the number of scattered sensor nodes in region A

R = the radio transmission range

$\mu(R)$ = the number of nodes within the transmission radius of each node in region A

- **Production cost.** Since sensor networks consist of a large number of sensor nodes, the cost of a single node is very important to justify the overall cost of the network. The cost of each sensor node has to be kept low. The cost of a sensor node should be much less than US\$1 in order for it to be feasible.

- **Hardware constraints.** A sensor node is made up of four basic components (see Figure 20.3): a sensing unit, a processing unit, a transceiver unit, and a power unit. There may also be additional application-dependent components such as a location finding system, power generator, and mobilizer. Sensing units are usually composed of two subunits: sensors and analog-to-digital converters (ADCs). The analog signals produced by the sensors based on the observed phenomenon are converted into digital signals by ADC, and then fed into the processing unit. The processing unit, which is generally associated with a small storage unit, manages the procedures that make the sensor node collaborate with other nodes to carry out the assigned sensing tasks. A transceiver unit connects the node to the network. One of the most important components of a sensing node is the power unit. Power units may be supported by power scavenging units such as solar cells. There are also other subunits that are application-dependent. Most of the sensor network routing techniques and sensing tasks require knowledge of location with high accuracy. Thus, it is common that a sensor node has a location finding system. A mobilizer may sometimes be needed to move sensor nodes when it is required to carry out the assigned tasks. All of these

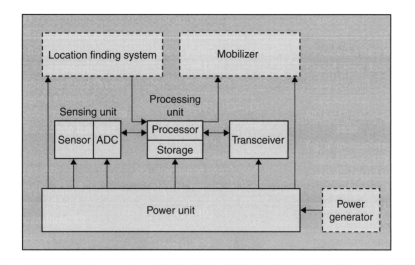

Figure 20.3   The components of a sensor node.

subunits may need to fit into a matchbox-sized module. The required size may be even smaller than a cubic centimeter. Apart from size, there are some other stringent constraints for sensor nodes. These nodes must consume extremely low power, operate in high volumetric densities, have low production cost, be dispensable and autonomous, operate unattended, and be adaptive to the environment.

- **Sensor network topology.** Hundreds to several thousands of nodes are deployed throughout the sensor field. Node densities may be as high as 20 nodes/m³. Deploying a high number of nodes densely requires careful handling of topological maintenance.

- **Operating environment.** Sensor nodes are densely deployed either very close to or directly inside the phenomenon to be observed. They usually work unattended in remote geographic areas. They may work in the interior of large machinery, at the bottom of an ocean, in a biologically or chemically contaminated field, in a battlefield beyond the enemy lines, or in a home or large building.

- **Transmission media.** In a multihop sensor network, communication nodes are linked by a wireless medium. These links are formed by radio, infrared, or optical media. Both infrared and optical media require a line-of-sight between sender and receiver. To enable the global operation of sensor networks, the chosen transmission medium must be available worldwide.

- **Power consumption.** The wireless sensor node can only be equipped with a limited power source (<0.5 Ah, 1.2 V). Sensor node lifetime shows a strong dependency on battery lifetime. In a multihop sensor network, each node plays the dual role of data originator and data router. The disfunctioning of a few nodes can cause significant topological changes and might require rerouting of packets and reorganization of the network. Hence, power conservation and power management take on additional importance. Power consumption can be divided into three domains: sensing, communication, and data processing. Sensing power varies with the nature of applications. Sporadic sensing might consume less power than constant event monitoring. The complexity of event detection also plays a crucial role in determining energy expenditure. Higher ambient noise levels might cause significant corruption and increase detection complexity. The energy in data communication involves both data transmission and reception. In the following formulation for radio power consumption (P_c) is given [18]:

$$P_c = N_T \left[P_T \left(T_{on} + T_{st} \right) + P_{out} \left(T_{on} \right) \right] + N_R \left[P_R \left(R_{on} + R_{st} \right) \right] \qquad (20.3)$$

where:

P_T = power consumed by transmitter
P_R = power consumed by receiver
P_{out} = the output power of transmitter
T_{on} = transmitter on time; this can be further broken down as T_{onL} and T_{onR} (L is the packet size and R is the data rate)
R_{on} = receiver on time
T_{st} = transmitter start-up time
R_{st} = receiver start-up time
N_T, N_R = transmitter and receiver switched on per unit time, which depends on the task and media access control (MAC) scheme

Today's state-of-the-art low-power radio transceiver has typical P_T and P_R values around 20 dBm and P_{out} close to 0 dBm. The pico-radio aims at a P_c value of −20 dBm.

Energy expenditure in data processing is much less compared to data communication. Local data processing is crucial in minimizing power consumption in a multihop sensor network. A sensor node must have built-in computational abilities and be capable of interacting with its surroundings. Further limitations of cost and size lead us to the choice of complementary metal oxide semiconductor (CMOS) technology for a microprocessor. The power consumption in data processing (P_p) can be given as [19]:

$$P_p = CV^2_{dd}f + \text{power loss due to leakage current} \qquad (20.4)$$

where:

C = total switching capacitance

V_{dd} = voltage swing

f = switching frequency

20.5 Sensor Network Protocol Stack

The protocol stack combines power and routing awareness, integrates data with networking protocols, communicates power efficiently through the wireless medium, and promotes cooperative efforts of sensor nodes. The protocol stack consists of the physical layer, data link layer, network layer, transport layer, application layer, power management plane, mobility management plane, and task management plane (see Figure 20.4).

20.5.1 Physical Layer

The physical layer addresses the needs of simple but robust modulation, transmission, and receiving techniques. It is responsible for frequency selection,

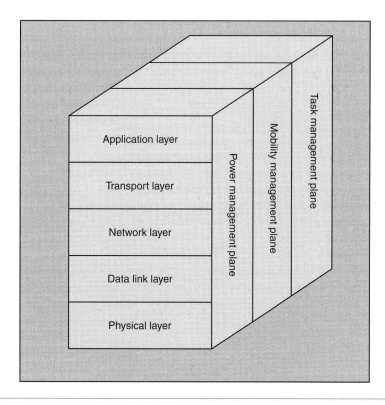

Figure 20.4 The sensor network protocol stack.

carrier frequency generation, signal detection, modulation, and data encryption. The 915 MHz ISM band has been widely suggested for sensor networks.

Modulation depends on the transceiver and hardware design constraints, which aim for simplicity, low power consumption, and low cost per unit. Binary modulation schemes are simpler to implement and thus are deemed to be more energy efficient. Low transmission power and simple circuity make ultra wideband (UWB) an attractive candidate because of baseband transmission (i.e., intermediate or carrier frequency), use of pulse position modulation, resilience to multipath and low transmission power, and simple transceiver circuitry. Energy consumption minimization is of paramount importance when designing the physical layer for a sensor network in addition to the usual effects such as scattering, shadowing, reflection, diffraction, multipath, and fading.

20.5.2 Data Link Layer

The data link layer is responsible for multiplexing data streams, data frame detection, medium access, and error control. It ensures reliable point-to-point and point-to-multipoint connections in a communication network. The MAC protocol in a wireless multihop self-organizing sensor network should achieve two goals. The first is the creation of the network infrastructure. Since thousands of sensor nodes are densely scattered in a sensor field, MAC must establish communication links for data transfer. This forms the basic infrastructure needed for wireless communication hop to hop and gives the sensor network self-organizing ability. The second objective is to fairly and efficiently share communication resources between sensor nodes. Since the environment is noisy and sensor nodes can be mobile, the MAC protocol must be power-aware and be able to minimize collision with neighbors' broadcasts. Table 20.1 provides a qualitative overview of MAC protocols for sensor networks.

Regardless of which type of MAC scheme is used for sensor networks, it certainly must have built-in power-saving mechanisms and strategies for proper management of node mobility or failure. The most obvious means of power conservation is to turn the transceiver off when it is not required. Although this power-saving method seemingly provides significant energy gains, an important point that must not be overlooked is that sensor nodes communicate using short data packets. The shorter the packets, the more the dominance of startup energy. Operation in a power-saving mode is energy efficient only if the time spent in that mode is greater than, a certain threshold. There can be a number of such useful modes of operation for the wireless sensor node, depending on the number of states of the microprocessor, memory, A/D converter, and transceiver. Each of these modes should be characterized by its power consumption and latency overhead, which is the transition power to and from that mode. The main features of sensor MAC are periodic listen and sleep, collision and overhearing avoidance, and message passing. The duration of the sleep and awake cycles are application-dependent and they are set the same for all nodes.

Table 20.1 MAC protocols for sensor networks.

MAC protocol	Channel access mode	Sensor network specifics	Power conservation
Self-organizing Media Access Control for Sensor networks (SMACS) and Eaves-drop-and-Register (EAR) Algorithm [20]	Fixed allocation of duplex time slots at fixed frequency	Exploitation of large available bandwidth compared to sensor data rate	Random wake-up during setup and turning radio off while idle
Hybrid TDMA/FDMA [18]	Centralized frequency and time division	Optimum number of channels calculated for minimum system energy	Hardware-based approach for system energy minimization
CSMA-based [21]	Contention-based random access	Application phase shift and pretransmit delay	Constant listening time for energy efficiency

Since a sensor node has limited power resources, forward error correction (FEC) is more feasible than automatic repeat request (ARQ), which in a multihop sensor network environment is limited by additional retransmission energy cost and overhead.

20.5.3 Network Layer

The network layer takes care of routing the data supplied by the transport layer. The network layer of a sensor network is usually designed according to the following principles:

- Power efficiency is always an important consideration.
- Sensor networks are mostly data-centric.
- Data aggregation is useful only when it does not hinder the collaborative effort of the sensor nodes.
- An ideal sensor network has attribute-based addressing and location awareness.

Energy-efficient routes can be found based on the available power in the nodes or energy required for transmission in the links along the routes. An energy-efficient route is selected by one of the following approaches:

- The maximum power available (PA) route is the route that has maximum total PA.
- The minimum energy (ME) route is the route that consumes minimum energy to transmit the data packets between the sink and sensor nodes.

- The minimum hop (MH) route is the route that makes the minimum hop to reach the sink.
- The minimum PA node route is the route along which the minimum PA is larger than the minimum PAs of the other routes.

Data aggregation combines data from many sensor nodes into a more compact form before forwarding to a location for processing. Data aggregation is needed to handle the large amount of data generated in sensor networks.

Another important issue is that routing may be based on the *data-centric* approach. In data-centric routing, interest dissemination is performed to assign the sensing tasks to the sensor nodes. There are two approaches used for interest dissemination: sinks broadcast the interest and sensor nodes broadcast an advertisement for the available data and wait for a request from the interested nodes.

In sensor networks, *attribute-based addressing* may be required. The user is more interested in querying an attribute of the phenomenon, rather than querying an individual node. Attribute-based addressing is used to carry out queries by using the attributes of the phenomenon. Other features include localization to determine the physical location of the sensors, fault tolerance, and clock synchronization.

An overview of the network layer protocols proposed for the sensor networks is given in Table 20.2. These protocols need to address higher topology changes and higher scalability.

Table 20.2 An overview of network layer schemes for sensor networks.

Network layer scheme	Description
Small Minimum Energy Communication Network (SMECN) [15]	Creates a subgraph of sensor network that contains the minimum energy path
Flooding	Broadcasts data to all neighbor nodes regardless of whether they receive it before or not
Gossiping [7]	Sends data to one randomly selected neighbor
Sensor Protocols for Information via Negotiation (SPIN) [8]	Sends data to sensor nodes only if they are interested; has three types of messages (i.e., ADV, REQ, and DATA)
Sequential Assignment Routing (SAR) [20]	Creates multiple trees where the root of each tree is one hop neighbor from the sink; select a tree for data to be routed back to sink according to energy resources and additive QoS metric
Low-Energy Adaptive Clustering Hierarchy (LEACH) [9]	Forms clusters to minimize energy dissipation
Directed diffusion [14]	Set up gradients for data flow from source to sink during interest dissemination

20.5.4 Transport Layer

The transport layer helps to maintain the flow of data if the sensor network application requires it. The transport layer is needed when a system is planned to be accessed through the Internet or other external networks. Transmission control protocol (TCP) with its current transmission window mechanisms matches the extreme characteristics of the sensor network environment. An approach such as TCP splitting may be needed to make sensor networks interact with other networks such as the Internet. In this approach, TCP connections are ended at sink nodes, and a special transport layer protocol can handle the communication between the user and sink nodes. Communication between the user and sink node is by user datagram protocol (UDP) or TCP via the Internet or satellite. On the other hand, communication between the sink and sensor nodes may be purely by UDP-type protocols, because each sensor node has limited memory.

Event-to-sink reliable transport (ESRT) has been proposed to achieve reliable event detection (at the sink node) with a protocol that is energy aware and has congestion control mechanisms.

The end-to-end communication schemes in sensor networks are not based on global addressing. These schemes use attribute-based naming to indicate the destinations of the data packets. Factors such as power consumption and scalability, and characteristics such as data-centric routing mean sensor networks need different handling in the transport layer. Thus, these requirements stress the need for new types of transport layer protocols.

20.5.5 Application Layer

Depending on the sensing tasks, different types of application software can be built and used on the application layer. Three possible application layer protocols for sensor networks are sensor management protocol (SMP), task assignment and data advertisement protocol (TADAP), and sensor query and data dissemination protocol (SQDDP). An application layer management protocol makes the hardware and software of the lower layers transparent to the sensor network management applications. System administrators interact with sensor networks by using SMP, which provides the software operations needed to perform the following tasks:

- Introducing the rules related to data aggregation, attribute-based naming, and clustering to sensor nodes
- Exchanging data related to the location finding algorithms
- Time synchronization of sensor nodes
- Moving sensor nodes
- Turning the sensor nodes on and off
- Querying the sensor network configuration and status of nodes, and re-configuring the sensor network
- Authentication, key distribution, and security in data communications

20.5.6   Power, Mobility, and Task Management Planes

The power, mobility, and task management planes monitor the power, movement, and task distribution among sensor nodes. These planes help the sensor nodes coordinate the sensing task and lower overall power consumption. The power management plane manages how a sensor node uses its power. When the power level of a sensor node is low, the sensor node broadcasts to its neighbors that it is low in power and cannot participate in routing messages. The remaining power is reserved for sensing. The mobility management plane detects and registers the movement of sensor nodes, so a route back to the user is always maintained, the sensor nodes can keep track of who their neighbor sensor nodes are, and the sensor nodes can balance their power and task usage. The task management plane balances and schedules the sensing tasks given to a specific region. Not all sensor nodes in that region are required to perform the sensing task at the same time. As a result, some sensor nodes perform the task more than others, depending on their power level. These management planes are needed so that sensor nodes can work together in power-efficient ways, route data in a mobile sensor network, and share resources between sensor nodes.

Cross layers design is also suggested to avoid conflicting behavior, remove unnecessary layers, and provide a new paradigm.

20.6   ZigBee Technology

The low rate (LR) wireless personal access network (WPAN) (IEEE 802.15.4/LR-WPAN) is intended to serve a set of industrial, residential, and medical applications with very low power consumption, low cost requirement, and relaxed needs for data rate and QoS [10,11]. The low data rate enables the LR-WPAN to consume little power.

ZigBee technology is a low data rate, low power consumption, low cost, wireless networking protocol targeted toward automation and remote control applications. The IEEE 802.15.4 committee and ZigBee Alliance worked together and developed the technology commercially known as ZigBee. It is expected to provide low-cost and low-power connectivity for devices that need battery life as long as several months to several years but does not require data transfer rates as high as those enabled by Bluetooth. ZigBee can be implemented in mesh (peer-to-peer) networks larger than is possible with Bluetooth. ZigBee-compliant wireless devices are expected to transmit 10–75 minutes, depending on the RF environment and power output consumption required for a given application, and operate in the unlicensed RF worldwide (2.4 GHz global, 915 MHz America, or 868 MHz Europe) bands. The data rate is 250 kbps at 2.4 GHz, 40 kbps at 915 MHz, and 20 kbps at 868 MHz.

The IEEE 802.15.4 committee is focusing on the specifications of the lower two layers of the protocol (the physical and data link layers). On the other hand,

ZigBee Alliance aims to provide the upper layers of the protocol stack (from the network to the application layer) for interoperable data interworking, security services, and a range of wireless home and building control solutions. ZigBee Alliance provides interoperability compliance testing, marketing of the standard, and advanced product engineering for the evolution of the standard. This will assure consumers to buy products from different manufacturers with confidence that those products will work together.

ZigBee often uses a basic master-slave configuration suited to static star networks of many infrequently used devices that talk via small data packets. It allows up to 254 nodes. Other network topologies such as peer-to-peer and cluster tree are also used. When ZigBee node is powered down, it can wake up and get a packet in around 15 msec.

20.6.1 ZigBee Components and Network Topologies

A ZigBee system consists of several components. The most basic is the device. A device can be a full-function device (FFD) or reduced-function device (RFD). A network includes at least one FFD, operating as the personal area network (PAN) coordinator. The FFD can operate in three modes: a PAN coordinator, a coordinator, or a device. An RFD is intended for applications that are extremely simple and do not need to send large amounts of data. An FFD can talk to reduced-function or full-function devices, while an RFD can only talk to an FFD.

ZigBee supports three types of topologies: star topology, peer-to-peer topology, and cluster tree (see Figure 20.5).

In the *star topology*, communication is established between devices and a single central controller, called the PAN coordinator. The PAN coordinator may be powered by mains while the devices will most likely be battery powered. Applications that benefit from this topology are home automation, personal computer (PC) peripherals, toys, and games.

After an FFD is activated for the first time, it may establish its own network and become the PAN coordinator. Each star network chooses a PAN identifier, which is not currently used by any other network within the radio sphere of influence. This allows each star network to operate independently.

In the *peer-to-peer topology*, there is also one PAN coordinator. In contrast to star topology, any device can communicate with any other device as long as they are in range of one another. A peer-to-peer network can be ad hoc, self-organizing, and self-healing. Applications such as industrial control and monitoring, wireless sensor networks and asset and inventory tracking would benefit from such a topology. It also allows multiple hops to route messages from any device to any other device in the network. It can provide reliability by multipath routing.

The *cluster-tree topology* is a special case of a peer-to-peer network in which most devices are full-function devices and an RFD may connect to a cluster-tree network as a leaf node at the end of a branch. Any of the full-function devices can

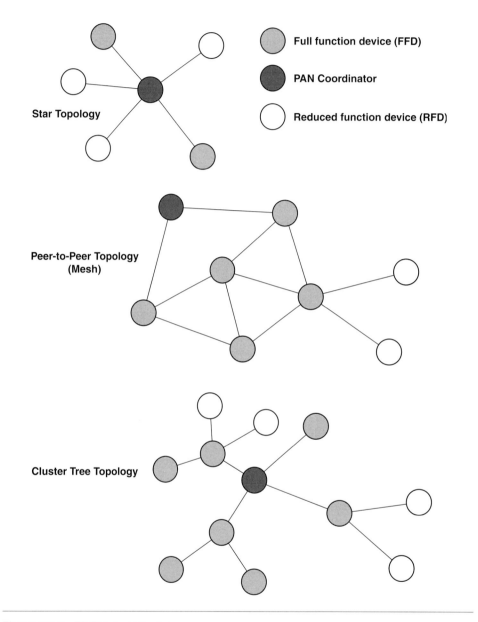

Figure 20.5   ZigBee topologies.

act as a coordinator and provide synchronization services to other devices and coordinators. However, only one of these coordinators is the PAN coordinator.

The PAN coordinator forms the first cluster by establishing itself as the cluster head (CLH) with a cluster identifier (CID) of zero, choosing an unused PAN identifier, and broadcasting beacon frames to neighboring devices. A candidate device receiving a beacon frame may request to join the network at the cluster head. If the PAN coordinator permits the device to join, it will add this new device to its neighbor list. The newly joined device will add the cluster head as its parent in its neighbor list and begin transmitting periodic beacons such that other candidate devices may then join the network at that device. Once application or network requirements are met, the PAN coordinator may instruct a device to become the cluster head of a new cluster adjacent to the first one. The advantage of the clustered structure is the increased coverage at the cost of increased message latency.

20.7   IEEE 802.15.4 LR-WPAN Device Architecture

Figure 20.6 shows an LR-WPAN device. The device comprises a physical layer (PHY), which contains the RF transceiver along with its low-level control mechanism. A MAC sublayer provides access to the physical channel for all types of transfer. The upper layers consist of a network layer, which provides network configuration, manipulation, and message routing, and an application layer, which provides the intended function of a device. An IEEE 802.2 logical link control (LLC) can access the MAC through the service specific convergence sublayer (SSCS).

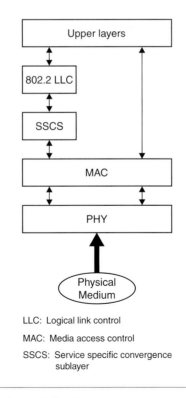

Figure 20.6   LR-WPAN device architecture.

20.7.1 Physical Layer

The PHY (IEEE 802.15.4) provides two services: the PHY data service and PHY management service interfacing to the physical layer management entity (PLME). The PHY data service enables the transmission and reception of PHY protocol data units (PPDUs) across the physical radio channel. The features of the PHY are activation and deactivation of the radio transceiver, energy detection (ED), link quality indication (LQI), channel selection, clear channel assessment (CCA) and transmitting as well as receiving packets across the physical medium.

The standard provides two options based on the frequency band. Both are based on direct sequence spread spectrum (DSSS). The data rate is 250 kbps at 2.4 GHz, 40 kbps at 915 MHz, and 20 kbps at 868 MHz. The higher rate at 2.4 GHz is attributed to a higher-order modulation scheme. Lower frequency provides longer range due to lower propagation losses. Low rate can be translated into better sensitivity and larger coverage area. Higher rate means higher throughput, lower latency, or lower duty cycle. Table 20.3 summarizes the information.

There is a single channel between 868 and 868.6 MHz, 10 channels between 902 and 928 MHz, and 16 channels between 2.4 and 2.4835 GHz (see Table 20.4). Several channels in different frequency bands enable the ability to relocate within the spectrum. The standard also allows dynamic channel selection. MAC includes a scan function that steps through a list of supported channels in search of a beacon, while the PHY contains several lower-level functions, such as receiver energy detection, link quality indication, and channel switching. These functions are used by the network to establish its initial operating channel and to change channels in response to a prolonged outage. Table 20.4 shows IEEE 802.15.4 channel frequencies.

Table 20.3 Frequency bands and data rates.

PHY (MHz)	Freq. band (MHz) and number of channels	Spreading parameters			Data parameters	
		Chip rate (kcps)	Modulation	Bit rate (kbps)	Symbol rate (ksps)	Symbol
868/915	868–868.6 (1 channel)	300	BPSK	20	20	Binary
	902–928 (10 channels)	600	BPSK	40	40	Binary
2450	2400–2483.5 (16 channels)	2000	OQPSK	250	62.5	16-ary orthogonal

Receiver sensitivities are −85 dBm for 2.4 GHz and −92 dBm for 868/915 MHz. The achievable range is a function of receiver sensitivity as well as transmit power. The standard specifies that each device should be capable of transmitting 1 mW, but, depending on the application needs, the actual transmit power may be lower (within regulatory limits). Typical devices (1 mW) are expected to cover a 10–20 m range, with good sensitivity and a moderate increase in transmit power. A star topology can provide complete home coverage. For applications allowing more latency, mesh network topologies provide an attractive alternative for home coverage since each device needs only enough power (and sensitivity) to communicate with its neighbor.

To maintain a common simple interface with MAC, both PHY share a single packet structure (see Figure 20.7). Each PPDU contains a synchronization header (preamble plus start of packet delimiter), a PHY header to indicate the packet length, and the payload, or PHY service data unit (PSDU). The 32-bit preamble is designed for the acquisition of symbol and chip timing, and in some cases may be used for coarse frequency adjustment. Channel equalization is not required for either PHY due to the combination of small coverage area and relatively low chip

Table 20.4 IEEE 802.15.4 channel frequencies.

Channel number	Channel center frequency (MHZ)
k = 0	868.3
k = 1, 2, . . ., 10	906 + 2(k − 1)
k = 11, 12, . . ., 26	2405 + 5(k − 11)

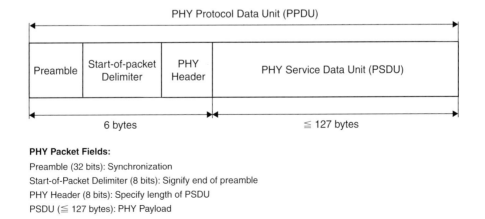

PHY Packet Fields:

Preamble (32 bits): Synchronization

Start-of-Packet Delimiter (8 bits): Signify end of preamble

PHY Header (8 bits): Specify length of PSDU

PSDU (≦ 127 bytes): PHY Payload

Figure 20.7   IEEE 802.15.4 PHY packet structure.

rates. Within the PHY header, 7 bits are used to specify the length of the payload (in bytes). This supports packets of length 0–127 bytes, although, due to MAC layer overhead, zero-length packets will not occur in practice. Typical packet sizes for home applications such as monitoring and control security, lighting, air conditioning, and other appliances are expected to be of the order of 30–60 bytes, while more demanding applications such as interactive games and computer peripherals, or multihop applications with more address overhead, may require larger packet sizes. Adjusting transmission rates in each frequency band, the maximum packet durations are 4.25 ms for the 2.4 GHz band, 26.6 ms for the 915 MHz band, and 53.2 ms for the 868 MHz band.

The 868/915 MHz PHY uses a simple DSSS approach in which each transmitted bit is represented by a 15-chip maximum length sequence (m-sequence, see Chapter 11). Binary data is encoded by multiplying each m-sequence by +1 or −1, and the resulting chip sequence is modulated onto the carrier using binary phase shift keying (BPSK). Differential data encoding is used prior to modulation to allow low-complexity differential coherent reception.

The 2.4 GHz PHY uses a 16-ary quasi-orthogonal modulation technique based on DSSS methods. Binary data is grouped into 4-bit symbols, and each symbol specifies one of sixteen nearly orthogonal 32-chip, pseudo-random noise (PN) sequences for transmission. PN sequences for successive data symbols are concatenated, and the aggregate chip sequence is modulated onto the carrier using, offset-quadrature phase shift keying (OQPSK).

20.7.2   Data Link Layer

The data link layer (IEEE 802.15.4) is divided into two sublayers, the MAC and LLC sublayers. The logical link control is standardized in IEEE 802.2 and is common among all IEEE 802 standards. The IEEE 802.15.4 MAC provides services to an IEEE 802.2 type logical link control through the service-specific convergence sublayer (SCCS), or a proprietary LLC can access the MAC services directly without going through the SCCS. The SCCS ensures compatibility between different LLC sublayers and allows the MAC to be accessed through a single set of access points.

The features of the IEEE 802.15.4 MAC are association and disassociation, acknowledged frame delivery, channel access mechanism, frame validation, guaranteed time slot management, and beacon management. The MAC provides two services to higher layers that can be accessed through two service access points (SAPs). The MAC data service is accessed through the MAC common part sublayer (MCPS-SAP), and the MAC management services are accessed through the MAC layer management entity (MLME-SAP). These two services provide an interface between the SCCS or another LLC and the physical layer.

The MAC management service has 26 primitives, making it very suitable for intended low-end applications. The MAC frame structure is very flexible to accommodate the needs of different applications and network topologies while maintaining a simple protocol. The general format of a MAC frame is shown in Figure 20.8.

The MAC protocol data unit (MPDU) consists of the MAC header (MHR), MAC service data unit (MSDU), and MAC footer (MFR). The first field of the MAC header is the frame control field, which indicates the type of MAC frame being transmitted, specifies the format of the address field, and controls the acknowledgment. The frame control field specifies how the rest of the frame looks and what it contains. The size of the address field may vary between 0 and 20 bytes. The flexible structure of the address field helps to increase the efficiency of the protocol by keeping the packet shorts.

The payload field is variable in length; however, the complete MAC frame may not exceed 127 bytes in length. The data contained in the payload is dependent on the frame type.

The IEEE 802.15.4 MAC has four different frame types. These are the *beacon frame*, *data frame*, *acknowledgment frame* and *MAC command frame*. Only the data and beacon frames actually contain information sent by higher layers; the acknowledgment and MAC command frames originate in the MAC and are used for MAC peer-to-peer communication.

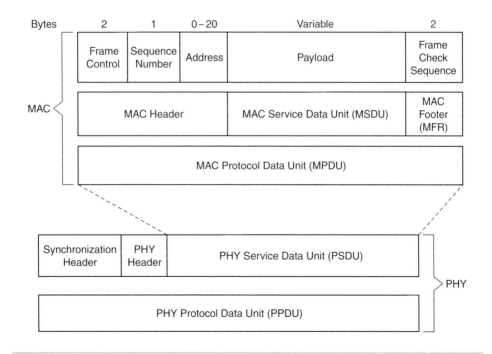

Figure 20.8   General MAC frame format.

Other fields in the MAC frame are the sequence number and frame check sequence (FCS). The sequence number in the MAC header matches the acknowledgment frame with the previous transmission. The FCS helps to verify the integrity of the MAC frame. The FCS in an IEEE 802.15.4 MAC frame is a 16-bit International Telecommunication Union—Telecommunication Standardization Sector (ITU-T) cyclic redundancy check (CRC).

Superframe Structure

Some applications may require a dedicated bandwidth to achieve low latencies. To accomplish these low latencies, the IEEE 802.15.4 LR-WPAN can operate in an optional superframe mode. In a superframe (see Figure 20.9), a dedicated PAN coordinator transmits superframe beacons in predetermined intervals. These intervals can be as short as 15 ms or as long as 245 seconds. The time between two beacons is divided into 16 equal time slots independent of the duration of the superframe. The beacon frame is sent in the first slot of each superframe. The beacons are used to synchronize the attached devices, to identify PAN, and describe the structure of superframes. A device can transmit at any time during the slot, but must complete its transaction before the next superframe beacon. The channel access in time slots is contention based; however, the PAN coordinator may assign time slots to a single device that requires a dedicated bandwidth or low latency transmissions. These assigned time slots are called guaranteed time slots (GTSs) and together form a contention-free period (CFP) located immediately before the next beacon. The size of the CFP may vary depending on the demand by the associated network devices; when guaranteed time slots are used, all devices must complete their contention-based transactions before the CFP begins. The beginning of the CFP and duration of the superframe are communicated to the attached

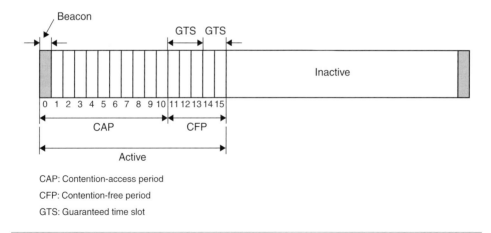

Figure 20.9   IEEE 802.15.4 superframe structure.

network devices by the PAN coordinator. The PAN coordinator may allocate up to seven of the GTSs and a GTS can occupy more than one slot period.

Other MAC Features

Depending on network configuration, one of two channel access mechanisms can be used. In the beacon-enabled network with superframes, a slotted carrier sense multiple access-collision avoidance (CSMA-CA) mechanism is used. In networks without beacons, unslotted CSMA-CA is used. In a beacon-enabled network, any device wishing to transmit during the contention-access period (CAP) waits for the beginning of the next time slot and then determines whether another device is currently transmitting in the same slot. If another device is already transmitting in the slot, the device backs off for a random number of slots or indicates a transmission failure after some retries. In slotted CSMA-CA, the backoff period boundaries of every device in the PAN are aligned with the superframe slot boundaries of the PAN coordinator. In unslotted CSMA-CA, the backoff period of a device does not need to be synchronized to the backoff period of another device.

An important function of the MAC is to confirm successful reception of a received frame. Successful reception and validation of a data or MAC command frame is confirmed with an acknowledgment. If the receiving device is unable to handle the incoming message for any reason, the receipt is not acknowledged. The frame control field in MAC indicates whether or not an acknowledgment is expected. The acknowledgment frame is sent immediately after successful validation of the received frame.

Three levels of security are provided: no security of any type; access control lists (noncryptographic security); and symmetric key security, using AES-128.

20.7.3 The Network Layer

The network layer of Zigbee (IEEE 802.15.4) is responsible for topology construction and maintenance as well as naming and binding services, which include the tasks of addressing, routing, and security. The network layer should be self-organizing and self-maintaining to minimize energy consumption and total cost. IEEE 802.15.4 supports multiple network topologies, including star, peer-to-peer, and cluster tree. The topology is an application design choice.

Routing Protocol

Routing protocols for ad hoc networks can be divided into two groups: *table-driven* (proactive) and *source-initiated on-demand-driven* (reactive) [16,17]. The table-driven approach has low latency and high overhead, and is more suitable when time constraints are significant. On the other hand, the source-initiated on-demand-driven approach has high latency and low overhead. It is more suitable for a mobile environment with a limited bandwidth capacity.

The table-driven routing protocols attempt to maintain consistent, up-to-date routing information from each node to every node in the network. These protocols require each node to maintain one or more tables to store routing information, and they respond to changes in network topology by propagating route updates throughout the network to maintain a consistent view. The destination sequenced distance vector (DSDV), wireless routing protocol (WRP), and cluster switch gateway routing (CSGR) protocol belong to this category.

The source-initiated on-demand-driven routing protocols create routes only when desired by a source node. When a node requires a route to a destination, it initiates a route discovery process within the network. This process is completed once a route is found or all possible route permutations have been examined. The ad hoc on-demand distance vector (AODV), dynamic source routing (DSR), temporally ordered routing algorithm (TORA), and cluster based routing protocol (CBRP) belong to this category.

The ZigBee routing algorithm is based on a hierarchical routing strategy with table-driven optimizations applied where possible. Below we provide summaries of the AODV and cluster-tree algorithm.

The AODV is a pure on-demand route acquisition algorithm in which nodes that do not lie on active paths neither maintain nor participate in any periodic routing table exchanges. Also, a node does not have to discover and maintain a route to another node until the two need to communicate, unless the former node is offering services as an intermediate forwarding station to maintain connectivity between two other nodes. The primary objectives of the algorithm are to broadcast discovery packets only when necessary in order to distinguish between local connectivity management and general topology maintenance and to disseminate information about changes in local connectivity to those neighboring mobile nodes that are likely to need information.

When a source node needs to communicate with another node for which it has no routing information in its table, the *path discovery* process is initiated. Every node maintains two separate counters: *sequence number* and *broadcast id*. The source node starts path discovery by broadcasting a route request (RREQ) packet to its neighbors, which includes *source address; source sequence number; broadcast id; destination address; destination sequence number;* and *hop count*. The pair *source address, broadcast id* uniquely identify an RREQ, where *broadcast id* is incremented whenever the source issues a new RREQ. When an intermediate node receives an RREQ, and if it has already received this RREQ with the same *broadcast id* and *source address*, it drops the redundant RREQ and does not rebroadcast it. Otherwise, it rebroadcasts it to its own neighbors after increasing *hop count*. Each node keeps the following information: *destination address, source address, broadcast id, expiration time for reverse path route entry*, and *source node sequence number*. As the RREQ travels from a source to a destination, it automatically sets up a *reverse path* from all nodes back to source. To set up a reverse path, a node records the address of the neighbor from which it received the first copy of RREQ. These reverse path route entries are maintained for at least enough time for the RREQ to traverse the network and produce a reply to the sender.

When the RREQ arrives at a node, possibly the destination itself, that possesses a current route to the destination, the receiving node first checks that the RREQ was received over a bidirectional link. If this node is not the destination, but has the route to the destination, it determines whether the route is current by comparing the destination sequence number in its own route entry to the destination sequence number in the RREQ. If the RREQ's sequence number for the destination is greater than that recorded by the intermediate node, the intermediate must not use this route to respond to the RREQ, but instead rebroadcast the RREQ.

If the route has a destination sequence number that is greater than that contained in the RREQ or equal to that contained in the RREQ but a smaller hop count, it can unicast a route reply (RREP) packet back to its neighbor from which it received the RREQ. A RREP contains the following information: *source address, destination address, destination sequence number, hop count,* and *lifetime.* As the RREP travels back to the source, each node along the path sets up a forward pointer to the node from which the RREP came, updates its time out information for route entries to the source and destination, and records the latest destination sequence number from the requested destination.

Nodes that are along the path determined by the RREP will time out after the route request expiration timer and delete the reverse pointers since they are not on the path from the source to the destination (see Figure 20.10). The value

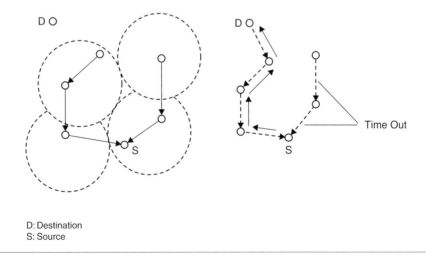

D: Destination
S: Source

Figure 20.10   Reverse and forward path formation in the AODV protocol.

of this time-out time depends on the size of the ad hoc network. Also, there is the routing caching time-out that is associated with each route entry to show the time after which the route is considered to be invalid. Each time a route entry is used to transmit data from a source toward a destination, the time-out for the entry is reset to the current time plus active route time-out.

The source node can begin data transmission as soon as the first RREP is received, and can later update its routing information if it learns of a better route. Each routing table entry includes the following fields: *destination address*, *next hop number of hops*, *destination sequence number*, *active neighbor for this route*, and *expiration time for the route table entry*.

For path maintenance, each node keeps the address of active neighbors through which packets for the given destination are received. The neighbor is considered active if it has originated or relayed at least one packet for that destination within the last active time-out period. Once the next hop on the path from the source to the destination becomes unreachable, the node upstream of the break propagates an unsolicited RREP with a fresh sequence number and hop count of infinity to all active upstream nodes. This process continues until all active source nodes are notified. Upon receiving the notification of a broken link, the source node can start the discovery process if it still requires a route to the destination. If it decides that it would like to rebuild the route to the destination, it sends out an RREQ with a destination sequence number of one greater than the previously known sequence number, to ensure that it builds a new, viable route and no nodes reply if they still regard the previous route as valid.

The cluster tree is a protocol of the logical link and network layer. It uses link-state packets to form either a single cluster network or a potentially larger cluster tree network. The network is self-organized and supports network redundancy to attain a degree of fault resistance and self-repair.

Nodes select a cluster head (CH) and form a cluster according to the self-organized manner. The self-developed clusters connect to each other using the designated device (DD) (see Figure 20.11).

The cluster formation process begins with CH selection. After a CH is elected, the CH expands links with other member nodes to form a cluster. The CH can be selected based on stored parameters of each node, like transmission range, power capacity, computing ability, or location information. After becoming the CH, the node broadcasts a periodic HELLO message that contains a part of the cluster head MAC address and node ID 0 that indicates the CH. The nodes that receive this message send a CONNECTION REQUEST message to the CH. When the CH receives it, it responds to the node with a CONNECTION RESPONSE message that contains a node ID for the node. The node that is assigned a node ID replies with an ACK message to the CH.

If all nodes are located in the range of the CH, the topology of connection becomes a star and every member node is connected to the CH with one hop.

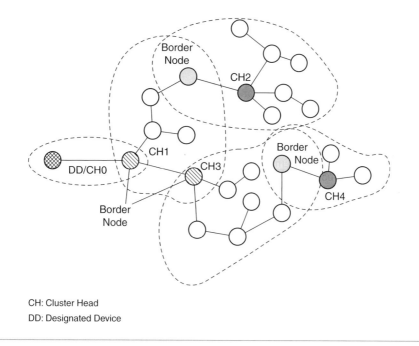

CH: Cluster Head
DD: Designated Device

Figure 20.11   Cluster tree protocol.

A cluster can expand into a multihop structure when each node supports multiple connections.

To form a multicluster network, a designated device (DD) is needed. The DD has the responsibility of assigning a unique cluster ID to each CH. This cluster ID combined with the node ID that the CH assigns to each node within a cluster forms a logical address that is used to route packets. Another role of the DD is to calculate the shortest route from the cluster to the DD and distribute it to all nodes within the network.

When the DD joins the network, it acts as the CH of cluster 0 and starts to send a HELLO message to the neighborhood. If a CH has received this message, it sends a CONNECTION REQUEST message and joins the cluster 0. After that, the CH requests a cluster ID (CID) to the DD. In this case, the CH is a border node that has two logical addresses. One is for a member of the cluster 0 and the other is for a CH. When the CH gets a new CID, it informs its member nodes by the HELLO message. The border node requests a connection and joins the cluster 0 as its member node.

The clusters not bordering cluster 0 use intermediate clusters to get a CID. The CH either becomes the border node to its parent cluster or it names a member node as the border to its parent cluster.

Each member node of the cluster has to record its parent cluster, child/lower clusters, and border node IDs associated with the parent and child clusters. The DD should store the whole structure of the clusters. The CHs report their link state information to the DD.

A backup DD can be prepared to prevent network down-time due to the DD trouble. Inter-cluster communication is realized by routing. The border nodes act as routers that connect clusters and relay packets between clusters. When a border node receives a packet, it examines the destination address, then forwards it to the next border node in the adjacent cluster or to the destination node within its own cluster.

Only the DD can send a message to all nodes within the network. The message is forwarded along the tree route of clusters. The border node should forward the broadcast packet from the parent cluster to the child cluster. Table 20.5 lists the features of the routing protocols.

20.7.4   Applications

The IEEE 802.15.4 has been designed to be useful in a wide variety of applications, including sensing and location determination at disaster sites; automotive sensing, such as tire pressure monitoring; smart badges and tags; and precision agriculture, such as the sensing of soil moisture, pesticide, herbicide, and pH levels. The largest application opportunities for the IEEE 802.15.4 are in home automation and networking including heating, ventilation, air conditioning, security, and lighting; the control of objects such as curtains, windows, doors, and locks; health monitoring, including sensors, monitors, and diagnostics; and toys and games.

The maximum required data rate for these applications is expected to range from 115.2 kbps for some PC peripherals to less than 10 kbps for some home

Table 20.5 Features of routing protocols for ad hoc networks.

Features	AODV	DSDV	TORA	DSR	CBRP
Loop free	Yes	Yes	Yes	Yes	Yes
Multicast	Yes	No	No	No	No
Distributed	Yes	Yes	Yes	Yes	Yes
Reactive	Yes	No	Yes	Yes	Yes
Bidirectional link	Yes	Yes	Yes	No	Yes
Multiple routes	No	No	Yes	Yes	Yes
Security	No	No	No	No	No
Power conservation	No	No	No	No	No

automation and consumer electronics. Message latency is expected to range from approximately 15 ms for PC peripherals to 100 ms or more for home automation applications.

20.8   IEEE 802.15.3a—Ultra WideBand

For wireless personal area networks, which only transmit over small distances in which signal propagation loss is small and less variable, greater capacity can be achieved through greater bandwidth. The use of ultra wideband (UWB) (see Figure 20.12) under FCC guidelines [3] offers tremendous capacity potential over short ranges (<10 m) at low radiated power (mean effective isotropic radiated power [EIRP] of −41.3 dBm/MHz). The FCC defines UWB signals as having a fractional bandwidth (the ratio of baseband bandwidth to RF carrier frequency) of greater than 0.2, or a UWB bandwidth greater than 500 MHz. UWB bandwidth is defined as the frequency band bounded by the points that are 10 dB below the highest radiated emission [6].

The growing demand for higher quality multimedia services and faster content delivery is driving the quest for higher data rates in communication networks. The IEEE 802.15.3a (high rate (HR) WPAN) defined a very high data rate alternative physical layer for IEEE 802.15.3—the current high data rate WPAN standard—to deliver data rates from 20 to 55 Mbps over short ranges (less than 10 m).

The FCC approval of UWB devices has prompted many companies to consider UWB radio when designing physical layers of IEEE 802.15.3a. The FCC ruling allows UWB communication devices to operate in an unlicensed spectrum from 3.1 to 10.6 GHz. The low emission limits for UWB ensure that UWB devices do not cause harmful interference to licensed services and other important radio

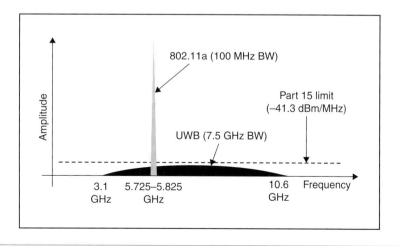

Figure 20.12   Spectrum of ultrawide band signal compared to Wi-Fi 802.11a signal.

operations. Table 20.6 summarizes the guidelines relevant to the use of UWB technology in WPAN devices.

The IEEE 802.15.3a has set out to develop a flexible standard which enables high data rate WPAN (110 Mbps at 10 m, 200 Mbps at 4 m, and 480 Mbps at 2 m) using a cost-effective architecture. The standard enables a broad range of applications including the wireless transmission of images and video. The IEEE 802.15.3a is only concerned with physical layer alternatives and uses the same MAC sublayer as IEEE 802.15.3. The IEEE 802.15.3a task group established technical requirements and selection criteria for the physical layer (see Table 20.7). The task group set forth goals for low-power consumption and low-cost devices to ensure that the standard could be implemented in CMOS technology.

Presently, the IEEE 802.15.3a task group is left with two proposals: (1) an orthogonal frequency division multiplexing (OFDM)-based multiband approach, which uses 528 MHz channels (three mandatory lower band channels and four optional upper band channels) supported by multiband OFDM, and (2) a dual-band impulse radio (IR) spread spectrum approach, where there is a high band (above 5.2–5.8 GHz unlicensed band) and a lower band (from 3.1 GHz to just below the 5.2–5.8 GHz unlicensed band), and which exploits all of the UWB spectrum allocation.

In a channelized UWB system, the UWB frequency band from 3.1 to 10.6 GHz is divided into several smaller bands. Each of these bands must have a bandwidth greater than 500 MHz to comply with the FCC definition of UWB. Frequency hopping between these bands is used to facilitate multiple access.

There are three group A bands which are used for standard operation. The group C bands are allocated for optional use in areas where simultaneous

Table 20.6 FCC guidelines for UWB technology in WPAN devices.

Operating Frequency Range 3.1 to 10.6 GHz	
Average Radiated Emission Limits	
Frequency range (MHz)	EIRP in dBm/MHz (indoor environment for handheld device)
960–1610	−75.3/−75.5
1610–1900	−53.3/−63.3
1900–3100	−51.3/−61.3
3100–10,600	−41.3/−41.3
Above 10,600	−51.3/−61.3
Peak emission level in band: 60 dB above average emission level	
Maximum unacknowledged transmission period: 10 seconds	

Table 20.7 Summary of technical requirements and selection criteria of IEEE 802.15.3a physical layer.

Parameter	Details
Data rates	110, 220, and 480 (optional) Mbps
Range	10 m, 4 m and 2 m
Power consumption	100 mW and 250 mW
Power management modes	Capabilities such as power saving, wake-up, etc.
Co-located piconets	4
Interference susceptibility	Robust to IEEE systems, PER <8% for a 1024 byte packet
Co-existence capability	Reduced interference to IEEE systems, interfering average power at least 6 dB below the minimum sensitivity level of non-802.15.3a device
Cost	Similar to Bluetooth
Location awareness	Location information to be propagated to a suitable management entity
Scalability	Backward compatibility with IEEE 802.15, adaptable to various regulatory regions
Signal acquisition	<20 μs for acquisition from the beginning of the preamble to the beginning of the header
Antenna practicality	Size and form factor consistent with original device

operating piconets are in close proximity. Groups B and D are reserved for future use. Each band uses frequency hopping-orthogonal frequency division multiplexing (FH-OFDM). Because of the increased length of the OFDM symbol period, this modulation scheme can successfully reduce the effects of inter-symbol interference (ISI). However, this increases the transceiver complexity.

In the impulse radio spread-spectrum approach, very short duration (sub-nano second) pulses transmit over each band, with the bandwidth in excess of 1 GHz used. The root raised cosine pulse shaping provides the spectrum roll-off that is required. The phase of the pulse is inverted to give simple BPSK modulation. Code sequences are used to transmit user data. The codes are selected to be as orthogonal as possible. The code sequences are designed to appear as pseudo-random strings. The design uses a direct sequence spread spectrum with 24 chips per symbol.

Tables 20.8 and 20.9 provide overviews of two UWB approaches. Impulse radio, the traditional approach to UWB communication, has greater precision for position location, and realizes better spectrum efficiency compared to multiband UWB.

Table 20.8 Overview of multiband-OFDM physical layer proposal (supported by multiband OFDM coalition).

	Spectrum allocation
Number of bands	3 (first generation bands); 10 optional bands
Bandwidth	528 MHz
Frequency range	Group A: 3.168–4.752 GHz; Group B: 4.752–6.072 GHz; Group C: 6.072–8.184 GHz; Group D 8.184–10.296 GHz
Modulation scheme	FH-OFDM (with 128-point FFT), QPSK
Coexistence method	Null band for WLAN (~5 GHz)
Multiple access method	Time frequency interleaving
Number of simultaneous piconets	4
Error correction codes	Convolution codes
Code rates	11/32 @ 110 Mbps; 5/8 @ 200 Mbps; 3/4 @ 480 Mbps
Link margin	5.3 dB @ 10 m @110 Mbps; 10 dB @ 4 m @ 200 mbps; 11.5 dB @ 2 m @ 480 Mbps
Multipath mitigation	1-tap (robust to 60.6 ns delay spread)

Table 20.9 Overview of impulse radio (IR) spread spectrum UWB.

	Spectrum Allocation
Number of bands	2
Bandwidths	1.368 GHz, 2.736 GHz
Frequency ranges	3.2–5.15 GHz; 5.825–10.6 GHz
Modulation scheme	BPSK, QPSK, DSSS
Coexistence method	Null band for WLAN (~5 GHz)
Multiple access method	CDMA
Number of simultaneous piconets	8
Error correction codes	Convolution codes and Reed-Solomon codes
Code rates	1/2 @ 110 Mbps; RS (255,223) @ 200 Mbps; RS (255,233) @ 480 Mbps
Link margin	6.7 dB @ 10 m @ 114 Mbps; 11.9 dB @ 4 m @ 200 Mbps; 1.7 dB @ 2 m @ 600 Mbps
Chip duration	731 ps (low band); 365.5 ps (high band)
Multipath mitigation	Decision feedback equalizer and Rake receiver

However, it has less flexibility with respect to foreign spectral regulation and may be too broadband if foreign governments choose to limit their UWB spectral allocations to smaller ranges than authorized by the FCC. The other advantages of the multiband UWB scheme are:

- More scalable and adaptive than single band designs
- Better co-existence characteristics with systems such as IEEE 802.11a due to adaptive selection of the bands
- Lower risk of implementations because it leverages more traditional radio design techniques.

In the multiband UWB system information is independently encoded in the different bands instead of being transmitted in a single pulse, this process results in very high bit rate systems realized with relatively low signaling rates. Since a much lower signaling rate than that for a single band UWB system is required for the same data rate, most of the electronics can run at lower speed and, therefore, the complexity, cost, and integration concerns are reduced.

The multiband OFDM (MB-OFDM) UWB is supported by the *WiMedia Alliance*, whereas direct sequence UWB (DS-UWB) is supported by the UWB forum. The WiMedia Alliance is a not-for-profit open industry association that promotes and enables the rapid adoption, regulation, standardization, and multivendor interoperability of UWB worldwide. WiMedia UWB is optimized for wireless personal area networks delivering high-speed (480 Mbps and beyond), low-power multimedia capabilities for PC, consumer electronics, mobile, and automotive market segments. WiMedia UWB has been selected by the Bluetooth Special Interest Group (SIG) and USB Implementers Forum as the foundation radio of their high-speed wireless specifications for the next generation consumer electronics, mobile, and computer applications.

20.9 Radio Frequency Identification

Radio frequency identification (RFID) is an automatic identification method, relying on storing and remotely retrieving data using devices called RFID tags or transponders. An RFID tag is an object that can be attached to or incorporated into a product, animal, or person for the purpose of identification using radio waves. Chip-based RFID tags contain silicon chips and antennas. Passive tags require no internal power source, whereas active tags require a power source. RFID is also called *dedicated short range communication* (DSRC).

In a typical RFID system, individual objects are equipped with a small, inexpensive tag. The tag contains a transponder with a digital memory chip that is given a unique electronic product code. The transponder emits messages with an identification number that is retrieved from a database and acted upon accordingly. The writable memory is used to transmit information among RFID readers in different locations.

The interrogator, an antenna packaged with a transceiver and decoder, emits a signal activating the RFID tag so it can read and write data to it. When an RFID tag passes through the electromagnetic zone, it detects the reader's activation signal. The reader decodes the data encoded in the tag's integrated circuit (silicon chip) and the data is passed to the host computer. The application software on the host processes the data, and may perform various filtering operations to reduce the numerous often redundant reads of the same tag to a smaller and more useful data set.

The following are the RFID components and their characteristics:

- Tags
  - ◆ Active (with watch-sized battery)
  - ◆ Passive (without battery)
  - ◆ Semi-active (with battery)
  - ◆ Size (varies from less than 1 square inch to many square inches)
    - — Dependent on power and frequency (13.56 MHz, 433 MHz, 900 MHz, 2.4 GHz with power from 1 mW to 1 W)
    - — Memory can be read only, write one and read many with 1 byte to 512 Kilo-bytes storage.
- Reader
  - ◆ Receive data
  - ◆ Validate data
  - ◆ Send data to tag
- Middleware
- Host data management software applications

The following are the key features of RFID:

- **No line-of-sight.** RFID tags do not need to be visible to read or write.
- **Robust.** Because RFID systems do not need to be visible, they can be encased within rugged material protecting them from the environment in which they are being used. This means they can be used in harsh fluid and chemical environments and rough handling situations.
- **Read speed.** Tags can be read from significant distances and can also be read very quickly—for example, on a conveyor.
- **Reading multiple items.** A number of tagged items can be read at the same time within an RF field. This cannot be done easily with visual identifiers.
- **Security.** Because tags can be enclosed, they are much more difficult to tamper with. A number of tag types now also come programmed with a unique

identifier (serial identification) which is guaranteed to be unique throughout the world.

- **Programmability.** Many tags are read/write capable, rather than read only. This means that information can be written to the tag, perhaps to show that the item being tagged has gone through a particular process, or that its condition or status has changed somehow.

RFID systems can be used just about anywhere, from clothing tags to missiles to pet tags to food—anywhere that a unique identification system is needed. The tag can carry information as simple as a pet owner's name and address or the cleaning instructions on a sweater to as complex as instructions on how to assemble a car. Some auto manufacturers use RFID systems to move cars through an assembly line. At each successive stage of production, the RFID tag tells the computers what the next step of the automated assembly is. The following are some of the applications of RFID systems:

- **Automotive.** Auto makers have added security and convenience to automobiles by using RFID technology for anti-theft immobilizers and passive-entry systems.

- **Animal tracking.** Ranchers and livestock producers use RFID technology to meet export regulations and optimize livestock value. Wild animals are tracked in ecological studies, and many pets who are tagged are returned to their owners.

- **Assets tracking.** Hospitals and pharmacies meet tough product accountability legislation with RFID; libraries limit theft and keep books in circulation more efficiently; and sports and entertainment entrepreneurs find that "smart tickets" are their ticket to a better bottom line and happier customers.

- **Contactless commerce.** Blue-chip companies such as American Express, Exxon Mobile, and MasterCard use innovative form factors enabled by RFID technology to strengthen brand loyalty and boost revenue per customer.

- **Supply chain.** Wal-Mart, Target, Best Buy, and other retailers have discovered that RFID technology can keep inventories at the optimal level, reduce out-of-stock losses, limit shoplifting, and speed customers through check-out lines.

RFID tags are often envisioned as a replacement for bar codes, having a number of important advantages over bar code technology. One of the key differences between RFID and bar code technology is RFID eliminates the need of line-of-sight reading that bar coding depends on. Also, RFID scanning can be done at greater distances than bar code scanning. High frequency RFID systems

(850–950 MHz, 2.4–2.5 GHz) offer transmission ranges more than 90 feet. Bar codes are fixed at the time of printing and can be rendered useless by defacement or smudging. Bar codes can be spoofed or easily defeated by any malicious individual having a laser printer at their disposal.

20.10 Summary

In this chapter we first focused on wireless sensor networks and then discussed the IEEE 802.15.3, 802.15.3a, and 802.15.4 standards. The IEEE 802.15.4 is the approved low-rate standard for wireless sensor networks, whereas the IEEE 802.15.3a is the high-rate wireless personal area network standard that deals with ultra wideband (UWB) network devices. We discussed the uses of wireless sensor networks, wireless sensor network models, wireless sensor network protocol stacks, and Zigbee technology. We then presented details of the physical and data link layers in LR-WPAN devices. The chapter concluded by presenting requirements for HR-WPAN, the current status of the IEEE 802.15.3a standard, and radio frequency identification (RFID) techniques.

Problems

20.1   What is a wireless sensor network?

20.2   Discuss various uses of wireless sensor networks.

20.3   What is the ZigBee technology? Discuss briefly.

20.4   Discuss different types of network topology that are supported in ZigBee.

20.5   The IEEE standard provides two options for physical layers in LR-WPAN devices. Discuss them.

20.6   List the features of the IEEE 802.15.4 MAC.

20.7   What are the IEEE 802.15.4 MAC management services?

20.8   Discuss the routing protocols used for ad hoc networks.

20.9   Briefly discuss the AODV and cluster-tree routing protocol used in ZigBee.

20.10   What is a UWB device according to the FCC?

20.11   Discuss technical requirements and selection criteria of the IEEE 802.15.3a.

20.12   Briefly discuss the two proposals which the IEEE 802.15.3a task group is considering for approval.

20.13   What is RFID? Discuss some of its applications.

References

1. Akyildiz, I., et al. A Survey on Sensor Networks. *IEEE Communication Magazine*, August 2002, pp. 102–114.

2. Akyildiz, I., et al. Wireless Sensor Network: a Survey. *Computer Network* 38 (2002) 393–422.

3. FCC, First Report and Order 02-48, February 2002.

4. Gorday, P., et al. "IEEE 802.15.4 Overview." IEEE 802.15-01/509r0, November 12, 2001.

5. Gilb, J. P. K. "Overview of Draft Standard 802.15.3." IEEE 802.15-01/509r0, November 14, 2001.

6. Gutierrez, J. A., Callaway, E. H., and Barrett, R. L. *IEEE 802.15.4 Low-rate Wireless Personal Area Networks: Enabling Wireless Sensor Network*. Standard Information Network, IEEE Press, ISBN 0-7381-3557-7.

7. Hedetniemi, S., et al. A Survey of Gossiping and Broadcasting in Communication Networks. *Networks*, vol. 18, 1988.

8. Heinzelman, W., et al. Adaptive Protocols for Information Dissemination in Wireless Sensor Networks. Proceedings of ACM MobiCom '99, Seattle, WA, 1999, pp. 174–185.

9. Heinzelman, W., et al. Energy-efficient Communication Protocol for Wireless Microsensor Networks. IEEE Proceedings of, Hawaii International Conference, Sys. Sci., January 2000, pp. 1–10.

10. http://www.palowireless.com/Zigbee/whatis.asp.

11. http://www.silabs.com/ZigBee.

12. http://www.cens.ucla.edu/sensys03/sensys03-callaway.pdf.

13. IEEE standard 802.15.1TM, June 14, 2002, http://standard.ieee.org/getieee802/download/802.15.1-2002_sectionone.pdf.

14. Intanagonwiwat, C., et al. Directed-Diffusion: A Scalable and Robust Communication Paradigm for Sensor Networks. Proceedings of ACM MobiCom '00, Boston, MA, 2000, pp. 56–67.

15. Li, L., and Halpern, J. Minimum-Energy Mobile Wireless Network Revisited. ICC '01, Helsinki, Finland, June 2001.

16. Mandke, K., et al. The Evolution of Ultra Wide Band Radio for Wireless Personal Area Network. *High Frequency Electronics*, September 2003.

17. Royer, E. M., and Toh, C.-K. A Review of Current Routing Protocols for Ad Hoc Mobile Wireless Networks. *IEEE Magazine Personal Communications*, 17(8), 1999, pp. 46–55.

18. Shih, E., et al. Physical Layer Driven Protocol and Algorithm Design for Energy-efficient Wireless Sensor Networks. Proceedings of ACM MobiCom '01, Rome, Italy, July 2001, pp. 272–286.

19. Sinha, A., and Chandrakasan. Dynamic Power Management in Wireless Sensor Networks. *IEEE Design Test Computer*, March/April 2001.

20. Sohrabi, K. Protocols for Self-organization of a Wireless Sensor Network. *IEEE Personal Communications*, October. 2000, pp. 16–27.

21. Toh, C.-K. *Ad hoc Mobile Wireless Network — Protocols and Systems*. Upper Saddle River, NJ: Prentice Hall, 2002.

22. Woo, A., and Culler, D. A. Transmission Control Scheme for Media Access in Sensor Networks. Proceedings of ACM MobiCom '01, Rome, Italy, July 2001, pp. 221–235.

23. Zhao, F., and Guibas, L. *Wireless Sensor Networks: An Information Processing Approach*. Morgan Kaufmann, 2004.

CHAPTER 21

Wireless Local Area Networks

21.1 Introduction

With the success of wired local area networks (LANs), the local computing market is moving toward wireless LAN (WLAN) with the same speed of current wired LAN. WLANs are flexible data communication systems that can be used for applications in which mobility is required. In the indoor business environment, although mobility is not an absolute requirement, WLANs provide more flexibility than that achieved by the wired LAN. WLANs are designed to operate in industrial, scientific, and medical (ISM) radio bands (see Table 21.1) and unlicensed-national information infrastructure (U-NII) bands. In the United States, the Federal Communications Commission (FCC) regulates radio transmissions; however, the FCC does not require the end-user to buy a license to use the ISM or U-NII bands. Currently, WLANs can provide data rates up to 11 Mbps, but the industry is making a move toward high-speed WLANs. Manufacturers are developing WLANs to provide data rates up to 54 Mbps or higher. High speed makes WLANs a promising technology for the future data communications market [7].

The IEEE 802.11 committee is responsible for WLAN standards. WLANs include IEEE 802.11a (WiFi 5), IEEE 802.11b (WiFi), IEEE 802.11g and IEEE 802.11n (see Figure 21.1). The deployment of WLANs can provide connectivity in homes, factories, and hot-spots. The IEEE 802.16 group is responsible for wireless metropolitan area network (WMAN) standards. This body is concerned with fixed broadband wireless access systems, also known as "last mile" access networks. In this chapter, we focus on different types of WLANs and introduce IEEE 802.16 standards including WiMAX (high speed WLAN) [1,18,19]. Table 21.2 lists all subgroups of IEEE 802.11.

Table 21.1 Industrial, scientific, and medical (ISM) bands.

No.	Band (GHz)	Bandwidth (MHz)	Power level	Spread spectrum
1	0.902–0.928	26	1 W	FHSS, DSSS
2	2.4–2.4835	83.5	1 W	FHSS, DSSS
3	5.725–5.850	125	1 W	FHSS, DSSS
4	24.0–24.5	250	50 mW/m @ 3 m	Not Applicable

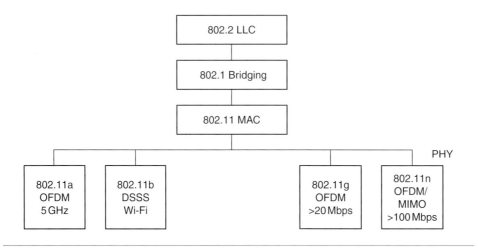

Figure 21.1   IEEE 802.11 WLAN standards.

Table 21.2 IEEE 802.11 subgroups.

802.11a	High speed physical layer in 5 GHz band
802.11b	Higher speed physical layer extension of wireless in 2.4 GHz band
802.11d	Local and metropolitan area wireless
802.11g	Broadband wireless
802.11i	Security
802.11n	Wideband service

WLANs are flexible data communications systems implemented as an extension or as an alternative for wired LANs. Using radio frequency (RF) technology, WLANs transmit and receive data over the air, minimizing the need for wired connections. Thus, WLANs combine data connectivity with user mobility [3–6,8,14–16].

Recently, manufacturers have deployed WLANs for process and control applications. Retail applications have expanded to include wireless point of sale (WPOS). The health-care and education industry are also fast-growing markets for WLANs. WLANs provide high-speed, reliable data communications in a building or campus environment as well as coverage in rural areas. WLANs are simple to install. Figure 21.2 provides application fields of WLANs.

In WLANs, the connection between the client and the user is accomplished by the use of a wireless medium such as RF or Infrared (IR) communications instead of a cable. This allows a remote user to stay connected to the network while mobile or not physically attached to the network. The wireless connection

Figure 21.2 WLAN application.

is usually accomplished by the user having a hand-held terminal or laptop that has an RF interface card installed inside the terminal or through the PC card slot of the laptop. The client connection from the wired LAN to the user is made through an access point (AP) that can support multiple users simultaneously. The AP can reside at any node on the wired network and acts as a gateway for wireless users' data to be routed onto the wired network.

The range of a WLAN depends on the actual usage and environment of the system. It may vary from 100 feet inside a solid walled building to several thousand feet in an outdoor environment with direct line-of-sight. Much like cellular phone systems WLANs are capable of roaming from the AP and reconnecting to the network through other APs residing at other points in the network. This can allow the wired LAN to be extended to cover a much larger area than the existing coverage by the use of multiple APs.

An important feature of WLANs is that they can be used independently of wired networks. They can be used as stand-alone networks anywhere to link multiple computers together without having to build or extend wired networks. The network communications take place in a part of the radio spectrum that is designed as *license free*. In this band, 2.4–2.5 GHz, users can operate without a license so as long as they use equipment that has been of the type approved for use in the license-free band. The 2.4–2.5 GHz band has been designated as license free by the international telecommunications Union (ITU), and is available as license free in most countries of the world.

Standard WLANs are capable of operating at speeds in the range of 1–2 Mbps depending on the actual system; both of these speeds are supported by the standard for WLANs defined by the IEEE. The fastest WLANs use 802.11b high-rate standard to move data through air at a maximum speed of 11 Mbps. The IEEE established the 802.11b standard for wireless networks and the wireless compatibility Ethernet alliance to assure that WLAN products are interoperable from manufacturer to manufacturer. Any LAN application, network operating systems, or protocol, including transmission control protocol/Internet protocol (TCP/IP) will run on 802.11b-compliant WLANs as easily as they run over the Ethernet.

The following are a few advantages of deploying WLANs:

- Mobility improves productivity with real-time access to information, regardless of worker location, for faster and more efficient decision making
- Cost-effective network setup for hard-to-wire locations such as older buildings and solid wall structures
- Reduced cost of ownership, particularly in a dynamic environment requiring frequent modification due to minimal wiring and installation costs per device and per user

However, there are several issues that should be considered in deploying the WLAN including:

- *Frequency allocation:* Operation of a wireless network requires that all users operate in a common frequency band. The frequency band must be approved in each country.
- *Interference and reliability:* In a wired LAN, one hears only the terminals connected to the network. In a WLAN, interference is caused by simultaneous transmission of information in the shared frequency band and by multipath fading. The reliability of a communication channel is measured by bit error rate (BER). Automatic repeat request (ARQ) and forward error correction (FEC) techniques are used to increase reliability.
- *Security:* Radio waves are not confined to the boundary of buildings or campuses. There exists the possibility of eavesdropping and intentional interference. Data privacy over a radio medium is usually accomplished by using encryption.
- *Power consumption:* WLANs are typically related to mobile applications. In these applications, battery power is a scarce resource. Therefore, the devices must be designed to be energy efficient.
- *Mobility:* One of the advantages of a WLAN is the freedom of mobility. The devices should accommodate handoff at transmission boundaries to route data calls to mobile users.
- *Throughput:* To support multiple transmissions simultaneously, spread spectrum techniques are often used.

21.2   WLAN Equipment

There are three main links that form the basis of the wireless network. These are:

- *LAN adapter:* Wireless adapters are made in the same basic form as their wired counterparts: PCMCIA, Card bus, PCI, and USB. They also serve the same function, enabling end-users to access the network. In a wired LAN, adapters provide an interface between the network operating system and the wire. In a WLAN, they provide the interface between the network operating system and an antenna to create a transparent connection to the network.

- *Access point (AP):* The AP is the wireless equivalent of an LAN hub. It receives, buffers, and transmits data between the WLAN and the wired network, supporting a group of wireless user devices. An AP is typically connected with the backbone network through a standard Ethernet cable, and communicates with wireless devices by means of an antenna. The AP or antenna connected to it is generally mounted on a high wall or on the ceiling. Like cells in a cellular network, multiple APs can support handoff from one AP to another as the user moves from area to area. APs have a range from 20 to 500 meters. A single AP can support between 15 to 250 users, depending on technology, configuration, and use. It is relatively easy to scale a WLAN by adding more APs to reduce network congestion and enlarge the coverage area. Large networks requiring multiple APs deploy them to create overlapping cells for constant connectivity to the network. A wireless AP can monitor movement of a client across its domain and permit or deny specific traffic or clients from communicating through it.

- *Outdoor LAN bridges:* Outdoor LAN bridges are used to connect LANs in different buildings. When the cost of buying a fiber optic cable between buildings is considered, particularly if there are barriers such as highways or bodies of water in the way, a WLAN can be an economical alternative. An outdoor bridge can provide a less expensive alternative to recurring leased-line charges. WLAN bridge products support fairly high data rates and ranges of several miles with the use of line-of-sight directional antennas. Some APs can also be used as a bridge between buildings of relatively close proximity.

21.3 WLAN Topologies

WLANs can be built with either of the following topologies:

- Peer-to-peer (ad hoc) topology
- Access point-based topology
- Point-to-multipoint bridge topology

In peer-to-peer topology, client devices within a cell communicate directly to each other as shown in Figure 21.3.

AP-based technology uses access points to bridge traffic onto a wired (Ethernet or Token Ring) or a wireless backbone as shown in Figure 21.4. AP enables a wireless client device to communicate with any other wired or wireless device on the network. AP-based topology is more commonly used and demonstrates that

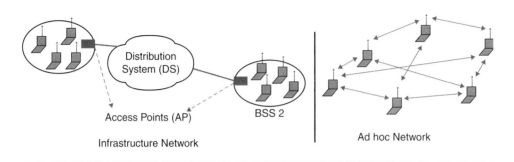

Figure 21.3   Peer-to-peer topology (Ad hoc Network).

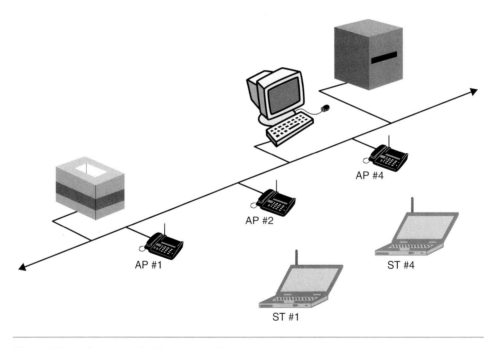

Figure 21.4   Access point-based topology.

the WLAN does not replace the wired LAN, it extends connectivity to mobile devices.

Another wireless network topology is the point-to-multipoint bridge. Wireless bridges connect LANs in one building to LANs in another building even if the buildings are miles apart. These conditions receive a clear line-of-sight between buildings. The line-of-sight range varies based on the type of wireless bridge and antenna used as well as environmental conditions.

21.4   WLAN Technologies

The technologies available for use in a WLAN include infrared, UHF (narrowband), and spread spectrum implementation. Each implementation comes with its own set of advantages and limitations.

21.4.1   Infrared Technology

Infrared is an invisible band of radiation that exists at the lower end of the visible electromagnetic spectrum. This type of transmission is most effective when a clear line-of-sight exists between the transmitter and the receiver.

Two types of infrared WLAN solutions are available: diffused-beam and direct-beam (or line-of-sight). Currently, direct-beam WLANs offer a faster data rate than the diffused-beam networks. Direct-beam is more directional since diffused-beam technology uses reflected rays to transmit/receive a data signal. It achieves lower data rates in the 1–2 Mbps range.

Infrared is a short-range technology. When used indoors, it can be limited by solid objects such as doors, walls, merchandise, or racking. In addition, the lighting environment can affect signal quality. For example, loss of communication may occur because of the large amount of sunlight or background light in an environment. Fluorescent lights also may contain large amounts of infrared. This problem may be solved by using high signal power and an optimal bandwidth filter, which reduces the infrared signals coming from an outside source. In an outdoor environment, snow, ice, and fog may affect the operation of an infrared-based system. Table 21.3 gives considerations for choosing infrared technology.

21.4.2   UHF Narrowband Technology

UHF wireless data communication systems have been available since the early 1980s. These systems normally transmit in the 430 to 470 MHz frequency range, with rare systems using segments of the 800 MHz range. The lower portion of this band — 430–450 MHz — is referred to as the unprotected (unlicensed), and 450–470 MHz is referred to as the protected (licensed) band. In the unprotected band, RF licenses are not granted for specific frequencies and anyone is allowed to use any frequencies, giving customers some assurance that they will have complete use of that frequency.

Table 21.3   Considerations for choosing infrared technology.

Advantages	No government regulations controlling use
	Immunity to electro-magnetic (EM) and RF interference
Disadvantages	Generally a short-range technology (30–50 ft radius under ideal conditions)
	Signals cannot penetrate solid objects
	Signal affected by light, snow, ice, fog
	Dirt can interfere with infrared

Because independent narrowband RF systems cannot coexist on the same frequency, government agencies allocate specific RFs to users through RF site licenses. A limited amount of unlicensed spectrum is also available in some countries. In order to have many frequencies that can be allocated to users, the bandwidth given to a specific user is very small.

The term *narrowband* is used to describe this technology because the RF signal is sent in a very narrow bandwidth, typically 12.5 kHz or 25 kHz. Power levels range from 1 to 2 watts for narrowband RF data systems. This narrow bandwidth combined with high power results in larger transmission distances than are available from 900 MHz or 2.4 GHz spread spectrum systems, which have lower power levels and wider bandwidths. Table 21.4 lists the advantages and disadvantages of UHF technology.

Many modern UHF systems are synthesized radio technology. This refers to the way channel frequencies are generated in the radio. The crystal-controlled products in legacy UHF products require factory installation of unique crystals for each possible channel frequency. Synthesized radio technology uses a single, standard crystal frequency and drives the required channel frequency by dividing the crystal frequency down to a small value, then multiplying it up to the desired channel frequency. The division and multiplication factors are unique for each desired channel frequency, and are programmed into digital memory in the radio at the time of manufacturing. Synthesized UHF-based solutions provide the ability to install equipment without the complexity of hardware crystals. Common equipment can be purchased and specific UHF frequency used for each device can be tuned based upon specific location requirements. Additionally, synthesized UHF radios do not exhibit the frequency drift problem experienced in crystal controlled UHF radios.

Modern UHF systems allow APs to be individually configured for operation on one of the several preprogrammed frequencies. Terminals are programmed with a list of all frequencies used in the installed APs, allowing them to change frequencies when roaming. To increase throughput, APs may be installed with overlapping coverage but use different frequencies.

Table 21.4 Considerations for choosing UHF technology.

Advantages	Longest range
	Low cost solution for large sites with low to medium data throughput requirements
Disadvantages	Large radio and antennas increase wireless client size
	RF site license required for protected bands
	No multivendor interoperability
	Low throughput and interference potential

21.4.3   Spread Spectrum Technology

Most WLANs use spread spectrum technology, a wideband radio frequency technique that uses the entire allotted spectrum in a shared fashion as opposed to dividing it into discrete private pieces (as with narrowband). The spread spectrum system (see Chapter 11) spreads the transmission power over the entire usable spectrum. This is obviously a less efficient use of the bandwidth than the narrowband approach. However, spread spectrum is designed to trade off bandwidth efficiency for reliability, integrity, and security. The bandwidth trade-off produces a signal that is easier to detect, provided that the receiver knows the parameters of the spread spectrum signal being broadcast. If the receiver is not tuned to the right frequency, a spread spectrum signal looks like background noise.

By operating across a broad range of radio frequencies, a spread spectrum device could communicate clearly despite interference from other devices using the same spectrum in the same physical location. In addition to its relative immunity to interference, spread spectrum makes eavesdropping and jamming inherently difficult.

In commercial applications, spread spectrum techniques currently offer data rates up to 2 Mbps. Because the FCC does not require site licensing for the bands used by spread spectrum systems, this technology has become the standard for high-speed RF data transmission. Two modulation schemes are commonly used to encode spread spectrum signals: direct sequence spread spectrum (DSSS) and frequency hopping spread spectrum (FHSS).

FHSS uses a narrowband carrier that changes frequency in a pattern known to both transmitter and receiver. Properly synchronized, the net effect is to maintain a single logical channel. To an unintended receiver, FHSS appears to be a short-duration impulse noise.

DSSS generates a redundant bit pattern for each bit to be transmitted. This bit pattern is called a *spreading code*. The longer the code, the greater the probability that the original data can be recovered (and, of course the more bandwidth will be required). To an unintended receiver DSSS appears as low-power, wideband noise and is rejected by most narrowband receivers.

21.5   IEEE 802.11 WLAN

In 1997 the IEEE developed an international standard for WLANs: IEEE 802.11-1997. This standard was revised in 1999. Like other IEEE 802 standards, the 802.11 standard focuses on the bottom two layers of the OSI model, the physical layer (PHY) and data link layer (DLL). Because of the common interface provided to upper layers, any LAN application, network operating system, or protocol including TCP/IP and Novell Netware will run on an 802.11-compliant WLAN. The objective of the IEEE 802.11 standard was to define an medium access control (MAC) sublayer, MAC management protocols and services, and three PHYs for wireless connectivity of fixed, portable, and moving devices within a local area. The three physical layers are an IR baseband PHY, an FHSS radio in the 2.4 GHz band, and a DSSS radio in the 2.4 GHz. All three physical layers support both 1 and 2 Mbps operations.

WLANs support asynchronous data transfers that refer to the traffic that is relatively insensitive to time delays such as electronic mail and file transfers. Optionally WLANs can also support the traffic, which is bounded by the specified time delay, to achieve an acceptable quality of service (QoS), such as packetized voice and video.

21.5.1 IEEE 802.11 Architecture

The architecture of the IEEE 802.11 WLAN is designed to support a network where most decision making is distributed to mobile stations. This type of architecture has several advantages. It is tolerant of faults in all of the WLAN equipment and eliminates possible bottlenecks a centralized architecture would introduce. The architecture is flexible and can easily support both small, transient networks and large, semipermanent or permanent networks. In addition, the architecture and protocols offer significant power saving and prolong the battery life of mobile equipment without losing network connectivity.

Two network architectures are defined in the IEEE 802.11 standard:

- *Infrastructure network:* An infrastructure network is the network architecture for providing communication between wireless clients and wired network resources. The transition of data from the wireless to wired medium occurs via an AP. An AP and its associated wireless clients define the coverage area. Together all the devices form a *basic service set* (see Figure 21.5).
- *Point-to-point (ad hoc) network:* An ad hoc network is the architecture that is used to support mutual communication between wireless clients. Typically, an ad hoc network is created spontaneously and does not support access to wired networks. An ad hoc network does not require an AP.

IEEE 802.11 supports three basic topologies for WLANs: the independent basic service set (IBSS), the basic service set, and the extended service set (ESS). The MAC layer supports implementations of IBSS, basic service set, and ESS configurations.

- The IBSS configuration is referred to as an independent configuration or an ad hoc network. An IBSS configuration is analogous to a peer-to-peer office network in which no single node is required to act as a server. IBSS WLANs include a number of nodes or wireless stations that communicate directly with one another on an ad hoc, peer-to-peer basis. Generally, IBSS implementations cover a limited area and are not connected to any large network. An IBSS is typically a short-lived network, with a small number of stations, that is created for a particular purpose.
- The basic service set configuration relies on an AP that acts as the logical server for a single WLAN cell or channel. Communications between station 1 and

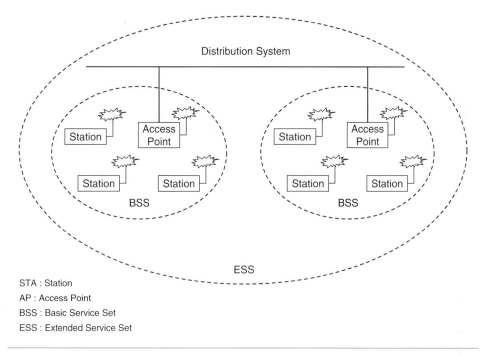

STA : Station
AP : Access Point
BSS : Basic Service Set
ESS : Extended Service Set

Figure 21.5   BSS and ESS configuration of IEEE 802.11 WLAN.

station 4 actually flow from station 1 to AP1 and then from AP1 to AP2 and then from AP2 to AP4 and finally AP4 to station 4 (refer to Figure 21.4). An AP performs a bridging function and connects multiple WLAN cells or channels, and connects WLAN cells to a wired enterprise LAN.

- The ESS configuration consists of multiple basic service set cells that can be linked by either wired or wireless backbones called a distributed system. IEEE 802.11 supports ESS configurations in which multiple cells use the same channel, and configurations in which multiple cells use different channels to boost aggregate throughput. To network the equipment outside of the ESS, the ESS and all of its mobile stations appear to be a single MAC-layer network where all stations are physically stationary. Thus, the ESS hides the mobility of the mobile stations from everything outside the ESS (see Figure 21.5).

21.5.2   802.11 Physical Layer (PHY)

At the physical layer, IEEE 802.11 defines three physical characteristics for WLANs: diffused infrared (baseband), DSSS, and FHSS. All three support a 1 to 2 Mbps data rate. Both DSSS and FHSS use the 2.4 GHz ISM band (2.4–2.4835 GHz). The physical layer provides three levels of functionality. These include: (1) frame exchange between the MAC and PHY under the control of the physical layer convergence procedure (PLCP) sublayer; (2) use of signal carrier and spread spectrum (SS) modulation to transmit data frames over the media under the control of the physical medium dependent (PMD) sublayer; and (3) providing a carrier sense indication back to the MAC to verify activity on the media (see Figure 21.6).

Each of the physical layers is unique in terms of the modulation type, designed to coexist with each other and operate with the MAC. The specifications for IEEE 802.11 meet the RF emissions guidelines of FCC, ETSI, and the Ministry of Telecommunications.

DSSS PHY

In the DSSS PHY, data transmission over the media is controlled by the PMD sublayer as directed by the PLCP sublayer. The PMD sublayer takes the binary information bits from the PLCP protocol data unit (PPDU) and converts them into RF signals by using modulation and DSSS techniques (see Figure 21.7). Figure 21.8 shows the PPDU frame, which consists of a PLCP preamble, PLCP header, and MAC protocol data unit (MPDU). The PLCP preamble and PLCP header are always transmitted at 1 Mbps, and the MPDU can be sent at 1 or 2 Mbps.

The start of frame delimiter (SFD) contains information that marks the start of the PPDU frame. The SFD specified is common for all IEEE 802.11 DSSS radios.

The *signal field* indicates which modulation scheme should be used to receive the incoming MPDU. The binary value in this field is equal to the data rate multiplied by 100 kbps. Two modulation schemes, differential binary phase shift keying (DBPSK)—for 1 Mbps—and differential quadrature phase shift keying (DQPSK)—for 2 Mbps—are available.

The *service field* is reserved for future use. The *length field* indicates the number of microseconds necessary to transmit the MPDU. The MAC layer uses this field to determine the end of a PPDU frame.

The *CRC field* contains the results of a calculated frame check sequence from the sending station. The ITU CRC-16 error detection algorithm is used to protect the signal, service, and length field.

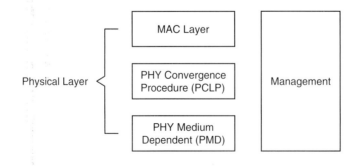

Figure 21.6 OSI model for IEEE 802.11 WLAN.

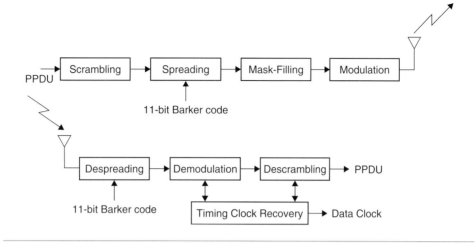

Figure 21.7 Transmit and receive DSSS PPDU.

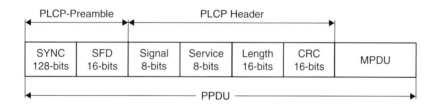

Figure 21.8 DSSS PHY PPDU frame.

The *SYNC field* is 128 bits (symbols) in length and contains a string of 1s which are scrambled prior to transmission. The receiver uses this field to acquire the incoming signal and to synchronize the receiver's carrier tracking and timing prior to receiving the SFD. The SFD field contains information to mark the start of the PPDU frame. The SFD specified is common for all IEEE 802.11 DSSS radios.

All information bits transmitted by the DSSS PMD are scrambled using a self-synchronizing 7-bit polynomial. An 11-bit Barker code $(1, -1, 1, 1, -1, 1, 1, 1, -1, -1, -1)$ is used for spreading. In the transmitter, the 11-bit Barker code is applied to a modulo-2 adder together with each of the information bits in the PPDU. The output of the modulo-2 adder results in a signal with a data rate that is 10 times higher than the information rate. The result in the frequency domain is a signal that is spread over a wide bandwidth at a reduced RF power level. Every station in the IEEE 802.11 network uses the same 11-bit sequence. At the receiver, the DSSS signal is convolved with the same 11-bit Barker code and correlated. The minimum requirement for processing gain (G_p) in North America and Japan is 10 dB.

Each DSSS PHY channel occupies 22 MHz of bandwidth and allows for three noninterfering channels spaced 25 MHz apart in the 2.4 GHz frequency band (see Figure 21.9). With this channel arrangement, a user can configure multiple DSSS networks to operate simultaneously in the same area. Table 21.5 lists the DSSS channels used in different parts of the world. Fourteen frequency

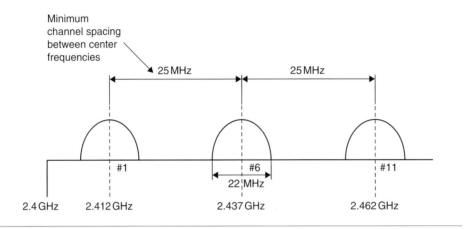

Figure 21.9   Channel spacing for IEEE 802.11 DSSS networks.

Table 21.5 DSSS channels for different parts of the world.

Channel number	Frequency GHz	North America	Europe	Spain	France	Japan
1	2.412	x	x			
2	2.417	x	x			
3	2.422	x	x			
4	2.427	x	x			
5	2.432	x	x			
6	2.437	x	x			
7	2.442	x	x			
8	2.447	x	x			
9	2.452	x	x			
10	2.457	x	x	x	x	
11	2.462	x	x	x	x	
12	2.467		x		x	
13	2.472		x		x	
14	2.483					x

Table 21.6 Maximum allowable transmit power.

Country	Power
North America	1000 mW
Europe	100 mW
Japan	10 mW/Hz

channels are defined for operation across the 2.4 GHz frequency band. In North America 11 frequencies are used ranging from 2.412 to 2.462 GHz. In Europe 13 frequencies are allowed between 2.412 and 2.472 GHz. In Japan only channels at the 2.483 GHz frequency are permitted.

The maximum allowable radiated power for DSSS PHY varies from region to region (refer to Table 21.6). The transmit power is directly related to the range that a particular implementation can achieve.

Example 21.1

A QPSK/DSSS WLAN is designed to transmit in the 902–928 MHz ISM band. The symbol transmission rate is 0.5 Mega symbols/sec. An orthogonal code with 16 symbols is used. A bit error rate of 10^{-5} is required. How many users can be supported by the WLAN? A sector antenna with a gain of 2.6 is used. Assume interference factor $\beta = 0.5$ to account for the interference from users in other cells and power control efficiency $\alpha = 0.9$. What is the bandwidth efficiency?

Solution

$$B_w = 928 - 902 = 26 \, \text{MHz}$$

$$R_b = R_s \log_2 16 = 2 \, \text{Mbps}$$

$$G_p = \frac{26}{2} = 13$$

$$P_b = 10^{-5} = \frac{1}{2} \, \text{erfc} \sqrt{\frac{E_b}{N_0}}$$

$$\therefore \frac{E_b}{N_0} = 10 \, \text{dB}$$

$$M = \frac{G_p}{(E_b/N_0)} \cdot \frac{1}{1 + \beta} \cdot \alpha \cdot \lambda = \frac{13}{10} \times \frac{1}{1 + 0.5} \times 2.6 \times 0.9 \approx 2$$

$$\eta = \frac{2 \times 2}{26} = 0.16 \, \text{bps/Hz}$$

FHSS PHY

In FHSS PHY, data transmission over media is controlled by the FHSS PMD sublayer as directed by the FHSS PLCP sublayer. The FHSS PMD takes the binary information bits from the whitened PSDU and converts them into RF signals by using carrier modulation and FHSS techniques (see Figure 21.10).

The format of the PPDU is shown in Figure 21.11. It consists of the PLCP preamble, PLCP header, and PLCP service data unit (PSDU). The PLCP preamble is used to acquire the incoming signal and synchronize the receiver's demodulator. The PLCP header contains information about PSDU from the sending physical layer. The PLCP preamble and header are transmitted at 1 Mbps.

The *sync field* contains a string of alternating 0s and 1s pattern and is used by the receiver to synchronize the receiver's packet timing and correct for frequency offsets.

The *SFD field* contains information marking the start of a PSDU frame. FHSS radios use a 0 0 0 0 1 1 0 0 1 0 1 1 1 1 0 1 bit pattern. The leftmost bit is transmitted first.

The *PLCP length word (PLW) field* specifies the length of the PSDU in octets and is used by the MAC layer to detect the end of a PPDU frame.

The *PLCP signaling field (PSF)* identifies the data rate of the whitened PSDU ranging from 1 to 4.5 Mbps in increments of 0.5 Mbps.

The *header error check field* contains the results of a calculated frame check sequence from the sending station. The ITU CRC-16 error detection algorithm is used to protect the PSF and PLW fields.

Data whitening is used for the PSDU before transmission to minimize DC bias on the data if long strings of 1s or 0s are contained in the PSDU. The PHY stuffs a special symbol every 4 octets of the PDSU in a PPDU frame. A 127-bit

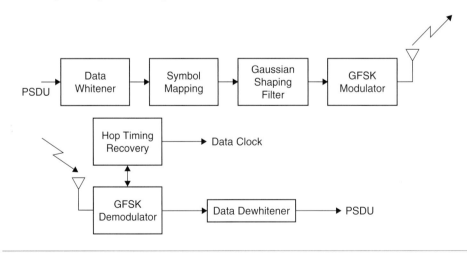

Figure 21.10   FHSS PHY transmitter and receiver.

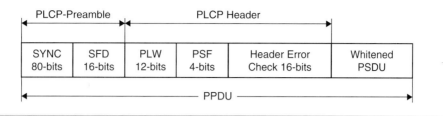

Figure 21.11 FHSS PHY PPDU frame.

sequence generator using the polynomial $x^7 + x^4 + 1$ and a 32/33 bias-suppression encoding algorithm are used to randomize and whiten the data.

The FHSS PMD uses two-level Gaussian frequency shift key (GMSK) modulation to transmit the PSDU at the basic rate of 1 Mbps. Four-level GFSK is an optimal modulation scheme defined in the standard that enables the whitened PSDU to be transmitted at a higher rate. The value in the PSF field of the PLCP header is used to determine the data rate of the PSDU.

In GFSK modulation, the frequency is shifted on either side of the carrier hop frequency depending on whether the binary symbol from the PSDU is either a 1 or 0. For two-level GFSK, a binary 1 represents the upper deviation frequency $(f_c + \Delta f)$ from the hopped carrier, and binary 0 represents the lower deviation frequency $(f_c - \Delta f)$, where f_c is the carrier hopped frequency and Δf the deviation frequency. The deviation frequency should be greater than 110 kHz for IEEE 802.11 FHSS radios. Four-level GFSK is similar to 2-level GFSK and is used to achieve a data rate of 2 Mbps in the same occupied frequency bandwidth. The symbol pairs (1 0, 1 1, 0 1, 0 0) generate four frequency deviations from the hopped carrier frequency, two upper and two lower.

A set of hop sequences is defined in IEEE 802.11 for use in the 2.4 GHz frequency band. The channels are evenly spaced across the band over a span of 83.5 MHz. In North America and Europe (except France and Spain) the number of hop channels is 79. The number of hop channels for Spain and France are 23 and 35, respectively. In Japan the required number of hop channels is 23. The hop channels are spaced uniformly across the 2.4 GHz frequency band occupying a bandwidth of 1 MHz. In North America and Europe (except Spain and France) the hop channels operate from 2.402 to 2.480 GHz and in Japan, 2.473 to 2.495 GHz. In Spain the hop channels operate from 2.447 to 2.473 GHz, and in France from 2.448 to 2.482 GHz. Channel 2 is the first hop channel located at the center frequency of 2.402 GHz and channel 95 is the last hop frequency channel at 2.495 GHz in the 2.4 GHz band.

Channel hopping is controlled by FHSS PMD. The FHSS PMD transmits the whitened PSDU by hopping from channel to channel in a pseudo-random fashion using one of the hopping sequences. In the United States, FHSS radios hop a minimum 2.5 hops per second for a minimum hop distance of 6 MHz.

The hopping sequences for IEEE 802.11 are grouped in hopping sets for worldwide operation: Set 1, Set 2, and Set 3. The sequences are selected when FHSS basic service set is configured for a WLAN. The hopping sets are designed to minimize interference between neighboring FHSS radios in a set. The following hop sets are valid IEEE 802.11 hopping sequences in North America and most of Europe (except Spain and France).

- **Set 1**: (0, 3, 6, 9, 12, 15, 18, 21, 24, 27, 30, 33, 36, 39, 42, 45, 48, 51, 54, 57, 60, 63, 66, 69, 72, 75)
- **Set 2**: (1, 4, 7, 10, 13, 16, 19, 22, 25, 28, 31, 34, 37, 40, 43, 46, 49, 52, 55, 58, 61, 64, 67, 70, 73, 76)
- **Set 3**: (2, 5, 8, 11, 14, 17, 20, 23, 26, 29, 32, 35, 38, 41, 44, 47, 50, 53, 56, 59, 62, 65, 68, 71, 74, 77)

802.11a—Orthogonal Frequency Division Multiplexing (OFDM)

The OFDM PHY provides the capability to transmit PSDU frames at multiple data rates up to 54 Mbps for a WLAN where the transmission of multimedia content is a consideration. The OFDM PHY defined for IEEE 802.11a is similar to the OFDM PHY specification of ETSI-HIPERLAN 2.

The PPDU is unique to the OFDM PHY. The PPDU frame consists of a PLCP preamble and signal and data fields (see Figure 21.12). The receiver uses the PLCP preamble to acquire the incoming OFDM signal and synchronize the demodulator. The PLCP header contains information about the PSDU from the sending OFDM PHY. The PLCP preamble and the signal fields are always transmitted at 6 Mbps, BPSK-OFDM modulated using a convolutional encoding rate $R = \frac{1}{2}$.

PLCP preamble field is used to acquire the incoming signal and train and synchronize the receiver. The PLCP preamble consists of 12 symbols, 10 of which are short symbols, and 2 long symbols. The short symbols are used to train the receiver's automatic gain control (AGC) and obtain a coarse estimate of the carrier frequency and the channel. The long symbols are used to fine-tune the frequency and channel estimates. Twelve subcarriers are used for the short symbols and 53 for the long. The training of an OFDM is accomplished in 16 μs. The PLCP preamble is BPSK-OFDM modulated at 6 Mbps.

The *signal* is a 24-bit field that contains information about the rate and length of the PSDU. The signal field is convolutional encoded rate $R = \frac{1}{2}$, BPSK-OFDM modulated. Four bits are used to encode the rate, eleven bits are used to define the length, one reserved bit, a parity bit, and six 0 tail bits (see Table 21.7).

The *length field* is an unsigned 12-bit integer to indicate the number of octets in the PSDU.

The *data field* contains the service field, PSDU, tail bits, and pad bits. A total of six tail bits containing 0s are appended to the PPDU to ensure that the convolutional encoder is brought back to the zero state.

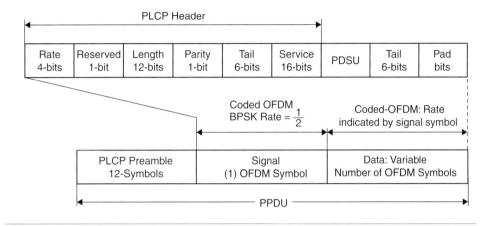

Figure 21.12   OFDM PLCP preamble, header, and PSDU.

Table 21.7 PSDU data rate.

Rate (Mbps)	Modulation	Coding rate	Signal bits
6	BPSK	$R = {}^1\!/_2$	1101
9	BPSK	$R = {}^3\!/_4$	1111
12	QPSK	$R = {}^1\!/_2$	0101
18	QPSK	$R = {}^3\!/_4$	0111
24	16-QAM	$R = {}^1\!/_2$	1001
36 (optional)	16-QAM	$R = {}^3\!/_4$	1011
48 (optional)	64-QAM	$R = {}^2\!/_3$	0001
54 (optional)	64-QAM	$R = {}^3\!/_4$	0011

All the bits transmitted by the OFDM PMD in the data portion are scrambled using a frame-synchronous 127-bit sequence generator. Scrambling is used to randomize the service, PSDU, pad bit, and data patterns, which may contain long strings of binary 1s or 0s. The tail bits are not scrambled.

All information contained in the service, PSDU, tail, and pad are encoded using convolutional encoding $R = {}^1\!/_2, {}^2\!/_3,$ or ${}^3\!/_4$ corresponding to the desired data rate. Puncture codes are used for the higher data rates.

In OFDM modulation, the basic principal of operation is to divide a high-speed binary signal to be transmitted into a number of lower data rate subcarriers. There are 48 data subcarriers and 4 carrier pilot subcarriers for a total of 52 nonzero subcarriers defined in IEEE 802.11a. Each lower data rate bit stream is used to modulate a separate subcarrier from one of the channels in the 5 GHz band. Intersymbol interference is generally not a concern for a lower speed carrier; however, the subchannels may be subjected to frequency selective fading. Therefore, bit interleaving and convolutional encoding is used to improve the BER performance. The scheme uses integer multiples of the first subcarrier, which are orthogonal to each other. Prior to transmission, the PPDU is encoded using a convolutional coded rate $R = {}^1/_2$, and the bits are reordered and bit interleaved for the desired data rate. Each bit is then mapped into a complex number according to the modulation type and subdivided into 48 data subcarriers and 4 pilot subcarriers. The subcarriers are combined using an inverse fast Fourier transform and transmitted. At the receiver, the carrier is converted back to a multicarrier lower data rate form using an fast frequency transform (FFT). The lower data subcarriers are combined to form the high rate PPDU. Figure 21.13 shows IEEE 802.11a OFDM PMD.

The 5 GHz frequency band is segmented into three 100 MHz bands for operation in the United States. The lower band ranges from 5.15–5.25 GHz, the middle band ranges from 5.25–5.35 GHz, and the upper band ranges from 5.725–5.825 GHz. The lower and middle bands accommodate 8 channels in a total bandwidth of 200 MHz and the upper band accommodates 4 channels in a 100 MHz bandwidth. The frequency channel center frequencies are spaced 20 MHz apart. The outermost channels of the lower and middle bands are centered 30 MHz from the outer edges. In the upper band the outermost channel centers are 20 MHz from the outer edges.

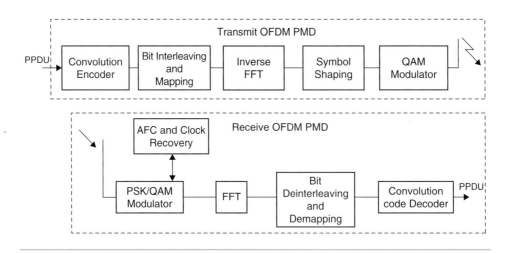

Figure 21.13   IEEE 802.11a transmit and receive OFDM PMD.

Table 21.8 Channel frequencies and channel numbers in the United States.

Regulatory domain	Frequency band	Channel number	Center frequencies
USA	Lower Band 5.15–5.25 GHz	36	5.180 GHz
		40	5.220 GHz
		44	5.220 GHz
		48	5.240 GHz
USA	Middle Band 5.25–5.35 GHz	52	5.26 GHz
		56	5.280 GHz
		60	5.300 GHz
		64	5.320 GHz
USA	Upper Band	149	5.745 GHz
		153	5.765 GHz
		157	5.785 GHz
		161	5.805 GHz

Table 21.9 Transmit power levels for North American operation.

Frequency band	Max. transmit power with 6 dBi antenna gain
5.150–5.250 GHz	40 mW (2.5 mW/MHz)
5.250–5.350 GHz	200 mW (12.5 mW/MHz)
5.725–5.825 GHz	800 mW (50 mW/MHz)

The channel frequencies and numbering defined in 802.11a start at 5 GHz and each channel is spaced 5 GHz apart (see Table 21.8).

Three transmit RF power levels are specified: 40 mW, 200 mW, and 800 mW (see Table 21.9). The upper band defines RF transmit power levels suitable for bridging applications while the lower band specifies a transmit power suitable for short-range indoor home and small office environments.

Example 21.2

Consider an FH/MFSK WLAN system, in which a pseudo-random noise (PN) generator is defined by a 20-stage linear feedback shift register with a maximal length sequence. Each state of the register dictates a new center frequency within the hopping band. The minimum step-size between center frequencies (hop to hop) is 200 Hz. The register clock rate is 2 kHz. 8-ary FSK modulation is used and the data rate is 1.2 kbps. (a) What is the hopping bandwidth? (b) What is the chip-rate? (c) How many chips are there in each data symbol? (d) What is the processing gain?

Solution

$\Delta f = 200\,\text{Hz}$

Minimum number of chips $= 20$

Number of tones contained in spreading bandwidth $(B_{ss}) = 2^{20} = 1,048,576$

$$1,048,576 = \frac{B_{ss}}{200}$$

$$\therefore B_{ss} = 209,715,200 = 209.715\,\text{MHz}$$

$$\text{Chip rate} = 1.2 \times 20 = 24\,\text{kchip/sec}$$

$$\text{Processing gain }(G_p) = \frac{209.715 \times 10^6}{1.2 \times 10^3} = 174,762.5$$

$$\text{Symbol rate} = \frac{1.2 \times 10^3}{3} = 400\,\text{symbols/sec}$$

$$\text{Number of chips per symbol} = \frac{24 \times 10^3}{400} = 60$$

Example 21.3

The IEEE 802.11a WLAN uses a 64-subchannel implementation of multicarrier modulation (MCM, i.e., OFDM). Forty-eight subcarriers are used for information transmission, 4 subcarriers for pilot tones are used for synchronization, and 12 are reserved. Each subchannel has a symbol rate of 250 kilo symbols per second (ksps). The occupied bandwidth is 20 MHz. Find the bandwidth of a subchannel. What is modulation efficiency? What is a user symbol rate? If 16-QAM modulation is used, what is the user data rate if the information bits are encoded with a rate of $^3/_4$? If the guard time between two transmitted symbols is 800 ns, what is the time utilization efficiency of the system?

Solution

- Total number of subcarriers $= 48 + 12 + 4 = 64$

- Bandwidth of subchannels $= \dfrac{20 \times 10^6}{64} = 312.5\,\text{kHz}$

- Modulation efficiency $= \dfrac{250}{312.5} = 0.8\,\text{symbols/sec/Hz}$

- User symbol transmission rate $= 48 \times 250 = 1.2\,\text{Msps}$

- User bit per symbol $= 4$ for 16-QPSK modulation

- User data rate $= \frac{3}{4} \times 4 \times 12 = 36\,\text{Mbps}$

- Symbol duration $= \dfrac{1}{250 \times 10^3} = 4000\,\text{ns}$

- Time utilization efficiency $= \dfrac{4000}{4800} = 0.83$

Example 21.4

A WLAN is required to have the minimum E_b/N_0 of 10 dB in an indoor office environment. The background noise at 2.4 GHz is -120 dBM. If the transmit power is 20 mW and there is no line-of-sight (NLOS) between the AP and mobile terminals, what is the coverage of the AP? Assume a data rate of 1 Mbps and channel bandwidth of 0.5 MHz. Use the joint technical commitee (JTC) path loss model in which for NLOS conditions in the indoor office environment, path loss at the first meter (A) is 37.7 dB and path loss exponent, γ, is 3.3. Assume power loss from the floors to be 19 dB and a shadow effect of 10 dB.

Solution

- Required $\dfrac{S}{N} = \dfrac{E_b}{N_0} \times \dfrac{R}{B_w} = 10 \times \dfrac{1}{0.5} = 20\,\text{dB}$
- Transmit power $= 10 \log 20 = 13\,\text{dBm}$

- Receiver sensitivity $= -120 + 20 = -100\,\text{dBm}$

- Maximum allowable path loss $= 13 - (-100) = 113\,\text{dB}$

- $L_p = A + 10\gamma \log d + L_f(n) + X_\sigma$

where:

A = path loss at the first meter
γ = path loss exponent
L_f = function relating power loss with number of floors, n
X_σ = lognormally distributed random variable representing the shadow effect
d = distance between AP and mobile terminal

$$113 = 37.7 + 33 \log d + 19 + 10$$

$$\therefore d = 25.3 \text{ meters}$$

21.5.3 IEEE 802.11 Data Link Layer

The data link layer within 802.11 consists of two sublayers: logical link control (LLC) and media access control (MAC). 802.11 uses the same 802.2 LLC and 48-bit addressing as the other 802 LAN, allowing for simple bridging from wireless to IEEE wired networks, but the MAC is unique to WLAN. The sublayer above MAC is the LLC, where the framing takes place. The LLC inserts certain fields in the frame such as the source address and destination address at the head end of the frame and error handling bits at the end of the frame.

The 802.11 MAC is similar in concept to 802.3, in that it is designed to support multiple users on a shared medium by having the sender sense the medium before accessing it. For the 802.3 Ethernet LAN, the carrier sense multiple access with collision detection (CSMA/CD) protocol regulates how Ethernet stations establish access to the network and how they detect and handle collisions that occur when two or more devices try to simultaneously communicate over the LAN. In an 802.11 WLAN, collision detection is not possible due to the *near/far* problem (see Chapter 11). To detect a collision, a station must be able to transmit and listen at the same time, but in radio systems the transmission drowns out the ability of a station to hear a collision.

21.5.4   IEEE 802.11 Medium Access Control

Wireless local area networks operate using a shared, high bit rate transmission medium to which all devices are attached and information frames relating to all calls are transmitted. MAC sublayer defines how a user obtains a channel when he or she needs one.

MAC schemes include random access, order access, deterministic access, and mixed access. The random access MAC protocols are: ALOHA (asynchronous, slotted), carrier-sense multiple-access (CSMA) (CSMA/collision-detection (CD), CSMA/collision-avoidance (CA), non-persistent, and p-persistent). The maximum throughput of slotted ALOHA protocol is about 36% of the data rate of the channel (see Chapter 5). It is simple, but not very efficient. Most WLANs implement a random access protocol, CSMA/CA with some modification, to deal with the *hidden node* problem. The CSMA peaks at about 60%. When the traffic becomes heavy, it degrades badly. The way of dealing with that situation is to use p-persistent. Most mobile data networks also use random access protocol, usually one that is simpler than CSMA, namely slotted ALOHA. Table 21.10 provides a comparison of MAC schemes for wireless networks.

Deterministic MAC schemes improve throughput and response time when traffic is heavy. They offer the guaranteed bandwidth for isochronous traffic. In mixed cases such as CSMA/TDMA, the frame is divided into a random access part and a reserved part. When the traffic is light, it is left to be mostly random. When the traffic is heavy and throughput is in danger of declining or if a node requires isochronous bandwidth, the control point allocates bandwidth deterministically. CSMA/TDMA approaches CSMA performance under light traffic, so it has fast access time. It approaches TDMA performance when the traffic becomes heavy, so its throughput can rise close to 100% of the data rate.

IEEE 802.11 uses a modified protocol known as *carrier sense multiple access with collision avoidance* (CSMA/CA) or distributed coordination function (DCF).

Table 21.10 Comparison of MAC access schemes in wireless networks.

Access	Protocols	Characteristics
Random	CSMA	• Under light load—fast response time
		• Under heavy load—throughput declines
		• Simple to implement
Deterministic	FDMA	• Able to provide guaranteed bandwidth
	TDMA	• Larger average delay compared to random access
	CDMA	• Smaller delay variance
Mixed	CSMA/TDMA	• Under light load—fast response time
		• Under heavy load—throughput approaches TDMA
		• Higher overhead compared to random and deterministic access

CSMA/CA attempts to avoid collisions by using *explicit packet acknowledgment* (ACK), which means an ACK packet is sent by the receiving station to confirm that the data packet arrived intact.

The CSMA/CA protocol is very effective when the medium is not heavily loaded since it allows stations to transmit with minimum delay. But there is always a chance of stations simultaneously sensing the medium as being free and transmitting at the same time, causing a collision. These collisions must be identified so that the MAC layer can retransmit the packet by itself and not by the upper layers, which would cause significant delay. In the Ethernet with CSMA/CD the collision is recognized by the transmitting station, which goes into a retransmission phase based on an exponential random backoff algorithm. While these collision detection mechanisms are a good idea on a wired LAN, they cannot be used on a WLAN environment for two main reasons:

- Implementing a collision detection mechanism would require the implementation of a full duplex radio capable of transmitting and receiving at the same time, an approach that would increase the cost significantly.

- In a wireless environment we cannot assume that all stations hear each other (which is the basic assumption of the collision detection scheme), and the fact that a station wants to transmit and senses the medium as free does not necessarily mean that the medium is free around the receiver area.

To overcome these problems, the 802.11 uses a CA mechanism together with a positive ACK. The MAC layer of a station wishing to transmit senses the medium. If the medium is free for a specified time, called *distributed inter-frame space* (DIFS), then the station is able to transmit the packet; if the medium is busy (or becomes busy during the DIFS interval) the station defers using the *exponential backoff algorithm*.

This scheme implies that, except in cases of very high network congestion, no packets will be lost because retransmission occurs each time a packet is not acknowledged. This entails that all packets sent will reach their destination in sequence.

The 802.11 MAC layer provides for two other robustness features: *cycle redundancy check (CRC) checksum* and *packet fragmentation*. Each packet has a CRC checksum calculated and attached to ensure that the data was not corrupted in transmit. This is different from the Ethernet, where higher-level protocols such as TCP handle error checking.

Packet fragmentation allows large packets to be segmented into smaller units when sent over the medium. This is useful in very congested environments or when interference is a factor, since large packets have a better chance of being corrupted. This technique reduces the need for retransmission in many cases and improves overall wireless network performance. The MAC layer is responsible for reassembling fragments received, rendering the process transparent to higher-level protocols. The following are some of the reasons it is preferable to use smaller packets in a WLAN environment:

- Due to higher BER of a radio link, the probability of a packet getting corrupted increases with packet size.

- In case of corrupted packets (either due to collision or interference), smaller packets cause less overhead.

- On an FHSS system the medium is interrupted periodically for hopping. With smaller packets the chance that the transmission will be postponed after dwell time is reduced.

A simple *send-and-wait* algorithm is used at the MAC sublayer. In this mechanism the transmitting station is not allowed to transmit a new packet until one of the following happens:

- Receives an ACK for the packet, or

- Decides that packet was retransmitted too many times and drops the whole frame.

Exponential Backoff Algorithm

Backoff is a scheme commonly used to resolve contention problems among different stations wishing to transmit data at the same time. When a station goes into the backoff state, it waits an additional, randomly selected number of time slots (in 802.11b a slot has a 20 μs duration and the random number must be greater than 0 and smaller than a maximum value referred to as the contention window (CW)). During the wait, the station continues sensing the medium to check whether it remains free or another transmission begins. At the end of its contention window, if the medium is still free the station can send its frame. If during the contention window another station begins transmitting data, the back-off counter is frozen and counting down starts again when the channel returns to the idle state.

There is a problem related to the CW dimension. With a small CW, if many stations attempt to transmit data at the same time it is very possible that some of them may have the same backoff interval. This means that there will continu-ously be collisions, with serious effects on the network performance. On the other hand, with a large CW, if few stations wish to transmit data they will likely have long backoff delays resulting in the degradation of the network performance. The solution is to use an exponentially growing CW size. It starts from a small value ($CW_{min} = 31$) and doubles after each collision, until it reaches the maximum value CW_{max} ($CW_{max} = 1023$). In 802.11 the backoff algorithm must be executed in three cases:

- When the station senses the medium is busy before the first transmission of a packet
- After each retransmission
- After a successful transmission

This is necessary to avoid a single host wanting to transmit a large quantity of data, occupying the channel for too long a period, and denying access to all other stations. The backoff mechanism is not used when the station decides to transmit a new packet after an idle period and the medium has been free for more than the DIFS (see Figure 21.14).

$$\text{The transmission time for a data frame} = \left(\text{PLCP} + \frac{D}{R}\right)\mu s$$

where:

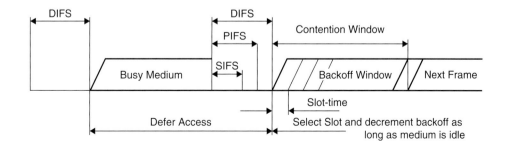

Figure 21.14   CSMA/CA in IEEE 802.11b.

PLCP = the time required to transmit the physical layer convergence protocol (PLCP)

D = the frame size

R = the channel bit rate

$$\text{CSMA/CA packet transmission time} = \text{BO} + \text{DIFS} + 2\text{PLCP} + \frac{D}{R} + \text{SIFS} + \frac{A}{R}\,\mu s$$

where:

A = the ACK frame size

BO = the backoff time

DIFS = the distributed inter-frame space

SIFS = the short inter-frame space

The loss of performance strongly depends on the packet size and data rate, but a 30% loss is more than likely to occur. The smaller the packets, the larger will be the impact of CSMA/CA on network performance. To evaluate the performance impact of CSMA/CA it is important to know how the various inter-frame spaces are defined. The 802.11 standard defines the following four inter-frame spaces to provide different priorities.

- Short inter-frame space (SIFS): It is used to separate transmissions belonging to a single dialog (e.g., fragment-ACK), and is the minimum inter-frame space. There is always at most one single station to transmit at any given time, therefore giving it priority over all other stations. This value is fixed per PHY and is calculated in such a way that the transmitting station will be able to switch back to receive mode and be capable of decoding the incoming packet. For the 802.11 DSSS PHY the value is $10\,\mu s$.

- Point coordinate inter-frame space (PIFS): It is used by the AP to gain access to the medium before any other station. This value is SIFS plus a slot time (i.e., $30\,\mu s$).

- Distributed inter-frame space (DIFS): It is the inter-frame space used for a station willing to start a new transmission. It is calculated as PIFS plus one slot time (i.e., $50\,\mu s$).

- Extended inter-frame space (EIFS): It is the longer inter-frame space used by a station that has received a packet which it could not understand. This is required to prevent the station (which could not understand the duration information for the virtual carrier sense) from colliding with a future packet belonging to the current dialog.

Hidden and Exposed Node Problem

Another major MAC layer problem specific to a WLAN is the *hidden node* issue, in which two stations on opposite sides of an AP can both hear activity from an AP, but not from each other, usually due to distance or an obstruction (see

Figure 21.15a). To solve this problem, 802.11 specifies an optional *request to send/clear to send* (RTS/CTS) protocol at the MAC layer. When this feature is in use, a sending station transmits an RTS and waits for the AP to reply with a CTS. Since all stations in the network can hear the AP, the CTS causes them to delay any intended transmissions, allowing the sending station to transmit and receive a packet acknowledgment without any chance of collision. Since RTS/CTS adds additional overhead to the network by temporarily reserving the medium, it is typically used only on the largest-sized packets, for which transmission would be expensive from a bandwidth standpoint. This mechanism reduces the probability of a collision on the receiver area by a station that is hidden from the transmitter to the short duration of the RTS transmission, because all stations hear the CTS and make the medium busy until the end of the transaction. The duration information on the RTS also protects the transmitter area from collisions during the ACK (from stations that are out of range of the acknowledged station). It should also be noted that, due to the fact that RTS and CTS are short frames, the mechanism also reduces the overhead of collisions, since these frames are recognized faster than if the whole packet were to be transmitted. The mechanism is controlled by a parameter called RTS threshold, which, if used, must be set on both the AP and the client side.

The time required to transmit a frame, taking into account the RTS/CTS four-way handshake is given as:

$$BO + DIFS + 4PLCP + \frac{RTS + CTS + D + A}{R} + 3SIFS\,\mu s$$

where:

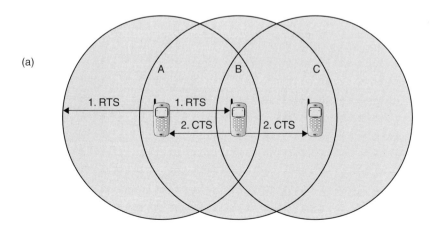

(a)

Figure 21.15(a)   Hidden and exposed node problem.

     BO = backoff time (μs)
  DIFS = distributed inter-frame space (50 μs)
  PLCP = time required to transmit physical layer convergence protocol (μs)
   RTS = request to send frame size (bits)
   CTS = clear to send frame size (bits)
      D = frame size (bits)
      A = acknowledgement frame size (bits)
      R = channel bit rate (bits per second)
  SIFS = short inter-frame space (10 μs)

We refer to Figure 21.15b and assume that node B and C intend to transmit data only without receiving data. When node C is transmitting data to node D, node B is aware of the transmission. This is because node B is within the radio coverage of node C. Without exchanging RTS and CTS frames, node B will not initiate data transmission to node A because it will detect a busy medium. The transmission between node A and node B, therefore, is blocked even if both of them are idle. This is referred as the *exposed node problem*. To alleviate this problem, a node must wait a random backoff time between the two consecutive new packet transmission times.

21.5.5   IEEE 802.11 MAC Sublayer

In IEEE 802.11, the MAC sublayer is responsible for asynchronous data service (e.g., exchange of MAC service data units (MSDUs)), security service (confidentiality, authentication, access control in conjunction with layer management), and MSDU ordering.

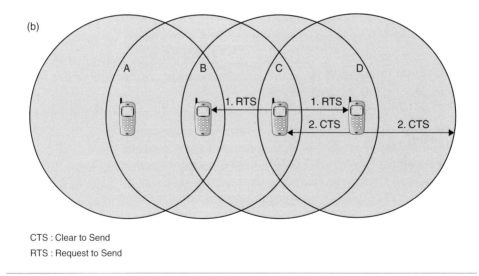

CTS : Clear to Send
RTS : Request to Send

Figure 21.15(b)   Hidden and exposed node problem.

The MAC sublayer accepts MSDUs from higher layers in the protocol stack to send them to the equivalent layer of the protocol stack in another station. The MAC adds information to the MSDU in the form of headers and trailers to generate a MAC protocol data unit (MPDU). The MPDU is then passed to the physical layer to be sent over the wireless medium to other stations. The MAC may fragment MSDUs into several frames to increase the probability of each individual frame being delivered successfully. The MAC frame contains addressing information, information to set the network allocation vector (NAV), and a frame check sequence to verify the integrity of the frame. The general IEEE 802.11 MAC frame format is shown in Figure 21.16.

The MAC frame format contains four address fields. Any particular frame type may contain one, two, three, or four address fields. The address format in IEEE 802.11-1997 is a 48-bit address, used to identify the source and destination of MAC addresses contained in a frame, as IEEE 802.3. In addition to source address (SA) and destination address (DA), three additional address types are defined: the transmitter address, the receiver address (RA), and the basic service set identifier (BSSID). The BSSID is a unique identifier for a particular basic service set of the IEEE 802.11 WLAN. In an infrastructure basic service set, the BSSID is the MAC address of the AP.

The *transmitter address* is the address of the MAC that transmitted the frame onto the wireless medium. This address is always an individual address. The transmitter address is used by stations receiving a frame to identify the station to which any responses in the MAC frame exchange protocol will be sent.

The *receiver address* (*RA*) is the address of the MAC to which the frame is sent over the wireless medium. This address may be either an individual or group address.

The *source address* (*SA*) is the address of the MAC that originated the frame. This address is always an individual address. This address does not always match the address in the transmitter address field because of the indirection that is performed by the distribution system of an IEEE 802.1 WLAN. It is the SA field that should be used to identify the source of a frame when indicating that a frame has been received to higher layer protocols.

The *destination address* (*DA*) is the address of the final destination to which the frame is sent. This address may be either an individual or group address. This address does not always match the address in the RA field because of the indirection that is performed by the DS.

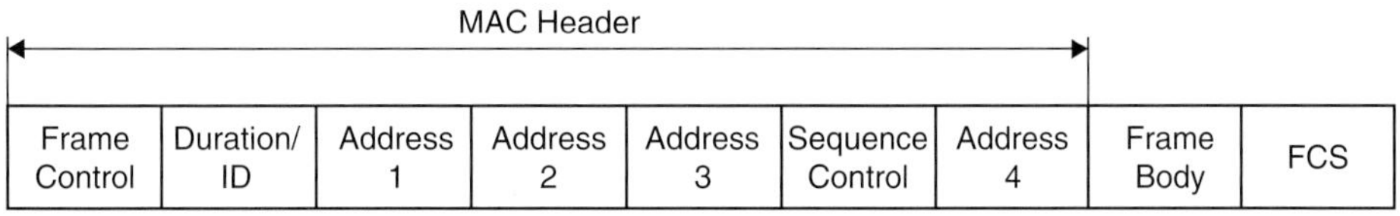

Figure 21.16   IEEE 802.11 MAC frame format.

The *sequence control field* is a 16-bit field that consists of two subfields. The subfields are a 4-bit fragment number and a 12-bit sequence number. This field is used to allow a receiving station to eliminate duplicate received frames. The *sequence number subfield* contains numbers assigned sequentially by the sending station to each MSDU. This sequence number is incremented after each assignment and wraps back to zero when incremented from 4095. The sequence number for a particular MSDU is transmitted in every data frame associated with the MSDU. It is constant over all transmissions and retransmissions of the MSDU. If the MSDU is frangmented, the sequence number of the MSDU is sent with each frame containing a fragment of the MSDU. The *fragment number subfield* contains a 4-bit number assigned to each fragment of an MSDU. The first, or only, fragment of an MSDU is assigned a fragment number of zero. Each successive fragment is assigned a sequentially incremented fragment number. The fragment number is constant in all transmissions or retransmissions of a particular fragment.

The *frame body field* contains the information specific to the particular data or management frames. This field is variable in length. It may be as long as 2034 bytes without encryption, or 2312 bytes when the frame body is encrypted. The value of 2304 bytes as the maximum length of this field was chosen to allow an application to send 2048-byte pieces of information, which can be encapsulated by as many as 256 bytes of upper layer protocol headers and trailers.

The *frame check sequence (FCS) field* is 32 bits in length. It contains the result of applying the C-32 polynomial to the MAC header and frame body.

The original 802.11 standard suffers from some serious limitations which prevent it from becoming a leading technology and a serious alternative to wired LAN. The following are some of the problems:

- Low data rate: The 802.11 protocol imposes very high overhead to all packets that reduce real data rate significantly
- No QoS guarantees

Several extensions to the basic 802.11 standard have been introduced by IEEE to provide higher data rates or QoS guarantees. 802.11a, 802.11b, and 802.11g focus on higher data rates whereas 802.11e is aimed at providing QoS guarantees.

21.6   Joining an Existing Basic Service Set

The 802.11 MAC sublayer is responsible for how a station associates with an AP. When an 802.11 station enters the range of one or more APs, it chooses an AP to associate with (also known as joining a basic service set), based on signal strength and observed packet error rates. Once accepted by the AP, the station tunes to the radio channel to which the AP is set. Periodically it surveys all 802.11 channels in order to access whether a different AP would provide it with better performance characteristics. If it determines that this is the case, it reassociates with the new AP, tuning to the radio channel to which that AP is set. Reassociating usually occurs because the wireless station has physically moved away from the original AP, causing the signal to be weakened. In other cases, reassociating occurs due to changes in radio characteristics in the building, or due to high network traffic on the original AP. In the latter case this function is known as load balancing, since its primary function is to distribute the total WLAN load most efficiently across the available wireless infrastructure.

The process of dynamically associating and reassociating with APs allows network managers to set up WLANs with very broad coverage by creating a series of overlapping 802.11b cells throughout a building or across a campus. To be successful, the IT manager ideally will employ channel reuse, taking care to set up each access point on an 802.11 DSSS channel that does not overlap with a channel used by a neighboring AP (see Figure 21.17).

As noted above, while there are 14 partially overlapping channels specified in 802.11 DSSS, there are only 3 channels that do not overlap at all and these are the best to use for multicell coverage (refer to Table 21.5). If two APs are in range of one another and are set to the same or partially overlapping channels, they may cause some interference for one another, thus lowering the total available bandwidth in the area of overlap.

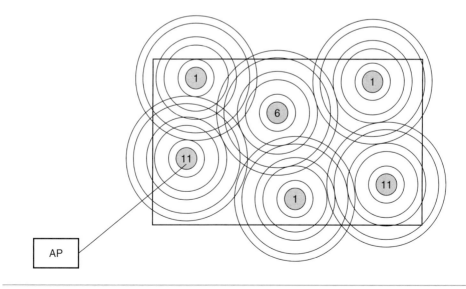

Figure 21.17 DSSS channel without overlap with a channel used by neighbor AP.

When a station wishes to access an existing basic service set, it needs to get synchronization information from the AP. The station can get this information in one of two ways:

- *Passive scanning:* In this case the station waits to receive a beacon frame from the AP. The beacon frame is a frame sent out periodically by the AP containing synchronization information.

- *Active scanning:* In this case the station tries to locate an AP by transmitting *probe request frame*, and waits for *probe response* from the AP.

A method is chosen according to the power consumption/performance trade-off. Once the station has located an AP, and decides to join its basic service set, it goes through the authentication process. This is the interchange of information between the AP and the station, where each side proves the knowledge of a given password. This is necessary because WLANs have limited physical security to prevent unauthorized access. The goal of authentication is to provide access control equal to a wired LAN. The authentication service provides a mechanism for one station to identify another station. Without this proof of identity, the station is not allowed to use the WLAN for data delivery. All 802.11 stations, whether they are part of an independent basic service set or extended service set (ESS) network, must use the authentication process prior to communicating with another station. IEEE 802.11 uses authentication services defined in IEEE 802.11i.

Once the station is authenticated, it then starts the association process. It is used to make a logical connection between a mobile station and an AP and to exchange information about the station and basic service set/capabilities, which allows the distribution system service (DSS) to know about the current position of the station. This is necessary so that the AP can know where and how to deliver data to the mobile station. A station is allowed to transmit data frames through the AP only after the association process is completed.

When a station determines that the existing signal is poor, it begins scanning for another AP. This can be done by passively listening or actively probing each channel and waiting for a response. Once information has been received, the station selects the most appropriate signal and sends an association request to the new AP. If the new AP sends an association response, the client connects to the new AP. This feature is known as *roaming* and is similar to the cellular handover, with two main differences:

- On a packet-based LAN system, the transition from cell to cell may be performed between packet transmissions as opposed to a cellular system where the transition may occur during a phone conversation. This makes WLAN roaming a little easier.

- On a voice system, a temporary disconnection may not affect the conversation, while in a packet-based data system it significantly reduces performance because retransmission is performed by the upper layer protocols.

The 802.11 standard does not define how roaming should be performed, but defines the basic tools including active/passive scanning, and a re-association process, in which a station roaming from one AP to another becomes associated with the new AP.

The 802.11 standard also provides a mechanism to remove a station from the basic service set. The process is called de-authentication. De-authentication is used to prevent a previously authenticated station from using the network any further. Once a station is de-authenticated, it is no longer able to access the WLAN without performing the authentication process again. De-authentication is a notification and cannot be refused. When a station wishes to be removed from a basic service set, it can send a de-authentication management frame to the associated AP. An AP could also de-authenticate a station by sending a de-authentication frame to the station.

21.7 Security of IEEE 802.11 Systems

The IEEE 802.11 provides for MAC access control and encryption mechanisms. Earlier, the *wireline equivalent privacy* (WEP) algorithm was used to encrypt messages. WEP uses a Rivest Cipher 4 (RC4) pseudo-random number generator with two key structures of 40 and 128 bits. Because of the inherent weaknesses of the WEP, the IEEE 802.11i committee developed, a new encryption algorithm (see Chapter 13) and worked on the enhanced security and authentication mechanisms for 802.11 systems.

For access control, ESSID (also known as a WLAN service area ID) is programmed into each AP and is required knowledge in order for a wireless client to associate with an AP. In addition, there is provision for a table of MAC addresses called an *access control list* to be included in the AP, restricting access to stations whose MAC addresses are not on the list.

Beyond layer 2, 802.11 WLANs support the same security standards supported by other 802 LANs for access control (such as network operating system logins) and encryption (such as IPSec or application-level encryption). These higher-level technologies can be used to create end-to-end secure networks encompassing both wired LAN and WLAN components, with the wireless piece of the network gaining additional security from the IEEE 802.11i feature set.

21.8 Power Management

Power management is necessary to minimize power requirements for battery powered portable mobile units. The standard supports two power-utilization modes, called *continuous aware mode* and *power save polling mode*. In the former, the radio is always on and draws power, whereas in the latter, the radio is dozing with the AP and is queuing any data for it.

A power saver mode or sleep mode is defined when the station is not transmitting in order to save battery power. However, critical data transmissions cannot be missed. Therefore APs are required to have buffers to queue messages.

Sleeping stations are required to periodically wake up and retrieve messages from the AP. Power management is more difficult for peer-to-peer IBSS configurations without central AP. In this case, all stations in the IBSS must be awakened when the periodic beacon is sent. Stations randomly handle the task of sending out the beacon. An announcement traffic information message window commences. During this period, any station can go to sleep if there is no announced activity for it during this short period.

21.9 IEEE 802.11b—High Rate DSSS

In September 1999 IEEE ratified the 802.11b *high rate* amendment to the standard, which added two higher speeds (5.5 and 11 Mbps) to 802.11. The key contribution of the 802.11b addition to the WLAN standard was to standardize the physical layer support to two new speeds, 5.5 and 11 Mbps. To accomplish this, DSSS was selected as the sole physical layer technique for the standard, since frequency hopping cannot support the higher speeds without violating current FCC regulations. The implication is that the 802.11b system will interoperate with 1 Mbps and 2 Mbps 802.11 DSSS systems, but will not work with 1 Mbps and 2 Mbps FHSS systems.

The original version of the 802.11 specifies in the DSSS standard an 11-bit chipping, called a *Barker sequence*, to encode all data sent over the air. Each 11-chip sequence represents a single data bit (1 or 0), and is converted to a waveform, called a symbol, that can be sent over the air. These symbols are transmitted at a 1 million symbols per second (Msps) rate using binary phase shift keying (BPSK). In the case of 2 Mbps, a more sophisticated implementation based on quadrature phase shift keying (QPSK) is used. This doubles the data rate available in BPSK, via improved efficiency in the use of the radio bandwidth.

To increase the data rate in 802.11b standard, advanced coding techniques are employed. Rather than the two 11-bit Barker sequences, 802.11b specifies complementary code keying (CCK). CCK allows for multichannel operation in the 2.4 GHz band by using existing 1 and 2 Mbps DSSS channelization schemes. CCK consists of a set of sixty-four 8-bit code words. As a set, these code words have unique mathematical properties that allow them to be correctly distinguished from one another by a receiver even in the presence of substantial noise and multipath interference. The 5.5 Mbps rate uses CCK to encode 4 bits per carrier, while the 11 Mbps rate encodes 8 bits per carrier. Both speeds use QPSK modulation and a signal at 1.375 Msps. This is how the higher data rates are obtained. Table 21.11 lists the specifications.

To support very noisy environments as well as extended ranges, 802.11b WLANs use dynamic rate shifting, allowing data rates to be automatically adjusted to compensate for the changing nature of the radio channel. Ideally, users connect at a full 11 Mbps rate. However, when devices move beyond the optimal range for 11 Mbps operation, or if substantial interference is present, 802.11b devices will

Table 21.11 802.11b data rate specification.

Data rate	Code length	Modulation 6-1	Symbol rate	Bits/symbol
1 Mbps	11 (Barker Sequence)	BPSK	1 Msps	1
2 Mbps	11 (Barker Sequence)	QPSK	1 Msps	2
5.5 Mbps	8 (CCK)	QPSK	1.375 Msps	4
11 Mbps	8 (CCK)	QPSK	1.375 Msps	8

transmit at lower speeds, falling back to 5.5, 2, and 1 Mbps. Likewise, if a device moves back within the range of a higher-speed transmission, the connection will automatically speed up again. Rate shifting is a physical layer mechanism transparent to the user and upper layers of the protocol stack.

21.10   IEEE 802.11n

In response to growing market demand for higher-performance WLANs, the IEEE formed the task group 802.11n. The scope of this task group is to define modifications to the physical and MAC layer to deliver a minimum of 100 Mbps throughput at the MAC service access point (SAP).

802.11n employs an evolutionary philosophy reusing existing technologies where practical, while introducing new technologies where they provide effective performance improvements to meet the needs of evolving applications. Reuse of legacy technologies such as OFDM, forward error correction coding, interleaving, and quadrature amplitude modulation mapping have been maintained to keep costs down and ease backward compatibility.

There are three key areas that need to be considered when addressing increases in WLAN performance. First, improvements in radio technology are needed to increase the physical transfer rate. Second, new mechanisms implementing the effective management of enhanced physical layer performance modes must be developed. Third, improvements in data transfer efficiency are needed to reduce the improvements achieved with an increase in physical transfer rate.

The emerging 802.11n specification differs from its predecessors in that it provides for a variety of optional modes and configurations that dictate different maximum raw data rates. This enables the standard to provide baseline performance parameters for all 802.11n devices, while allowing manufacturers to enhance or tune capabilities to accommodate different applications and price points. WLAN hardware does not need to support every option to be compliant with the standard.

The first requirement is to support an OFDM implementation that improves upon the one employed in 802.11a/g standards, using a higher maximum code rate and slightly wider bandwidth. This change improves the highest attainable raw data rate to 65 Mbps from 54 Mbps in the existing standards.

Multi-input multi-output (MIMO) technology (see Chapter 23) is used in 802.11n to evolve the existing OFDM physical interface presently implemented with legacy 802.11a/g. MIMO harnesses multipath with a technique known as space division multiplexing (SDM). The transmitting WLAN device splits a data stream into multiple parts, called spatial streams, and transmits each spatial stream through separate antennas to corresponding antennas on the receiving end. The current 802.11n provides for up to four spatial streams, even though compliant hardware is not required to support that many.

Doubling the number of spatial streams from one to two effectively doubles the raw data rate. There are trade-offs, however, such as increased power consumption and, to a lesser extent, cost. The 802.11n specification includes an MIMO power-save mode, which mitigates power consumption by using multiple paths only when communication would benefit from the additional performance. The MIMO power-save mode is a required feature in the 802.11n specification.

There are two features in the specification that focus on improving MIMO performance: (1) beam-forming and (2) diversity. Beam-forming is a technique that focuses radio signals directly on the target antenna, thereby improving range and performance by limiting interference. Diversity exploits multiple antennas by combining the outputs of or selecting the best subset of a larger number of antennas than required to receive a number of spatial streams. The 802.11n specification supports up to four antennas.

Another optional mode in the 802.11n effectively doubles data rates by doubling the width of a WLAN communications channel from 20 to 40 MHz. The primary trade-off is fewer channels available for other devices. In the case of the 2.4-GHz band, there is enough room for three nonoverlapping 20-MHz channels. A 40-MHz channel does not leave much room for other devices to join the network or transmit in the same air space. This means intelligent, dynamic management is critical to ensuring that the 40-MHz channel option improves overall WLAN performance by balancing the high-bandwidth demands of some clients with the needs of other clients to remain connected to the network.

One of the most important features in the 802.11n specification to improve mixed-mode performance is aggregation. Rather than sending a single data frame, the transmitting client bundles several frames together. Thus, aggregation improves efficiency by restoring the percentage of time that data is being transmitted over the network.

The 802.11n specification was developed with previous standards in mind to ensure compatibility with more than 200 million Wi-Fi (802.11b) devices currently in use. An 802.11n access point will communicate with 802.11a devices on the 5-GHz band as well as 802.11b and 802.11g hardware on 2.4-GHz frequencies. In addition to basic interoperability between devices, 802.11n provides for greater network efficiency in mixed mode over what 802.11g offers.

Table 21.12 lists the major components of 802.11n. Table 21.13 compares the primary IEEE 802.11 specifications.

Table 21.12 Major components of IEEE 802.11n.

Feature	Definition	Status
Better OFDM	Supports wider bandwidth and higher code rate to bring maximum data rate to 65 Mbps	Mandatory
Space-Division Multiplexing (SDM)	Improve performance by parsing data into multiple streams transmitted through multiple antennas	Optional for up to four antennas
Diversity	Exploits the existence of multiple antennas to improve range and reliability. Typically employed when the number of antennas on the receiving end is higher than the number of streams being transmitted.	Optional for up to four antennas
MIMO Power Save	Limits power consumption penalty of MIMO by utilizing multiple antennas only on as-needed basis.	Required
40 MHz Channels	Effectively doubles data rates by doubling channel width from 20 MHz to 40 MHz	Optional
Aggregation	Improves efficiency by allowing transmission bursts of multiple data packets between overhead communication	Required
Reduced Inter-frame Spacing (RIFS)	Designed to improve efficiency by providing a shorter delay between OFDM transmissions than in 802.11a or g.	Required
Greenfield Mode	Improves efficiency by eliminating support for 802.11a/b/g devices in an all 802.11n network	Optional

Table 21.13 Primary IEEE 802.11 specifications and their comparisons.

	802.11a	802.11b	802.11g	802.11n
Approval date	July 1999	July 1999	June 2003	August 2006
Maximum data rate	54 Mbps	11 Mbps	54 Mbps	600 Mbps
Modulation	OFDM	DSSS or CCK	DSSS or CCK or OFDM	DSSS or CCK or OFDM
RF band	5 GHz	2.4 GHz	2.4 GHz	2.4 GHz or 5 GHz
Number of spatial streams	1	1	1	1, 2, 3, or 4
Channel width	20 MHz	20 MHz	20 MHz	20 MHz or 40 MHz

Because it promises far greater bandwidth, better range, and reliability, 802.11n is advantageous in a variety of network configurations. And as emerging networking applications take hold in the home, a growing number of consumers will view 802.11n not just as an enhancement to their existing network, but as a necessity. Some of the current and emerging applications that are driving the need for 802.11n are voice over IP (VoIP), streaming video and music, gaming, and network attached storage.

21.11 Other WLAN Standards

The high performance radio local area network (HIPERLAN) committee in ETSI, referred to as Radio and Equipment Systems (RES) 10, worked with the European Conference of Postal and Telecommunications Administration (CEPT, a committee of PTT and other administration representatives) to identify its target spectrum. The CEPT identified the 5.15–5.25 GHz band (this allocation allows three channels), with an optional expansion to 5.30 GHz (extension to five channels). Any country in the CEPT area (which covers all of Europe, as well as other countries that implement CEPT recommendations) may decide to implement this recommendation. Most of the CEPT countries permit HIPERLAN systems to use this 5 GHz band. In the United States, the U.S. Federal Communications Commission (FCC) followed the European model roughly. The unlicensed National Information Infrastructure band (U-NII) covers approximately 300 MHz in three different bands between 5.1 and 5.8 GHz. The regulators in Japan are likely to align with the 5 GHz band also. In addition, a second band from 17.1 to 17.3 GHz was identified by CEPT but so far no systems have been defined to use this band.

21.11.1 HIPERLAN Family of Standards

HIPERLAN/1 is aligned with the IEEE 802 family of standards and is very much like a modern wireless Ethernet. HIPERLAN/1, a standard completed and ratified in 1996, defines the operation of the lower portion of the OSI reference model, namely the data link layer and physical layer [9,10]. The data link layer is further divided into two parts, the channel access control (CAC) sublayer and media access control (MAC) sublayer. The CAC sublayer defines how a given channel access attempt will be made depending on whether the channel is busy or idle and at what priority level an attempt will be made. The HIPERLAN MAC layer defines the various protocols which provide the HIPERLAN/1 features of power conservation, security, and multihop routing (i.e., support for forwarding), as well as the data transfer service to the upper layers of protocols. HIPERLAN/1 uses the same modulation technology that is used in GSM, Gaussian minimum shift keying (GMSK). It has an over air data rate of 23.5 Mbps and maximum user data rate (per channel) of over 18 Mbps. The range in a typical indoor environment is 35 to 50 meters. HIPERLAN/1 provides quality of service (QoS), which lets critical traffic be prioritized.

HIPERLAN/2 has many characteristics of IEEE 802.11 WLAN. HIPER-LAN/2 has three basic layers: physical layer (PHY), data link control layer (DLC), and convergence layer (CL) (see Figure 21.18). The protocol stack is divided into a control plane and a user plane. The user plane includes functions for transmission of traffic over established connections, and the control plane performs functions of connection establishment, release, and supervision. The transmission format on the physical layer is a burst consisting of a preamble part and a data part. The data part originates from each of the transport layers within DLC. A key feature of the physical layer is to provide several modulation and coding schemes according to current radio link quality and meet the requirements for different physical layer modes as defined by transport channels within DLC. Table 21.14 provides a comparison of the IEEE 802.11 WLAN with HIPERLAN/2.

The faster HIPERLANs include the high performance radio access (HIPER-ACCESS) and high performance metropolitan area network (HIPERMAN). Both HIPERACCESS and HIPERMAN are designed for broadband speeds and greater ranges than HIPERLAN/2. HIPERACCESS provides up to 100 Mbps in the 40.5–43.5 GHz band whereas HIPERMAN is designed for a WMAN in 2 GHz and 11 GHz bands.

The DLC layer constitutes the logical link between an access point (AP) and mobile terminals (MTs). The DLC includes functions for medium access and transmission as well as terminal/user connection handling. The DLC layer consists of medium access control (MAC), error control (EC), radio link control (RLC), DLC connection control (DCC), radio resource control (RRC) and association control function (ACF) (see Figure 21.19). Compared to IEEE 802.11 WLAN, medium access in HIPERLAN/2 is based on the time division duplex/time division

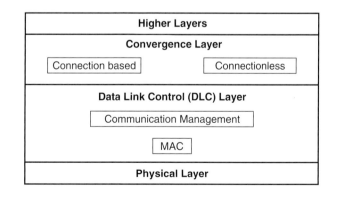

Figure 21.18 Protocol stack of HIPERLAN/2.

Table 21.14 A comparison of HIPERLAN/2 and IEEE 802.11.

Characteristic	IEEE 802.11	HIPERLAN/2
Spectrum	2.4 GHz	5 GHz
Max. physical rate	2 Mbps	54 Mbps
Max. data rate, layer 3	1.2 Mbps	32 Mbps
Medium access control/ Media sharing	CSMA/CA	Central resource control, TDMA/TDD
Access scheme	DCF/PCF	Elimination yield-non preemptive priority multiple access
Connectivity	Connectionless	Connection oriented
Multicast	Yes	Yes
QoS support	PCF	ATM/802.1p/Rsource reSerVation Protocol/Differential service (full control)
Frequency selection	Frequency hopping or DSSS	Single carrier with dynamic frequency selection
Authentication	No	Network access identifier/IEEE address/X.509
Encryption	40-bit RC4	Data Encryption Standard (DES), triple DES
Handover support	No	No
Fixed network support	Ethernet	Ethernet, IP, ATM, UMTS, Firewire, PPP
Management	802.11 MIB	HIPERLAN/2 MIB
Radio link quality control	No	Link adaptation

multiple access (TDD/TDMA) and uses a MAC frame of 2 ms duration. An AP provides centralized control and informs the mobile terminals at which point in time in the MAC frame they are allowed to transmit their data. Time slots are allocated dynamically depending on the need for transmission resources.

HIPERLAN/2 operates as a connection-oriented wireless link. It supports different QoS levels required for the transmission of various traffic types. The convergence layer (CL) between the data link layer and network layer provides QoS. The role of the convergence layer is two-fold—it maps the service requirements of the higher layer to the service offered by the data link control layer, and converts packets received from the core network to the format expected at the lower layers. There are two types of convergence layer. One is cell based and the other is packet based. We focus only on the packet based convergence layer, which can be further divided into a common part and a service specific part. The packet based service specific convergence sublayer (SSCS) is for switched Ethernet

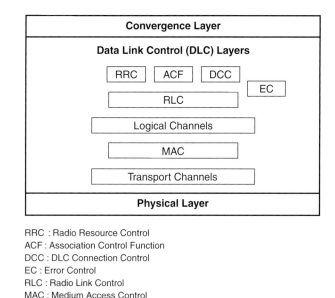

RRC : Radio Resource Control
ACF : Association Control Function
DCC : DLC Connection Control
EC : Error Control
RLC : Radio Link Control
MAC : Medium Access Control

Figure 21.19 Relation between logical and transport channels in HIPERLAN/2.

and IEEE 1394 Firewire. Broadband radio access for IP-based networks (BRAIN) focuses on the specifications of an innovative SSCS dedicated to provide support to IP traffic in a mobile environment. The architecture of the convergence layer makes HIPERLAN/2 suitable as a radio access for different types of fixed networks, e.g., Ethernet, IP, ATM, UMTS, etc. The main function of the common part is to segment packets received from the SSCS, and reassemble segmented packets received from the data link control layer before they are handed over to the SSCS. The Ethernet SSCS makes the HIPERLAN/2 network look like a wireless segment of a switched Ethernet.

HIPERLAN/2 supports two QoS schemes: the best effort scheme and IEEE 802.1p-based priority scheme. The connection-oriented nature of HIPERLAN/2 allows implementation of QoS. Each connection is assigned a specific QoS in terms of bandwidth, delay, jitter, bit error rate, and so on. Also a simple approach is used where each connection is assigned a priority level relative to other connections. The QoS support with high data rate facilitates the transmission of many different types of data streams, e.g., video, voice, and data.

QoS in HIPERLAN/2

Quality of service deals with the ability of a network to provide service for specific network traffic over various underlying wireline or wireless technologies. Compared to end-to-end IP network QoS, wireless network may provide QoS only for one or two hops of an end-to-end connection. The cellular radio access in the BRAIN uses HIPERLAN/2 and supports QoS on a per connection basis.

An IP convergence layer is used to provide the functions required for mapping the QoS requirements of the individual connection to the QoS parameters available in DLC connections. The convergence layer offers a QoS interface to support different IP QoS schemes. By using IP QoS parameters, the convergence layer establishes DLC connections in which IP QoS parameters are mapped into DLC connections for priority, radio bandwidth reservation, appropriate ARQ scheme, and handover strategy. This procedure is realized by mapping IP packets into different DLC connection queues based on respective code point and destination address fields. The convergence layer associates a specific link scheduling priority, discarding time and/or bandwidth reservation to each DLC queue. The convergence layer segments IP traffic to fixed length packets. The segmentation and reassembly causes extra complexity in the convergence layer but enables a better bandwidth reservation policy.

MAC enables full rescheduling in every 2 ms and dynamically adjusts uplink and downlink capacity. Radio link control (RLC) provides connection-oriented secured link service to the convergence layer. There are up to 63 unicast data connections per terminal that can be supported with various QoS parameters. Error correction (EC) provides a selective repeat ARQ mode for each connection. Alternatively, for delay intolerant and multicast services a repetition mode can be used. Thus, DLC provides means for executing several IP QoS techniques such as prioritizing, on-demand-based bandwidth reservation, and delay guarantee. DLC also provides dynamic frequency selection (DFS), link adaptation, power control, and power saving.

The physical layer uses orthogonal frequency division multiplexing (OFDM) to combat frequency selective fading and randomize the burst errors caused by a wide band fading channel. There are seven modes with different coding and modulation schemes available in physical layer; all of them can be adapted dynamically by a link adaptation scheme. There is a strong interaction between the PHY modes, retransmission load and utilization of the radio link, delay, and overall throughput.

The OFDM transmits broadband, high data rate information by dividing the data into several interleaved, parallel bit streams, and lets each one of them modulate on a separate subcarrier. Various coding and modulation schemes are used by a link adaptation mechanism. This is to adapt to current radio link quality and meet the requirements for different physical layer properties as defined by the transport channels within DLC. Table 21.15 provides the different PHY modes and their transmission rates. The seven PHY modes use BPSK, QPSK, and 16-QAM as mandatory subcarrier modulation schemes whereas 64-QAM is optional. Forward error correction is performed by a convolutional code of a rate of 1/2 and a constraint length of 7. Other code rates of 3/4 and 9/16 can be obtained by puncturing.

Table 21.15 Physical modes and transmission rates of HIPERLAN/2.

Mode	Modulation	Coding rate R	Nominal bit rate (Mbps)	Coded bit rate per subcarrier	Coded bits per OFDM symbol	Data bit per OFDM symbol
1	BPSK	$^1/_2$	6	1	48	24
2	BPSK	$^3/_4$	9	1	48	36
3	QPSK	$^1/_2$	12	2	96	48
4	QPSK	$^3/_4$	18	2	96	72
5	16-QAM (HIPERLAN/2 only)	$^9/_{16}$	27	4	192	108
6	16-QAM (IEEE only)	$^1/_2$	24	4	192	96
7	16-QAM	$^3/_4$	36	4	192	144
8	64-QAM	$^3/_4$	54	6	288	216

The DLC scheduling algorithm deals with the properties of the HIPERLAN/2 radio access that are dependent on ARQ and link adaptation. ARQ reacts on transmission errors and initiates retransmission. When the error check bit detects error(s) in the transmission, ARQ sends a request for retransmission of the error packet data unit (PDU). Thus, for a poor radio link, retransmission will cause large transmission delay. Selective repeat ARQ uses a transmission window at the transmitter and receiver. The receiver notifies the transmitter of the sequence number below which all PDUs are received correctly and points out which PDU is not correct.

Based on the current radio link conditions, the link adaptation in the DLC layer assigns a specific PHY mode to the PDU dedicated to a connection. Each connection and its direction are addressed individually and the assignment varies from one MAC frame to another. The link adaptation scheme adapts the PHY mode based on link quality measurements. ARQ and link adaptation reduce packet loss rate but introduce additional overhead and delay to the radio access system.

Total system throughput, transmission delay, and bit error rate are the important parameters in determining the performance of the HIPERLAN/2 radio access. There is a strong interaction between PHY modes and these parameters.

The MAC protocol functions are used for organizing the access and transmission of data on the radio link. Since HIPERLAN/2 uses a central resource controlled TDD/TDMA scheme, MAC frame (MF) is allowed to simultaneously communicate via a number of DLC connections in both downlink and uplink directions. Each MF allocates time slots for broadcast channel (BCH), frame channel, access feedback channel (ACH), random channel (RCH), downlink (DL) phase, uplink (UL) phase and directlink (DiL) phase. Data is grouped as PDU trains. There are two kinds of PDU, one is long PDU (LCH PDU) of 54 bytes and another is short PDU (SCH PDU) of 9 bytes. The PDU error ratio refers to the error rate of LCH PDU.

21.11.2   Multimedia Access Communication—High Speed Wireless Access Network

Multimedia access communication (MMAC)—high speed wireless access network (HiSWAN)—started in Japan in 1996. It uses two frequency bands: 5 GHz for HiSWANa and 25 GHz for HiSWANb. The HiSWAN uses the 5 GHz license-free frequency band and is closely aligned with the HIPERLAN/2. HiSWAN uses the OFDM physical layer to provide a standard speed of 27 Mbps and 6 to 36 Mbps by link adaptation. However, MMAC-HiSWANa differs from the HIPERLAN/2 in radio network functions due to the differences in regional frequency planning and regulations. Instead of dynamic frequency selection in HIPERLAN/2, carrier sense functions of access points are mandatory in MMAC-HiSWANa. Also, inter-access point synchronizations are specified to avoid interference among access points and to use four available channels in Japan for wide coverage. Table 21.16 summaries various WLANs.

Example 21.5

Consider the HIPERLAN/2 that uses BPSK and $R = {}^3/_4$ codes for 9 Mbps information transmission and 16-QAM with the same coding for the actual payload data transmission rate of 36 Mbps. Calculate the coded symbol transmission rate per subcarrier for each of the two modes. What is the bit transmission rate per subcarrier for each of the two modes?

Solution

- User data transmission rate per carrier with $R = {}^3/_4$ convolution encoder (refer to Table 21.11).

$$\text{Mode I (9 Mbps)} = \frac{9 \times 10^6}{48} = 187.5 \text{ kbps}$$

$$\text{Mode II (36 Mbps)} = \frac{36 \times 10^6}{48} = 750 \text{ kbps}$$

- Carrier transmission rate with $R = {}^3/_4$ convolutional encoder

$$\text{Mode I} = \frac{187.5}{(3/4)} = 250 \text{ kbps}$$

$$\text{Mode II} = \frac{750}{(3/4)} = 1000\,\text{kbps}$$

- Carrier symbol rate

$$\text{Mode I (BPSK)} = 250\,\text{ksps}$$

$$\text{Mode II (16-QAM)} = \frac{1000}{4} = 250\,\text{ksps}$$

Example 21.6

What is the user data rate for HIPERLAN/2 with 64-QAM modulation with $R = {}^3/_4$ covolutional coder?

Solution

Carrier symbol rate $= 250\,\text{ksps}$; bits per symbol for 64-QAM $= 6$

User data rate $= 250 \times \frac{3}{4} \times 6 \times 48 = 54\,\text{Mbps}$

21.12 Performance of a Bluetooth Piconet in the Presence of IEEE 802.11 WLANs

Due to its global availability, the 2.4 GHz ISM unlicensed band is a popular frequency band to low-cost radios. Bluetooth and the IEEE 802.11 WLAN both operate in this band. Therefore, it is anticipated that some interference will result from both these systems operating in the same environment. Interference may lead to significant performance degradation. In this section, we evaluate Bluetooth

Table 21.16 Comparisons of various WLAN standards.

	IEEE 802.11	IEEE 802.11b	IEEE 802.11a	IEEE 802.11g	HIPERLAN/1	HIPERLAN/2	MMAC HiSWAN
Rectification	June 1997	Sept. 1999	Sept. 1999	June 2003	early 1993	Feb. 2000	April 1997
RF bandwidth (GHz)	2.4	2.4	5.0	2.4	5	5	5
Max. data rate (Mbps)	2	11	54	54	23.5	54	27
Physical layer (PHY)	FHSS, DSSS, IR	DSSS	OFDM	OFDM	GMSK	OFDM	OFDM
Range (m)	50–100	50–100	50–100	50–100	50	50 indoor, 300 outdoor	100–150

MAC layer performance in the presence of neighboring Bluetooth piconets and neighboring IEEE 802.11 WLANs.

A packet collision occurs when a desired Bluetooth packet [11,12,17] overlaps the interfering packets in time and frequency. In Bluetooth, the duration of a single slot packet is 366 ms and the duration of the slot is 625 ms. The time between the end of the transmission of the packet and the start of the next slot is the idle time. Similarly, the duration of one 802.11 packet traffic time includes packet transmission time and a backoff period.

To simplify the analysis, we make the following assumptions:

- The link is continuously established and the collocated systems are sufficiently close to each other such that the Bluetooth packet will be corrupted completely by the interference packets even if they overlap by a single bit.

- The desired Bluetooth packet won't be destroyed by another piconet if it is hit during the idle time.

- The desired Bluetooth packet won't be destroyed by an IEEE 802.11 network during the IEEE 802.11 backoff period.

- In Bluetooth, the hopping patterns are 100% uncorrelated.

- For a long enough observation time, a given transmitter uses the 79 hopping channels equally.

- There are also 79 channels spaced 1 MHz apart in the IEEE 802.11 frequency hopping (FH) system.

- Each station's signal hops from one modulating frequency to another in a predetermined pseudo-random sequence.

The collision probability of Bluetooth to the IEEE 802.11 FH system is 1/79. In the IEEE 802.11 direct sequence (DS), the data stream is converted into a symbol stream which spreads over a relatively wide band channel of 22 MHz, so the interference on a Bluetooth packet from IEEE 802.11 DS system is much higher than that from the 802.11 FH system. It is because the bandwidth of a channel in DS is 22 times as wide as Bluetooth one channel. The collision probability of Bluetooth to the IEEE 802.11 DS system is 22/79.

21.12.1 Packet Error Rate (PER) from N Neighboring Bluetooth Piconets

- **Synchronous mode:** The probability that one interfering Bluetooth piconet selects the same channel as the desired piconet is 1/79. The probability for a Bluetooth packet to be transferred successfully for synchronous mode is $(1 - 1/79) = 78/79$

- **Asynchronous mode:** The duration of a single slot packet is 366 µs and the duration of a slot is 625 µs. If r is equal to 366/625, there is a probability of $2(1 - r)$ that only the preceding or the following slot is vulnerable and a probability of $(2r - 1)$ that both preceding and following slots are vulnerable. Hence the probability of the desired piconet not being disturbed by one advisory piconet is given as:

$$P_s^{UR} = 2(1 - r) \cdot \left(\frac{78}{79}\right) + (2r - 1) \cdot \left(\frac{78}{79}\right)^2 \tag{21.1}$$

The PER due to N neighboring Bluetooth piconets for synchronous and asynchronous modes are given as:

$$(\text{PER})_{\text{syn}} = 1 - \left(\frac{78}{79}\right)^N \tag{21.2}$$

$$(\text{PER})_{\text{asyn}} = 1 - \left[2(1 - r) \cdot \left(\frac{78}{79}\right) + (2r - 1) \cdot \left(\frac{78}{79}\right)^2\right]^N = 1 - (P_s^{UR})^N \tag{21.3}$$

21.12.2 PER from *M* Neighboring IEEE 802.11 WLANs*

Under IEEE 802.11 FH, the probability of M IEEE 802.11 induced collisions on a Bluetooth packet is given as [17]:

$$\text{PER} = 1 - \left\{\left[1 - \frac{|G|}{L}\right] \cdot \left(\frac{78}{79}\right)^{\lceil H/L \rceil} + \frac{|G|}{L} \cdot \left(\frac{78}{79}\right)^{\lceil H/L \rceil - G/|G|}\right\}^M \tag{21.4}$$

Under IEEE 802.11 DS, the probability of M IEEE 802.11 induced collisions on a Bluetooth packet is given as:

$$\text{PER} = 1 - \left\{\left[1 - \frac{|G|}{L}\right] \cdot \left(\frac{57}{79}\right)^{\lceil H/L \rceil} + \frac{|G|}{L} \cdot \left(\frac{57}{79}\right)^{\lceil H/L \rceil - G/|G|}\right\}^M \tag{21.5}$$

where:
H = the duration of a Bluetooth packet
L = the dwell period of an IEEE 802.11 transmission
T_w = the packet duration of IEEE 802.11
G = $\lceil H/L \rceil L - T_w - H$
$\lceil x \rceil$ = the least integer greater than or equal to x

The collision probability of a Bluetooth piconet where there are N neighboring Bluetooth piconets and M IEEE 802.11 WLANs will be:

$$(\text{PER})_{\text{syn}} = 1 - \left[\left(\frac{78}{79}\right)^N \cdot (P_s)^M\right] \tag{21.6}$$

---

* Refer to Reference [17] for the derivation of equations and details.

$$(PER)_{asyn} = 1 - (P_s^{UR})^N (P_s)^M \tag{21.7}$$

21.12.3 Aggregated Throughput

The aggregated throughput of successfully transmitted packets in all piconets can be expressed as $S_a(n) = nP_s(n,m)$, where n is the number of Bluetooth piconets, m is the number of IEEE 802.11 WLANs, and $P_s(n,m)$ is the probability that a Bluetooth packet is free from other piconets and 802.11 network collisions. The aggregate throughput for different cases are given in Table 21.17.

Example 21.7

Determine the collision probability of 1000 bytes IEEE 802.11 frequency hopping (FH) packet at 2 Mbps and Bluetooth. Assume a dwell period of an 802.11 transmission to be 3 ms. What is the PER with an IEEE 802.11 direct spread (DS) packet? Duration of Bluetooth packet = 0.625 ms.

Solution

$$T_w = \frac{1000 \times 8}{2 \times 10^6} = 4\,\text{ms}$$

$$\left\lceil \frac{H}{L} \right\rceil = \left\lceil \frac{0.625}{3} \right\rceil = 1$$

$$G = 3 - 4 - 0.625 = -1.625\,\text{ms}$$

$$|G| = 1.625\,\text{ms}$$

$$(PER)_{FH} = 1 - \left\{ \left[1 - \frac{1.625}{3}\right]\left(\frac{78}{79}\right) + \frac{1.625}{3} \cdot \left(\frac{78}{79}\right)^{1+1} \right\} = 0.0195 \approx 2\%$$

$$(PER)_{DS} = 1 - \left\{ \left[1 - \frac{1.625}{3}\right] \cdot \left(\frac{57}{79}\right) + \frac{1.625}{3} \cdot \left(\frac{57}{79}\right)^2 \right\} = 0.3873 \approx 38.73\%$$

Note the collision probability with 802.11 DS is much higher than with 802.11 FH.

Table 21.17 Aggregated throughputs for different cases.

Case	n	m	$P_s(n,m)$	$S_a(n)$
Synchronous	$N + 1$	0	$(78/79)^N$	$(N + 1)(78/79)^N$
	$N + 1$	M	$(78/79)^N P_s^M$	$(N + 1)(78/79)^N P_s^M$
Asynchronous	$N + 1$	0	$(P_s^{UR})^N$	$(N + 1)(P_s^{UR})^N$
	$N + 1$	M	$(P_s^{UR})^N P_s^M$	$(N + 1)(P_s^{UR})^N P_s^M$

21.13   Interference between Bluetooth and IEEE 802.11

Interference range is the distance between two devices in order to interfere, if they operate at the same frequency and at the same time. The interference range depends on propagation characteristics of the environment, processing gains of receivers, and transmitted power from different devices. Figure 21.20 shows an interference scenario between a transmitting Bluetooth (BT-1) device and a receiving IEEE 802.11 FH device (MS) collocated in the same area [15]. Since IEEE 802.11 AP is usually located on the wall to provide better coverage, it is less likely to be interfered with by the BT. Interference occurs when the MS receives information from the AP, and BT-1 transmits information to BT-2; or when the MS transmits and BT-1 receives. We assume the interference from the AP to the BT devices and the interference of the BT-2 to the IEEE 802.11 device are negligible. When the MS is receiving and BT-1 is transmitting, the signal-to-interference ratio (SIR) at the MS will be:

$$\text{SIR} = \frac{S}{I} = \frac{KP_{\text{AP}}d^{-\gamma}}{KP_{\text{BT}}r^{-\gamma}} = \frac{P_{\text{AP}}}{P_{\text{BT}}} \cdot \left(\frac{r}{d}\right)^{\gamma} \tag{21.8}$$

$$r_{\max} = d[(\text{SIR})_{\min} \cdot (P_{\text{BT}}/P_{\text{AP}})]^{1/\gamma} \tag{21.9}$$

where:

P_{AP} and P_{BT} = the transmitted power by the AP and Bluetooth devices, respectively
γ = path loss exponent
d = distance between AP and IEEE 802.11 device
r = distance between Bluetooth device and IEEE 802.11 device

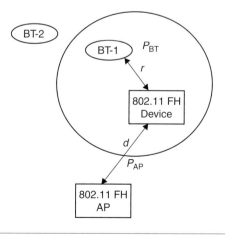

Figure 21.20   First interference scenario between Bluetooth and IEEE 802.11 FH device.

r_{max} = range of interference between Bluetooth and 802.11 device
$(SIR)_{min}$ = minimum signal-to-interference ratio

In the second scenario, the MS transmits to the AP and BT-1 receives from BT-2 (see Figure 21.21), then

$$r_{max} = d[(SIR)_{min}(P_{MS}/P_{BT})]^{1/\gamma} \tag{21.10}$$

where:
P_{MS} is the transmitted power of the IEEE 802.11 device

If instead of the IEEE 802.11 FH system, the IEEE 802.11 DS system is used in the scenario of Figure 21.20, the minimum required received SIR at the mobile terminal is reduced by the processing gain, G_p, of the DSSS. The interference range r_{max} will be:

$$r_{max} = d[(SIR)_{min}(P_{BT}/\{P_{AP}G_P\})]^{1/\gamma} \tag{21.11}$$

Similarly, the interference range for the second scenarios of Figure 21.21 with the IEEE 802.11 DSSS system will be:

$$r_{max} = d[(SIR)\{P_{MS}/(P_{BT}G_P\})]^{1/\gamma} \tag{21.12}$$

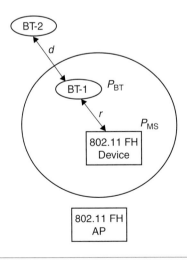

Figure 21.21   Second interference scenario between IEEE 802.11 FH device and Bluetooth.

Example 21.8

Consider the Bluetooth and IEEE 802.11 FH interference scenario (refer to Figure 21.20):

(a) Assuming that the acceptable error probability for the mobile terminal is 10^{-5}, find $(\text{SIR})_{\min}$ that supports this error rate, and (b) using $(\text{SIR})_{\min}$ from (a) calculate $r_{\max}$ for $d = 10$ m, $\gamma = 4$, $P_{BT} = 20$ dBm and $P_{AP} = 40$ dBm

Solution

$$P_e = 0.5_e^{-0.5E_b/N_0}$$
$$10^{-5} = 0.e^{-0.5E_b/N_0}$$
$$\frac{E_b}{N_0} = 21.6 = 13.35\,\text{dB}$$
$$r_{\max} = 10\left[\frac{21.6 \times 20}{40}\right]^{1/4} = 18.13\,\text{m}$$

Example 21.9

Calculate $r_{\max}$ for the interference scenarios (see Figure 21.21) using $S_{\min}$ from Example 21.8 and $d = 10$ m, $\gamma = 4$, $P_{BT} = 20$ dBm and $P_{MS} = 40$ dBm

Solution

$$r_{\max} = 10\left[\frac{21.6 \times 40}{20}\right] = 25.6\,\text{m}$$

Example 21.10

Repeat Problems 21.8 and 21.9, if the IEEE 802.11 FH device is replaced by the IEEE 802.11 DS device ($G_p = 11$).

Solution

$$r_{\max} = 10\left[\frac{21.6 \times 20}{40 \times 11}\right]^{1/4} = 9.95\,\text{m}$$

$$r_{\max} = 10\left[\frac{21.6 \times 20}{20 \times 11}\right]^{1/4} = 14.08\,\text{m}$$

Note the interference ranges are smaller for the IEEE 802.11 DS device compared to the IEEE 802.11 FH device.

21.14   IEEE 802.16

The IEEE 802.16 standard delivers performance comparable to traditional cable, DSL, or T1 offerings. The principal advantages of systems based on 802.16 are multifold: faster provisioning of service, even in areas that are hard for wired infrastructure to reach; lower installation cost; and ability to overcome the physical limitations of the traditional wired infrastructure. 802.16 technology provides a flexible, cost-effective, standard-based means of filling gaps in broadband services not envisioned in a wired world. For operators and service providers, systems built upon the 802.16 standard represent an easily deployable "third pipe" capable of delivering flexible and affordable last-mile broadband access for millions of subscribers in homes and businesses throughout the world [18,19].

The 802.16a is an extension of the 802.16 originally designed for 10–66 GHz. It covers frequency bands between 2 and 11 GHz and enables non line-of-sight (NLOS) operation, making it an appropriate technology for last-mile applications where obstacles such as trees and buildings often present and where base stations may need to be unobtrusively mounted on the roofs of homes or buildings rather than towers on mountains.

The 802.16a has a range of up to 30 miles with a typical cell radius of 4 to 6 miles. Within the typical cell radius NLOS performance and throughputs are optimal. In addition, the 802.16a provides an ideal wireless backhaul technology to connect 802.11 WLAN and commercial 802.11 hotspots with the Internet. Table 21.18 provides a road map of IEEE 802.16 standards.

Applications of the 802.16 are cellular backhaul, broadband on-demand, residential broadband, and best-connected wireless service (see Figure 21.22). The 802.16 delivers high throughput at long ranges with a high spectral efficiency. Dynamic adaptive modulation allows base stations to trade off throughput for range. The 802.16 supports flexible channel bandwidths to accommodate easy cell planning in both licensed and unlicensed spectra worldwide. The 802.16 includes robust security features and QoS needed to support services that require low latency, such as voice and video. The 802.16 voice service can either be TDM voice or voice over IP (VoIP). Privacy and encryption features are also included to support secure transmission and data encryption.

The *worldwide interoperability for microaccess inc.* (WiMAX) forum, an industry group, focused on creating system profiles and conformance programs

Table 21.18 Road-map of IEEE 802.16 standard.

Standard	Features
802.16 (2001)	Air interface for fixed broadband wireless access system, MAC and PHY specification for 10–66 GHz (LOS)
802.16a (January 2003)	Amendment to 802.16; MAC modifications and additional PHY specifications for 2–11 GHz (NLOS); three physical layers—OFDM, OFDMA, single carrier; additional MAC functions; mesh topology support; ARQ
802.16d (July 2004)	Combine 802.16 and 802.16a, some modification to MAC and PHY
802.16e (December 2005)	Amendment to 802.16d, MAC modifications for limited mobility

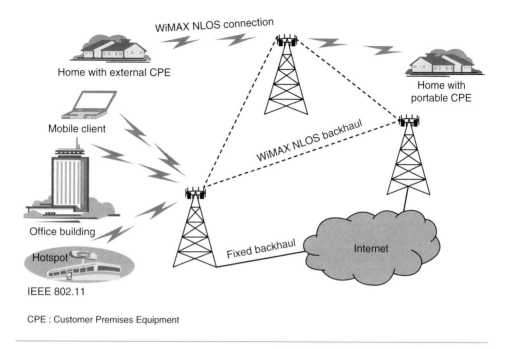

Figure 21.22   Applications of IEEE 802.16 (WiMax).

to ensure operability among devices based on the 802.16 standard from different manufacturers. For more details see references [18] and [19].

21.15 World Interoperability for MicroAccess, Inc. (WiMAX)

WiMAX is an advanced technology solution based on an open standard, designed to meet the need for very high speed wide area Internet access, and to do so in a low-cost, flexible way. It aims to provide business and consumer broadband service on the scale of the metropolitan area network (MAN). WiMAX networks are designed for high-speed data and will spur innovation in services, content, and new mobile devices. WiMAX is optimized for IP-based high-speed wireless broadband which will provide for a better mobile wireless broadband Internet experience.

The WiMAX product certification program ensures interoperability between WiMAX equipment from vendors worldwide. The certification program also considers interoperability with high performance radio metropolitan area network (HIPERMAN), the European telecommunication standards institute's MAN standard.

With its large range and high transmission rate, WiMAX can serve as a backbone for 802.11 hotspots for connecting to the Internet. Alternatively, users can connect mobile devices such as laptops and handsets directly to WiMAX base stations without using 802.11. Mobile devices connected directly can achieve a range of 4 to 6 miles, because mobility makes links vulnerable. The WiMAX technology can also provide fast and cheap broadband access to markets that lack infrastructure (fiber optics or copper wire), such as rural areas and unwired countries. WiMAX can be used in disaster recovery scenes where the wired networks have broken down. It can be used as backup links for broken wired links.

WiMAX can typically support data rates from 500 kbps to 2 Mbps. WiMAX also has clearly defined QoS classes for applications with different requirements such as VoIP, real-time video streaming, file transfer, and web traffic. A cellular architecture similar to that of mobile phone systems can be used with a central base station controlling downlink/uplink traffic (see Figure 21.22).

WiMAX is a family of technologies based on IEEE 802.16 standards. There are two main types of WiMAX today, *fixed WiMAX* (IEEE 802.16d—2004), and *mobile WiMAX* (IEEE 802.16e—2005). Fixed WiMAX is a point-to-multipoint technology, whereas mobile WiMAX is a multipoint-to-multipoint technology, similar to that of a cellular infrastructure. Both solutions are engineered to deliver ubiquitous high-throughput broadband wireless service at a low cost. Mobile WiMAX uses orthogonal frequency division multiple access (OFDMA) technology which has inherent advantages in latency, spectral efficiency, advanced antenna performance, and improved multipath performance in, an NLOS environment. Scalable OFDMA (SOFDMA) has been introduced in IEEE 802.16e to support scalable channel bandwidths from 1.25 to 20 MHz. Release 1 of mobile WiMAX will cover 5, 7, 8.75, and 10 MHz channel bandwidths for licensed worldwide spectrum allocations in 2.3 GHz, 2.5 GHz, 3.3 GHz, and 3.5 GHz frequency bands. Also, next generation 4G wireless technologies (see Chapter 23) are evolving toward OFDMA and IP-based networks as they are ideal for delivering cost-effective high-speed wireless data services.

The WiMAX specification improves upon many of the limitations of the Wi-Fi standard (802.11b) by providing increased bandwidth and stronger encryption. Table 21.19 provides comparisons of Wi-Fi and WiMAX.

The 802.16 standard was designed mainly for point-to-multipoint topologies, in which a base station distributes traffic to many subscriber stations that are mounted on rooftops. The point-to-multipoint configuration uses a scheduling mechanism that yields high efficiency because stations transmit in their scheduled slots and do not contend with one another. WiMAX does not require stations to listen to one another because they encompass a larger area. This scheduling design suits WiMAX networks because subscriber stations might aggregate traffic from several computers and have steady traffic, unlike terminals in 802.11 hotspots, which usually have bursty traffic. The 802.16 also supports a mesh mode, where subscriber stations can communicate directly with one another. The mesh mode can help relax the line-of-sight requirement and ease the deployment costs for high frequency bands by allowing subscriber stations to relay traffic to one another. In this case, a station that does not have line-of-sight with the base station can get its traffic from another station (see Figure 21.23).

Table 21.19 Comparison of Wi-Fi and WiMAX.

Wi-Fi	WiMAX
802.11a—OFDM, maximum rate = 54 Mbps 802.11b—DSSS, maximum rate = 11 Mbps 802.11g—OFDM, maximum rate = 54 Mbps	802.16—OFDM, maximum rate = 50 Mbps 802.16e—OFDM, maximum rate ∼ 30 Mbps
Range <100 m	A few km's non-line-of-sight, more with line of sight
Indoor environment	Outdoor environment
No admission control, no load balancing	Admission control and load balancing
No quality of service (QoS)	Five QoS classes enforced by base station

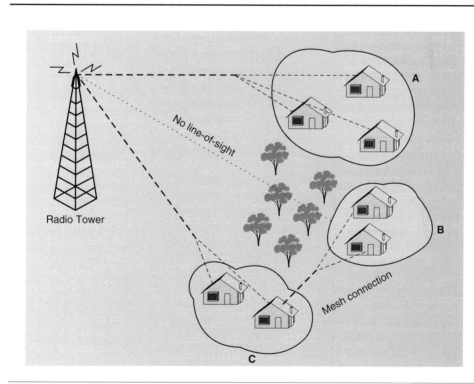

Figure 21.23   Mesh mode in IEEE 802.16 (WiMAX).

Mobile WiMAX systems offer scalability in both radio access technology and network architecture, thus providing a great deal of flexibility in network deployment options and service offerings. Some of the salient features supported by WiMAX are:

- **High data rates.** The inclusion of multi-input multi-out (MIMO) antenna techniques along with flexible sub-channelization schemes, advanced coding and modulation all enable the mobile WiMAX technology to support peak downlink data rates of 63 Mbps per sector and peak uplink data rates of up to 28 Mbps per sector in a 10 MHz channel.

- **Quality of service (QoS).** The fundamental premise of the IEEE 802.16 MAC architecture is QoS. It defines service flows which can map to DiffServ code points or MPLS flow labels that enable end-to-end IP based QoS. Additionally, sub-channelization and MAP-based signaling schemes provide a flexible mechanism for optimal scheduling of space, frequency, and time resources over the air interface on a frame-by-frame basis.

- **Scalability.** Mobile WiMAX is designed to be able to scale to work in different channelization from 1.25 to 20 MHz to comply with varied worldwide requirements as efforts proceed to achieve spectrum harmonization in the longer term.

- **Security.** Support for a diverse set of user credentials exists including SIM/USIM cards, smart cards, digital certificates, and user name/password schemes based on the relevant extensible authentication protocol (EAP) methods for the credential type.

- **Mobility.** Mobile WiMAX supports optimized handoff schemes with latencies less than 50 ms to ensure that real-time applications such as VoIP can be performed without service degradation. Flexible key management schemes assure that security is maintained during handoff.

21.15.1   WiMAX Physical Layer (PHY)

The 802.16 PHY supports TDD and full and half duplex FDD operations; however, the initial release of mobile WiMAX only supports TDD. Other advanced PHY features include adaptive modulation and coding (AMC), hybrid automatic repeat request (HARQ) and fast channel feedback (CQICH) to enhance coverage and capacity of WiMAX in mobile applications.

For the bands in the 10–66 GHz range, 802.16 defines one air interface called Wireless MAN—SC. The PHY design for the 2–11 GHz range (both licensed and unlicensed bands) is more complex because of interference. Hence, the standard supports burst-by-burst adaptability for modulation and coding schemes and specifies three interfaces. The adaptive features at the PHY allow trade-off between robustness and capacity. The three air interfaces for the 2–11 GHz range are:

- Wireless MAN—SCa uses single carrier modulation.
- Wireless MAN—OFDM uses a 256-carrier OFDM. This air interface provides multiple access to different stations through time-division-multiple access.
- Wireless MAN—OFDM uses a 2048-carrier OFDM scheme. The interface provides multiple access by assigning a subset of the carriers to an individual receiver.

Support for QPSK, 16-QAM, and 64-QAM are mandatory in the downlink with mobile WiMAX. In the uplink 64-QAM is optional. Both convolutional code and turbo code with variable code rate and repetition coding are supported. The combinations of various modulation and code rates provide a fine resolution of data rates. The frame duration is 5 ms. Each frame has 48 OFDM symbols with 44 OFDM symbols available for data transmission.

The base station (BS) scheduler determines the appropriate data rate for each burst allocation based on the buffer size, channel propagation conditions at the receiver, etc. A channel quality indicator (CQI) channel is used to provide channel state information from the user terminals to the BS scheduler.

WiMAX provides signaling to allow fully asynchronous operation. The asynchronous operation allows variable delay between retransmissions which gives more flexibility to the scheduler at the cost of additional overhead for each retransmission. HARQ combined with CQICH and AMC provides robust link adoption in the mobile environment at vehicular speeds in excess of 120 km/h.

21.15.2   WiMAX Media Access Control (MAC)

The IEEE 802.16 MAC is significantly different from that of IEEE 802.11b Wi-Fi MAC. In Wi-Fi, the MAC uses contention access—all subscribers wishing to pass data through an access point compete for the access point's (AP's) attention on a random basis. This can cause distant nodes from the AP to be repeatedly interrupted by less sensitive, closer nodes, greatly reducing their throughput. This makes services, such as VoIP or IPTV which depend on a determined level of QoS, difficult to maintain for large numbers of users.

The MAC layer of 802.16 is designed to serve sparsely distributed stations with high data rates. Subscriber stations are not required to listen to one another because this listening might be difficult to achieve in the WiMAX environment. The 802.16 MAC is a scheduling MAC where the subscriber only has to compete once (for initial entry into the network). After that it is allocated a time slot by the base station. The time slot can enlarge and constrict, but it remains assigned to the subscriber, meaning that other subscribers are not supposed to use it but take their turn. This scheduling algorithm is stable under overload and oversubscription. It is also more bandwidth efficient. The scheduling algorithm allows the base station to control QoS by balancing the assignment among the needs of subscribers.

Duplexing, a station's concurrent transmission and reception, is possible through time division duplex (TDD) and frequency division duplex (FDD). In TDD, a station transmits then receives (or vice versa) but not at the same time. This option helps reduce subscriber station costs, because the radio is less complex. In FDD, a station transmits and receives simultaneously on different channels.

The 802.16 MAC protocol is connection-oriented and performs link adaptation and ARQ functions to maintain target bit error rate while maximizing the data throughput. It supports different transport technologies such as IPv4, IPv6,

Ethernet, and ATM. This lets service providers use WiMAX independently of the transport technology they support.

The recent WiMAX standard, which adds full mesh networking capabilities, enables WiMAX nodes to simultaneously operate in "subscriber" and "base station" mode. This blurs the initial distinction and allows for widespread adoption of WiMAX based mesh networks and promises widespread WiMAX adoption. Mobile WiMAX with OFDMA and scheduled MAC allows wireless mesh networks to be much more robust and reliable.

21.15.3 Spectrum Allocation for WiMAX

The IEEE 802.16 specification applies across a wide swath of RF spectrum. There is no uniform global licensed spectrum for WiMAX in the United States. The biggest segment available is around 2.5 GHz, and is already assigned—primarily to Sprint Nextel. Elsewhere in the world, the most likely bands used will be around 3.5 GHz, 2.3/2.5 GHz, or 5 GHz, with 2.3/2.5 GHz probably being most important in Asia.

There is some prospect that some of a 700 MHz band might be made available for WiMAX in the United States, but it is currently assigned to analog TV and awaits the complete rollout of HD digital TV before it can become available, likely by 2009. There are several variants of 802.16, depending on local regulatory conditions and thus of which spectrum is used.

Mobile WiMAX based on the 802.16e standard will most likely be in 2.3 GHz and 2.5 GHz frequencies—low enough to accommodate the NLOS conditions between the base station and mobile devices. The key technologies in 802.16e on PHY level are OFDMA and SOFDMA. OFDMA uses a multicarrier modulation in which the carriers are divided among users to form subchannels (see Figure 21.24). For each subchannel, the coding and modulation are adapted separately, allowing channel optimization on a smaller scale (rather than using the same parameters for the whole channel). This technique optimizes the use of spectrum resources and enhances indoor coverage by assigning a robust scheme to vulnerable links. SOFDMA is an enhancement of OFDMA that scale the number of subcarriers in a channel with possible values of 128, 512, 1024, and 2048.

802.16e includes power-saving and sleep modes to extend the battery life of mobile devices. 802.16e also supports hard and soft handoff to provide users with seamless connections as they move across coverage areas of adjacent cells. Other improvements for mobile devices include a real-time polling service to provide QoS, HARQ scheme to retransmit erroneous packets, and private key management schemes to help with the distribution of encryption keys.

21.16 Summary

In this chapter we provided an overview of the wireless local area networks (WLANs). We divided WLAN standards into IEEE 802.11 connectionless and

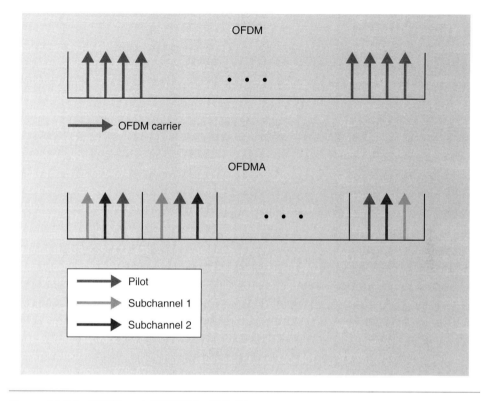

Figure 21.24   OFDM and OFDMA in 802.16e.

ETSI HIPERLAN/2 connection-oriented networks. HIPERLAN/2 is useful for variable QoS support. IEEE 802.11 WLAN uses FHSS and DSSS techniques to offer data rates up to 11 Mbps.

Bluetooth is a digital wireless data transmission standard in the 2.4 GHz ISM band aimed at providing a short-range, wireless link between laptops, mobile phones, and other devices. The IEEE 802.11 WLANs also operate in this frequency band. Since both systems operate in the same frequency band, interference between them is caused. We evaluated the performance of the Bluetooth MAC layer when Bluetooth radio operates in close proximity of other Bluetooth piconets and IEEE 802.11 WLANs. A probabilistic approach was suggested to obtain the packet error rate (PER) of a Bluetooth piconet and the aggregated throughput of N collocated piconets.

The chapter concluded by providing a brief description of the IEEE 802.16 standard for metropolitan area networks (MAN) and highlighting the features of the WiMAX.

Problems

21.1   What are the application fields of a WLAN?

21.2   Discuss WLAN topologies.

21.3   What are the two network architectures for the WLAN defined in the IEEE 802.11 standard? Discuss them briefly.

21.4   What is CSMA/CA protocol?

21.5   What are the hidden and exposed node problems in a WLAN? Discuss them briefly.

21.6   What is a Wi-Fi? Discuss briefly.

21.7   What is IEEE 802.11n? Discuss briefly.

21.8   Discuss the exponential backoff algorithm used in a WLAN.

21.9   What is the HiSWAN? Discuss briefly.

21.10 In the HIPERLAN/2 PHY layer, 64-subchannel implementation is used. 48 subcarriers are used for information transmission, 4 subcarriers for pilot tones used for synchronization, and 12 subcarriers reserved for other purposes. The HIPERLAN/2 standard uses 16-QAM with $R = 9/16$ coding for an actual payload data transmission rate of 27 Mbps. Calculate the coded symbol transmission rate per subcarrier. What is the bit transmission rate per subcarrier?

21.11 In a DSSS WLAN how many simultaneous data users at 38.4 kbps can be supported by an AP so that an average bit error data rate of 10^{-3} is maintained for each user? Assume the following data: spreading rate = 3.84 Mcps, power control efficiency = 0.8, interference factor from other terminals = 0.67, 3-sector antenna gain = 2.55, and cell loading = 70%. Assuming that each user transmits on average 8 packets per second, what is the average packet length that can be used in the system?

21.12 An FHSS WLAN system uses a 50 kHz channel over a continuous 20 MHz spectrum. Fast frequency hopping is used, in which two hops occur for each bit. If binary frequency shift keying (BFSK) is the modulation scheme used in the system, determine (a) number of hops per second if each user transmits 25 kbps data, and (b) probability of error for a single user operating at E_b/N_0 equal to 10 dB.

21.13 If we switch from 36 Mbps mode to 9 Mbps mode in WLAN, how much more (in dB) of the path loss can be afforded? If the system covers 50 m with 36 Mbps, what would be the coverage with 9 Mbps?

21.14 Consider the Bluetooth and IEEE 802.11 FH interference scenario (refer to Figure 21.20). If the acceptable error probability for the mobile terminal is 10^{-4}, find (a) $(\text{SIR})_{\min}$ that supports this error rate, and (b) calculate $r_{\max}$ for $d = 12\,\text{m}$, $\gamma = 4.3$, $P_{BT} = 23\,\text{dBm}$ and $P_{AP} = 43\,\text{dBm}$. What is $r_{\max}$ if the IEEE 802.11 FH device is replaced by the IEEE 802.11 DS device?

21.15 Find the collision probability of 1200 bytes IEEE 802.11 FH packet at 2 Mbps and the Bluetooth. Assume dwell period of an 802.11 transmission to be 3 ms and Bluetooth packet duration 0.625 ms. What is PER with IEEE 802.11 DS?

21.16 Discuss WiMax.

21.17 What are the main differences between the IEEE 802.11b (Wi-Fi) and WiMax?

References

1. Abichar, Z., Peng, Y., and Chang, M. "WiMAX: The Emergence of Wireless Broadband." *IT Pro* July/August 2006, pp. 44–48.

2. Amre El-Hoiydi. Interference Between Bluetooth Network-Upper Bound on the Packet Error Rate. *IEEE Communication Letters*, vol. 5, no. 6, June 2001.

3. Breeze Wireless Communications, Inc. "Network Security in a Wireless LAN." http://www.breezecom.com/pdfs/security.pdf.

4. Bantz, D. F. Wireless LAN Design Alternative. *IEEE Network,* March/April 1994, pp. 43–53.

5. IEEE. "Wireless Medium Access Control (MAC) and Physical Layer (PHY) Specifications." P. 802.11D6.2, July 1998.

6. Intel Corporation. "IEEE 802.11b High Rate Wireless LAN." http://www.intel.com/network/white.paper/wireless lan/.

7. Intermec Technologies Corporation. "Guide to Wireless Technologies." http://www.intermec.com/datactr/wlan_wp.pdf.

8. Jain, R. "Wireless Local Area Network Recent Developments." Wireless Seminar Series Electrical and Computer Engineering Department. Ohio State University, February. 19, 1998.

9. Jonsson, M. "HiperLan2—The Broadband Radio Transmission Technology Operating in the 5 GHz Frequency Band." HIPERLAN-2 Global Forum White Paper.

10. JTC Technical Report on RF Channel Characterization and Deployment Model. Air Interface Standards, September. 1994.

11. Haartsen, J. C., and Mattisson, S. Bluetooth—A New Low-Power Radio Interface Providing Short Range Connectivity. *Proceedings of IEEE*, vol. 88, no. 10, October 2000.

12. Muller, N. J. *Bluetooth Demystified*. New York: McGraw Hill, 2000.

13. NDC Communications, Inc. "Wireless LAN Systems-Technology and Specifications." http://networking.ittoolbox.com/peer/.

14. Pahlavan, K. Trends in Local Wireless Networks. *IEEE Communications Magazine*, March 1995, pp. 88–95.

15. Pahlavan, K., and Krishnamurthy, P. *Principle of Wireless Networks—A Unified Approach*. Upper Saddle River, NJ: Prentice Hall, 2002.

16. Proxim, Inc. "What is a Wireless LAN." http://www.proxim.com/wireless/whiteppr/whatwlan.shtml.

17. Wang, F., Nallanathan, A., and Garg, H. K. "Performance of a Bluetooth Piconet in the Presence of IEEE 802.11 WLANs,"

18. www.ieee802.org/16.

19. www.wimaxforum.org.

Traffic Tables

This appendix provides traffic tables (Tables A.1 through A.9) for a variety of blocking probabilities and channels.[*] The blocked-calls-cleared (Erlang-B) call model is used. In Erlang B, we assume that, when traffic arrives in the system, it either is served, with probability from the table, or is lost to the system. A customer attempting to place a call therefore either will see a call completion or will be blocked and abandon the call. This assumption is acceptable for low blocking probabilities. In some cases, the call will be placed again after a short period of time. If too many calls reappear in the system after a short delay, the Erlang-B model will no longer hold.

In Tables A.8 and A.9, where the number of channels is high (greater than 250 channels), linear interpolation between two table values is possible. We provide the deltas for one additional channel to assist in the interpolation.

---

[*]The data in the tables was supplied by V. H. MacDonald.

Table A.1 Offered loads (in Erlangs) for various blocking objectives: according to the Erlang-B model—system capacity from 1–20 channels.

P(B) = Trunks	0.01	0.015	0.02	0.03	0.05	0.07	0.1	0.2	0.5
1	0.010	0.015	0.020	0.031	0.053	0.075	0.111	0.250	1.000
2	0.153	0.190	0.223	0.282	0.381	0.471	0.595	1.000	2.732
3	0.455	0.536	0.603	0.715	0.899	1.057	1.271	1.930	4.591
4	0.870	0.992	1.092	1.259	1.526	1.748	2.045	2.944	6.501
5	1.361	1.524	1.657	1.877	2.219	2.504	2.881	4.010	8.437
6	1.913	2.114	2.277	2.544	2.961	3.305	3.758	5.108	10.389
7	2.503	2.743	2.936	3.250	3.738	4.139	4.666	6.229	12.351
8	3.129	3.405	3.627	3.987	4.543	4.999	5.597	7.369	14.318
9	3.783	4.095	4.345	4.748	5.370	5.879	6.546	8.521	16.293
10	4.462	4.808	5.084	5.529	6.216	6.776	7.511	9.684	18.271
11	5.160	5.539	5.842	6.328	7.076	7.687	8.487	10.857	20.253
12	5.876	6.287	6.615	7.141	7.950	8.610	9.477	12.036	22.237
13	6.607	7.049	7.402	7.967	8.835	9.543	10.472	13.222	24.223
14	7.352	7.824	8.200	8.803	9.730	10.485	11.475	14.412	26.211
15	8.108	8.610	9.010	9.650	10.633	11.437	12.485	15.608	28.200
16	8.875	9.406	9.828	10.505	11.544	12.393	13.501	16.807	30.190
17	9.652	10.211	10.656	11.368	12.465	13.355	14.523	18.010	32.181
18	10.450	11.024	11.491	12.245	13.389	14.323	15.549	19.215	34.173
19	11.241	11.854	12.341	13.120	14.318	15.296	16.580	20.424	36.166
20	12.041	12.680	13.188	14.002	15.252	16.273	17.614	21.635	38.159

Table A.2 Offered loads (in Erlangs) for various blocking objectives: according to the Erlang-B model—system capacity from 20–39 channels.

P(B) = Trunks	0.005	0.01	0.015	0.02	0.03	0.05	0.07	0.1
20	11.092	12.041	12.680	13.188	14.002	15.252	16.273	17.614
21	11.860	12.848	13.514	14.042	14.890	16.191	17.255	18.652
22	12.635	13.660	14.352	14.902	15.782	17.134	18.240	19.693
23	13.429	14.479	15.196	15.766	16.679	18.082	19.229	20.737
24	14.214	15.303	16.046	16.636	17.581	19.033	20.221	21.784
25	15.007	16.132	16.900	17.509	18.486	19.987	21.216	22.834
26	15.804	16.966	17.758	18.387	19.395	20.945	22.214	23.885
27	16.607	17.804	18.621	19.269	20.308	21.905	23.214	24.939
28	17.414	18.646	19.487	20.154	21.224	22.869	24.217	25.995
29	18.226	19.493	20.357	21.043	22.143	23.835	25.222	27.053
30	19.041	20.343	21.230	21.935	23.065	24.803	26.229	28.113
31	19.861	21.196	22.107	22.830	23.989	25.774	27.239	29.174
32	20.685	22.053	22.987	23.728	24.917	26.747	28.250	30.237
33	21.512	22.913	23.869	24.629	25.846	27.722	29.263	31.302
34	22.342	23.776	24.755	25.532	26.778	28.699	30.277	32.367
35	23.175	24.642	25.643	26.438	27.712	29.678	31.294	33.435
36	24.012	25.511	26.534	27.346	28.649	30.658	32.312	34.503
37	24.852	26.382	27.427	28.256	29.587	31.641	33.331	35.572
38	25.694	27.256	28.322	29.168	30.527	32.624	34.351	36.643
39	26.539	28.132	29.219	30.083	31.469	33.610	35.373	37.715

Table A.3 Offered loads (in Erlangs) for various blocking objectives: according to the Erlang-B model—system capacity from 40–60 channels.

P(B) = Trunks	0.005	0.01	0.015	0.02	0.03	0.05	0.07	0.1
40	27.387	29.011	30.119	30.999	32.413	34.597	36.397	38.788
41	28.237	29.891	31.021	31.918	33.359	35.585	37.421	39.861
42	29.089	30.774	31.924	32.838	34.306	36.575	38.447	40.936
43	29.944	31.659	32.830	33.760	35.255	37.565	39.473	42.012
44	30.801	32.546	33.737	34.683	36.205	38.558	40.501	43.088
45	31.660	33.435	34.646	35.609	37.156	39.551	41.530	44.165
46	32.521	34.325	35.556	36.535	38.109	40.545	42.559	45.243
47	33.385	35.217	36.468	37.463	39.063	41.541	43.590	46.322
48	34.250	36.111	37.382	38.393	40.019	42.537	44.621	47.401
49	35.116	37.007	38.297	39.324	40.976	43.535	45.654	48.481
50	35.985	37.904	39.214	40.257	41.934	44.534	46.687	49.562
51	36.856	38.802	40.132	41.190	42.893	45.533	47.721	50.644
52	37.728	39.702	41.052	42.125	43.853	46.533	48.756	51.726
53	38.601	40.604	41.972	43.061	44.814	47.535	49.791	52.808
54	39.477	41.507	42.894	43.999	45.777	48.537	50.827	53.891
55	40.354	42.411	43.817	44.937	46.740	49.540	51.864	54.975
56	41.232	43.317	44.742	45.877	47.704	50.544	52.902	56.059
57	42.112	44.224	45.667	46.817	48.689	51.548	53.940	57.144
58	42.993	45.132	46.594	47.759	49.636	52.553	54.979	58.229
59	43.875	46.041	47.522	48.701	50.603	53.559	56.018	59.315
60	44.759	46.951	48.451	49.645	51.570	54.566	57.058	60.401

Table A.4 Offered loads (in Erlangs) for various blocking objectives: according to the Erlang-B model—system capacity from 61–80 channels.

P(B) = Trunks	0.005	0.01	0.015	0.02	0.03	0.05	0.07	0.1
61	45.644	47.863	49.381	50.590	52.539	55.573	58.099	61.488
62	46.531	48.776	50.311	51.535	53.509	56.581	59.140	62.575
63	47.418	49.689	51.243	52.482	54.479	57.590	60.181	63.663
64	48.307	50.604	52.176	53.429	55.450	58.599	61.224	64.750
65	49.197	51.520	53.110	54.377	56.422	59.609	62.266	65.839
66	50.088	52.437	54.044	55.326	57.395	60.620	63.309	66.927
67	50.980	53.355	54.980	56.276	58.368	61.631	64.353	68.016
68	51.874	54.273	55.916	57.226	59.342	62.642	65.397	69.106
69	52.768	55.193	56.853	58.178	60.316	63.654	66.442	70.196
70	53.663	56.113	57.791	59.130	61.292	64.667	67.487	71.286
71	54.560	57.035	58.730	60.083	62.268	65.680	68.532	72.376
72	55.457	57.957	59.670	61.036	63.244	66.694	69.578	73.467
73	56.356	58.880	60.610	61.991	64.222	67.708	70.624	74.558
74	57.255	59.804	61.551	62.945	65.199	68.723	71.671	75.649
75	58.155	60.729	62.493	63.901	66.178	69.738	72.718	76.741
76	59.056	61.654	63.435	64.857	67.157	70.753	73.765	77.833
77	59.958	62.581	64.379	65.814	68.136	71.769	74.813	78.925
78	60.861	63.508	65.322	66.772	69.116	72.786	75.861	80.018
79	61.765	64.435	66.267	67.730	70.097	73.803	76.909	81.110
80	62.669	65.364	67.212	68.689	71.078	74.820	77.958	82.203

Table A.5 Offered loads (in Erlangs) for various blocking objectives: according to the Erlang-B model—system capacity from 81–100 channels.

P(B) = Trunks	0.005	0.01	0.015	0.02	0.03	0.05	0.07	0.1
81	63.574	66.293	68.158	69.648	72.059	75.838	79.007	83.297
82	64.481	67.223	69.104	70.608	73.042	76.856	80.057	84.390
83	65.387	68.153	70.051	71.568	74.024	77.874	81.107	85.484
84	66.295	69.085	70.999	72.529	75.007	78.893	82.157	86.578
85	67.204	70.016	71.947	73.491	75.991	79.912	83.207	87.672
86	68.113	70.949	72.896	74.453	76.975	80.932	84.258	88.767
87	69.023	71.882	73.846	75.416	77.959	81.952	85.309	89.861
88	69.933	72.816	74.796	76.379	78.944	82.972	86.360	90.956
89	70.844	73.750	75.746	77.342	79.929	83.993	87.411	92.051
90	71.756	74.685	76.697	78.306	80.915	85.014	88.463	93.146
91	72.669	75.621	77.649	79.271	81.901	86.035	89.515	94.242
92	73.582	76.557	78.601	80.236	82.888	87.057	90.568	95.338
93	74.496	77.493	79.553	81.202	83.875	88.079	91.620	96.434
94	75.411	78.431	80.506	82.167	84.862	89.101	92.673	97.530
95	76.326	79.368	81.460	83.134	85.850	90.123	93.726	98.626
96	77.242	80.307	82.414	84.101	86.838	91.146	94.779	99.722
97	78.158	81.245	83.368	85.068	87.827	92.169	95.833	100.819
98	79.075	82.185	84.323	86.036	88.815	93.193	96.887	101.916
99	79.993	83.125	85.279	87.004	89.805	94.217	97.941	103.013
100	80.911	84.065	86.235	87.972	90.794	95.240	98.995	104.110

Table A.6 Offered loads (in Erlangs) for various blocking objectives: according to the Erlang-B model—system capacity from 105–200 channels.

P(B) = Trunks	0.005	0.01	0.015	0.02	0.03	0.05	0.07	0.1
105	85.518	88.822	91.030	92.823	95.747	100.371	104.270	109.598
110	90.147	93.506	95.827	97.687	100.713	105.496	109.550	115.090
115	94.768	98.238	100.631	102.552	105.680	110.632	114.833	120.585
120	99.402	102.977	105.444	107.426	110.655	115.772	120.121	126.083
125	104.047	107.725	110.265	112.307	115.636	120.918	125.413	131.583
130	108.702	112.482	115.094	117.195	120.622	126.068	130.708	137.087
135	113.366	117.247	119.930	122.089	125.615	131.222	136.007	142.593
140	118.039	122.019	124.773	126.990	130.612	136.380	141.309	148.101
145	122.720	126.798	129.622	131.896	135.614	141.542	146.613	153.611
150	127.410	131.584	134.477	136.807	140.621	146.707	151.920	159.122
155	132.106	136.377	139.337	141.724	145.632	151.875	157.230	164.636
160	136.810	141.175	144.203	146.645	150.647	157.047	162.542	170.152
165	141.520	145.979	149.074	151.571	155.665	162.221	167.856	175.668
170	146.237	150.788	153.949	156.501	160.688	167.398	173.173	181.187
175	150.959	155.602	158.829	161.435	165.713	172.577	178.491	186.706
180	155.687	160.422	163.713	166.373	170.742	177.759	183.811	192.227
185	160.421	165.246	168.602	171.315	175.774	182.943	189.133	197.750
190	165.160	170.074	173.494	176.260	180.809	188.129	194.456	203.273
195	169.905	174.906	178.390	181.209	185.847	193.318	199.781	208.797
200	174.653	179.743	183.289	186.161	190.887	198.508	205.108	214.323

Table A.7 Offered loads (in Erlangs) for various blocking objectives: according to the Erlang-B model—system capacity from 205–245 channels.

P(B) = Trunks	0.005	0.01	0.015	0.02	0.03	0.05	0.07	0.1
205	179.407	184.584	188.192	191.116	195.930	203.700	210.436	219.849
210	184.165	189.428	193.099	196.073	200.976	208.894	215.765	225.376
215	188.927	194.276	198.008	201.034	206.023	214.089	221.096	230.904
220	193.694	199.127	202.920	205.997	211.073	219.287	226.427	236.433
225	198.464	203.981	207.836	210.963	216.125	224.485	231.760	241.963
230	203.238	208.839	212.754	215.932	221.180	229.686	237.094	247.494
235	208.016	213.700	217.675	220.902	226.236	234.887	242.430	253.025
240	212.797	218.564	222.598	225.876	231.294	240.090	247.766	258.557
245	217.582	223.430	227.524	230.851	236.354	245.295	253.103	264.089

Table A.8 Offered loads (in Erlangs) for various blocking objectives: according to the Erlang-B model—system capacity from 250–600 channels.

P(B) = Trunks	0.005	0.01	0.015	0.02	0.03	0.05	0.07	0.1
250	222.370	228.300	232.452	235.828	241.415	250.500	258.441	269.622
delta	0.961	0.977	0.988	0.998	1.015	1.042	1.069	1.107
300	270.410	277.144	281.853	285.707	292.142	302.617	311.866	324.961
delta	0.966	0.980	0.991	1.001	1.017	1.044	1.070	1.108
350	318.698	326.155	331.424	335.738	342.995	354.836	365.359	380.384
delta	0.969	0.984	0.994	1.005	1.018	1.045	1.071	1.109
400	367.163	375.334	381.128	385.963	393.895	407.096	418.890	435.813
delta	0.972	0.989	0.998	1.004	1.020	1.046	1.071	1.109
450	415.779	424.774	431.022	436.178	444.877	459.408	472.456	491.263
delta	0.975	0.987	0.997	1.006	1.021	1.047	1.072	1.109
500	464.518	474.130	480.890	486.480	495.919	511.759	526.049	546.730
delta	0.977	0.989	0.999	1.007	1.022	1.048	1.072	1.110
550	513.361	523.600	530.843	536.846	547.012	564.142	579.663	602.208
delta	0.979	0.991	1.000	1.008	1.023	1.048	1.073	1.110
600	562.292	573.142	580.859	587.267	598.145	616.552	633.295	657.697

Table A.9 Offered loads (in Erlangs) for various blocking objectives: according to the Erlang-B model—system capacity from 600–1050 channels.

P(B) = Trunks	0.005	0.01	0.015	0.02	0.03	0.05	0.07	0.1
600	562.292	573.142	580.859	587.267	598.145	616.552	633.295	657.697
delta	0.983	0.992	1.001	1.009	1.023	1.049	1.073	1.110
650	611.418	622.748	630.927	637.732	649.313	668.982	686.941	713.193
delta	0.981	0.993	1.002	1.010	1.024	1.049	1.073	1.110
700	660.462	672.410	681.042	688.238	700.511	721.432	740.598	768.697
delta	0.982	0.994	1.003	1.011	1.024	1.049	1.073	1.110
750	709.586	722.119	731.196	738.777	751.735	773.896	794.266	824.206
delta	0.984	0.995	1.004	1.011	1.025	1.050	1.074	1.110
800	758.762	771.872	781.386	789.346	802.981	826.375	847.943	879.719
delta	0.985	0.996	1.004	1.012	1.025	1.050	1.074	1.110
850	807.987	821.662	831.608	839.942	854.247	878.865	901.627	935.236
delta	0.985	0.996	1.005	1.012	1.026	1.050	1.074	1.110
900	857.256	871.487	881.857	890.561	905.530	931.365	955.317	990.757
delta	0.986	0.997	1.005	1.013	1.026	1.050	1.074	1.110
950	906.565	921.343	932.132	941.202	956.829	983.875	1009.013	1046.281
delta	0.987	0.998	1.006	1.013	1.026	1.050	1.074	1.111
1000	955.910	971.226	982.430	991.862	1008.142	1036.393	1062.715	1101.808
delta	0.988	0.998	1.006	1.014	1.027	1.050	1.074	1.111
1050	1005.289	1021.136	1032.748	1042.539	1059.468	1088.918	1116.420	1157.337

Acronyms

A

AAA	Authentication, Authorization, and Accounting
AAD	Additional Authentication Data
AAL	ATM Adaptation Layer
ABR	Answer-Busy Ratio
ABS	Average Busy Season
AC	Authentication Center
ACCH	Associated Control Channel
ACELP	Algebraic Code Excitation Linear Prediction
ACF	Association Control Function
ACH	Access Channel
ACI	Adjacent Channel Interference
ACK	Acknowledgment
ACKCH	Acknowledgment Channel
ACL	Asynchronous Connectionless
ACM	Address Complete Message
ACSE	Association Control Service Element
ADC	Analog-to-Digital Converter
ADPCM	Adaptive Differential Pulse Code Modulation
AES	Advanced Encryption Standard
AGC	Automatic Gain Control
AGCH	Access Grant Channel
AICH	Acquisition Indicator Channel
AK	Anonymity Key
AKA	Authentication and Key Agreement
A-Key	Authentication Key
AL	Application Layer
ALT	Automatic Link Transfer
ALCAP	Access Link Control Application Part
AMC	Adaptive Modulation and Coding
AMPS	Advanced Mobile Phone System
AMR	Adaptive Multi-Rates
AODV	Ad hoc On-demand Distance Vector

AP	Access Point
APBBS	Antipodal Baseband Signaling
APC	Automatic Power Control
API	Application Programming Interface
APN	Access Point Name
ARIB	Association of Radio Industries and Business
ARP	Address Resolution Protocol
ARQ	Automatic Repeat Request
ASIC	Application Specific Integrated Circuit
ASK	Amplitude Shift Keying
ASR	Answer-Seizure Ratio
AT	Attention
ATC	Adaptive Transform Coding
ATDPICH	Auxiliary Transmit Diversity Pilot Channel
ATIS	Alliance for Telecommunications Industry Solutions
ATM	Asynchronous Transmission Mode
AuC	Authentication Center
AWGN	Additive White Gaussian Noise

B

BCC	Blocked Call Clear
BCCH	Broadcast Control Channel
BCD	Blocked Call Delayed
BCH	Broadcast Channel
BER	Bit Error Rate
BH	Busy Hour
BHCA	Busy Hour Call Attempt
BLER	Block Error Rate
BPSK	Binary Phase Shift Keying
BS	Base Station
BSIC	Base Station Identity Code
BSC	Base Station Controller
BSS	Base Station Subsystem
BSSGP	Base System Station GPRS Control
BSSID	Basic Service Set Identifier
BTS	Base-station Transceiver
BWA	Broadband Wireless Access

C

C	Code
CA	Collision Avoidance

CAC	Channel Access Control
CACH	Common Assignment Channel
CAMEL	Customized Applications for Mobile Enhanced Logic
CAP	Contention-Access Period
CASE	Common Application Service Elements
CATV	Community Antenna Television
CAVE	Cellular Authentication and Voice Encryption
CAP	Contention Access Period
CBC-MAC	Cipher Block Chaining-Message Authentication Code
CBRP	Cluster Based Routing Protocol
CBCH	Cell Broadcast Channel
CC	Call Control
CCA	Clear Channel Assessment
CCH	Control Channel
CCCH	Common Control Channel
CCIP	Committee on Communications and Information Policy
CCK	Complementary Code Keying
CCS	Centrum Call Seconds
CCMP	Counter-mode/block Chaining Message Authentication Code Protocol
CCR	Commitment Concurrency and Recovery
CD	Collision Detection
CDG	CDMA Development Group
CDM	Code Division Multiplex
CDMA	Code Division Multiple Access
CDPD	Cellular Digital Packet Data
CELP	Code-Excited Linear Predictive
CEPT	Conference Européene des administrations des Postes et des Télécommunications
CFP	Contention-Free Period
CGI	Cell Global Identity
CH	Cluster Head
CHAP	Challenge Handshake Authentication Protocol
CHCNT	Call Counter
cHTML	Compact HTML
CI	Cell Identification
CID	Cluster Identification
CK	Ciphering Key
CL	Convergence Layer
CLH	Cluster Head
CLIP	Connectionless Interworking Protocol
CMOS	Complementary Metal Oxide Semiconductor

CN	Core Network
CNB	Core Network Bearer
CoA	Care-of Address
CPCH	Common Packet Channel
CPCCH	Common Power Control Channel
CPICH	Common Pilot Channel
CPHCH	Common Physical Channel
CPM	Continuous Phase Modulation
CPU	Central Processing Unit
CQI	Channel Quality
CQICH	Channel Quality Channel
CR	Cognitive Radio
CRC	Cyclic Redundancy Check
CS	Circuit Switched
CSCF	Cell State Control Function
csch	Common Signaling Channel
CSD	Circuit Switched Data
CSMA	Carrier Sense Multiple Access
CSS	Cascading Style Sheet
CSGR	Cluster Switch Gateway Routing
CT2	Cordless Telephony 2
CTCH	Common Traffic Channel
CTS	Clear to Send
CW	Contention Window

D

DA	Destination Address
DAB	Digital Audio Broadcast
DAMPS	Digital Advanced Mobile Phone System
DAPICH	Dedicated Auxiliary Pilot Channel
dB	Decibel
DBPSK	Differential Binary Phase Shift Keying
DC	Dedicated Control
DCC	DLC Connection Control
DCH	Dedicated Channel
DCCH	Dedicated Control Channel
DCF	Distributed Coordination Function
DD	Designated Device
DECT	Digital Enhancements of Cordless Technology
DES	Data Encryption Standard
DFS	Dynamic Frequency Selection
DH	Diffie-Hellman

DHCP	Dynamic Host Configuration Protocol
DID	Direct Inward Dialing
DiffServ	Differentiated Services
DiL	Directlink Phase
DIFS	Distributed Inter-Frame Space
DL	Downlink
DLC	Data Link Control
DLCI	Data Link Control Identity
DLL	Data Link Layer
dmch-control	dedicated MAC channel control
DNS	Domain Name Server
DO	Data Only
DPCH	Dedicated Physical Channel
DPCCH	Dedicated Physical Control Channel
DPCM	Delta Pulse Code Modulation
DPDCH	Dedicated Physical Data Channel
DQPSK	Differential Quadrature Phase Shift Keying
DRC	Datarate Request Channel
DRNC	Drift Radio Network Controller
DS	Direct Sequence
DS-CDMA	Direct Sequence Code Division Multiple Access
dsch	Dedicated Signaling Channel
DSCH	Downlink Shared Channel
DSP	Digital Signal Processor
DSR	Dynamic Source Routing
DSRC	Dedicated Short Range Communication
DSS	Distribution System Service
DSSS	Direct Sequence Spread Spectrum
DSDV	Destination Sequenced Distance Vector
DTx	Discontinuous Transmission
dtch	Dedicated traffic channel
DV	Data and Voice
DVB	Digital Video Broadcast

E

EACH	Enhanced Access Channel
EAP	Extensible Authentication Protocol
EAPOL	Extensive Authentication Protocol over LANs
EAPOW	EAP-Over WLAN
EAR	Eavesdrop-And-Register
EC	Error Control
ECC	Elliptic Curve Cryptography

ECN	Explicit Congestion Notification
ECSD	Enhanced Circuit Switched Data
ED	Energy Detection
EDN	External Data Network
EDGE	Enhanced Data rates for GSM Evolution
EEPROM	Electrically Erasable PROM
EFR	Enhanced Full Rate
EGPRS	Enhanced General Packet Radio Service
EIB	Erasure Indicator Bit
EIFS	Extended Inter-Frame Space
EIR	Equipment Identity Register
EIRP	Effective Isotropic Radiated Power
ELN	Explicit Loss Notification
EMR	Electro Magnetic Interference
EMS	Element Manager System
EPROM	Erasable Programmable Read Only Memory
ERP	Effective Radiated Power
ESN	Electronic Serial Number
ESS	Extended Service Set
ESRT	Event-to-Sink Reliable Transport
ESSID	Enhanced Service Set ID
ETSI	European Telecommunications Standards Institute
EV	Enhanced Version
EV-DO	Evolution for Data-Only
EV-DV	Evolution for Data and Voice
EVRC	Enhanced Variable Rate Codec

F

FA	Foreign Agent
FACCH	Fast Associated Control Channel
FACH	Forward Access Channel
FAUSCH	Fast Uplink Signaling Channel
FBI	FeedBack Information
FCC	Federal Communications Commission
FCCH	Frequency Correction Channel
FCH	Fundamental Channel
FCS	Frame Check Sequence
FD	Full Duplex
FDD	Frequency Division Duplex
FDM	Frequency Division Multiplexing
FDMA	Frequency Division Multiple Access
FEC	Forward Error Correction

FER	Frame Error Rate
FFD	Full Function Device
FFPC	Fast Forward Power Control
FFT	Fast Fourier Transform
FH	Frequency Hopping
FH-DFDM	Frequency Hopping DFDM
FHMA	Frequency Hopping Multiple Access
FHSS	Frequency Hopping Spread Spectrum
FIPS	Federal Information Processing Standard
FM	Frequency Modulation
FOMA	Freedom Of Mobile Access
F-PDCCH	Forward-Packet Data Control Channel
FPDCH	Forward Packet Data Channel
FQDN	Fully Qualified Domain Name
FR	Frame Relay
FSK	Frequency Shift Keying
FT	Frequency Time
FTP	File Transfer Protocol

G

1G	First-Generation
2G	Second-Generation
3G	Third-Generation
4G	Fourth-Generation
GBN	Go-Back-N
GBS	GPRS Backbone System
GC	General Control
GCI	Global Cell Identity
GFSK	Gaussian Frequency Shift Keying
GGSN	Gateway GPRS Support Node
GIF	GPRS Interworking Function
GMM	GPRS Mobility Management
GMSC	Gateway Mobile Switching Center
GMSK	Gaussian Minimum Shift Keying
GoS	Grade of Service
GPRS	General Packet Radio Service
GPS	Global Positioning System
GRE	Generic Routing Encapsulation
GSM	Groupe Special Mobile/Global System for Mobile Communications
GSN	GPRS Support Node
GT	Global Title

GTP	GPRS Tunneling Protocol
GTS	Guaranteed Time Slot

H

HA	Home Agent
HARQ	Hybrid Automatic Repeat Request
HCI	Host Controller Interface
HD	High Day
HDR	High Data Rate
HIPERLAN	High Performance Radio Local Area Network
HIPERMAN	High Performance Metropolitan Area Network
HLR	Home Location Register
HoA	Home Address
HSCSD	High-Speed Circuit Switched Data
HS-DSCH	High-Speed Downlink Shared Channel
HSDPA	High-Speed Downlink Packet Access
HS-DPCCH	High-Speed Dedicated Physical Control Channel
HS-SCCH	High Speed Shared Control Channel
HR	High Rate
HTML	Hyper Text Markup Language
HTTP	Hyper Text Transfer Protocol
HUMAN	High-speed Unlicensed Metropolitan Area Network

I

IAM	Initial Address Message
IBSS	Independent Basic Service Set
IC	Interference Cancellation
ICMP	Internet Control Message Protocol
IDEA	International Data Encryption Algorithm
IETF	Internet Engineering Task Force
IF	Intermediate Frequency
IFFT	Inverse Fast Frequency Transform
IGMP	Internet Group Management Protocol
IHL	Internet Header Length
IKE	Internet Key Exchange
IM	Inter Modulation
IMEI	International Mobile Equipment Identity
IMS	IP Multimedia Service
IMSI	International Mobile Subscriber Identity
IMT-DS	IMT Direct Spread
IMT-FT	IMT Frequency Time

IMT-MC	IMT Multicarrier
IMT-SC	IMT Single Carrier
IMT-2000	International Mobile Telecommunications-2000
IN	Intelligent Network
IN/CAMEL	Intelligent Network/Customized Applications for Mobile Enhanced Logic
IntServ	Integrated Services
IOS	Interoperability Specifications
IP	Internet Protocol
IPCP	Internet Protocol Control Protocol
IPSec	Internet Protocol Security
IR	Impulse Radio
IrDA	Infrared Data Association
ISDN	Integrated Services of Digital Network
ISI	Inter-Symbol Interference
ISM	Industrial, Scientific, and Medicine
ISMA	Idle Signal Casting Multiple Access
ISP	Internet Service Provider
ISUP	ISND User Part
ITU	International Telecommunications Union
I-TCP	Indirect TCP
ITU-T	International Telecommunications Union-Technical
IWF	Inter Working Function
IWMSC	Inter Working MSC

J

JTC	Joint Technical Committee

K

KPI	Key Performance Indicator

L

L2CAP	Logical Link Control and Adaptation Protocol
LA	Location Area
LAC	Link Access Channel
LAC	Location Area Code
LAI	Location Area Identity
LAN	Location Area Network
LAPD	Link Access Protocol D
LCR	Level Crossing Rate

LCP	Link Control Protocol
LD	Low Delay
LDAP	Lightweight Directory Access Protocol
LD-CELP	Long delay-CELP
LEACH	Low-Energy Adaptive Cluster Hierarchy
LEO	Low Earth Orbit
LL	Link Layer
LP	Link Protocol
LLC	Logical Link Control
LMDS	Local Multipoint Distribution System
LML	Link Management Layer
LMP	Link Manager Protocol
LO	Local Oscillator
LOM	Lowest Order Modulation
LOS	Line of Sight
LP	Linear Predictor
LPAS	Linear-Prediction-Based Analysis-by-Synthesis
LPC	Linear Prediction Coding
LPF	Linear Pass Filter
LQI	Link Quality Indication
LR	Low Rate
LSP	Linear Spectral Pair
LT	Long Term
LTP	Long-Term Prediction
LU	Location Update

M

MAC	Media Access Control
MAGIC	Mobile multimedia, Anytime anywhere, Global mobility support, Integrated wireless solution, and Customised personal service.
MAHO	Mobile Assisted Handoff
MAN	Mobile Access Network
MANET	Mobile Ad hoc Network
MAP	Mobile Application Part
MB-DFDM	Multiband DFDM
MC	Message Center
MC	Multiple Carrier
MCC	Mobile Country Code
MC-CDMA	Multicarrier CDMA
MC-DS-CDMA	Multicarrier DS-CDMA
MCHO	Mobile Controlled Handoff
MCM	MultiCarrier Modulation

MCPS-SAP	MAC Common Port Sublayer-Service Access Point
MCTD	Multi-Carrier Transmit Diversity
MExE	Mobile Station Execution Environment
MF	MAC Frame
MFR	MAC Footer
MFSK	M-ary Frequency Shift Keying
MGCF	Media Gateway Control Function
MGW	Media Gateway
MHR	MAC Header
MIMO	Multiple Inputs, Multiple Outputs
MIN	Mobile Identification Number
MIP	Mobile Internet Protocol
MIPS	Millions of Instructions Per Second
MISO	Multiple Input, Single Output
ML	Maximum-Likelihood
MLME-SAP	MAC Layer Management Entity-SAP
MM	Mobility Management
MMAC	MultiMedia Access Communication
MMDS	Multichannel Multipoint Distribution System
MMS	MultiMedia Service
MN	Mobile Node
MNC	Mobile Network Code
MOS	Mean Opinion Score
MOU	Minutes of Use
MPDU	MAC Protocol Data Unit
M-PGW	Mobile Packet Gateway
M-QAM	Multilevel Quadrature Amplititude Modulation
MRF	Media Resource Function
MS	Mobile Station
MSC	Mobile Switching Center
MSDU	MAC Service Data Unit
MSIN	Mobile Subscriber Identification Number
MSK	Minimum Shift Keying
MSR	Mobile Support Router
MSRN	Mobile Subscriber Roaming Number
MSS	Maximum Segment Size
MT	Mobile Terminal
M-TCP	Mobile TCP
MT-CDMA	Multitone CDMA
MTP3-B	Message Transfer Part 3-B
MUD	Multi-User Detection
MUI	Multi-User Interference
MWAN	Mobile Wide-Area Network

N

NAM	Number Assignment Module
NAP	Network Access layer Protocol
NAPT	Network Address and Port Translation
NAT	Network Address Translation
NAV	Network Allocation Vector
NBAP	Node B Application Part
NCHO	Network Controlled Handoff
NE	Network Element
N-ISDN	National ISDN
NIST	National Institute of Standards and Technology
NL	Network Layer
NLOS	Non-Line-of-Sight
NMC	Network Management Center
NMSI	National Mobile Subscriber Identity Number
NMT	Nordic Mobile Telephone
NNI	Network-to-Network Interface
NRZ	Non-Return to Zero
NSS	Networking and Switching Subsystem
Nt	Notification

O

O + I	Originating plus Incoming
OBEX	Object Exchange
OCCCH	ODMA Common Control Channel
OCS	Online Changing System
ODMA	Opportunity Driven Multiple Access
ODCCH	ODMA Dedicated Control Channel
ODCH	ODMA Dedicated Channel
ODTCH	ODMA Dedicated Traffic Channel
OFDM	Orthogonal Frequency Division Multiple access
OMA	Open Mobile Alliance
OMC	Operations and Maintenance Center
OMSS	Operational and Maintenance Subsystem
OQPSK	Offset-Quadrature Phase Shift Keying
OS	Operations Systems
OSI	Open System Interconnection
OSPF	Open Shortest Path First
OTD	Orthogonal Transmit Diversity
OVSF	Orthogonal Variable Spreading Factor

P

PACS	Personal Access Communication System
PACCH	Packet Associated Control Channel
PAD	Packet Assembler/Disassembler
PAGCH	Packet Access Grant Channel
PAM	Pulse Amplitude Modulation
PAN	Personal Area Network
PBCCH	Packet Broadcast Control Channel
PC	Point Coordinate
PCF	Packet Control Function
PCH	Paging Channel
PCCH	Paging Control Channel
PCCCH	Packet Common Control Channel
PCCPCH	Primary Common Control Physical Channel
PCG	Power Control Group
PCM	Pulse Code Modulation
PCN	Personal Communication Network
PCPCH	Physical Common Packet Channel
PCPICH	Primary Common Pilot Channel
PCS	Personal Communications Services
PCU	Packet Control Unit
PDA	Personal Digital Assistants
PDC	Personal Digital Cellular
PDCCH	Packet Dedicated Control Channel
PDCH	Packet Data Channel
PDC-P	Personal Digital Cellular-Packet
PDF	Probability Density Function
PDG	Packet Data Gateway
PDN	Packet Data Network
PDP	Packet Data Protocol
PDSCH	Physical Downlink Shared Channel
PDSN	Packet Data Service Node
PDTCH	Packet Data Traffic Channel
PDU	Packet Data Unit
PER	Packet Error Rate
PGP	Pretty Good Privacy
PHS	Personal Handy Phone System
PHY	Physical layer
PICH	Page Indicator Channel
PICH	Pilot Channel
PIFS	Point coordinates Inter-Frame Space
PKI	Public Key Infrastructure
PLCP	Physical Layer Convergence Procedure

PLL	Phase-Locked Loop
PLME	Physical Layer Management Entity
PLDCF	Physical Layer Dependent Convergence Function
PLICF	Physical Layer Independent Convergence Function
PLMN	Public Land Mobile Network
PLW	PLCP Length Word
PMD	Physical Medium Dependent
PMK	Pairwise Master Key
PN	Pseudo-random Noise
PPCH	Packet Paging Channel
PNCH	Packet Notification Channel
POP	Post Office Protocol
POS	Personal Operating Space
PPDU	PLCP Protocol Data Unit
PPP	Point-to-Point Protocol
PRACH	Packet Random Access Channel
PRMA	Packet Reservation Multiple Access
PROM	Programmable Read Only Memory
PS	Packet Switched
PSC	Primary Synchronization Code
PSD	Power Spectral Density
PSDU	PHY Service Data Unit
PSF	PLCP Signaling Field
PSK	Phase Shift Keying
PSMM	Pilot Strength Measurement Message
PSPDN	Packet Switched Public Data Network
PSTN	Public Switched Telephone Network
PTCCH	Packet Timing advance Control Channel
PTCH	Packet Traffic Channel
PTM	Point-to-Multipoint
PTP	Point-to-Point
PTT	Post Telegraph and Telephone
PWT-E	Personal Wireless Telecommunications-Enhanced

Q

QAM	Quadrature Amplitude Modulation
QCELP	Qualcomm Code-Excited Linear Predictive
QOQAM	Quaternary Offset Quadrature Amplitude Modulation
QoS	Quality of Service
QPCH	Quick Paging Channel
QPSK	Quadrature Phase Shift Keying

R

RA	Receiver Address
RAB	Radio Access Bearer
RACE	Research in Advanced Communications Equipment
RACH	Random Access Channel
R-ACKCH	Reverse ACK Channel
RADIUS	Remote Authentication Dial-In User Service
RAM	Random Access Memory
RAN	Radio Access Network
RANAP	Radio Access Network Application Part
RAND	Random Number
RARP	Reverse Address Resolution Protocol
RAS	Remote Access Server
RAU	Routing Area Update
RBP	Radio Burst Protocol
RC	Rate Class
RCELP	Relaxed Code-Excited Linear Predictive
R-CQIC	Reverse Channel Quality
RES	Response
RFD	Reduced Function Device
RFID	Radio Frequency Identification
RLC	Radio Link Control
RLAC	Radio Link Access Control
RLP	Radio Link Protocol
RN	Radio Node
RNC	Radio Network Controller
RNCP	Radio Network Control Plane
RNL	Radio Network Layer
RNS	Radio Network Subsystem
RNSAP	Radio Network Subsystem Application Part
ROM	Read Only Memory
ROSE	Remote Operation Service Element
RPE	Regular Pulse Excited
RRC	Radio Resource Control
RREP	Route Reply
RREQ	Route Request
RRI	Reverse Rate Indicator
R-S	Reed-Solomon
RSA	Rivet-Shamir-Adelman
RSS	Radio Station Subsystem
RSS	Radio Signal Strength
RSV	Reed-Solomon Viterbi

RTO	Round Trip Time-Out
RTS	Request to Send
RTT	Radio Transmission Technology
Rx	Receiver

S

S	Slave
SA	Source Address
SAAL	Signaling Asynchronous transfer mode Adaptation Layer
SACK	Selective Acknowledgment
SACCH	Slow Associated Control Channel
SAP	Service Access Point
SAPI	Service Access Point Identity
SAR	Sequential Assignment Routing
SASE	Specifies Application Service Element
SBC	Sub-Band Coding
SC	Single Carrier
SCCH	Supplementary Code Channel
SCCP	Signaling Connection Control Part
SCCPCH	Secondary Common Control Physical Channel
SCH	Synchronization Channel
SCH	Secondary Channel
SCO	Synchronous Connection Oriented
SCP	Stream Control Protocol
SCPICH	Secondary Common Pilot Channel
SCTP	Simple Control Transmission Protocol
SDCCH	Stand-alone Dedicated Control Channel
SDM	Space Division Multiplexing
SDP	Service Discovery Protocol
SDR	Software Defined Radio
SDU	Service Data Unit
SF	Spreading Factor
SFD	Start of Frame Delimiter
SGML	Standard Generalized Markup Language
SGSN	Serving GPRS Support Node
SIFS	Short Inter-Frame Space
SIG	Special Interest Group
SIM	Subscriber Identity Module
SIMO	Single Input, Multiple Output
SINR	Signal-to-Interference-plus-Noise Ratio
SIP	Session Initiated Protocol
SIR	Signal-to-Interference Ratio

SISO	Soft Input, Soft Output
SM	Service Management
SMACS	Self-organized Media Access Control for Sensor network
SMECN	Small Minimum Energy Communication Network
SMP	Sensor Management Protocol
SMS	Short Message Service
SMTP	Simple Mail Transport Protocol
SNDCP	Sub-network Dependent Convergence Protocol
SNR	Signal-to-Noise Ratio
SOHO	Small Office Home Office
SOVA	Soft Output Viterbi Algorithm
SPIN	Sensor Protocol for Information via Negotiation
SQDDP	Sensor Query and Data Dissemination Protocol
SR	Spreading Rate
SRBP	Signaling Radio Burst Protocol
SRES	Signature Response
SRLP	Signaling Radio Link Protocol
SRNC	Serving RNC
SRNS	Serving RNS
SRP	Selective Repeat Protocol
SS	Spread Spectrum
SS7	Signaling System 7
SSC	Secondary Synchronization Code
SSCF	Service-Specific Coordinating Function
SSCS	Service Specific Convergence Sublayer
SSCOP	Service-Specific Connection Oriented Protocol
SSD	Shared Secret Data
SSID	Service Set Identifier
SSL	Secure Socket Layer
SSP	Sevice-Signaling Point
ST	Short Term
STD	Silence Descriptor
SYNCH	Sync Channel

T

T	Transparent
TA	Terminal Adaptor
TACS	Total Access Communication System
TACAS	Terminal Access Controller Access control System
TADAP	Task Assignment and Data Advertisement Protocol
TAG	Technical Ad hoc Group
TB	Transport Block

TCH	Traffic Channel
TCH/F	Traffic Channel/Full
TCH/H	Traffic Channel/Half
TCM	Trellis Coded Modulation
TCP	Transmission Control Protocol
TCS	Telephone Control Specification
TCS-1	Telephone Control Channel-1
TDD	Time Division Duplex
TDM	Time Division Multiplexing
TDMA	Time Division Multiple Access
TDPICH	Transmit Diversity Pilot Channel
TD-SCDMA	Time Division-Synchronous CDMA
TeleVAS	Telephony Value-Added Service
TEI	Terminal Equipment Identity
TFCI	Transport Format Combination Indicator
THSS	Time-Hop Spread Spectrum
TIA	Telecommunication Industries Association
TKIP	Temporal Key Integrity Protocol
TLLI	Temporary Logical Link Identifier
TLS	Transport Layer Security
TMSI	Temporary Mobile Subscriber Identity
TNCP	Transport Network Control Plane
TNL	Transport Network Layer
TORA	Temporally Ordered Routing Algorithm
TOS	Type of Service
TPC	Transmit Power Control
TP4	Transport Protocol Class 4
TRAU	Transponder Rate Adapter Unit
TS	Time Slot
TST	Time Slot Transfer
TTI	Transmission Time Interval
TTL	Time-to-Live
TTP	Traffic Termination Point
Tx	Transmitter

U

UCN	UMTS Core Network
UDI	Unrestricted Digital Information
UDP	User Datagram Protocol
UE	User Equipment
UHF	Ultra High Frequency
UMTS	Universal Mobile Telecommunications Services

U-NII	Unlicensed-National Information Infrastructure
UP	Uplink
UP	User Plane
URAN	UMTS Radio Access Network
URI	Uniform Resource Identifier
URL	Uniform Resource Locator
USF	Uplink State Flag
USIM	UMTS SIM
USSD	Unstructured Supplementary Service Data
UTRA	UMTS Terrestrial Radio Access
UTRAN	UMTS Terrestrial Radio Access Network
UWB	Ultra WideBand
UWCC	Universal Wireless Communication Consortium

V

V	Viterbi
VAD	Voice Activity Detector
VC	Virtual Circuit
VLR	Visitor Location Register
VLSI	Very Large System Integration
VM	Voice Mail
VMR	Variance-to-Mean Ratio
VOFDM	Vector Orthogonal Frequency Division Multiple-access
VoIP	Voice-over IP
VPN	Virtual Private Network
VSELP	Vector Self-Excited Linear Predictor

W

WACS	Wireless Access Communication System
WAE	Wireless Application Environment
WAF	WLAN Adaptation Function
WAG	WLAN Access Gateway
WAN	Wide-area Network
WAP	Wireless Application Protocol
WAP CSS	Wireless Application Protocol Cascading Style Sheet
W-APN	WLAN-Access Point Name
WCDMA	Wideband CDMA
WCMP	Wireless Control Message Protocol
WDP	Wireless Datagram Protocol
WEP	Wired Equivalent Privacy
WIG	Wireless Interworking Group

WiMAX	Worldwide interoperability for Microwave Access
WIN	Wireless Intelligent Networking
WLAN	Wireless Local Area Network
WLL	Wireless Local Loop
WMAN	Wireless Metropolitan Area Network
WML	Wireless Markup Language
WPA	Wi-Fi Protected Access
WPAN	Wireless Personal Area Network
WPOS	Wireless Point of Sale
WORM	Window-control Operation based on Reception Memory
WRP	Wireless Routing Protocol
WSP	Wireless Service Provider
WSP	Wireless Session Protocol
WTA	Wireless Telephony Application
WTAI	Wireless Telephony Application Interface
WTLS	Wireless Transport Layer Security
WTP	Wireless Transaction Protocol
WWAN	Wireless Wide-Area Network
WWW	World Wide Web

X

XHTML MP	Extensible HTML Mobile Profile
XML	Extensible Mark-up Language
XRES	Expected Response

Z

ZF	Zero-Forcing

Index